Blockchain and Artificial Intelligence for Secure Computer Vision Technologies and Applications

There are intrinsic vulnerabilities in traditional computer vision systems, such as data tampering and adversarial attacks. This book illustrates how blockchain's decentralized and immutable ledger can mitigate these risks. Further, it examines the role of proxy smart contracts in implementing security protocols and ensuring data integrity within computer vision approaches.

- Discusses the integration of emerging technologies such as blockchain and generative artificial intelligence for computer vision applications with service scenarios.
- Explores existing secure solutions for computer vision applications with blockchain and advanced technologies.
- Explains the fusion of blockchain and machine learning techniques for intrusion detection in computer vision.
- Covers real-world computer vision applications, such as smart healthcare, smart parking, and smart education, with secure environments as case studies.
- Examines security and privacy challenges in computer vision applications, discussing mitigation procedures, regulatory frameworks, and ethical contemplations crucial for protecting individuals' rights.

Blockchain and Artificial Intelligence for Secure Computer Vision Technologies and Applications is primarily written for senior undergraduates, graduate students, and academic researchers in electrical engineering, electronics and communications engineering, computer vision, artificial intelligence, and computer engineering.

Blockchain and Artificial Intelligence for Secure Computer Vision Technologies and Applications

Edited by
Sushil Kumar Singh, Himanshu Gupta, Rajendrasinh Jadeja, Ashish Khanna, and Bal Virdee

CRC Press
Taylor & Francis Group
Boca Raton London New York

CRC Press is an imprint of the
Taylor & Francis Group, an informa business

Front cover image: Summit Art Creations/Shutterstock

This edition published in 2027
by CRC Press
2385 NW Executive Center Drive, Suite 320, Boca Raton FL 33431

and by CRC Press
4 Park Square, Milton Park, Abingdon, Oxon, OX14 4RN

CRC Press is an imprint of Taylor & Francis Group, LLC

ISBN: 978-1-041-11263-1 (hbk)
ISBN: 978-1-041-11264-8 (pbk)
ISBN: 978-1-003-65911-2 (ebk)

DOI: 10.1201/9781003659112

Typeset in Sabon
by codeMantra

Contents

Preface

The rapid evolution of digital technologies has transformed the way visual data are captured, processed, analyzed, and secured. As computer vision increasingly permeates modern applications—ranging from smart healthcare and intelligent transportation to autonomous vehicles, industrial automation, and smart cities—the need for trustworthy, transparent, and secure visual information processing has become more critical than ever. Traditional computer vision systems, although powerful, face persistent challenges, including data tampering, adversarial attacks, privacy violations, and the absence of robust authentication mechanisms. These limitations highlight an urgent demand for innovative solutions that can strengthen the reliability and resilience of vision-based intelligent systems.

This book, *Blockchain and Artificial Intelligence for Secure Computer Vision Technologies and Applications*, presents a comprehensive exploration of two groundbreaking technologies—blockchain and generative artificial intelligence (AI)—and their transformative role in addressing security, privacy, and integrity issues within computer vision environments. The convergence of blockchain's decentralized, immutable ledger with generative AI's creative and predictive capabilities introduces a new paradigm that not only safeguards visual data but also enhances the intelligence and adaptability of modern vision systems.

Across 14 chapters, this edited volume brings together diverse perspectives from researchers and practitioners worldwide. This book begins by introducing foundational concepts in computer vision, layered neural network architectures, and the fundamentals of blockchain technologies and generative AI models such as generative adversarial networks (GANs) and VAEs. It further explores Internet of Things (IoT)-enabled vision systems, security and privacy vulnerabilities, cryptographic protocols, and secure infrastructure design. Emphasis is placed on practical deployments—ranging from smart parking, healthcare diagnostics, intelligent surveillance, and smart education to advanced communication-assisted secure environments. Special chapters present blockchain-driven intelligent architectures, intrusion detection using machine learning, emerging technology fusion, and future research challenges.

The contributors offer theoretical insights, design methodologies, implementation strategies, and real-world case studies, making the book suitable for researchers, graduate students, academicians, engineers, and industry professionals. Through its multidisciplinary approach, this book provides a unified platform for understanding how modern technologies can safeguard visual information, strengthen authentication, ensure transparency, and promote trustworthy decision-making.

We believe this volume will serve as a valuable reference for those seeking to advance research and development in secure computer vision environments. The editors gratefully acknowledge the efforts of all chapter contributors, reviewers, and supporting institutions. Their dedication has enabled the creation of a resource that we hope will inspire innovative research pathways and foster technological advancements across academia and industry.

Editors:

Dr. Sushil Kumar Singh
Marwadi University, Rajkot, Gujarat
sushilkumar.singh@marwadieducation.edu.in

Dr. Himanshu Gupta
Manipal Institute of Technology,
Manipal Academy of Higher Education, Manipal
himanshu.gupta@manipal.edu

Prof. Rajendrasinh Jadeja
Marwadi University, Rajkot, Gujarat
rajendrasinh.jadeja@marwadieducation.edu.in

Prof. Ashish Khanna
Maharaja Agrasen Institute of Technology, Delhi
ashishkhanna@mait.ac.in

Prof. Bal Virdee
London Metropolitan University, UK
b.virdee@londonmet.ac.uk

About the Editors

Sushil Kumar Singh (Member, IEEE) is an Associate Professor in the Department of Computer Engineering at Marwadi University, Rajkot, Gujarat, India. He has completed a PostDoc at London Metropolitan University, London, United Kingdom, under the supervision of Professor Bal Virdee. He received a Ph.D. degree from Seoul National University of Science and Technology, Seoul, South Korea, under the supervision of Professor Jong Hyuk Park. He received an M.Tech. degree in Computer Science and Engineering from Uttarakhand Technical University, Dehradun, India. He also received an M.E. degree in Information Technology from Karnataka State University, Mysore, India. He has also been the lab leader of the UCS Lab at the Department of Computer Science and Engineering, Seoul National University of Science and Technology, Seoul, South Korea. He has received the Best Lab Leadership Award from UCS Lab for 2019–2021. He is selected in the top 2% scientists for the 2022–2023, 2023–2024, and 2024–2025 years as per the list compiled by Stanford University and published by Elsevier. He has more than 15 years of experience teaching in the field of computer science. He has published more than 10 books on the following topics: Computer C Programming, Cyber Security, Big Data Analytics, Mobile Computing, Secure and Intelligent IoT-Enabled Smart Cities, and others. He attended the Erasmus+ Program at Pitesti University, Romania, in 2024 and delivered two essential sessions to foreign students. He has also published many high-quality papers (Q1, Top 10% JCR Rank) in international journals and conferences. He has already delivered international lectures in many countries. His research interests include blockchain, artificial intelligence, big data, Internet of Things, smart city security, and cyberphysical systems. He is an Associate/Guest Editor in the *Human-centric Computation and Information Sciences (HCIS) Journal*, *IEEE Journal of Biomedical and Health Informatics (IEEE JBHI)*, *IGI Global Publication*, and *Wiley Scrivener Publication*. He is a reviewer of the *IEEE Wireless Communication Magazine*, *IEEE Systems*, *IEEE Internet of Things*, *FGCS*, *TETT*, *EXSY*, *JISA*, *Computer Network*, *MDPI*, *CIE*, *HCIS*, *JIPS*, *Computing (COMP)*, *Multimedia Tools and Applications*, and *SCIS Journal*. He also organizes the Research Activities Club, which promotes quality research activities among young researchers at Marwadi University, Rajkot, Gujarat, India.

Himanshu Gupta is currently working as an Assistant Professor (Senior Scale) in the Department of Instrumentation and Control Engineering at Manipal Institute of Technology, Manipal Academy of Higher Education, Manipal, Karnataka, India. He has more than 14 years of academic experience. He received his B.Tech. degree from Uttar Pradesh Technical University, Lucknow, India, in 2006 and M.Tech. and Ph.D. degrees from Dr B R Ambedkar National Institute of Technology Jalandhar, India, in 2012 and 2023, respectively. He also worked as a team leader in several sponsored projects, including ISRO projects. He has published more than 35 research articles in peer-reviewed journals and conferences. He has been a potential reviewer of several international journals and conferences of high repute. He is also associated with many conference series, such as Soft Computing: Theories and Applications (SoCTA) and MAchine inTelligence for Research and Innovations (MAiTRI). He is a member of the Instrument Society of India, IISc Bangalore, STEM Research Society, and Automatic Control and Dynamic Optimization Society, India. His research interests include computer vision, drone technology, soft-computing approaches, and optimization.

Rajendrasinh B. Jadeja (Senior Member, IEEE) is working as a ProVost and Dean at the Faculty of Engineering, Marwadi University, Gujarat, India. He completed his B.E. in Electrical Engineering from L. D. College of Engineering (Gujarat University) in 1998, M.Tech. in Electrical Engineering from Nirma University in 2005, and Ph.D. from the M S University, Baroda, in 2010. He was awarded the title "ING.PAED.IGIP" in 2020. He is a Senior Member of IEEE. He worked as the Treasurer of IEEE Region-10 during 2021. He is currently working as the Chair, Membership Development Committee of IEEE Region-10. He worked as the Treasurer of IEEE India Council and as the Chair of IEEE Gujarat Section during 2017. His areas of research interest include control of power electronic converters, AC drives, pulse width-modulated power electronic converters, and power quality issues and their mitigation techniques. Various projects have been undertaken by him, and he has completed a project funded by Gujarat Council on Science and Technology (GUJCOST), and he is working on a project making "Deep Learning and AI skills" mainstream in India to fulfil trilateral needs of entrepreneurship, industry–academia partnership, and application-inspired engineering research. He actively participates and initiates Erasmus+ for staff mobility and has also completed such a program for teaching between partner countries. He is a recipient of the Young Professional Award for a travel grant to attend the IEEE Section Congress held on 22–24 August 2014, Amsterdam, Netherlands. He has been working as a member of the Board of Studies (BoS) of Electrical Engineering of the Gujarat state's largest public technical university, Gujarat Technological University, and as a BoS member at PDPU, Gandhinagar.

Ashish Khanna (IEEE Member 2019, Senior Member 2020, MACM) has expertise in teaching, entrepreneurship, and research and development with a specialization in computer science engineering subjects. He is part of the world's leading 2% research community list from Stanford University, USA. He is also part of the top research community, including the AD research community and research.com group. He has completed a PostDoc at INATEL Brazil and the University of Valladolid, Spain, Europe. He received his Ph.D. degree from the National Institute of Technology, Kurukshetra, in March 2017. He completed his M.Tech. in 2009 and B.Tech. from GGSIPU, Delhi, in 2004. He has around 200 accepted and published research papers and book chapters in reputed SCI, Scopus journals, conferences, and reputed book series, including 98 papers accepted and published in TOP SCI-indexed journals. He also has ten published patents and one granted German patent to his credit. Additionally, he has co-authored and edited around 50 books. He has also served as a keynote speaker resource person and has given several invitations. He was a resource person speaker and delivered a lecture at the ATAL program at JNU in 2020. He has worked as a Guest Editor in several reputed journals. He is also serving as a Series Editor in publishing houses like De Gruyter (Germany) of the "Intelligent Biomedical Data Analysis" series, Elsevier of "Intelligent Biomedical Data Analysis," and CRC Press of "Intelligent Techniques in Distributed Systems: Principles and Applications."

Bal Virdee (Senior Member, IEEE) received his B.Sc. and M.Phil. degrees in Communications Engineering from the University of Leeds, UK, and his Ph.D. degree in Electronic Engineering from the University of London, UK He has worked in the industry for various companies including Philips, UK, as a Research and Development Engineer, and Teledyne Defence and Space as a Future Products Developer in RF/microwave communications. He has taught at several academic institutions before joining London Metropolitan University, where he is currently a Senior Professor in Communications Technology with the School of Computing and Digital Media, where he is also the Head of the Communications Technology Research Center. His research, in collaboration with industry and academia, is in wireless communications encompassing mobile phones to satellite technology. He has chaired technical sessions at IEEE international conferences and published numerous research papers. He is an Executive Member of IET's Technical and Professional Network Committee on RF/Microwave Technology. He is a fellow of the IET.

Contributors

Kinnari Utsavkumar Acharya
Department of Computer Science
Ganpat University
Mehsana, Gujarat, India

Shivam Agarwal
Amity Institute of Information Technology
Amity University Uttar Pradesh
Lucknow, Uttar Pradesh, India

Saddala Anil
Department of CSE, School of CS&AI
SR University
Warangal, Telangana, India

Deepali Vishnu Bavlecha
Electrical Engineering Department
Government Engineering College
Rajkot, Gujarat, India

Rahul Bhandari
Department of CSE
Chandigarh University
Mohali, Punjab, India

Juhi Chaudhary
Department of Computer Science & Engineering
Shri Ram Murti Smarak College of Engineering & Technology
Bareilly, Uttar Pradesh, India

Jessica Chauhan
Amity Institute of Information Technology
Amity University Uttar Pradesh
Lucknow, Uttar Pradesh, India

Rushabh Doshi
Electrical Engineering Department
B. H. Gardi College of Engineering & Technology
Rajkot, Gujarat, India

Chetan R. Dudhagara
International Agribusiness Management Institute
Anand Agricultural University (AAU)
Anand, Gujarat, India

Shahnaz Fatima
Amity Institute of Information Technology
Amity University Uttar Pradesh
Lucknow, Uttar Pradesh, India

Abhishek Gaikwad
NVIDIA
Santa Clara, California

Krishnendu Ghosh
Brainware University
Kolkata, West Bengal, India

Sachinkumar Anandpal Goswami
Department of Computer Science
Ganpat University
Mehsana, Gujarat, India

Himanshu Gupta
Department of Instrumentation and Control Engineering
Manipal Institute of Technology, Manipal Academy of Higher Education
Manipal, Karnataka, India

Pradeep Gupta
Ajay Kumar Garg Engineering College
Ghaziabad, Uttar Pradesh, India

Sonam Gupta
Department of Computer Science and Engineering
Ajay Kumar Garg Engineering College
Ghaziabad, Uttar Pradesh, India

Utkarsh Gupta
Latent View Analytics
San Jose, California

Piali Haldar
Department of Management
Brainware University
Kolkata, West Bengal, India

A. Hanuman Prasad
Computer Science and Artificial Intelligence
SR University
Warangal, Telangana, India

Malik Shahzad Ahmed Iqbal
Department of Computer Science & Engineering
Acharya University
Karakul, Uzbekistan

Riaz K. Israni
School of Engineering
R K University
Rajkot, Gujarat, India

Neelesh Kumar Jain
Parul Institute of Engineering and Technology
Parul University
Vadodara, Gujarat, India

Yuvraj Kaushik Jani
Unitedworld Institute of Technology
Karnavati University
Gandhinagar, Gujarat, India

Amudhavel Jayavel
Faculty of Computing, Engineering and Media
De Montfort University
The Gateway House, Leicester, United Kingdom

Ashwini Kumar Jha
Parul Institute of Engineering and Technology
Parul University
Vadodara, Gujarat, India

Sudan Jha
Kathmandu University
Dhulikhel, Nepal

Weiwei Jiang
School of Information and Communication Engineering
Beijing University of Posts and Telecommunications
Beijing, China

Michael L. Johns
University of Western Australia
Crawley, Western Australia, Australia

R. Karthick Manoj
Department of Electrical and Electronics Engineering
Academy of Maritime Education and Training, Deemed to Be University
Chennai, Tamil Nadu, India

Ragini Karwayun
Department of Computer Applications
Inderprastha Engineering College
Ghaziabad, Uttar Pradesh, India

Rishabh Karwayun
Veeva Systems
Toronto, Canada

Suresh Kaswan
Department of CSE
Chandigarh University
Mohali, Punjab, India

Nanthini Kempaiyan
School of Computer Science and Information Technology
JAIN (Deemed-to-be University)
Bengaluru, Karnataka, India

Bright Keswani
Department of Computer Science & Engineering
Poornima University
Jaipur, India

Mohammedvaris S. Khan
Electrical Engineering Department
Government Polytechnic
Diu (U.T.), India

Ashish Khanna
Maharaja Agrasen Institute of Technology
PSP Area, Delhi, India

Manisha Koranga
Department of Computer Science and Engineering
Graphic Era Hill University
Haldwani, Uttarakhand, India

Tarun Kumar
Department of Allied Sciences
Meerut Institute of Engineering & Technology
Kumaon, Haldwani, Uttarakhand, India

Nitaigour Premchand Mahalik
Department of Industrial Technology
California State University
Fresno, California

Dev Kumar Mandal
Department of Management
Brainware University
Kolkata, West Bengal, India

Suresh Kumar Mandala
Computer Science and Artificial Intelligence
SR University
Warangal, Telangana, India

Darshan B. Mehta
Electrical Engineering Department
B. H. Gardi College of Engineering & Technology
Rajkot, Gujarat, India

Raj Pankajkumar Mehta
Electrical Engineering Department
Darshan University
Rajkot, Gujarat, India

Anand Kumar Mishra
School of Engineering & Technology (UIET)
CSJM University
Kanpur, Uttar Pradesh, India

Shivansh Mishra
Amity Institute of Information Technology
Amity University Uttar Pradesh
Lucknow, Uttar Pradesh, India

Anita Mohanty
Department of Electronics Engineering
Silicon University
Bhubaneswar, Odisha, India

Subrat Kumar Mohanty
Department of Electronics and Communication Engineering
Templecity Institute of Technology and Engineering
Bhubaneswar, Odisha, India

Ambarish G. Mohapatra
Department of Electronics Engineering
Silicon University
Bhubaneswar, Odisha, India

Naveen Kumar Nandhakumar
Department of Computer Applications
Madanapalle Institute of Technology & Science
Madanapalle, Andhra Pradesh, India

Sasmita Nayak
Department of Mechanical Engineering
Government College of Engineering
Kalahandi, Odisha, India

Olukayode A. Oki
Department of Information Technology
Walter Sisulu University
East London, South Africa

Devi Paianoor Prabhakar
Department of Artificial Intelligence and Machine Learning,
Vel Tech Rangarajan Dr. Sagunthala R&D Institute of Science and Technology
Chennai, Tamil Nadu, India

Akruti Pandwal
Parul Institute of Engineering and Technology
Parul University
Vadodara, Gujarat, India

Poonam Nanjibhai Parmar
Electrical Engineering Department
Government Engineering College
Rajkot, Gujarat, India

Kashyap C. Patel
Department of Computer Science
Ganpat University
Mehsana, Gujarat, India

Hiral Chaitanya Prajapati
Department of Computer Science
Ganpat University
Mehsana, Gujarat, India

Imran Qureshi
College of Computing and Information Sciences
University of Technology and Applied Science
Al-Musannah, Sultanate of Oman, Oman

Ranjana Rajnish
Amity Institute of Information Technology
Amity University Uttar Pradesh
Lucknow, Uttar Pradesh, India

Maruthamuthu Ramaswamy
Department of Computer Applications
Madanapalle Institute of Technology & Science
Madanapalle, Andhra Pradesh, India

S. Ramesh
Department of AI&DS
AKT Memorial College of Engineering and Technology
Kallakurichi, Tamil Nadu, India

Shivaratri Narasimha Rao
Department of Computer Science
VeeraNari Chakali llamma Women's University
Hyderabad, Telangana, India

Amarjit Seal
Parul Institute of Engineering and Technology
Parul University
Vadodara, Gujarat, India

Bhagyashree Rajeshkumar Shah
Department of AIML
Parul University
Vadodara, Gujarat, India

Yashasvi Shah
Electrical Engineering Department
B. H. Gardi College of Engineering & Technology
Rajkot, Gujarat, India

Mohammed Ali Shaik
School of Computer Science & Artificial Intelligence
SR University
Warangal, Telangana, India

Mohammad Shahnawaz Shaikh
Faculty of Engineering and Technology
Marwadi University
Rajkot, Gujarat, India
and
Parul Institute of Engineering and Technology
Parul University
Vadodara, Gujarat, India

Meha Sharma
School of Engineering & Technology (UIET)
CSJM University
Kanpur, Uttar Pradesh, India

Pradeep Kumar Sharma
Department of Computer Science & Engineering
Shri Ram Murti Smarak College of Engineering & Technology
Bareilly, Uttar Pradesh, India

Aditya Shinde
Indian Institute of Technology Hyderabad
Kandi, Telangana, India

Swati Shinde
Pimpri Chinchwad College of Engineering
Pune, India

V. Shobha Rani
Department of CSE, School of CS&AI
SR University
Warangal, Telangana, India

Daulat Sihag
Department of CSE
Chandigarh University
Mohali, Punjab, India

Arpita Singh
School of Engineering & Technology (UIET)
CSJM University
Kanpur, Uttar Pradesh, India

Sushil Kumar Singh
Department of Computer Science
Marwadi University
Rajkot, Gujarat, India

Swapna Singh
Department of CSE
Inderprastha Engineering College
Ghaziabad, Uttar Pradesh, India

Yogendra Singh
Parul Institute of Engineering and Technology
Parul University
Vadodara, Gujarat, India

Balamurugan Souprayen
School of Computer Science and Information Technology
JAIN (Deemed-to-be University)
Bengaluru, Karnataka, India

Disha Srivastava
Amity Institute of Information Technology
Amity University Uttar Pradesh
Lucknow, Uttar Pradesh, India

Ajimuse Muyiwa Sunday
Federal University Oye Ekiti
Ibadan, Nigeria

K. Suresh Kumar
Department of Computer Science and Engineering
Sri Eshwar College of Engineering
Coimbatore, Tamil Nadu, India

Om Prakash Suthar
Department of CSE
Marwadi University
Rajkot, Gujarat, India

Salman Ali Syed
Department of Computer Science, College of Computer and Information Sciences
Jouf University
Sakaka, Al-Jouf province, Kingdom of Saudi Arabia, Saudi Arabia

Umesh B. Tank
Vasave Business Solutions Ltd.
London, United Kingdom

Parul Verma
Amity Institute of Information Technology
Amity University Uttar Pradesh
Lucknow, Uttar Pradesh, India

Bal Virdee
London Metropolitan University
London, United Kingdom

Vikash Yadav
Department of Technical Education
Government Polytechnic Bighapur
Unnao, Uttar Pradesh, India

Chapter 1

Computer vision

Introduction and challenges

Himanshu Gupta and Sushil Kumar Singh

1.1 INTRODUCTION

In the world of digital transformation, the processing of visual data by machines has emerged as a key driving force of innovations across industries. From identifying faces in security systems and scanning barcodes at retail stores to driving cars autonomously and detecting diseases in medical images, computer vision (CV) has become a key component in a variety of new and emerging intelligent systems [1]. With the growing emphasis toward automation [2], Industry 4.0 [3], augmented reality [4], robotics [5], and digital transformation [6], the need for machine-level visual understanding has become very critical.

This need is further intensified by the rapid growth of visual data produced daily through social media, surveillance cameras, drones, satellite systems, etc. Moreover, a large volume of unstructured visual data has been produced in real-time due to the advent of Internet of Things (IoT) devices, edge cameras, and sensor-embedded hardware systems. However, raw pixels alone are meaningless and very difficult to interpret. CV turns pixels into structured knowledge, which allows the system to recognize patterns, objects, segment scenes, and provide real-world information to make decisions. For example, CV models can easily identify the cancerous lesions from radiographs and are now able to outperform human experts [7], track crop health and detect pests [8], and support real-time traffic management and face recognition-based access in smart cities [9]. The opportunities for transformational impact that CV possesses are underscored by its use in application frameworks.

The disruptive impact that CV technology has on market systems worldwide is profound. By 2030, according to reports by Grand View Research, the CV market is projected to surpass 58 billion USD, attributable to its 20% annual growth from the automotive, health, manufacturing, and security applications. Besides, the decreasing costs of sophisticated computers, combined with rapid improvements in deep learning systems and the abundance of development frameworks like OpenCV, TensorFlow, and Pytorch, have demystified and cheapened the cost of developing CV solutions.

In a nutshell, CV is a branch of artificial intelligence (AI) that enables computers and systems to automate human visual system processes. It refers to the process of automatically extracting useful information from images for purposeful output. In other words, CV aims to replicate in machines the capabilities of human vision and to automate tasks that require vision, such as object recognition, scene reconstruction,

DOI: 10.1201/9781003659112-1

motion estimation, and semantic segmentation. In general, CV performs three tasks: low level, mid level, and high level. The low-level tasks often include tasks related to per-pixel operations, such as denoising, sharpening, edge detection, and more simple enhancements. In contrast, tasks such as segmentation, recognizing contours, estimating motion, or recovering depth are mid level, whereas high-level vision tasks deal with semantic interpretation (object recognition, action understanding, scene 3D reconstruction, and interaction reasoning).

1.2 HISTORICAL BACKGROUND

CV started as a field in the 1960s, while even the AI community at that time was still in the nurturing stage. Earlier works were focused on simple tasks such as line drawing understanding, object-background segmentation, and hand-written character recognition, usually with synthetic images in controlled settings. Still, projects such as MIT's block-world studies were revolutionary, showing that visual understanding could be reduced to a computational problem [10].

During the 1970s and 1980s, CV matured into a discipline with foundational work; for example, edge detection has laid the foundation for decades of CV research [11,12]. In the 1990s, attention had shifted to Eigenfaces, hidden Markov models, Bayesian inference, and other mathematical approaches that went beyond hand-crafted geometry. Subsequently, feature engineering and feature descriptors, such as Scale-Invariant Feature Transform (SIFT) [13], Speeded-Up Robust Features (SURF) [14], and Histogram of Oriented Gradients (HOG) [15], have been popularly employed due to robust object detection on scale, rotation, and illumination. These approaches were used in applications such as object (image) stitching, 3D reconstruction, and pedestrian detection, usually in combination with machine learning classifiers, e.g., support vector machines (SVMs), to achieve state-of-the-art performance.

With the development of AlexNet, a deep convolutional neural network (CNN), the world has witnessed a significant advancement in the field of CV in the decade of 2010 [16]. It outperformed the classical approaches in the ImageNet Large-Scale Visual Recognition Challenge (ILSVRC) [17]. Over the next few years, most hand-engineered features were replaced by end-to-end training, and columns of networks learned feature hierarchies from raw pixel values. Subsequent architectural designs, such as VGGNet [18], ResNet [19], and Inception [20], have further enhanced classification accuracy and made complex tasks such as object detection, semantic segmentation, and action recognition more accurate and scalable.

Now in the 2020s, transformers and multimodal models have been increasingly employed for CV. Vision Transformers (ViTs) take inspiration from natural language processing (NLP) to model global relationships over the images [21]. Further, models, such as contrastive language–image pretraining (CLIP) bring vision and language together to perform zero-shot learning [22]. Meanwhile, neural radiance field (NeRF) and implicit 3D representations are revolutionizing scene reconstruction [23].

1.3 POPULAR DATASETS

Datasets and benchmarks have played an instrumental role in shaping the development of CV. Historically, in the absence of benchmark datasets, customized datasets have been used to validate the performance of CV algorithms. However, the models trained and compared on these do not have wide acceptance. Therefore, benchmark datasets such as PASCAL VOC, ImageNet, and COCO have been developed.

1.3.1 ImageNet

The ImageNet dataset is undoubtedly the most influential benchmark in the history of CV [24]. ImageNet contains over 14 million images organized into 1,000 top-level categories and over 20,000 labeled subsets. With its series of annual contests, the ILSVRC from the year 2010 to 2017 spurred rapid progress on image classification and detection. It remains a major touchstone for training image classification and other models for detection-based systems and continues to be used to set performance standards.

1.3.2 PASCAL VOC

The PASCAL Visual Object Classes (VOC) challenge was one of the earliest attempts at detection and segmentation of an object [25]. It was the first to be introduced and crucially benchmarked the annotations of an object's bounding box for the detection and segmentation of objects and action recognition along with rich object labels and bounding box annotations for 20 object categories. Through various evaluation processes, it, used to measure the mean average precision (mAP), is still the most used evaluation technique today.

1.3.3 COCO

The COCO dataset brought large-scale annotated images that account for real-world complexity. With more than 330,000 images and 1.5 million object instances with pixel-wise annotations across 80 object categories, COCO is a popular choice for a variety of tasks, including object detection, keypoint detection, instance segmentation, and image captioning [26]. Unlike previous datasets, COCO focuses on object detection in contexts, such as overlapping objects and occluded objects, and viewed from different angles, making it far more challenging. The current generation models (such as Faster R-CNN, YOLO, Mask R-CNN, and DETR) have been extensively evaluated against COCO to achieve state-of-the-art performance.

1.3.4 KITTI

The KITTI Vision Benchmark Suite has been core to autonomous driving research since its inception [27]. The database contains stereo images, optical flow, LiDAR point clouds, as well as GPS/IMU data. Its realistic driving scenes, such as cars, trucks, pedestrians, and cyclists, make it most suitable for autonomous vehicles and urban robotics. KITTI also attracted a great amount of research advances in multisensor fusion and 3D vision.

1.3.5 Cityscapes

The Cityscapes dataset was developed in 2016 with the aim of understanding urban scenes. It consists of 5,000 densely annotated images and 20,000 weakly annotated images acquired under various climatic and lighting conditions [28]. Cityscapes provides full-pixel annotations of 30 visual classes for semantic and instance segmentation, which are highly relevant in driving scenarios, like cars, pedestrians, roads, and buildings. It is widely used for segmentation models in relation to autonomous driving and smart-city scenes.

1.4 MAJOR APPLICATIONS OF COMPUTER VISION

CV has evolved to a cornerstone of modern intelligent systems that drives applications that have a direct impact on our daily lives and industrial processes. With the proliferation of data from IoT devices, surveillance cameras, drones, and satellites, the impact of CV technologies is being noticed in the transportation, healthcare, agriculture, retail, security, manufacturing, and robotics sectors. Further, as per Statista, the CV has been most extensively employed for manufacturing followed by healthcare and security, as depicted in Figure 1.1.

1.4.1 Autonomous vehicles

Self-driving cars constitute one of the most high-stakes applications of CV in which safety and accuracy are crucial. It is one of the key technologies that facilitates autonomous cars to be able to recognize the lanes, traffic signals, road signs, pedestrians, or other vehicles in real time [29]. Deep learning models, such as YOLO and Faster R-CNN, have been employed for obstacle detection, whereas SegNet-based semantic segmentation models are used to mark the boundaries of roads and drivable regions. Moreover, algorithms, such as Simultaneous Localization And Mapping (SLAM), have been employed to map the environment.

1.4.2 Healthcare

In medicine, CV has transformed diagnostic imaging, pathology, and surgical support. CNNs are used to process X-rays, CT scans, and MRIs to diagnose a range of

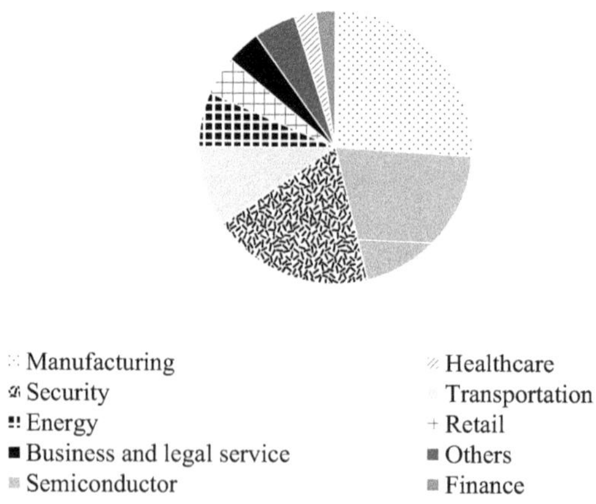

Figure 1.1 Computer vision market size across various verticals.

conditions including tumors, fractures, and pneumonia, with the accuracy often comparable with that of trained radiologists. High-resolution slide images are segmented in digital pathology to remove false detection and errors during diagnosis [30]. In the field of ophthalmology, applications from retinal scans for identifying early signs of diabetic retinopathy and glaucoma are widespread, while in the surgery domain, CV-assisted systems deliver real-time video guidance to vision-mediated robotic surgeries [31,32]. CV is enhancing both the quality and availability of healthcare worldwide by speeding up workflows, reducing costs, and extending the range of diagnostic tests carried out.

1.4.3 Surveillance and security

The CV technology has been highly recognized in surveillance and public safety because of its capability of automatic monitoring and active threat identification in real-time . Facial recognition systems have become increasingly common in airports, public transport, and access control, and deep CNN-based networks provide robust recognition across different illuminations and poses [33]. In addition to identity verification, CV-powered behavior analysis is used to recognize abnormality in the crowds, and forensic analysis is applied to make the low-quality video clearer for further investigations.

1.4.4 Industrial automation

In the world of industrial automation, CV has revolutionized, particularly with the advance of Industry 4.0. Automatic inspection systems utilize high-resolution images to identify surface anomalies (scratches, cracks, and misalignments) in the parts of the product line and provide uniform quality assurance for all parts of inspection. CV not only enhances the abilities of robotic arms to perform accurate tasks but also prevents costly breakdowns [2]. CV also helps in improving the cost-effectiveness and lowering manufacturing costs in smart factories.

1.4.5 Augmented reality and virtual reality

Augmented reality and virtual reality (AR and VR) systems utilize CV to offer engagement and interactivity with the virtual world. In AR systems, CV methods integrate virtual objects with the real world by accomplishing gesture recognition, scene mapping, and object tracking. VR systems, on the other hand, use CV mechanisms to capture and reconstruct 3D motions [34]. These technologies are applied to education, healthcare, and entertainment fields to foster an interaction with real and virtual environments.

1.4.6 Agriculture

CV drives the adoption of precision farming to ensure more sustainable and data-driven agriculture. Further, drones and ground robots equipped with cameras and CV algorithms have been used to monitor the physical conditions of fields, weed detection, and yield estimation [8]. Moreover, CV is also employed in agriculture to identify diseased plants, which helps in the proper calibration of spray equipment to optimize pesticide use on the field. This improves the total productivity as water and fertilizers are used more effectively [35]. Therefore, CV gives farmers the ability to maximize productivity with minimal environmental effects.

1.4.7 Retail and e-commerce

In the fields of retail and e-commerce, CV has improved customer experience and productivity in the supply chain. CV is utilized in automated checkout systems (e.g., Amazon Go) to monitor selected items by customers and perform payments without customer intervention. Similarly, robots can manage inventory warehouses. Online platforms utilize visual search engines that allow customers to search products by uploading images [36]. Therefore, these technologies help in optimizing store layout and transforming the retail industry.

1.4.8 Robotics

Computers receiving input from different sensors enable the robot to perceive and sense its environment. To navigate complex environments, SLAM is commonly employed. Whereas, visual odometry can perform mapping in complex scenarios [37]. For human–robot interaction, CV enables to perform recognition and follow natural interactions using gestures and face recognition. The incorporation of various technologies in robotics, such as 3D vision and reinforcement learning, enables the robot to work in a dynamically changing environment without human input. As a result, a rising number of robotics use-cases can be found in the manufacturing, healthcare, and service sectors.

1.5 CHALLENGES IN COMPUTER VISION

Nonetheless, CV, despite its flourishing success in multiple sectors, still suffers from a number of pressing challenges that it has yet to overcome to achieve more widespread use. These challenges stem from the complexity of visual data, insufficient advanced algorithms, and the practicality of the CV system in reference to its ethical use. These challenges are some of the most critical for the next generation of CV systems.

1.5.1 Illumination variations

Objects can look very different in images under different lighting conditions. Shadows, reflections, and low illuminations significantly affect the performance of the recognition system. For example, a face recognition model learned from well-illuminated indoor images may not perform well when used in an outdoor scenario with changing sunlight. Older techniques tried to normalize illumination by applying histogram equalization and color constancy methods; more recent deep learning models use data with a wider range of different illuminations. However, constructing illumination-invariant CV systems is still facing a significant challenge, particularly for safety-critical scenarios that involve autonomous driving.

1.5.2 Occlusion and clutter

Objects in real environments might be occluded (partially hidden) or may be affected by clutter, which poses great challenges to the detection and recognition of the objects. For example, it is very difficult to recognize a pedestrian behind a parked car or an item on a crowded shelf. The feature-based methods, such as SIFT and HOG, find it difficult to detect the objects in occluded regions, but deep learning models try to solve this by using context, attention mechanisms, or generative models to infer occluded parts. Despite these advances, robust occlusion handling is still a challenging task, particularly in the presence of dynamic clutter in busy environments such as traffic intersections or busy public places.

1.5.3 Real-time processing and scalability

CV applications, such as autonomous vehicles, robotics, surveillance, and AR/VR, need to analyze high-resolution visual data with low latency. However, deep networks with millions of parameters require high computational power, making it difficult to deploy on edge devices. Systems also encounter scalability challenges when they have to handle continuous streams from hundreds or thousands of cameras, as in smart-city scenarios. Solutions such as model compression, quantization, and dedicated hardware (GPUs, TPUs, and edge AI chips) have made it possible for some models; however, the trade-off between accuracy, latency, and energy still remains a major challenge.

1.5.4 Generalization and domain adaptation

CV models often struggle with generalization, which means that though these models perform well on benchmark datasets, they tend to fail when deployed to a new environment. This loss of generalization is due to the problem-specific bias: an image-based fruit recognition model trained on the laboratory dataset could have very poor performance when applied to a real farm due to its low tolerance for diversities in illumination, background, or species. Adversarial training, domain adaptation, and transfer learning approaches try to reduce the domain shift effects with fine-tuning data-limited models with respect to target data or with adversarial training to align the feature distributions between domains. But achieving robust generalization across complex real-world scenarios still remains an open question.

1.5.5 Interpretability and explainability

The black-box nature of the modern CV systems, in particular deep neural networks, makes it challenging to understand why a model made a specific decision. This lack of explainability is particularly limiting the applicability of CV in high-stakes domains, such as healthcare and public safety and law enforcement. Although efforts have been made to enhance the interpretability, they achieve only partial explainability. Therefore, there is an urgent requirement for the development of an accurate and transparent CV system because it has a direct impact on ethical adoption, regulatory approval, and user trust.

1.6 ETHICAL CONCERNS AND SOCIAL IMPLICATIONS

As their use becomes more ubiquitous, CVs raise pressing ethical and social issues. While CVs can improve the realms of health, safety, and automation, they raise deeper concerns regarding privacy, equity, and responsibility in surveillance and security, in addition to trust and autonomous decision-making systems.

1.6.1 Surveillance ethics and privacy

The fundamental problem with CVs is mass surveillance. The potential loss of privacy to surveillance systems and the use of CCTV and facial recognition at airports and train stations are alarming. These technologies aid security and law enforcement, yet they also endanger privacy and individual rights. There is a dispute regarding where surveillance and privacy begin and where they end. Therefore, balancing security and individual rights is one of the critical ethical concern.

1.6.2 Facial recognition bias and fairness

There has been a considerable amount of research regarding the difficulty and accuracy of recognition of particular classes in facial recognition systems. For example, models tend to be less accurate with classes of women, children, and darker-skinned individuals. Such discrepancies can lead to adverse outcomes, including wrongful arrests and other forms of social discrimination, including denial of service, employment, and protection under the law. The absence of fairness in automated decision-making diminishes user confidence and promotes unaccountability. Thus, fairness–bias in algorithmic decisions, unrepresentative dataset construction, or legal frameworks should be intensely contained.

1.6.3 Deepfakes and misinformation

There are other serious issues related to the advent of deepfake technology. The fusion of CV and deep learning can realistically alter a person's face and voice and even the entire scene. While the creative world can greatly benefit from such generative models, it can also be weaponized for the spread of misinformation, politically charged falsehoods, and other malicious intent. Deepfakes are extremely dangerous as they erode trust in visual media, making it nearly impossible to discriminate between real media and contents generated by a computer. Thus, issues of deepfake detection, watermarking, and the legal frameworks should be carefully adressed.

1.7 CONCLUSION

Since its inception, CV was imagined mainly as a concept for checking fundamental shapes. At that time, its possibilities were limited to bare primitives that CV was able to recognize. From that moment to now, the leap that the whole domain has achieved is hard to comprehend. Today, CV is able to handle describing features of objects, detecting, segmenting, as well as performing 3D reconstruction. We are able to define its usage and resume its wide usage, including autonomous machines, telemedicine, customer surveillance, machines, virtuality systems, field, shops, and intelligent assistants. Although we achieved so much, multiple CV systems still have major ethical issues to resolve. Also, the real-world scenarios impose challenges like illumination variation, occlusion, and clutter. In addition, CV systems often encounter difficulty in generalization across domains and lack interpretability, therefore limiting their trustworthiness in high-stakes scenarios.

The future of CV will not solely depend on the refinement of algorithms and hardware but also on the integration of fairness, transparency, and accountability in the design of these systems. Its further development holds the potential to transform business and society, if innovation is governed by responsibility.

REFERENCES

1. H. Gupta, O.P. Verma, Monitoring and surveillance of urban road traffic using low altitude drone images: A deep learning approach, Multimed. *Tools Appl.* (2021) 1–21. https://doi.org/10.1007/s11042-021-11146-x
2. I. Yousif, L. Burns, F. El Kalach, R. Harik, Leveraging computer vision towards high-efficiency autonomous industrial facilities, *J. Intell. Manuf.* 36 (2025) 2983–3008. https://doi.org/10.1007/s10845-024-02396-1
3. L. Yang, R. Kumar, R. Kaur, A. Babbar, G.S. Makhanshahi, A. Singh, R. Kumar, A. Bhowmik, A.H. Alawadi, Exploring the role of computer vision in product design and development: A comprehensive review, *Int. J. Interact. Des. Manuf.* 18 (2024) 3633–3680. https://doi.org/10.1007/s12008-024-01765-7
4. K. Sai Suryaa, C. Saravana Kumar, S. Sudharsan, K.M. Anandkumar, ARNAV: Computer vision and machine learning-based augmented reality-based Indoor navigation system, in: *2024 IEEE International Conference on Computing, Power and Communication Technologies (IC2PCT)*, vol. 5, 2024: pp. 519–524. https://doi.org/10.1109/IC2PCT60090.2024.10486552
5. J. Qi, L. Ma, Z. Cui, Y. Yu, Computer vision-based hand gesture recognition for human-robot interaction: A review, *Complex Intell. Syst.* 10 (2024) 1581–1606. https://doi.org/10.1007/s40747-023-01173-6
6. S. Fuentes, C. Gonzalez Viejo, E. Tongson, F.R. Dunshea, The livestock farming digital transformation: Implementation of new and emerging technologies using artificial intelligence, *Anim. Heal. Res. Rev.* 23 (2022) 59–71. https://doi.org/10.1017/S1466252321000177
7. S.B. Mukadam, H.Y. Patil, *Machine Learning and Computer Vision Based Methods for Cancer Classification: A Systematic Review*, Springer Netherlands, 2024. https://doi.org/10.1007/s11831-024-10065-y
8. S. Jain, R. Sahni, T. Khargonkar, H. Gupta, O.P. Verma, T.K. Sharma, T. Bhardwaj, S. Agarwal, H. Kim, Automatic rice disease detection and assistance framework using deep learning and a chatbot, *Electron.* 11 (2022). https://doi.org/10.3390/electronics11142110
9. G. Liu, H. Shi, A. Kiani, A. Khreishah, J. Lee, N. Ansari, C. Liu, M.M. Yousef, Smart traffic monitoring system using computer vision and edge computing, *IEEE Trans. Intell. Transp. Syst.* 23 (2022) 12027–12038. https://doi.org/10.1109/TITS.2021.3109481
10. L.G. Shapiro, Computer vision: The last 50 years, *Int. J. Parallel Emergent Distrib. Syst.* 5760 (2020). https://doi.org/10.1080/17445760.2018.1469018
11. D. Marr, E. Hildreth, Theory of edge detection, *Proc. R. Soc. London. Ser. B. Biol. Sci.* 207 (1980) 187–217. https://doi.org/10.1098/rspb.1980.0020
12. L. Ding, A. Goshtasby, On the Canny edge detector, *Pattern Recognit.* 34 (2001) 721–725.

13. D.G. Lowe, Object recognition from local scale-invariant features, in: *Proceedings of the Seventh IEEE International Conference on Computer Vision*, vol. 2, 1999: pp. 1150–1157. https://doi.org/10.1109/ICCV.1999.790410
14. H. Bay, T. Tuytelaars, L. Van Gool, SURF: Speeded up robust features, in: *European Conference on Computer Vision*, Berlin, Heidelberg: Springer Berlin Heidelberg, 2006: pp. 404–417.
15. N. Dalal, B. Triggs, D. Europe, Histograms of oriented gradients for human detection, in: *2005 IEEE Computer Society Conference on Computer Vision and Pattern Recognition (CVPR'05)*, 2005.
16. A. Krizhevsky, I. Sutskever, G.E. Hinton. Imagenet classification with deep convolutional neural networks. *Advances in Neural Information Processing Systems* 25 (2012).
17. O. Russakovsky, J. Deng, H. Su, J. Krause, S. Satheesh, S. Ma, Z. Huang, A. Karpathy, A. Khosla, M. Bernstein, A.C. Berg, L. Fei-Fei, ImageNet large scale visual recognition challenge, Int. J. Comput. Vis. 115 (2015) 211–252. https://doi.org/10.1007/s11263-015-0816-y
18. K. Simonyan, A. Zisserman, Very deep convolutional networks for large-scale image recognition, in: *3rd International Conference on Learning Representations, ICLR 2015 – Conference Track Proceedings*, 2014. https://doi.org/10.48550/arxiv.1409.1556
19. K. He, X. Zhang, S. Ren, J. Sun, Deep residual learning for image recognition, in: *Proceedings of the IEEE Conference on Computer Vision and Pattern Recognition*, IEEE Computer Society, 2016: pp. 770–778. https://doi.org/10.1109/CVPR.2016.90
20. C. Szegedy, W. Liu, Y. Jia, P. Sermanet, S. Reed, D. Anguelov, D. Erhan, V. Vanhoucke, A. Rabinovich, Going deeper with convolutions, 2014. https://arxiv.org/abs/1409.4842 (accessed August 26, 2022).
21. A. Dosovitskiy, L. Beyer, A. Kolesnikov, D. Weissenborn, X. Zhai, T. Unterthiner, M. Dehghani, M. Minderer, G. Heigold, S. Gelly, others, J. Uszkoreit, N. Houlsby, An image is worth 16x16 words: Transformers for image recognition at scale, ArXiv Prepr. ArXiv2010.11929, 2020. https://doi.org/https://doi.org/10.48550/arXiv.2010.11929
22. A. Radford, J. Wook, K. Chris, H. Aditya, R. Gabriel, G. Sandhini, G. Sastry, A. Askell, P. Mishkin, J. Clark, G. Krueger, I. Sutskever, Learning transferable visual models from natural language supervision, in: *International Conference on Machine Learning*, Long Beach Convention Center, Long Beach, California, 2019.
23. B. Mildenhall, P.P. Srinivasan, M. Tancik, J.T. Barron, R. Ramamoorthi, R. Ng, U.C. Berkeley, NeRF: Representing scenes as neural radiance fields for view synthesis, in: Lecture Notes in Computer Science, Springer International Publishing, 2020: pp. 405–421. https://doi.org/10.1007/978-3-030-58452-8
24. J. Deng, W. Dong, R. Socher, L. Li, K. Li, L. Fei-Fei, ImageNet: A large-scale hierarchical image database, in: *2009 IEEE Conference on Computer Vision and Pattern Recognition*, Miami, FL, USA, 2009: pp. 248–255.
25. M. Everingham, L. Van Gool, C.K.I. Williams, J. Winn, A. Zisserman, The Pascal Visual Object Classes (VOC) challenge, *Int. J. Comput. Vis.* 88 (2010) 303–338. https://doi.org/10.1007/s11263-009-0275-4
26. T.Y. Lin, M. Maire, S. Belongie, J. Hays, P. Perona, D. Ramanan, P. Dollár, C.L. Zitnick, Microsoft COCO: Common objects in context, in: *Lecture Notes in Computer Science (including subseries Lecture Notes in Artificial Intelligence and Lecture Notes in Bioinformatics)*, 8693 LNCS, 2014: pp. 740–755. https://doi.org/10.1007/978-3-319-10602-1_48
27. A. Geiger, P. Lenz, C. Stiller, R. Urtasun, Vision meets robotics : The KITTI dataset, *Int. J. Robot. Res.* 32(11) (2013), 1231–1237. https://doi.org/10.1177/0278364913491297
28. M. Cordts, M. Omran, S. Ramos, T. Rehfeld, M. Enzweiler, R. Benenson, U. Franke, S. Roth, B. Schiele, The cityscapes dataset for semantic urban scene understanding, in: *Proceedings of the IEEE Conference on Computer Vision and Pattern Recognition*, 2016: pp. 3213–3223. https://doi.org/10.1109/CVPR.2016.350

29. F. Qu, N. Dang, B. Furht, M. Nojoumian, Comprehensive study of driver behavior monitoring systems using computer vision and machine learning techniques, *J. Big Data.* 11 (2024). https://doi.org/10.1186/s40537-024-00890-0
30. Y. Matsuzaka, R. Yashiro, The diagnostic classification of the pathological image using computer vision, *Algorithms.* 18 (2025) 1–32. https://doi.org/10.3390/a18020096
31. M. Oulhadj, J. Riffi, C. Khodriss, A.M. Mahraz, A. Yahyaouy, M. Abdellaoui, I.B. Andaloussi, H. Tairi, Diabetic retinopathy prediction based on vision transformer and modified capsule network, *Comput. Biol. Med.* 175 (2024) 108523. https://doi.org/10.1016/j.compbiomed.2024.108523
32. A.Y. Tsai, S.R. Carter, A.C. Greene, Artificial intelligence in pediatric surgery, *Semin. Pediatr. Surg.* 33 (2024) 151390. https://doi.org/10.1016/j.sempedsurg.2024.151390
33. S. Kumar, D. Yadav, H. Gupta, M. Kumar, O.P. Verma, Towards smart surveillance as an aftereffect of COVID-19 outbreak for recognition of face masked individuals using YOLOv3 algorithm, *Multimed. Tools Appl.* 82 (2023) 8381–8403. https://doi.org/10.1007/s11042-021-11560-1
34. M. Rajendran, C.T. Tan, I. Atmosukarto, A.B. Ng, S. See, Review on synergizing the Metaverse and AI-driven synthetic data: Enhancing virtual realms and activity recognition in computer vision, *Vis. Intell.* 2 (2024). https://doi.org/10.1007/s44267-024-00059-6
35. S. Ghazal, A. Munir, W.S. Qureshi, Computer vision in smart agriculture and precision farming: Techniques and applications, *Artif. Intell. Agric.* 13 (2024) 64–83. https://doi.org/10.1016/j.aiia.2024.06.004
36. W. Villegas-Ch, A.M. Navarro, S. Sanchez-Viteri, Optimization of inventory management through computer vision and machine learning technologies, *Intell. Syst. with Appl.* 24 (2024) 200438. https://doi.org/10.1016/j.iswa.2024.200438
37. Y. Cohen, A. Biton, S. Shoval, Fusion of computer vision and AI in collaborative robotics: A review and future prospects, *Appl. Sci.* 15 (2025) 1–30. https://doi.org/10.3390/app15147905

Chapter 2

Latest trends in blockchain, generative AI, and computer vision

Shivaratri Narasimha Rao, Mohammed Ali Shaik, Imran Qureshi, and Salman Ali Syed

2.1 INTRODUCTION

2.1.1 Brief background on emerging technologies and their transformative role

The digital landscape of the last 20 years has experienced a paradigm shift due to the quick development of new technologies. Blockchain, generative artificial intelligence (AI), and computer vision (CV) are some of the innovations whose early experimental years are the past and the present the basis of transformation centers in a wide range of industries [1]. What they have in common is that they can transform a traditional business process, challenge a conventional business model, and open new sources of value creation.

The history of technological progress always refers to several different waves, which have stretched since the Industrial Revolution to the Digital Revolution. The present age in many ways can be described as the Fourth Industrial Revolution (Industry 4.0) and is dominated by the so-called blend of technologies diluting the borders between the physical, digital, and biological worlds. Blockchain, GAI and CV have been deemed a high-impact area that is transforming industries and the society so far [2]. The driving forces of such technologies include an exponential growth in computing power, widespread access to high-speed connections, vast amounts of data, and new classes of deep learning algorithms. The addition of cloud infrastructure, edge computing, and Internet of Things (IoT) devices has put this on the fast track, thus becoming commercially viable and available to a globalized audience.

The idea of a decentralized and unchangeable ledger was introduced when blockchain came to the minds of the people in 2008, when Bitcoin was launched. First purely used in cryptocurrencies, blockchain has now developed into a multifaceted platform. Contemporary uses of blockchain are also not limited to financial applications and are used in supply chain transparency, secure management of medical records, decentralized finance (DeFi), smart contracts, and nonfungible tokens (NFTs) to represent digital ownership [3]. Some of the early limitations of the technology have been overcome by recent trends, including energy-efficient consensus mechanisms (proof of stake, PoS) and layer-2 optimizations, providing scalability. The disruption that

 DOI: 10.1201/9781003659112-2

blockchain brings with it is a concept of trust redefined in the context of transparent and safeguarded transactions, when the intervention of centralized governance is not essential anymore.

GAI was initially a self-limiting field with models such as variational autoencoders (VAEs) in the early 2010s, but after the publication of generative adversarial networks (GANs) in 2014, it became much more widely studied. Most recently, there have been transformer architectures, GPT, and stable diffusion, which expand the scope of what is possible to generate by AI [4]. Current GAI has the ability to generate a near-photorealistic level of images, videos, text, and audio that is virtually impossible to detect or tell apart. It has various applications across diverse faculties like art, advertising, as well as entertainment sectors on the one hand and across scientific sectors like drug discovery and architectural designs on the other hand. The revolutionary side to the technology is the fact that it has changed creativity by making it democratized so that everyone can create high-quality content without being a specialist and has posed significant questions on authorship, authenticity, and ethics.

AI: CV is one of the sub-fields. It is concerned with the use of machines to explain visual information. Deep learning has seen CVCV develop at a greater pace. A watershed moment followed in 2012 when AlexNet won the ImageNet challenge, which led to a burst of inventions in object detection, image segmentation, and facial recognition. Autonomous vehicles, industrial inspection, medical imaging, and augmented reality (AR) are some of the applications in which real-time CVCV is used today [5]. Vision models (Vision Transformers or ViT) based on transformers have also increased accuracy and flexibility. CVCV has the transformational power of automating perception-based tasks, enabling computers to perform in manipulated environments and contexts that were previously the preserve of humans.

When considered separately, all these technologies are quite potent each in their way, but together, they form synergies that even multiply their effect. To give one example, one could prove the provenance and authenticity of AI-made media, allaying deepfake anxieties, with blockchain, and then use CVCV to strengthen those verifications using blockchain-based identity verification solutions. GAI and CVCV can achieve automated visual creation of content that is grounded in existing data via visual, so-called real-world data [6]. A combination of all three can then drive more complex systems, like the metaverse or secure digital identity frameworks, and personalized medicine platforms. It turns out that with such a conjunction, the cycles of innovation are accelerated, trust in the relationship on the digital front increases, and new economic opportunities are opened.

Blockchain, GAI, and CVCV integration and adoption can disrupt industries and government and shift cultural standards. They will be able to reduce innovation cycles, increase traceability, and make user experiences hyper-personalized. Nevertheless, their disruptive nature is mixed with the issues of ethics, privacy, bias, and regulation [7]. The challenge of fast technological changes is that quick and engaging policy-making and joint governance are needed to maintain equal opportunities and to prevent hazards. These technologies are the backbones of the new audiovisual era and must be managed with tremendous responsibility—ranging from economic events to human intelligence and machine sovereignty.

2.1.2 Overview of blockchain, generative AI, and CV as distinct yet interconnected domains

2.1.2.1 Blockchain technology

Blockchain represents one example of distributed ledger technology (DLT), which allows maintaining a secure and tamper-free record-keeping process without a central authority. Records are recorded in chronological order of cryptographically linked blocks of data, with each block having some number of transactions [8]. The popular consensus mechanism by which the transactional validity is agreed upon includes the PoW, PoS consensus mechanisms, and newer energy-saving protocols.

The technology can be described as decentralized, i.e., the elimination of the single points of critical failure; transparent, i.e., the ability to publicly check records; secure, i.e., the use of cryptographic algorithms; and immutable, i.e., retroactive resistance to change. In addition to cryptocurrencies, blockchains are used in a wide variety of applications such as keeping track of provenance in a supply chain, DeFi, digital identity, secure storage of medical records, and contracts between parties that are automated using smart contracts.

2.1.2.2 Generative artificial intelligence

GAI is an application of AI used to synthesize new examples of data that look similar to patterns in training data. It exploits the capabilities of GANs, VAEs and transformer-based diffusion models to produce realistic content in many forms, including text, pictures and audio, and 3D structures [9]. Prominent GAI functionalities are content generation to generate new multimedia output, style transfer to modify the esthetic content without losing meaning, and simulation and modeling to synthetically generate designed data in the case of scarce real-world data. Its applications extend to automated marketing content development, virtual environment design, drug discovery, the creation of synthetic media, and prototyping of personalized products. Although its creative possibilities are limitless, GAI also houses valuable ethical questions pertaining to deepfakes, misinformation, and intellectual property.

2.1.2.3 CV

CV is an area of AI that aims at bringing machines to be able to see, understand, and respond to what is seen in their surroundings through visual information. Early work has been incredibly dependent on handcrafted features, though deep learning and convolutional neural networks (CNNs) have led to a massive improvement in the field. Such developments have enabled CVCV systems to perform as humans do with regard to image classification, object detection, segmentation, and even achieving 3D reconstructions [10]. Examples of contemporary commercial applications of CVCV are self-driving vehicles, healthcare diagnostics, intelligent surveillance, manufacturing quality control, and augmented (AR) and virtual reality (VR) immersion systems [11]. More recently, the introduction of ViTs [12] has opened up these systems to even greater capacities, i.e., adaptation and ever-increasing accuracy in the performed visual understanding tasks.

2.1.2.4 Flow of connection between the domains

- Blockchain, GAI, and CV are different in their intent and underlying technical basis but also define powerful, synergistic combinations when combined.
- Combined, blockchain and GAI could be used to verify the credentials of AI-generated content, uphold intellectual property rights, and enable marketplaces for digitalizable creative items to run on a decentralized basis.
- Blockchain and CV help approve visual data securely and free of fraud, especially in critical fields such as biometric checks and medical imaging audit.
- GAI and CV combine forces to increase content production and visual data augmentation, creating realistic synthetic training datasets and increasing the ability to restore images.
- A combination of the three technologies can help develop groundbreaking solutions that can be seen as secure AI-driven visual analytics in healthcare, authenticated virtual spaces in the metaverse, and privacy-preserving digital ID systems.

2.1.2.5 Some uses of integration

The combination of these fields opens out a variety of high-impact applications. An example of this is NFT marketplaces featuring AI-generated art, which uses blockchain to record ownership alongside GAI as a content creation technology [13]. In facial recognition, the combination of blockchain identity management and CVCV to increase privacy and trust is done through decentralized facial recognition systems. In healthcare, GAI and CVCV in a format of AI-enhanced medical imaging can be safely saved in the blockchain network for higher reliability of diagnostics [14]. Lastly, the combination of the three technologies can be used to facilitate tamper-free AR/VR metaverse spaces, which can guarantee safe immersive digital experiences.

Table 2.1 shows the comparison of the three areas according to the main intent, technologies, data types, their main strong and weak areas, and an application example. Blockchain is a form of technology that enables trust in maintaining secure,

Table 2.1 Unique aspects of Blockchain, Generative AI, and Computer Vision

Feature	*Blockchain*	*Generative AI*	*Computer Vision*
Core purpose	Secure, decentralized, tamper-proof record-keeping	Create new data/content based on learned patterns	Interpret and understand visual data
Underlying technology	Distributed ledger and consensus algorithms	GANs, VAEs, Diffusion Models, and Transformers	CNNs, Vision Transformers, and deep learning
Data type	Transactional and structured data	Text, images, audio, and 3D models	Images, videos, and spatial data
Key strength	Trust, transparency, and security	Creativity, content synthesis, and simulation	Perception, recognition, and spatial awareness
Limitations	Scalability, energy use, and regulation	Ethical concerns, bias, and data misuse	Privacy concerns, data bias, and adversarial attacks
Example applications	DeFi, supply chain, and identity	Digital art, drug design, and synthetic media	Autonomous vehicles and healthcare imaging

Table 2.2 Examples of integrated use cases

Integration	*Example use case*	*Benefit*
Blockchain + GAI	NFT marketplaces for AI-generated art	Secure ownership and transparent sales history
Blockchain + CV	Decentralized facial recognition access control	Privacy-preserving authentication
GAI + CV	AI-enhanced medical imaging for rare diseases	Improved diagnostic accuracy with synthetic data
All three	Verified AR/VR metaverse environments	Secure, immersive, and tamper-proof digital experiences

decentralized, and impervious records at scale with the largest amount of transactional and structured data. GAI is best applied in the creation of new and human-like multimedia materials with the aid of highly advanced models such as GANs and diffusion networks, whereas CVCV is concerned with visual data interpretation and understanding using perception models that are based on deep learning. As seen in the table, although the technologies follow different data modalities and fulfil different purposes, none of them lacks the capacity to revolutionize numerous industries, as shown in Table 2.2.

As depicted in Table 2.2, the three technologies can be paired or combined in a combination to generate innovative applications. To give an example of how this can be done, through blockchain and GAI, we can create NFT marketplaces where AI-generated art will be stored and where we could verify the ownership and its history of transactions. Integration of CV and blockchain can facilitate a decentralized and privacy-preserved facial recognition system, and GAI and CV may be used to facilitate realistic generation of synthetic data to benefit medical imaging. The availability of all three technologies favors the evolution of safe, authorized, and immersive metaverse spaces. The table highlights the complementarity of these fields and unlocks high-value, cross-disciplinary solutions through their intersection.

2.1.3 Motivation for integrating these technologies

2.1.3.1 Dealing with limitations of separate technologies

Despite their high potential, the aforementioned technologies, blockchain, GAI, as well as CV, explicitly have certain issues when implemented individually but can be compensated, to some extent, by teamwork. Blockchain provides a level of trust, transparency, and immutability but is not able to create or analyze complex multimedia information in a manual way [12]. Gen-AI software may create very realistic output, and although it also has no built-in way to check the authenticity or source of its output, it is susceptible to manipulation into producing deepfakes. CV equips computers and systems with vision and the ability to analyze the world with imagery; however, the results of the processes can be altered due to the manipulation of the data or adversarial attacks on the processes.

2.1.3.2 Securing data and increasing data provenance, security, and trust

During the digital age, authenticity, integrity, and origin of the data are highly pertinent areas of concern. When combined with GAI and CV, blockchain can deliver strong ecosystems where any generated or captured data can be secured with cryptography, time-stamped, and made immutable. Blockchain may be used to provide a decentralized record of how AI images have been created or what outcome a CVCV-based surveillance captured, hence increasing accountability and minimizing the possibility of falsification.

2.1.3.3 Moving innovation in multimodal, multifaceted applications

The combination of these technologies allows the creation of completely new categories of applications that cannot be carried out by any single domain. To illustrate, immersive 3D environments can be generated with GAI, CVCV can offer real-time interaction and tracking of user behavior, and blockchain can define secure transactions and ownership of virtual assets, all within the metaverse. Likewise, in the precision medicine sector, blockchain may keep and organize safe patient records, GAI may model a drug effect, and CVCV may process medical image diagnostics. These are the multimodal applications that provide more engaging, interacting, and more credible experiences, which translates to an effective incentive that researchers, developers, and businesses would want to accomplish integration.

2.1.3.4 Supporting ethical and regulatory compliance

The rapid growth of GAI and CV raises ethical concerns about misinformation, privacy, and bias, while blockchain faces scrutiny over energy consumption and scalability. Integration allows these challenges to be managed more effectively. For instance, blockchain can be used to enforce digital rights management and traceability for AI-generated content, ensuring compliance with copyright laws. CV outputs used for surveillance or biometric verification can be securely logged on blockchain networks, enabling transparent auditing and regulatory oversight.

2.1.3.5 Unlocking economic and societal value

Integrating blockchain, GAI, and CV holds significant promise for creating new business models, reducing operational inefficiencies, and enhancing societal well-being. In industries like supply chain management, combining blockchain's traceability with CVCV's inspection capabilities and GAI's predictive modeling can result in unprecedented efficiency and cost savings. In cultural preservation, AI-generated reconstructions of historical sites can be authenticated through blockchain to prevent forgery, while CVCV aids in digitizing artifacts. The motivation, therefore, is not only solely technical but also economic and societal—these integrated systems have the potential to reshape how we work, create, transact, and interact in both physical and digital realms.

In Figure 2.1, the conceptual framework diagram shows how blockchain, GAI, and CV, which are different in their mechanisms, unite to reach common objectives and deliver many integrated life-changing results. Blockchain plays a role in security, authenticity, transparency, and governance, which guarantees that processes and data can be traced and verified as tamper-proof. GAI opens the door to increased creativity, innovation, and tracking of provenance of generated content, whereas CV provides a boost in automation, perception, and reliability of visual information. Such joint objectives as security and privacy, authenticity and provenance, and transparency

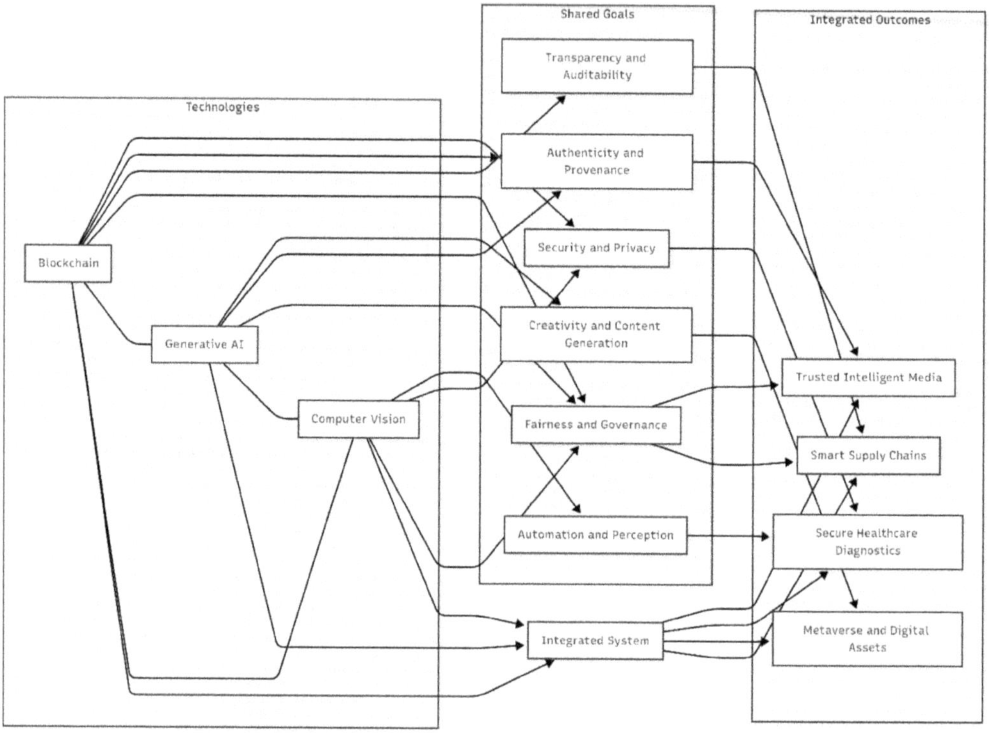

Figure 2.1 Conceptual framework.

and auditability are bridging grounds between the technologies. The combination of all three makes high-impact applications, such as smart healthcare diagnostics, secure trust-enabled media, smart supply chain, and metaverse environments with verified digital assets. The connected ecosystem of the mutually reinforcing application of these technologies emphasizes that, rather than ideally complementing each specific technology, collective usage creates more than the sum of its parts, in terms of functionality, security, and ethics of digital solutions.

2.1.4 Chapter objectives and scope

Objectives

This chapter will

- Study the recent innovations and development of the fields of blockchain, GAI, and CV as separate yet related fields of technology.
- Study the technical background and strengths and weaknesses of the respective technologies, giving a clear insight about the existing and emerging uses of the technologies.
- Research the integration capabilities of the blockchain, GAI, and CV to overcome the issue of data authenticity, privacy, scalability, and controlled governance.

- Identify use cases across domains and other inventive solutions where these technologies come together to build high-value solutions in domains including healthcare, supply chain, finance and creative arts, and the metaverse.
- Illustrate central issues, risks, and implications of ethics that need to be overcome in order to realize sustainable and responsible adoption of these integrated technologies.

Scope

This chapter is concerned with the following:

- The introduction to blockchain, GAI and CV, how it evolved, and what its main capabilities are as well as the state of the art in development.
- A timeline of key deep dive into industry applications and the ways in which these kinds of technologies are transforming such things as security, automation, and creativity in immersive environments.
- In order to examine how the combined disciplines may address current technological shortfalls to present new possibilities of innovation, a cross-disciplinary survey of the possibilities of the two fields will be discussed.
- Real-life and possible examples of implementing integrated solutions depicted using conceptual frameworks and case studies.
- Discussion of policy, regulatory, and ethical structures that must be in place to use these technologies in a responsible manner one day.

This chapter does not intend to offer weary implementation guidelines or programming tutorials but is a strategic-conceptual piece aimed at complementing the researcher, practitioner, and/or policymaker who wants to know more about its potential and relevance as per the current context of convergence among these technologies.

2.2 BLOCKCHAIN: EVOLVING BEYOND CRYPTOCURRENCY

Utilized as the underlying technology behind the Bitcoin blockchain in 2008, blockchain technology has shown explosive growth within a short time span to become a broader digital innovation capable of so much more than what the term was intended to suggest from the beginning. Fundamentally, blockchain is a DLS that can be used to create decentralized and tamperproof record-keeping across numerous nodes. Removal of centralized middlemen is a plus, whose benefits include increased transparency, security, and resilience in this architecture. Although in the initial stages, it was mostly the role of financial transactions and the central functions of the technology, such as being immutable, decentralized, and consensus-based validation, that have come to fit the needs of a large variety of industries.

Among the most dramatic developments in the history of blockchain, its growth into sector-specific and enterprise applications can be ranked among the major ones. With the supply chain business, the blockchain allows real-time tracking of products along the supply chain—increasing supply chain traceability and decreasing potential fraud. It is used in healthcare to provide patients' records with secure and interoperable storage so that your healthcare is safe and complies with privacy requirements but still allows sharing of data with other healthcare providers. There are also DeFi platforms, which provide blockchain-based substitutes to conventional banking services, allowing peer-to-peer lending, decentralized exchanges, and programmable financial contracts. Besides, smart contracts, self-executing contracts that are integrated into blockchains, enable terms to be executed trustlessly and automatically and have revolutionized law, insurance, and logistics.

2.2.1 Overview of blockchain technology

2.2.1.1 Core principles: decentralization, immutability, and transparency

Blockchain technology has several guiding principles that constitute its architecture, governance, and its trust model. When it is stated that blockchain has the principles of decentralization, immutability, and transparency, it is not the technical characteristics of the phenomenon but the main values that distinguish blockchain applications that make them different and more valuable than mainstream data management systems.

2.2.1.1.1 Decentralization

In traditional systems, there is a centralized power where the data are held by a bank, government agency, or corporate server, which verifies, holds, and performs transactions. Some of these weaknesses of this centralization include a potential single point of failure, susceptibility to cyberattacks, and inherent risk of unilateral control or censorship.

Blockchain does not give a single source of control because it disseminates the ledger on a peer-to-peer network of nodes, where each of the operators has an updated copy of the data. Consensus is reached by the network in order to validate transactions using a mechanism such as Proof of Work (PoW), Proof of Stake (PoS), or newer variations such as Delegated Proof of Stake (DPoS) and Practical Byzantine Fault Tolerance (PBFT). This makes it such that no individual actor is capable of tampering with the ledger without a majority consensus.

Decentralization also promotes resilience and uptimes, but more than that, it democratizes trust, and this aspect makes blockchain highly applicable to financial applications, government and governance, and all other facets of collaboration, where divisions of power are essential.

2.2.1.1.2 Immutability

Blockchain is characterized by one important aspect: it is almost impossible to change the information already saved in it and confirmed. This is done by cryptographic connections between blocks, where every block has its exclusive hash created using block content and a hash of a previous block connected to it. An attempt to manipulate a transaction would invalidate all subsequent blocks and warn the network because it changes its hash.

This property provides guarantee of data integrity, which is essential to programs with a strong need of records retention over long periods—land registries, intellectual property databases, or audit histories. Immutability further prevents frauds since a third party would have to access a substantial portion of the network computing resources (the so called 51% attack) to update the ledger, which is unfeasible to do in many cases in an effectively secured blockchain.

The principle has great consequence as part of the legal admissibility of records, adherence to audit rules, and tracing of history in multistakeholder ecosystems.

2.2.1.1.3 Transparency

One of the things is that blockchain is transparent since all participants within the network can access and confirm the history of transactions; there is no concealment or distortion of the data. With public blockchains, transaction information is not only pseudonymous but also transparent to anyone; the notion of independent verification of activities is possible without trusting an intermediary.

In even the permissioned or private blockchain, transparency is imposed by authorized stakeholders so that every stakeholder acts on a unified source of truth. This minimizes the conflicts, hastens reconciliation procedures, and enhances the responsibility.

Nonetheless, transparency should be countered by privacy-preserving, such as zero-knowledge proofs, confidential transactions and selective disclosure, in particular during a sensitive application in healthcare or financial services where the transaction content might have to be kept secret yet verifiable.

The outcome is a trust layer by which multiparty collaborations can be facilitated without having to believe all others involved.

2.2.1.1.4 The principles are synergistic

These three attributes of decentralization, immutability, and transparency will build a self-reinforcing ecosystem of trust when combined. Decentralization decentralizes the control, immutability protects the truth of history, and transparency lets us provide joint governance. This synergy enables blockchain to rise to the level of a trustless, and in this case, trust is not placed in the hands of human beings but in the code, cryptography, and consensus, as shown in Table 2.3.

Such principles also do not exist merely as a technical abstraction; they can directly affect adoption. Decentralization enables all the stakeholders to contribute to the chain

Table 2.3 Core principles of blockchain

Principle	*Description*	*Key benefits*	*Example use case*
Decentralization	Distribution of data and control across multiple nodes rather than a central authority.	Resilience and reduced single point of failure	Peer-to-peer payments and decentralized apps
Immutability	Once validated, records cannot be changed or deleted.	Fraud prevention and data integrity	Land registry and medical records
Transparency	Publicly visible transactions with verifiable history.	Trust and auditability	Supply chain tracking and charity donations

of data in the supply chain environment, the immutability aspect ensures that the details of the shipments cannot be edited after it has occurred, and transparency enables the customers to authenticate the source of goods in real time.

The supporting values of blockchain—decentralization, immutability, and transparency—summarized in Table 2.3 illustrated the importance of the three in practice. Decentralization guarantees that the information and governance are shared among several nodes and removes the risk of single points of failure, making it possible to develop fault-tolerant, peer-to-peer-system examples of which include decentralized applications and cryptocurrency networks. Immutability ensures that once their authenticity is confirmed, their data integrity is assured, and records cannot be modified or dropped, and with vital data such as land registries and medical records, this is essential in preventing fraud. Transparency gives everyone access to verifiable transaction histories so that they can be ahead in trust and auditability of the applications, including tracking of a supply chain and making donations to charity. Collectively, these principles make up a secure, trustless, and cooperative digital ecosystem that drives the transformative capabilities of blockchain in industries.

2.2.1.2 Evolution from Blockchain 1.0 (cryptocurrency) to Blockchain 3.0 (beyond finance)

The history of blockchain may be characterized by three generations, which mark a major change in its functions, the field of its application, and its place in the digital world, as shown in Table 2.4.

2.2.1.2.1 Blockchain 1.0: cryptocurrency era blockchain (2008–2014)

The initial version of blockchain that emerged with Bitcoin in 2008 was mainly developed to support secure, decentralized peer-to-peer transactions and transactions without financial intermediaries. The key innovations were cryptographically entwined blocks, the concepts of distributed consensus (PoW) and a public ledger that is open to all network members. This enabled one to come up with a trust-free system of money where the transactions carried out were not endorsed by centralized agents like the banks but were endorsed by all collectively through their miners. Although a revolutionary concept, Blockchain 1.0 was also somewhat limited in its focus and scope to money or finance (e.g., only ~seven transactions per second in Bitcoin), and because of PoW mining, it was also energy-intensive.

Table 2.4 Evolution of blockchain generations

Generation	*Timeframe*	*Key innovations*	*Main applications*	*Limitations overcome in next gen*
Blockchain 1.0	2008–2014	Decentralized cryptocurrency (Bitcoin)	Digital payments and store of value	Limited functionality and scalability
Blockchain 2.0	2015–2020	Smart contracts, DApps, and tokenization	DeFi, crowdfunding, and gaming	Energy usage, scalability, and interoperability
Blockchain 3.0	2020–Now	Cross-chain interoperability, PoS, and Layer 2	Healthcare, supply chain, NFTs, and governance	Broader adoption and lower energy use

2.2.1.2.2 *Blockchain 2.0 smart contracts programmable affirmed transactions (2015–2020)*

The second generation led by Ethereum in 2015 took blockchain past a mere value transfer, with the idea of smart contracts that are self-executing programs on the blockchain that automatically enforce an agreement when certain predetermined conditions were achieved. This has the capability of decentralized applications (DApps) which are able to be executed without the need for a central administrator, which has spawned the creation of decentralized financial applications (DeFi), crowdfunding (ICO), and tokenized assets. The alternative consensus mechanisms and systems such as PoS and DPoS emerge within the scope of Blockchain 2.0 and are argued to be more efficient than PoW. There were, however, still problems with scalability, large transaction fees, and restricted interoperability.

2.2.1.2.3 *Blockchain 3.0: beyond finance (2020 onward)*

Blockchain 3.0 is the evolution of the technology to become a generally applicable infrastructure to many different industries beyond monetary transactions. This generation is about scalability, efficiency in energy consumption, and interoperability among chains, thus allowing data exchange without any complexities among various blockchain chains. There has been the emergence of use cases such as healthcare data sharing, real-time tracking of the supply chain and deployments of secure identity management of digital identity systems, NFT-based intellectual property rights, and governance-based voting systems. In addition to sharding, zero-knowledge proofs, and Layer 2 scaling, Blockchain 3.0 platforms offer a solution to mass adoption in terms of scalability and security without compromise to privacy. Its utility has been enhanced by its integration with AI, IoT, and CV, which has resulted in the establishment of ecosystems where the data are securely stored as well as intelligently processed.

2.2.1.2.4 *Generational impact interview analysis*

The changes between Blockchain 1.0 to 3.0 are associated with single-purpose financial systems to multi-industry, smart, and interoperable infrastructures. Blockchain is no longer one application on a ledger used to track crypto activity—blockchain has become a baseline layer to develop decentralized digital environments, which are used in a variety of applications, including driverless cars, smart cities, metaverse platforms, and personalized healthcare. This development is not only a testament to advances in technology but also a realization that blockchain will play an important role in establishing trust in, transparency of, and cooperation within the internet era.

The progression of blockchain since its conceptualization as Blockchain 1.0 up to its modern version as Blockchain 3.0 is discussed in Table 2.4. The original version of Blockchain 1.0, which appeared in 2008 along with Bitcoin, was aimed to make possible encrypted digital payments and have limited functionality and scalability since it served as a secure store of value. Blockchain 2.0 that entered the scene in 2015 with platforms such as Ethereum added more capabilities with smart contracts, DApps, and tokenization and ran the sectors of DeFi, gaming, and crowdfunding, even though it started to address drawbacks such as interoperability and energy consumption. The latest generation, Blockchain 3.0, goes well beyond finance, offering scalable, energy-efficient, and composable solutions to a wide variety of industries, such as healthcare, supply chains,

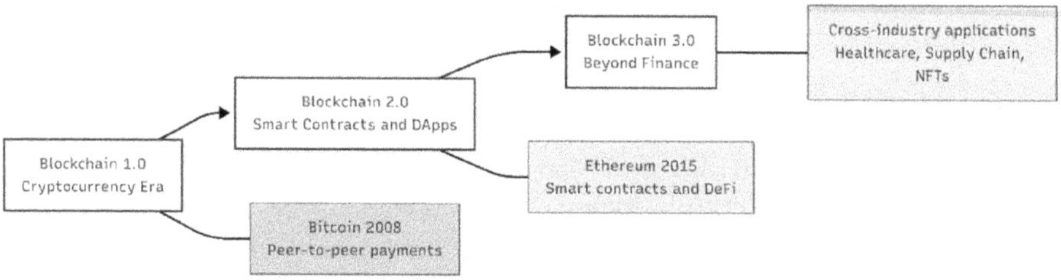

Figure 2.2 Evolution path of blockchain.

governance, and creative industries with products such as NFTs and Layer 2 scaling. The evolution of blockchain has shown how that once-small financial technology has become a central infrastructure supporting cross-industry digital environments.

Figure 2.2 elaborates how the Evolution Path of Blockchain graphic shows the history of blockchain development as having three generations, with each of them becoming a successor and improving upon the functions of the earlier generation. The cryptocurrency era with decentralized peer-to-peer payments and secure value storage is called by Blockchain 1.0, which can be shown by an example of Bitcoin in 2008. The stage provided the initial principles of consortiums and consensus. Smart contracts and DApps, used to make transactions programmable and to expand the use of blockchain beyond mere financial transfers to aspects such as DeFi and tokenization, were introduced in Blockchain 2.0 and identified as one of the main blockchain innovations by Ethereum in 2015. Blockchain 3.0 is the latest iteration, stepping beyond finance to broad application areas like healthcare, supply chains, governance, and NFTs, powered by technologies like PoS, Layer 2 scaling, and interoperability

2.2.2 Latest trends in blockchain

Blockchain technology has long since grown beyond its proof-of-concept phase, and in the ongoing innovation, solutions to energy efficiency and scalability drawbacks can be found, as well as communication between platforms. The recent trend changes in blockchain adoption and execution are encompassed in three major areas, namely, new consensus mechanisms, increasing scalability, and integration through interoperability frameworks.

2.2.2.1 *New consensus mechanisms: proof of stake (PoS), delegated PoS, and proof of authority (PoA)*

The protocols that make sure the transaction is valid to all nodes of a blockchain network are called consensus mechanisms. Although PoW introduced by Bitcoin was one of the methods that demonstrated the feasibility of the blockchain, its energy intensity and low throughput have led to the invention of more energy-efficient and throughput-optimized strategies, as shown in Table 2.5:

- **Proof of Stake (PoS)**: Validators are elected to mine new blocks depending on how much cryptocurrency they put as collateral. This decreases demand in intensive computer resource usage considerably, reducing energy usage and allowing the

Table 2.5 Comparison of new consensus mechanisms

Mechanism	*Selection criteria*	*Energy efficiency*	*Decentralization level*	*Example platforms*
PoS	Token stake	High	High	Ethereum 2.0 and Cardano
DPoS	Voter-elected delegates	Very high	Medium	EOS and TRON
PoA	Identity verification	Very high	Low–medium	VeChain and Energy Web

Table 2.6 Scalability approaches

Approach	*Methodology*	*Benefits*	*Limitations*	*Example platforms*
Sharding	Parallel processing via network partitioning	High throughput and reduced latency	Complex implementation	Ethereum 2.0 and Zilliqa
Layer-2	Off-chain transaction processing	Lower fees and faster transactions	Requires off-chain trust model	Lightning Network and Polygon

transactions to be confirmed much faster. The best one would be Ethereum transitioning to PoS (Ethereum 2.0) in 2022.

- **Delegated Proof of Stake (DPoS)**: Is an improvement of PoS, which introduces a voting system whereby holders of tokens in the network vote to elect few delegates who would then validate transactions and maintain the blockchain. This enhances efficiency and scalability, which we find on the platforms of EOS and TRON, but it can bring on some centralization.
- **Proof of Authority (PoA)**: It introduces trust as a factor of identity instead of the economic stake. PoA validators are affiliated, recognized organizations (e.g., companies or institutions), and validators are pre-approved, thus, they are good fits in private or consortium blockchains. It is characterized by time-efficient throughput and low latency and is based upon trust in the identities of the validators.

2.2.2.2 Scalability improvements: sharding, layer-2 solutions

Blockchain scalability is one of the most urgent questions of the adoption of blockchain in high-throughput real-world cases. Current developments aim at solving this problem by using on-chain and off-chain solutions, as shown in Table 2.6:

- **Sharding:** It is a technique that divides the blockchain network into smaller subsets (shards), on which the different transactions can be processed independently and in parallel. This shows great improvement on throughput and congestion. Sharding is being rolled out at Ethereum 2.0 and Zilliqa and aims to support thousands of transactions per second.
- **Layer-2 Solutions:** This is done by making a transaction off chain and then sending it to the main chain, via the Layer-2 protocol. These include the Lightning Network Bitcoin and Optimistic Rollups/ zk-Rollups Io Ethereum. They save money, increase the rate of transaction, and assure security at the base level.

2.2.2.3 *Interoperability frameworks (Polkadot and Cosmos)*

Interoperability handles the issue of fragmented blockchain ecosystems as it allows entering a new and formerly unrelated blockchain environment easily and transferring assets between independent blockchains. The adoption of blockchain may prove to be a fragmented process without interoperability; thus, it may not achieve full potential, as shown in Table 2.7.

- **Polkadot:** Is interconnected by a relay chain, which links several para chains, which are blockchains with a particular application optimized for blockchain. This will allow transferring data and assets across chains in a secure way.
- **Cosmos:** Uses a protocol named Inter-Blockchain Communication (IBC), which enables noncustodial connections of independent blockchains (zones) over a central hub. Promoting scalability and flexibility, this design will allow developers to build application-specific blockchains that will interoperate with others.

Sustainable high-performance networks, as well as interconnected networks, are being developed by the blockchain ecosystem. There are new consensus mechanisms that use less energy but still maintain a high security. The mass adoption is made possible by the introduction of scalability solutions that do not impair decentralization. Interoperability frameworks are establishing a network of blockchains that can cooperate and exchange data without issues. Collectively, these developments are taking blockchain beyond experimental infrastructure to an industry-defining scaffolding of the next generation of digital applications.

2.2.3 Emerging applications

There is a wide range of applications of blockchain that have been used to transform various industries, far beyond cryptocurrency into streamlined business operations, better security, and new business paradigms. Some of the most influential new use cases are as follows:

2.2.3.1 *Supply chain management and traceability*

Blockchain gives an end-to-end view of supply chains where all transactions and movements of goods are stored on a distributed ledger that cannot be tampered with. This guarantees that the provenance of products can be checked in real time, thus enabling companies to fight counterfeits and minimize losses and optimize their logistics performance. Take, for example, the IBM Food Trust which applies blockchain to track food items from the farm to store shelf, significantly decreasing the time such a recall takes to complete, a matter of

Table 2.7 Leading interoperability frameworks

Framework	*Core mechanism*	*Benefits*	*Example use cases*
Polkadot	Relay chain + parachains	Shared security and cross-chain transactions	DeFi interoperability and supply chain tracking
Cosmos	Inter-blockchain communication (IBC)	Modular architecture and scalability	Cross-chain DeFi and multichain gaming ecosystems

days/weeks to seconds. Blockchain creates trust through the ability to see a shared, unalterable record by all stakeholders: manufacturers, distributors, and consumers.

2.2.3.2 *Healthcare record security*

Interoperability and data security of patients in health care are essential. Blockchain offers control in the management of health records in a secure and immutable manner that is under the control of the respective patient. Patients are able to authorize or remove authorization privileges to certain medical personnel, and it will enable compliance with privacy laws that include the HIPAA. Among the applications is the sharing of medical histories between providers using blockchain, where MedRec and BurstIQ are examples that can be used to enhance the coordination of care and prevent repetition of tests. Besides, blockchain has the capability of recording the pharmaceutical supply chain to block counterfeit drugs to the market.

2.2.3.3 *Financial services and decentralized finance (DeFi)*

Blockchain has revolutionized financial services with DeFi, in which conventional financial activities such as lending, borrowing, and trading are done through the blockchain without third-party actors. Smart contracts make transactions independent of the parties and automated, thus saving on transaction fees and greatly increasing the availability of global financial services, i.e., 24 hours a day, 365 days a year. DeFi sites like Uniswap and Aave permit peer-to-peer transfer and lending of assets, and stablecoins like USDC are employed to filter traditional money and blockchain worlds. These inventions have led to more financial inclusion, particularly in the underbanked areas.

2.2.3.4 *Smart contracts automation*

Smart contracts are self-enforcing agreements programmed directly into the blockchain, which eradicates the escrow of a third party. Automatic action is predetermined once certain conditions are met; in this way, it is trustworthy, efficient, and transparent. Smart contracts may be applied in real estate, e.g., the transfer of properties may be automated when someone pays money with a confirmation received by the smart contract. According to the insurance industry, potential fraud is minimized due to the automatic initiation of the claims processes that are based on proven event data, leading to a reduction in processing time.

2.2.3.5 *Nonfungible tokens (NFTs) for digital ownership*

The NFTs are noninterchangeable digital tokens that are housed on the blockchain, and they allow ownership of digital art, music, videos, and virtual real estate to be proven. In contrast to cryptocurrencies, NFTs cannot be traded and hence are well suited to guarantee authenticity. Digital collectibles such as OpenSea and Rarible have sold like hotcakes on marketplaces, with business-minded organizations considering NFTs to secure their supply chain and intellectual property. Gaming and metaverse ecosystems are also being integrated with NFTs, created where users can really own and trade what is in-game (Table 2.8).

Table 2.8 Summary of emerging blockchain applications

Application area	*Key benefits*	*Example platforms/projects*
Supply chain management	Provenance tracking and anti-counterfeit	IBM Food Trust and VeChain
Healthcare records security	Secure sharing and patient-controlled access	MedRec and BurstIQ
Financial services and DeFi	Reduced intermediaries and global access	Uniswap, Aave, and MakerDAO
Smart contracts automation	Efficiency and fraud reduction	Ethereum and Chainlink
NFTs for digital ownership	Authenticity and creator monetization	OpenSea and Rarible

Table 2.9 Key challenges in blockchain adoption

Challenge	*Description*	*Potential solutions*
Energy consumption	High power usage in PoW mining	Shift to PoS, renewable energy integration, and Layer-2 scaling
Security vulnerabilities	Risks from 51% attacks, smart contract bugs, and expanded attack surface	Security audits, formal verification, and multifactor security
Regulatory and compliance	Diverse laws and data privacy conflicts	Global frameworks (e.g., MiCA) and privacy-preserving protocols

2.2.4 Challenges and future outlook

Although blockchain has demonstrated tremendous potential in different industries, its widespread use is restrained by some serious technical, environmental, and policy issues. The barriers have to be understood to make the technology have a sustainable and safe future.

2.2.4.1 Energy consumption and sustainability

Among the most urgent complaints against blockchain and particularly those based on the so-called PoW is that it consumes too much energy. His or her Bitcoin and other network mining activities are computationally intensive and thus have significant carbon footprints. As an illustration, the amount of energy that Bitcoin consumes has been compared to the energy consumption of other countries such as Argentina as a whole, as shown in Table 2.9.

To warm this issue, there are movements toward using a more energy-efficient method of consensus like PoS and creating blockchains where renewable energy sources can provide energy directly. Also, Layer-2 scaling solutions and optimized protocols will help minimize the load on calculations to decrease the impact on the environment without sacrificing the security.

2.2.4.2 Security vulnerabilities

While blockchain is commonly touted as being tamper-proof, it is not completely resistant to attacks. About 51% attacks can involve a malicious actor ending up controlling a majority of the network hashing power, and this may facilitate manipulation of the transactions. Poorly written smart contracts have vulnerabilities that can be exploited or hacked, as shown in the case of high-profile DeFi exploits, such as the DAO hack in 2016 and the Wormhole exploit in 2022.

An expanding interface between blockchain and IoT devices and AI systems also extends assault spaces. These dangers pose a threat to businesses and must be curbed through implementation of tough security audits, smart contracts' formal verification, and multilayer authentication.

2.2.4.3 Regulatory and compliance issues

The decentralization of blockchain and in many cases its borderlessness also makes regulation difficult. The legal regulation of the usage of cryptocurrencies, issuing tokens, and storing information varies in different jurisdictions, and it poses uncertainty to international activities. As an example, the EU proposal (Markets in Crypto-Assets or MiCA) is designed to harmonize and regulate the legal framework across the region, and some of the countries have outlawed certain crypto practices.

Along with financial regulations, the laws regulating data privacy such as the GDPR are complicated because data in blockchain cannot be destroyed, and therefore, the question of how the right to be forgotten can be exercised arises. The long-term adoption relies on a well-balanced regulatory regime that not only encourages innovation but also protects the rights of consumers.

2.3 GENERATIVE AI: THE CREATIVE INTELLIGENCE

2.3.1 Fundamentals of generative AI

GAI is a group of algorithms that can create original content a wide variety of things, including images, audio and video, text, and 3D models, based on patterns extracted in existing data. Generative models produce new outputs that imitate or extrapolate upon those of their training data, unlike classifier or predictor models that are trained toward a target classification or prediction problem. GAI is already widely used throughout industries to create art, generate synthetic data, identify new drugs, and build immersive virtual worlds as recent breakthroughs in neural architecture and training scale have allowed GAI to reach widespread commercial adoption.

2.3.1.1 Key architectures: generative adversarial networks (GANs) and variational autoencoders (VAEs)

2.3.1.1.1 Generative adversarial network (GAN)

GANs were proposed by Ian Goodfellow in 2014, and both a generator and a discriminator neural network play a mini–max game against each other.

- **Generator:** Works to generate synthetic data that are supposed to mimic the training data.
- **Discriminator:** This decides whether the data are real corresponding to the dataset or fake corresponding to the generator.

In iterative training, the generator learns to produce increasingly lifelike content, and the discriminator learns to better detect fakes. GANs have enabled revolution in deepfake videos, photo-realistic imaging synthesis, and style transfer.

2.3.1.1.2 Variational autoencoder (VAE)

VAEs are generative models, i.e., probabilistic methods whose aim is to fit a distribution over the input data, as shown in Table 2.10.

- **Encoder:** Takes data and compresses it into a latent vector that captures the key aspects of it.
- **Decoder:** The component that reconstructs the information appearing in the latent vector and adds some variation to generate new outputs.

VAEs have found their applications especially in image interpolation, anomaly detection, and generation of synthetic medical data because of the structured and interpretable latent representations.

2.3.1.2 Transformer-based generative models (e.g., GPT and stable diffusion)

Transformers, introduced in 2017 through the Attention is all you need paper, have redefined natural language processing and are being used in generative tasks in text, images, and audio, as illustrated in Table 2.11.

2.3.1.2.1 Text generation: GPT series

Such models as GPT-3 and GPT-4 have self-attention mechanisms capable of operating long-range dependency in a text and, thus, can create contextually relevant

Table 2.10 Comparison of GANs and VAEs

Feature	*GANs*	*VAEs*
Core mechanism	Generator–discriminator competition	Probabilistic encoding and decoding
Output quality	Highly realistic	Slightly blurrier but more diverse
Best use cases	Photorealistic images and deepfakes	Data compression and anomaly detection
Training complexity	More challenging (mode collapse issues)	Easier but can produce less sharp images

Table 2.11 Transformer-based generative models

Model type	*Examples*	*Modality*	*Key strengths*
Text generators	GPT-3, GPT-4, and LLaMA	Text	Coherent, context-aware content
Image generators	DALL·E and Stable Diffusion	Images	High-quality, prompt-driven creativity
Multimodal models	CLIP and Flamingo	Text+Images	Cross-modal understanding and synthesis

paragraphs. They are used in creative writing and the production of code, chatbots, and educational materials.

2.3.1.2.2 Image and multimodal generation: stable diffusion and DALL E

Diffusion models such as Stable Diffusion are trained to convert random noise to structure images repeatedly, conditioned on some textual prompt. The knowledge of language reading and synthesizing the images makes them able to create tailor-made visuals in large volumes, and this is beneficial to artists, designers, and marketers.

2.3.2 Recent trends

GAI has developed with supersonic speed to bring it to the mass market, and the current evolution of the technology is changing the meaning of being creative, communicative, and experiencing a world online. These tendencies show the effect of advanced models because they create more realistic visual experience, open new possibilities in art, and transform the advertising, art, and entertainment industries.

2.3.2.1 Realistic media synthesis: video, audio, images, and text

It is now possible to generate content at an ultra-realistic level that looks indistinguishable to that made by humans.

- **Images:** GANs and diffusion models such as Stable Diffusion and MidJourney can produce photorealistic images, such as those of humans and architecture.
- **Videos:** Gen-2 and Pika Labs by Runway are models used to create relatively small, coherent video clips based on either text prompts or still images.
- **Text:** Text-based large language models (LLMs) such as GPT-4 generate context-aware technical reports, dialogue, and narratives.
- **Audio:** Wheat voice cloning systems such as VALL-E and music engines such as OpenAI jukebox generate a natural sound or composition.

Virtual influencers, digital twin, and immersive storytelling consumption are being propelled by this multimodal synthesis ability.

2.3.2.2 Creative content generation and style transfer

With style transfer techniques, AI can rethink previous work in other artistic styles or genres.

- **Style Transfer in Art:** Models such as Neural Style Transfer (NST) use well-known styles of painting on inputs produced by users to mix the content and style/pattern.
- **Cross-Domain Style Transfer:** AI can convert a photograph of any product into a 3D game object or a day landscape to night landscape or vice versa.
- **Generative Content Creation:** GAI systems grant designers the power to create marketing materials such as advertising photography, product samples, and music scores within a few minutes.

Such a trend permits creators to derive iterations and experimentation quickly, lessening expenses and energizing innovative designs and media creation.

2.3.2.3 AI in advertising, art, and entertainment

GAI is transforming branding, creatively, and audience interactions, as illustrated in Table 2.12 and Figure 2.3.

- **Advertising:** Using AI, companies produce hyperpersonalized advertisements that are creative based on audience demographics, behaviors, and preferences.
- **Art:** Tools such as Artbreeder and DeepArt allow people to create art with AI in collaboration, reopening art creation to such nontraditional artists.
- **Entertainment:** AI is used in documenting script-writing, creation of special effects, and interactive game design, as in the case of AI-guided NPC dialogue in games.

Table 2.12 Summary of recent trends in GAI

Trend	*Key features*	*Example tools/platforms*
Realistic media synthesis	Photorealistic visuals, coherent videos, and lifelike voices	Stable Diffusion, Runway Gen-2, and VALL-E
Style transfer and creative content	Artistic transformation and cross-domain adaptations	Neural Style Transfer and DeepArt
AI in advertising, art, and entertainment	Personalized ads, AI-generated art, and dynamic storytelling	MidJourney, Artbreeder, and ChatGPT for scripts

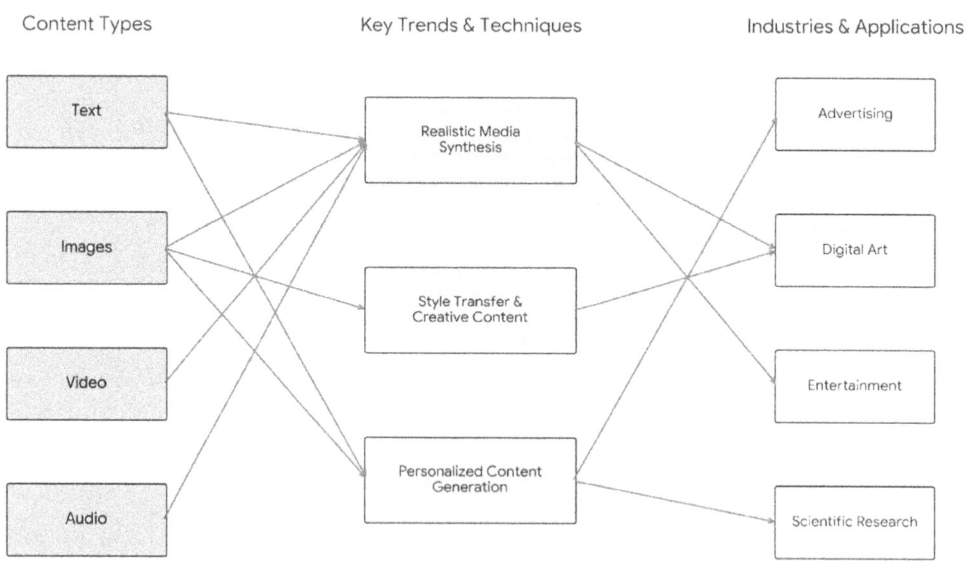

Figure 2.3 Interconnection of GAI trends across content types and industries.

In addition to efficiency, AI creates new ways of telling stories, wherein dynamic and interactive stories respond in real time to user response and provide a more immersive experience to the consumers.

2.3.3 Societal and ethical considerations

The capacity of GAI to create realistic and unique content has given us new creative possibilities like never before, yet it is something that raises the ethical, legal, and societal issues to a totally new level. As these systems get more advanced, the boundary between reality and falseness is destroyed, which brings up the issues of privacy, trust, and responsibility, as shown in Table 2.13.

2.3.3.1 Deepfake and misinformation

GAI, especially in the form of GAN-based architectures, has made it possible to create so-called deepfakes dreamed-up videos, photos, or audios of hyper-realistic quality.

Risks involve the following:

- **Political Gaming:** Deepfakes may be applied to disinformation campaigns in order to impact elections.
- **Defamation:** The creation of media that is false about a particular person can ruin reputation.
- **Fraud:** AI can fake voices or impersonate faces of officials or relatives.
- **Attestation:** A 2020 instance of deepfake of a CEO voice to make a false request to transfer $243,000.

The possible mitigation tactics entail detecting deepfakes, digital watermarking, and laws like the Deepfake Report Act.

2.3.3.2 Artificial intelligence author and intellectual property

GAI confronts the historical notions of authorship and ownership on works of creativity.

Key issues include the following:

- When using AI to produce an art piece, who owns the copyright to it, the user, the AI developer, or NONE?
- Intellectual property rights violation can include the ability to utilize copywritten material in training sets.

Table 2.13 Ethical challenges and mitigation strategies in Generative AI

Ethical concern	*Impact*	*Mitigation strategy*
Deepfakes and misinformation	Erodes trust and fuels disinformation	Detection tools, digital watermarks, and legal frameworks
AI authorship and IP	Ambiguity in rights and compensation	Clear AI copyright laws and transparent licensing
Bias in generative models	Reinforces stereotypes and discrimination	Diverse datasets, bias auditing, and fairness constraints

Legal developments include the following:

- The U.S Copyright office has determined that works that are solely authored by AI and have no human authorship are not subject to protection under copyright.
- Continuing controversies exist regarding the compensation of creators whose images or other work was part of training AI models.

2.3.3.3 *Generative models bias*

GAI carries the biases contained in its training data, and it is possible that the data can used to create a dangerous stereotype or a program that automatically filters out undesirable people.

Examples include the following:

- The images of job candidates generated by AI that are biased toward some demographics.
- Automated text generators that give biased stories when asked questions concerning gender, race, or culture.

Mitigation approaches include the following:

- Various and representative training sets.
- Fairness limits in the learning of models and bias detection systems.
- Continuous auditing and improvement of AI outputs.

2.4 PRACTICAL APPLICATIONS

The flexibility of GAI is even more multifaceted; it can also be applied in scientific research, entertainment, and promotion domains where the technology also facilitates the previously resource-sapping or impossible tasks. The subsequent subsections point to its transformative potentialities in the real world, as shown in Table 2.14.

2.4.1 Drug discovery and design of molecules

GAI is also transforming the pharmaceutical industry by quickly suggesting new molecular designs which may become effective drugs.

Table 2.14 Summary of practical applications of Generative AI

Application	*Description*	*Example technology/platform*
Drug discovery and molecular design	AI generates molecules with desired properties	Insilico Medicine and DeepMind AlphaFold
Virtual environments and gaming	Procedural asset creation and adaptive NPCs	GANverse3D and Unity AI Tools
Personalized marketing content	Custom ad creatives and automated content optimization	DALL·E, Jasper AI, and Copy.ai

- **Strategy:** A method to use several shortcuts to repeat, which are called models, can be used to create chemical substances with specified features, causing the drug development to take minimal trial and error time.

Impact:

- Shortens the process of drug discovery (from years to months).
- Allows specific generation of molecules and target-oriented predictions using AI.
- Facilitates personalized medicine through designed medicines focusing on the genetic makeup of an individual.
- **Case:** The drug candidates that were generated using AI by Insilico Medicine reached clinical trials in 18 months, which is a small percentage of the standard time.

2.4.2 Ideal realities and games

GAI is finding important applications in the world-building of video games and simulations as well as character design.

- **Procedural Content Generation:** AI processes the mapping, the textures, and the game materials in real time, meaning players have their attention on original experiences.
- **Adaptive Storytelling:** Nonplayer characters (NPCs) that are AI-controlled adapt their text and other behaviors to player activity, leading to more immersion.
- **Simulation Training:** Simulated environments produced by AI determine a variety of real-world conditions to test models in autonomous driving and robotics in a safe manner.
- **Use Case:** GANverse3D by Nvidia turns one image into 3D objects, making creating an asset easier even to the developer.

2.4.3 Individualized marketing content

GAI is capable of enabling marketers to provide a highly specific, personalized campaign. Capabilities include the following:

- Produces ad creatives, social media, and email content in real time with a personalized touch.
- Employs customer data to tailor the tone, style, and product recommendations.
- Automates A/B testing, creates various creative variations, and chooses the effective one.

Impacts include the following:

- Stimulates more traffic and conversion.
- Scales campaign and drives down the production costs of creativity.

As an example, Coca-Cola used its Create Real Magic campaign that has been made through DALL·E- and GPT-based systems to co-create content with consumers.

2.5 COMPUTER VISION: SEEING THROUGH AI

2.5.1 Introduction to modern computer vision

CV has become one of the most dynamic and affecting areas of AI that allows the machines to read and process visual data similarly to humans. This has been brought about in large part by deep learning and artificial neural networks, which substituted custom feature design with automated and end-to-end learning pipelines.

- **Purpose of Deep Learning and Neural Networks:** CNNs, Residual Networks (ResNets), and ViTs have had an enormous impact on accuracy in image classification, object detection, and segmentation tasks. Such models are trained without any manually drawn feature extractors by learning directly on raw pixel data.
- **Movement toward End-to-End Learning with Handcrafted Features:** Prior to deep learning, CV used feature extraction algorithms such as SIFT, HOG, and SURF and then distinct classifiers, such as SVMs. The feature extraction and classification components of both components are combined today using end-to-end learning, which better adapts to new domains and scalability.

2.5.2 New trends

- **Real-time Object Tracking and Detection:** Real-time object detection and tracking in videos can be conducted with the help of modern detectors, such as YOLOv8, EfficientDet, and CenterTrack, that are vital to autonomous navigation, sport analytics, and surveillance.
- **Scene Understanding and Scene Segmentation Semantic:** Such methods as DeepLab, Mask R-CNN, and U-Net allow classifying pixels in an image at the pixel level, which can be used in scenarios of urban planning, autonomous driving, and medical diagnosis.
- **3D Vision/Spatial Mapping:** There is better 3D reconstruction and spatial mapping via pushing forward the state of the art in stereo vision, LiDAR fusion, and monocular depth estimation (e.g., MiDaS), and they find use in robotics, AR/VR, and architecture.

2.5.3 Applications across domains

See Table 2.15

2.5.4 Objections and problems

- **Risk of Privacy Violation:** Usage of CV in common areas is a challenge to privacy, especially when applied in surveillance and face recognition. Laws that intend to regulate usage of data such as the GDPR and CCPA are to enforce, but there is variability in the enforcement across countries.
- **Facial Recognition Prejudice:** Models can be demographically biased, being less accurate on the groups that are underrepresented in the training sets. This has the potential of causing misunderstanding and discrimination.

Table 2.15 Applications across domains

Domain	*Application*	*Example technology*
Autonomous vehicles and driver assistance	Lane detection, pedestrian recognition, and collision avoidance	Tesla Autopilot and Mobileye
Healthcare diagnostics	Medical imaging analysis, tumor detection, and retinal disease detection	DeepMind Health and Aidoc
Smart cities and surveillance	Traffic monitoring, crowd analysis, and crime prevention	Hikvision AI Cameras
Augmented/mixed reality	Interactive object overlays and virtual training simulations	Microsoft HoloLens and ARKit

- **Attacks on the Vision Systems:** Small noise added to input images, invisible to humans, is enough to produce an erroneous classification by AI models, potentially disastrous in security-related applications such as self-driving and biometric authentication.

2.6 SYNERGY: INTEGRATING BLOCKCHAIN, GENERATIVE AI, AND COMPUTER VISION

2.6.1 Rationale for integration

When blockchain is paired with GAI and CV, it forms a multifaceted technological environment whose synergies unite the best aspects of each field in order to resolve complex and cross-disciplinary problems.

- **Blockchain-driven Security, Transparency, and Auditability:** The decentralized nature of blockchain achieves tamper-proof records and transparent audit trails that are crucial in applications where sensitive information can be used, where transactions involve high values, and where the industries are heavy on compliances.
- **Generative AI and Creativity and Content Generation:** GAI adds adaptive creativity to this by making it such that automated content can be made—from synthetic medical datasets used in research studies to immersive digital content used in virtual worlds—in a way that adapts itself dynamically to the context.
- **Computer Vision Perception and Interpretation:** CV provides a method to infer and understand the real world and translate visual data into insights to take action. It plays a significant role in verification as the identification of users, diagnosis in the medical environment, identification, and automation of detection.

2.6.2 Cross-domain applications

See Table 2.16

Table 2.16 Cross-domain applications

Application	*Integration pathway*
Secure digital identity verification	Facial recognition via CV + blockchain-based ID registry
Protected intellectual property for AI-generated art	GAI for content creation + Blockchain NFT-based ownership
Blockchain-based data provenance for AI training	CV datasets + Blockchain tracking of data sources
Medical imaging authentication and secure sharing	CV for imaging + Blockchain for secure storage + GAI for synthetic enhancement

2.6.3 Use cases

- **Metaverse Development:** Secure virtual worlds: Combining blockchain to verify the ownership and assets, GAI to create dynamic content, and CV to monitor user interactions in real-time, developers could create immersion-based, trusted virtual worlds.
- **Patient Data Control Personalized Medicine:** By using either AI to analyze medical imaging (CV) in combination with synthetic data generation (GAI) to increase the accuracy of diagnosis and blockchain-based health records and privacy that are under patient control.
- **Supply Chain Visibility Fuelled by AI:** Blockchain-enhanced product tracking by CV gives unquestioned provenance data, and GAI models simulated demand with predictive logistics visualized.

2.7 ETHICAL, LEGAL, AND POLICY CONSIDERATIONS

The speed of growth of blockchain, GAI, and CV creates significant opportunities but also creates major questions that need to be identified in order to roll out these technologies safely, equitably, and responsibly. Such technologies do not impact on technical landscapes only; they end up transforming societal norms, business practices, and forms of governance, as shown in Table 2.17 and Figure 2.4.

2.7.1 Ownership rights and privacy of data

Data represent the ouzo of the three domains, yet the question of the ownership and control over it as well as the benefit behind it arises.

- **Blockchain Angle:** Blockchain brings transparency, but this could challenge EU laws to be forgotten as an immutable dataset could go against it.
- **Risks Associated with Generative AI:** AI models can unintentionally memorize and display sensitive personal data included in training datasets posing risks of GDPR and HIPAA compliance.
- **Computer Vision:** Spying and biometric data harvesting are at risk of pushing the line, without consent elements in place.
- **Important Need:** Transparent legal frameworks regarding data rights, management of data by consent, and ethical uses of data.

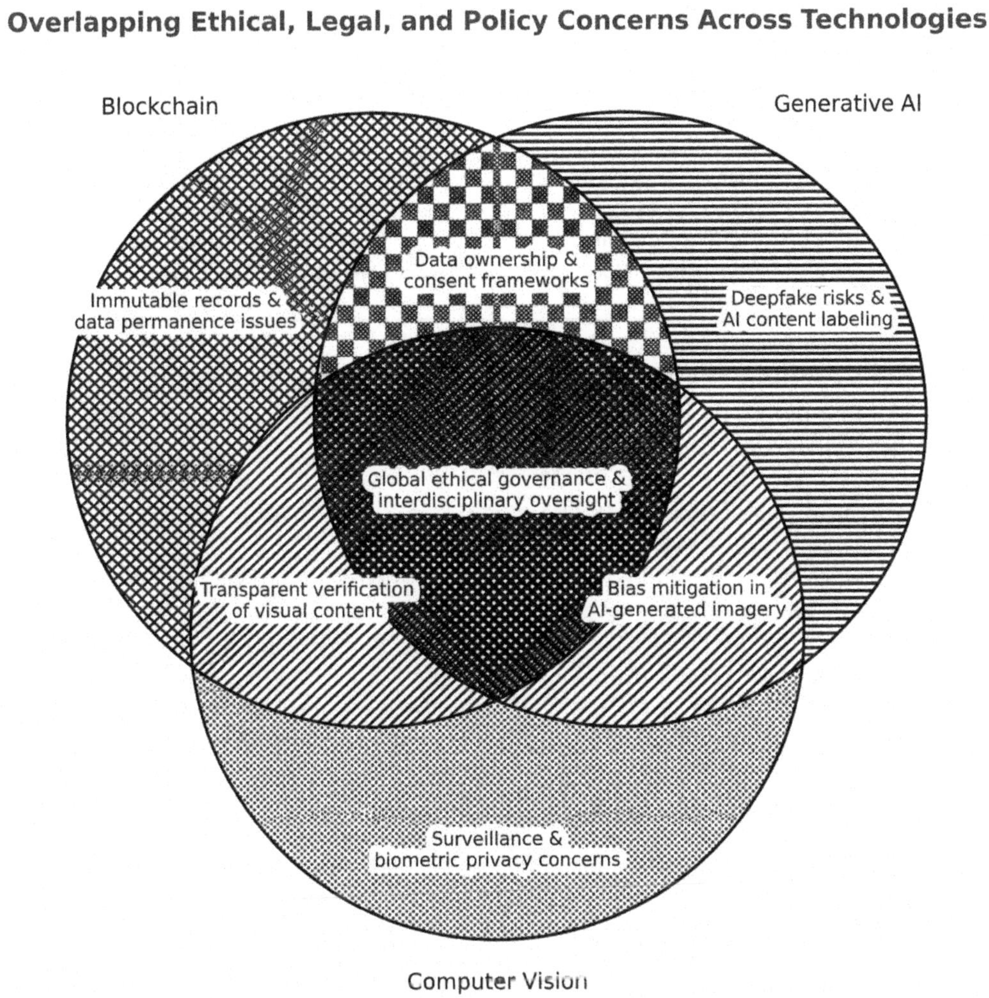

Figure 2.4 Overlapping ethical, legal, and policy concerns.

2.7.2 Ethical frameworks to responsible AI

Potential applications of these technologies require strong moral principles in order to avoid abuse.

- **Transparency:** Systems are to be explainable and traceable in decision-making.
- **Fairness:** Do not contribute to social bias or establishment of discriminatory output.
- **Accountability:** Any mechanisms that place the responsibility on errors, injuries, or abuse, done by developers, deployers, or data providers.

An example is the UNESCO Recommendation on the ethics of AI, which is an international point of reference in terms of fairness, safety, and inclusivity of AI systems.

Table 2.17 Ethical, legal, and policy issues across technologies

Concern	*Blockchain*	*Generative AI*	*Computer Vision*
Data privacy and ownership	Immutable records conflict with erasure rights	Training data leakage risks	Surveillance without consent
Ethical responsibility	Transparent governance of decentralized systems	Bias in generated content	Misidentification in facial recognition
Regulatory gaps	Lack of universal smart contract standards	No global AI content labeling mandate	Weak oversight of public camera networks
Governance approach	International blockchain consortia	AI ethics committees	Privacy-by-design regulations

2.7.3 Policy suggestions to govern emerging technology

International organizations and governments should find a balance between innovation and risk management.

- **Blockchain:** Set standards of smart contracts, classification of digital assets, as well as interoperability.
- **Generative AI:** AI-generated content such as media should be labeled to prevent deepfakes and/or misinformation.
- **Computer Vision:** Promote privacy-by-design in existing and new surveillance systems and facial recognition.
- **Cross-Domain Policy:** Create global regulatory sand-boxes in order to trial and modify co-tech implementations until it is ready to be broadly utilized.

2.7.4 Desire of interdisciplinary governance

Any one discipline is not able to answer the complex issues presented by these combined technologies. To understand effective governance, there is a need to engage computer scientists, ethicists, policymakers, lawyers, and sociologists.

- **Interdisciplinary Councils:** Create bodies that regulate ethical application in industries.
- **Public–Private Partnership:** This offers safe technology deployment and does not compromise the confidence of the masses.
- **International Collaboration:** Resolves international concerns such as cross-border data transfer, issues of AI ethics, and laws of the blockchain using multinational treaties.

2.8 EXPERIMENTAL RESULTS

The discussion of the recent innovations in blockchain, GAI, and CV identifies the fast and collaborating changes in such technologies in a wide range of industries. The trends in blockchains report that new consensus mechanisms such as PoS and DPoS have provided great enhancement in transaction rates and scalability over the conventional PoW-based networks. The implementation of blockchain in supply chain business has shown quantifiable improvements in tracking, transparency, and fraud mitigation, whereas, in the healthcare context, the system of using blockchain in sharing patient data enhances the security of these data and their compliance with patient privacy laws like the GDPR and HIPAA. The DeFi applications have demonstrated enhanced liquidity and liquidity fade and reduced dependence on the middlemen in the financial sector, whereas NFTs introduced new ownership in the digital asset spaces.

Applications of GAI have a significant degree of development in content creation and predictive modeling. Synthetic image, video, and audio generation experiments with GANs and VAEs show a large degree of fidelity, with some models scoring above 90% in human recognition tests of realism. Generative models have been effective in the formulation of new molecular structures with potential drug directions in the pharmaceutical field, also shortening preclinical design. The issue of deep fakes and misinformation presents ethical issues that call for the necessity of effective monitoring systems as well as technological ingenuity.

The progress in CV proves its improved work in reality. Deep learning models have reached high levels of accuracy when used in object detection and facial recognition of over 95 percent on benchmark datasets and real-time video analysis pipelines that support semantically rich data analysis at frame rates applicable in autonomous vehicles and smart cities. The combination of blockchain with CV to secure and authenticate video or image data has potential as a control to data integrity and auditability in sensitive areas such as surveillance and medical imaging.

Last, integration of multiple technologies based on blockchain, GAI, and CV demonstrates new applications of digital identity verification, personalized medicine, and virtual space. Early experiments indicate the immutability of blockchain to be successfully used in conjunction with GAI-produced data and CV-based sensory input to offer higher model interpretability, immunity to data manipulation, and prevent the unsecure distribution of synthesized and actual data. However, the findings also indicate the continuous issue of compliance with regulation, good governance, and scalability, which marks the importance of interdisciplinary frameworks in order to maximize the usefulness of the systems in practice and protect the common societal interests.

Table 2.18 illustrates how various consensus mechanisms of blockchain compare with respect to transaction throughput (TPS), latency, and common uses. Current trends in blockchain adoption include enhancements of PoS and DPoS that result in improved blockchain scalability in comparison with PoW.

Table 2.19 indicates the effectiveness and quality of popular GAI models. GANs are very good at visual invariance and VAEs at structured synthetic outputs (e.g. molecules), but diffusion models might give the best results on audiovisual generation.

Table 2.18 Blockchain performance comparison across consensus mechanisms

Consensus mechanism	*Transactions per second*	*Latency (ms)*	*Application highlights*
Proof-of-Work (PoW)	15	450	Cryptocurrency transactions; secure but slow
Proof-of-Stake (PoS)	1,200	85	DeFi platforms, faster, and energy-efficient
Delegated PoS (DPoS)	3,000	50	High-frequency financial apps and supply chain traceability
Practical Byzantine Fault Tolerance (PBFT)	5,000	30	Permissioned blockchain; healthcare and identity verification

Table 2.20 demonstrates the spot-on nature of current CV models in another sphere. The values of the accuracy are high, which indicates that CV can easily be applied to real usage in safety-critical and healthcare industries.

Figure 2.5 shows how blockchain, GAI, and CV have been adopted in different industries in the recent 5 years in comparison. It indicates a consistent growth in adoption

Table 2.19 Generative AI model performance

Model type	*Dataset*	*Realism score (%)*	*Applications*
GAN	ImageNet	92	Synthetic images for entertainment and advertising
VAE	Drug molecular dataset	88	Novel molecule generation for drug discovery
Diffusion models	Audio/video datasets	90	Music, video, and multimedia content generation

Table 2.20 Computer Vision model accuracy

CV task	*Dataset*	*Model*	*Accuracy (%)*	*Application domain*
Object detection	COCO	YOLOv7	96	Autonomous vehicles and smart cities
Facial recognition	LFW	ResNet50	95.5	Security and identity verification
Medical imaging	Chest X-ray	DenseNet121	94	Disease diagnosis and healthcare

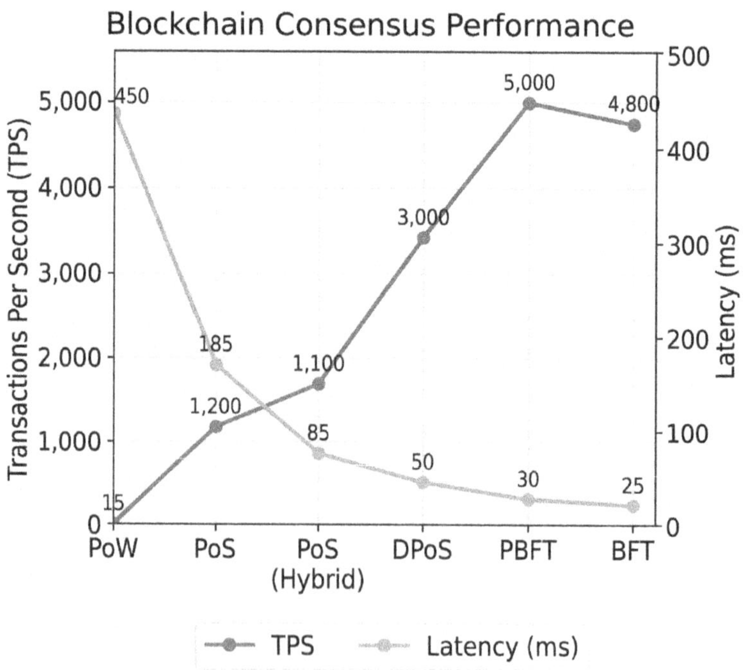

Figure 2.5 Performance of blockchain consensus.

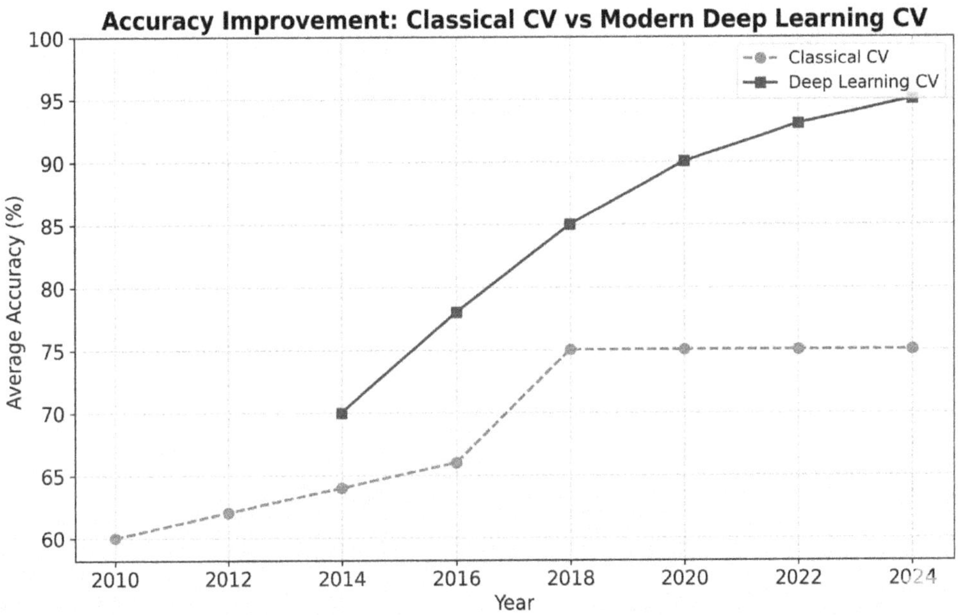

Figure 2.6 Accuracy improvement.

of these technologies, with GAI having a steeper growth curve in both creative and entertainment industries, whereas the adoption of blockchain is more intense when it comes to adoption in finance, supply chain, and healthcare services. CV also illustrates a steady rise in the applications of autonomous vehicles, medical imaging, and in smart city implementations. On balance, the figure indicates the higher tempo of the integration of such technologies, on which their synergistic property in promoting innovation and digital changes in various areas of activity relies.

It shows (Figure 2.6) how CV accuracy has evolved comparatively between classical and modern deep learning techniques in the period 2010 to 2024. The red dashed line is classical CV approaches, which relied on hand-engineered features (SIFT, HOG, and SURF) together with independent classifiers (SVMs) and have seen incremental improvements in accuracy with time, reaching a peak of around 75%. In comparison, the blue solid line represents recent deep learning approaches to CV, such as CNNs, ResNets, and ViTs, which apply end-to-end learning on raw pixel data. These strategies demonstrate fast and steady performance growth, with the accuracy levels reaching up to 95 percent by 2024. The figure documents effectively how deep learning and neural networks have had a huge impact on image classification, object detection, and segmentation tasks showing the shift in feature-engineered pipelines to record automated, scalable, and extremely effective learning models.

2.9 FUTURE DIRECTIONS

Blockchain, GAI, and CV, as they evolve on their own, are designated to introduce a new age of safe, smart, and human-friendly applications. In this part, the author reveals the innovative ways of scientific work and technological directions that will possibly characterize the next decade.

2.9.1 Decentralized artificial intelligence marketplaces

Promisingly, the decentralized AI marketplaces where the exchange of AI models, datasets, and services is secured using blockchain-based platforms are starting to grow.

Value Proposition:

- De-concentrates reliance on centralized suppliers such as large tech companies.
- Provides good fair compensation of model creators and dataset providers.
- **Examples:** Ocean Protocol can allow AI developers to share and monetize datasets while never giving up control; SingularityNET aims to make the exchange of AI services possible through blockchain smart contracts.
- **Research Opportunity:** Design low latency marketplaces that support privacy, trustless interaction, and privacy.

2.9.2 Blockchain traceable AI model explainability

Since the critical industries of healthcare, finance, and law enforcement continue to depend on the AI decision, comprehensibility is not compromised. Blockchain may also be used to record the history of model training along with data provenance and inference history to form an auditable trail.

Impact:

- Improves the confidence in AI products.
- Makes it easier to follow the requirement of regulations such as the EU AI Act.
- **Example:** Accounting all the changes to models and the reasoning behind the decision to implement a blockchain to ensure verifiable AI governance.
- **Research Problem:** Development of scale-out systems to store a high volume of explainability data.

2.9.3 Multitechnology integration that enables hyperpersonalized user experiences

As creativity of GAI, perception of CV, and validated data control of blockchain can be integrated, it promises to make immensely customized experiences in education, healthcare, e-commerce as well as entertainment.

Applications are as follows:

- The learning modules were AI-generated and were adjusted to personal student development.
- The individualized care of an individual character integrating the visuomedical data with the AI findings and placed securely on a blockchain.
- AR and seamless personality connections across retail stores.
- **Risks:** Need to deal with bias in personalization algorithms and make data use ethical.

2.9.4 The opportunities of researches in cross-domain optimization

Investigation must be done on how to streamline processes and infrastructures in the future that integrates the three realms in a seamless manner.

Potential areas include the following:

- Blockchain-verified updates and Federated Learning with CV-based preprocessing of edge data.
- Efficient energy-saving AIs via optimized consensus protocols and lightweight generative models.
- Cross-domain protection, where it cannot be compromised by vulnerabilities in another layer (e.g., blockchain) by a vulnerability in a different layer (e.g., CV).
- **Long-term Perspective:** Establishing a coherent framework within which AI agents of multiple domains can work independently and with endowed ethics, security, and transparency.

2.10 CONCLUSION

The fast development of blockchain, GAI, and CV is transforming both the industries as well as social organizations. In this chapter, the unique capabilities of each of these areas, their most recent advances, and the synergies that arise when these technologies are combined have been discussed. Blockchain technology provides security, transparency, and trust, GAI is a source of creativity and adaptive intelligence, and CV introduces perception and situational awareness into the digital space.

The trends that we have discussed may be considered as follows: the evolution of Blockchain 1.0 (cryptocurrency) to Blockchain 3.0 (cross-industry applications), unprecedented development of GAI in content creation and synthesis of data, and the revolutionized CV powered by deep learning and real-time analysis. The combination of these technologies has the potential to facilitate cross-domain solutions, including safe medical imaging, AI-generated art secured by blockchain, and identity verification mechanisms that are valuable and reliable in terms of their operation.

Socially speaking, the proper incorporation of such technologies may result in the emergence of safer, more transparent, and highly personalized systems in the spheres of healthcare and education, commerce, and even governance. But this pledge should be accompanied by consideration of ethical, legal, and policy concerns, such as right to privacy, mitigation of bias, explicability, and cross-border regulation.

Finally, the future of the discussed technologies will be based on the ability to strike an appropriate balance between innovation and regulation. There are certain risks of overregulation (there is also anti-progression that could be caused by it), and there are also dangers of under-regulation (which could also lead to negative misuse). To attain this balance, technologists, policymakers, ethicists, and end-users must work together. The encompassing of the blockchain, GAI, and CV as an idea can turn the digital and physical space into an intelligent, safe, and human-friendly environment, which would be one of the biggest technological changes of the 21st century, in case pursued in a reflective manner.

REFERENCES

1. Zeng, H., Cao, J., Zhang, K., Chen, Y., Luong, H., & Philips, W. (2024). Unmixing Diffusion for Self-Supervised Hyperspectral Image Denoising. *2024 IEEE/CVF Conference on Computer Vision and Pattern Recognition (CVPR)*, Seattle, WA, USA, pp. 27820–27830. https://doi.org/10.1109/CVPR52733.2024.02628
2. Bandara, W. G. C., Nair, N. G., & Patel, V. M. (2022). DDPM-CD: Remote Sensing Change Detection Using Denoising Diffusion Probabilistic Models. *arXiv preprint arXiv*:2206.11892.
3. Bandara, W. G. C., & Patel, V. M. (2022). HyperTransformer: A Textural and Spectral Feature Fusion Transformer for Pansharpening. *CVPR*, New Orleans, LA, USA, pp. 1767–1777.
4. Bodrito, T., Zouaoui, A., Chanussot, J., & Mairal, J. (2021). A Trainable Spectral-Spatial Sparse Coding Model for Hyperspectral Image Restoration. *NeurIPS*, 34, pp. 5430–5442.
5. Cao, X., Fu, X., Xu, C., & Meng, D. (2021). Deep Spatial-Spectral Global Reasoning Network for Hyperspectral Image Denoising. *IEEE Transactions on Geoscience and Remote Sensing*, 99, 1–14.
6. Chu, X., Tian, Z., Wang, Y., Zhang, B., Ren, H., Wei, X., Xia, H., & Shen, C. (2021). Twins: Revisiting the Design of Spatial Attention in Vision Transformers. *NeurIPS*, 34, pp. 9355–9366.
7. He, W., Yao, Q., Li, C., Yokoya, N., Zhao, Q., Zhang, H., & Zhang, L. (2020). Non-Local Meets Global: An Iterative Paradigm for Hyperspectral Image Restoration. *IEEE Transactions on Pattern Analysis and Machine Intelligence*, 44(4), pp. 2089–2107.
8. Li, M., Fu, Y., & Zhang, Y. (2023). Spatial-Spectral Transformer for Hyperspectral Image Denoising. *Proceedings of the AAAI Conference on Artificial Intelligence*, 37, pp. 1368–1376.
9. de Diego, S., & Gutiérrez-Aguero, I. (2025). Decentralized Digital Product Passport Building Blocks for Enhancing Supply Chain Sovereignty and Circular Economy Practices. *IEEE Access*, 13, pp. 137973–137985. https://doi.org/10.1109/ACCESS.2025.3594826
10. Zohrabi, N., Jones, J. C., Keegan, B., Adhikari, S., Verrelli, B. C., & Abdelwahed, S. (2025). Addressing Urban Food Insecurity through Data-Driven and Community-Centric Smart City Frameworks. *IEEE Access*, 13, pp. 133847–133868. https://doi.org/10.1109/ACCESS.2025.3589963
11. Kim, S., Lee, D.-Y., Koh, J.-S., & Cho, K.-J. (2025). Programmable Shape-Changing System through Origami Pop-Up Cells. *IEEE Robotics and Automation Letters*, 10(9), pp. 9136–9143. https://doi.org/10.1109/LRA.2025.3588796
12. Chen, W., Aslan, Y., Litjens, R., & Yarovoy, A. (2025). System-Level Assessment of Shaped Elevation Beam Patterns for Hybrid Beamforming in mm-Wave 5G Networks With Spatially Heterogeneous Traffic. *IEEE Access*, 13, pp. 125933–125943. https://doi.org/10.1109/ACCESS.2025.3588913
13. Kwon, D., & Hong, S. (2025). Side-Channel Attack on ARADI in Non-Profiling Scenarios. *IEEE Access*, 13, pp. 122628–122635. https://doi.org/10.1109/ACCESS.2025.3585559
14. Xue, Y., et al. (2025). Advancements of Large Language Models for Enhancing Carbon Capture Technologies: A Comprehensive Review. *CHAIN*, 2(2), pp. 131–147. https://doi.org/10.23919/CHAIN.2025.000010

Chapter 3

Layered architecture of computer vision neural networks

Saddala Anil and V. Shobha Rani

3.1 INTRODUCTION

Increased use of computer vision (CV) as a central building block in contemporary artificial intelligence (AI) systems has been mainly propelled by advances in deep neural networks with clearly defined layered architectures. Such architectures are the structural foundation of today's vision systems, allowing them to attain human levels of performance in tasks like object recognition, scene understanding, medical imaging, and autonomous navigation. With visual data expanding exponentially, scalable and efficient learning architectures that can learn meaningful representations from images with little supervision are urgently needed.

Hierarchical architectures in CV neural networks are meant to model the hierarchical processing of the human visual system. They usually start with low-level feature detectors like edges or corners and move toward capturing higher-level semantic information like object parts and categories. Each of the layers in the architecture has a specific and complementary function. Convolutional layers are experts in extracting spatial features, pooling layers shrink spatial dimensions while retaining the key features, normalization layers make the learning process stable, and fully connected layers function as decision-making units by combining features for classification or regression.

The goal of this chapter is to provide a comprehensive understanding of how various types of layers help ensure the success of CV models. We discuss not just the structure and operation of these layers but also how they communicate within popular architectures like LeNet, AlexNet, VGG, ResNet, and EfficientNet. Stress is put on both the theoretical design and practical implementation of layered models in actual applications. Additionally, we touch upon general issues with training deep networks, like vanishing gradients and overfitting, and look at advanced solutions like depthwise separable convolutions and skip connections that have developed as main innovations.

Finally, this chapter is intended to be a knowledge base for students, researchers, and practitioners who want to design solid, efficient, and interpretable CV systems. By knowing what role each layer plays in the processing pipeline, readers will be in a better position to choose, adapt, or come up with novel architectures optimized for certain vision tasks and deployment settings.

DOI: 10.1201/9781003659112-3

3.2 RELATED WORK

The exploration of stacked architectures in CV neural networks has come a long way since the initial creation of artificial neural networks. Traditional methods like LeNet-5 established the foundation for hierarchical learning with the use of convolutional and pooling layers as the basis for building blocks. These initial models showed that multilayered representation could be viable for classifying handwritten digits, establishing a precedent for the next wave of research into deep learning.

With the proliferation of large-scale data such as ImageNet and the provision of computing power through graphics processing units (GPUs), deeper and more intricate architectures became a possibility. AlexNet was a milestone in terms of greatly increasing the depth of the network architecture and utilizing ReLU activations, dropout regularization, and data augmentation strategies. This design triggered a rapid eruption of activity, breaking the barriers of network depth and performance.

Later models like VGGNet (Simonyan & Zisserman, 2014) followed a simple and effective strategy with smaller 3×3 convolution filters along the entire network with depth enhancement. GoogLeNet (Szegedy et al., 2015) came up with the Inception module, combining various convolutional operations in a single layer, so there was improved parameter efficiency with greater feature diversity. While this was happening, ResNet (He et al., 2016) solved the degradation issue with deep networks by adding residual connections, which made it possible to train networks of more than 100 layers and was a building block of many subsequent models.

New architectures like EfficientNet (Tan & Le, 2019) investigate compound scaling strategies that scale network depth, width, and resolution uniformly. These models deliver state-of-the-art performance with fewer parameters and computational demands, rendering them ideal for both high-performance systems and deployment on mobile devices.

Also, effort has gone into improving layer operations themselves. Examples include batch normalization, which prevents internal covariate shift, depthwise separable convolutions (employed by MobileNet) decreasing computation without compromising accuracy, and attention mechanisms (such as Vision Transformers) trying out a different way from the strictly layered convolutional approach also discussed by Howard et al. (2017).

Together, these contributions establish the significance of layer design as a way to determine a network's generalization capacity, computational efficiency, and adaptability across a variety of vision tasks. The ongoing refinement of these architectural elements is the foundation for the layered analysis described in this chapter.

3.3 LAYERED PIECES AND THEIR PURPOSES

Deep CV neural networks are based on a structured setup of specialized layers, where each layer is engineered to accomplish a certain transformation or computation on the input information. This chapter describes the primary layer types usually deployed in deep vision models and outlines each's individual purpose, mathematical operation, and connections in the framework of visual comprehension.

3.3.1 Convolutional layers

Convolutional layers are the workhorse of the majority of CV models. Their job is mainly to learn local features of input images through a collection of learnable filters or kernels. Each filter sweeps over the image, detecting spatial hierarchies like edges, textures, and patterns.

Algebraically, the value of the output of a convolution operation at position (i,j) in feature map k is as follows:

$$y_{i,j}^{(k)} = \sum_{m}\sum_{n} x_{i+m,j+n} \cdot w_{m,n}^{(k)} + b^{(k)}$$

where x is the input, $w^{(k)}$ is the kernel for feature map k, and $b^{(k)}$ is bias. The sparse connectivity and weight-sharing property of convolutions confer to them computational efficiency and translation invariance.

3.3.2 Pooling layers

Pooling layers compress the spatial dimensions (height and width) of feature maps but preserve the most significant information. This downsampling not only reduces computational expenses but also allows for some spatial invariance to transpositions and distortions.

The most popularly used pooling operations are as follows:

Max Pooling: Returns the maximum value within a given window.
Average Pooling: Returns the mean value across the pooling window.

A standard pooling layer employs a window (e.g., 2×2) and a stride (e.g., 2) to downsample features, easing overfitting and promoting generalization.

3.4 LAYER DESIGN STRATEGIES IN STATE-OF-THE-ART ARCHITECTURES

The last decade saw the development of a number of benchmark convolutional neural network (CNN) architectures, each providing new layer organization and feature learning strategies. This section investigates the architectural advances in state-of-the-art models like LeNet, AlexNet, VGG, ResNet, and EfficientNet, highlighting their distinctive design philosophies, layer configurations, and practical implications on performance and scalability.

3.4.1 LeNet: the foundational design

LeNet-5, built by Yann LeCun in the late 1990s for handwritten digit recognition (MNIST dataset), is one of the first CNN architectures. It has two convolutional layers followed by pooling and developed by Dosovitskiy et al. (2021), followed by fully connected layers for classification.

Important design practices:

Application of 5×5 filters for feature extraction.
Average pooling over max pooling.
Clear distinction between feature extraction and classification blocks.
Tanh activation functions.

While shallow by today's benchmark, LeNet proved the strength of layered structures and end-to-end training for vision tasks.

3.4.2 AlexNet: deep learning breakthrough

AlexNet revolutionized deep learning by defeating the ImageNet Large Scale Visual Recognition Challenge (ILSVRC) in 2012 by a landslide margin.

Architectural highlights:

Five convolutional layers, preceded by three fully connected layers.
Application of ReLU activation to allow quicker convergence.
Overlapping max pooling for improved generalization.
Dropout in fully connected layers to prevent overfitting.
Split computation across two GPUs for scalability.

AlexNet confirmed that deeper models, when trained on large datasets with GPUs, could perform better.

3.4.3 VGG: similarity in layer design

The VGG architecture, especially VGG-16 and VGG-19, focused on simplicity and depth.

Design philosophies:

Nested small 3×3 convolutional filters.
Fixed-size max pooling layers to downsample.
Iteration of convolution-pooling blocks for deeper hierarchical feature extraction.

Advantages:

Less parameters per filter because of small kernels.
Easy modular extension.
Uniformity of layer structure made transfer learning easier.
Even with its accuracy, VGG models are computationally intensive and memory-hungry.

3.5 COMPARATIVE EVALUATION AND TRAINING CHALLENGES

With advancements in CNN architectures, comparative evaluation and assessment of their performance, robustness, and trainability become pertinent. Although deep and wide architectures typically yield improved accuracy, such improvements are made at the cost of computational complexity, training dynamics, and generalization strength. This section compares the state-of-the-art architectures presented above, along with significant challenges encountered during training.

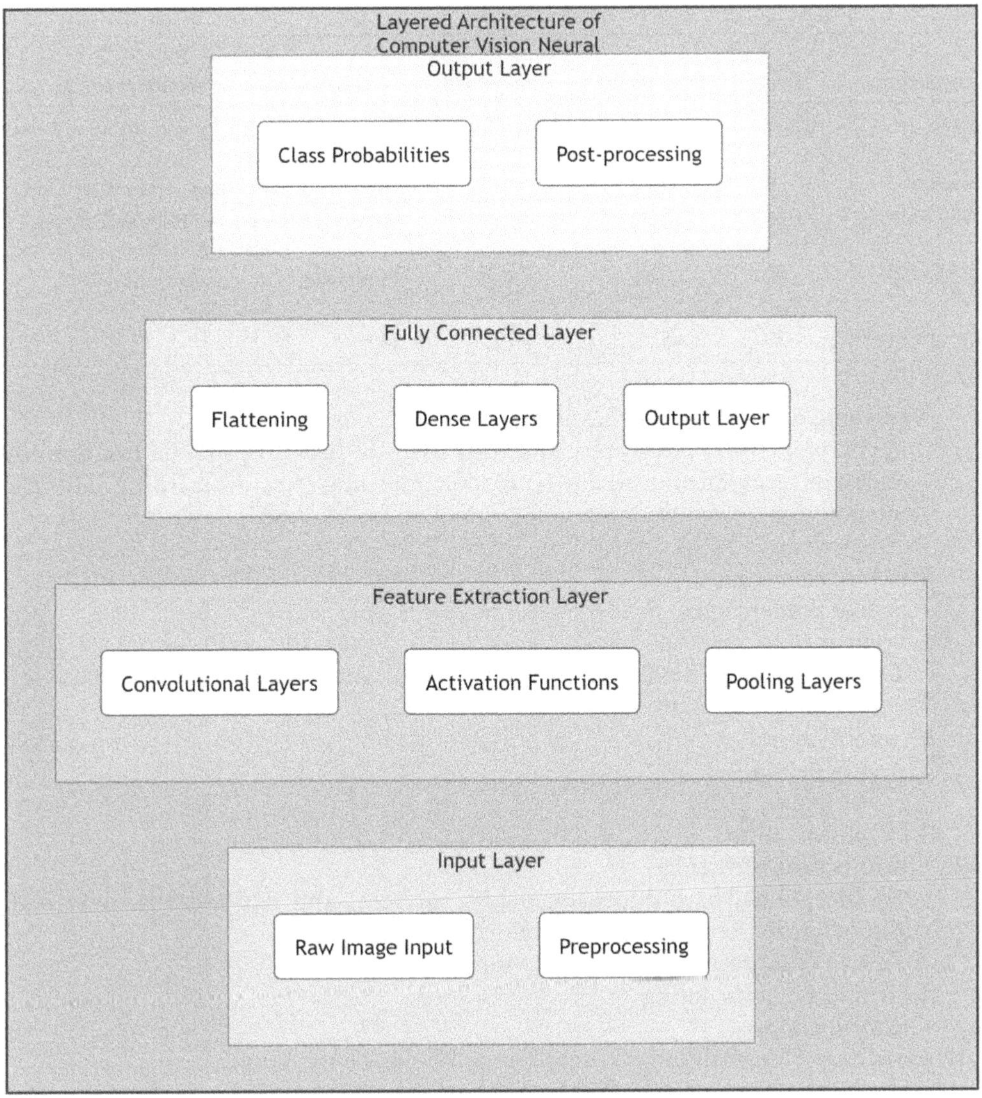

Figure 3.1 Layered architecture of computer vision neural networks.

3.5.1 Comparative performance analysis

A summary of the relative strengths of different architectures on baseline benchmarks like ImageNet is given as follows.

Although deeper architectures (EfficientNet and ResNet) usually perform better, they tend to need sophisticated optimization methods and hardware accelerators.

3.5.2 Training challenges in deep networks

Training deep neural networks introduces a range of challenges that are tightly linked to architectural design:

Table 3.1 Comparison of EfficientNet and ResNet

Architecture	*Depth*	*Parameters*	*Top-1 accuracy (ImageNet)*	*FLOPs*	*Notable strength*
LeNet	7	~60K	~99% (MNIST)	Very low	Simplicity and traditional use case
AlexNet	8	~60M	~57%	High	GPU usage and use of ReLU
VGG-16	16	~138M	~71%	Very high	Uniform filters and deeper learning
ResNet-50	50	~25M	~76%	Moderate	Skip connections and scalability
EfficientNet-B0	~20	~5M	~77%	Low	

i. **Vanishing and Exploding Gradients**

As networks become deeper, gradients used for backpropagation can diminish (vanish) or grow uncontrollably (explode), impeding effective learning. This phenomenon was a key limitation before innovations like the following:

ReLU activation.

Batch normalization.

Residual connections.

ii. **Overfitting**

Large, complex models are susceptible to overfitting on small data sets. Prevention strategies are as follows:

Dropout layers.

Data augmentation.

Early stopping.

L2 regularization.

iii. **Optimization Difficulty**

Saddle points, high-dimensional loss surfaces, and bad local minima make training harder. Solutions are as follows:

Adaptive optimizers (Adam and RMSprop).

Learning rate scheduling.

Gradient clipping.

iv. **Hardware Constraints**

Training time and memory usage grow exponentially with width and depth. EfficientNet and MobileNet are able to address this through light-weight modules, but older models such as VGG need high-end GPUs.

3.5.3 Interpretability and debugging

When CNNs get deeper, their decision-making process becomes more obscure. Some major methods to mitigate this include the following:

Activation map visualization (e.g., Grad-CAM).

Layer-wise relevance propagation (LRP).

Filter analysis and pruning for reduction.

Training challenges highlight the need to balance architectural complexity with interpretability, stability, and efficiency.

3.5.4 Model optimization and deployment

Model optimization and deployment are vital processes in taking CV models from research prototypes to efficient, trustworthy, and usable real-world applications. This process is centered on maximizing computational efficiency, minimizing resource utilization, and making sure that they are easy to integrate into target environments.

3.5.4.1 *Optimization techniques*

Pruning and Quantization of Parameters: Eliminating excess weights and transforming floating-point parameters to lower precision data types (e.g., INT8) in order to save memory while minimizing accuracy loss.

Knowledge Distillation: Supervising smaller "student" networks with predictions from a larger, high-performance "teacher" network.

Efficient Architectures: Utilizing light-weight models such as MobileNet, ShuffleNet, or EfficientNet for edge deployment.

3.5.4.2 *Deployment platforms*

Cloud Deployment: Deploying models on cloud-hosted services (AWS, Azure, and Google Cloud) to support scalable and globally accessible application programming interfaces (APIs).

Edge and On-Device Deployment: Executing models locally on embedded devices, smartphones, or Internet of Things (IoT) sensors for low-latency, real-time processing.

Hybrid Systems: Merging edge processing for real-time inference with cloud processing for heavy analytics and retraining.

3.5.4.3 *Security and privacy considerations*

Using secure model packaging to avoid reverse-engineering.

Implementing federated learning for model updates without data movement.

Following data protection laws (e.g., GDPR and HIPAA).

3.5.4.4 *Monitoring and maintenance*

Establishing performance dashboards to track accuracy drift and system delay.

Periodically retraining the model when there are notable performance declines due to domain shift or data drift.

3.6 LAYER INNOVATION BREAKTHROUGHS

The last few years have seen the emergence of a tremendous breakthrough in neural network layer architectures, overcoming long-standing limitations of classical designs and improving both performance and efficiency. They help make contemporary vision systems faster, more accurate, and frugal. This part discusses some of the most influential breakthroughs: depthwise separable convolutions, attention mechanisms, and transformer-based layers.

3.6.1 Depthwise separable convolutions

Depthwise separable convolutions are a significant innovation toward minimizing computational cost in convolutional networks. This method breaks a normal convolution operation into two individual steps:

Depthwise Convolution: For each input channel (feature map), a single filter is applied.
Pointwise Convolution: 1×1 convolutions are applied for summing up the outputs of depthwise convolution over channels.

This decoupling decreases the parameters and computations by a huge amount with minimal loss of accuracy. Architectures like MobileNet and Exception have effectively used this mechanism, allowing deployment on embedded and mobile devices.

As an example, MobileNet saves computation by almost 8–9 times from the traditional convolutions while the classification performance remains high on tasks like ImageNet.

3.6.2 Attention mechanisms

Attention mechanisms enable neural networks to dynamically concentrate on the most pertinent areas of input data. In CV tasks, self-attention allows models to learn global dependencies among various spatial positions in an image. Some of the popular attention-based methods are as follows:

Squeeze-and-Excitation (SE) Blocks: Re-weight feature channels by learning interdependencies.
Convolutional Block Attention Module (CBAM): Inserts both channel and spatial attention sequentially.
Nonlocal Blocks: Expose long-range dependencies, particularly beneficial for video understanding.

Focus has emerged as a key ingredient in enhancing interpretability and performance, particularly in medical imaging and fine-grained recognition.

3.6.3 Vision Transformer layers

Transformers, introduced for natural language processing (NLP) tasks, have transformed CV with models such as Vision Transformer (ViT) and DETR. They eschew the conventional convolution layers completely and rely exclusively on self-attention to represent image data.

Important features of transformer-based vision models:

Patch-based Representation: Divide an image into fixed-sized patches (e.g., 16×16) where each is considered a token.
Positional Encoding: Adds spatial information to the sequence of image patches.

Global Context Modeling: Enables attention computation over the whole image.

Even with their computational expense and data demands, ViTs have obtained state-of-the-art performance in various vision benchmarks and are regarded as an exciting direction for the future.

3.6.4 Hybrid architectures

In order to take advantage of the strengths of both CNNs and transformers, hybrid models like ConvNeXt, Convolutional vision Transformer (CoaT), and Swin Transformers blend convolution local feature extraction with attention layer global modeling capabilities.

These hybrid models are designed to achieve the following:

Improved scalability.
More efficient inductive biases (from CNNs).
Better generalization on complicated vision tasks.

3.6.5 Self-supervised and few-shot learning layers

Self-supervised and few-shot learning have become revolutionary paradigms of contemporary CV, allowing models to perform competitively with very little labeled data. These techniques play a direct role in shaping the design of new neural network layers capable of learning richer, more generalized representations from sparse supervision.\

3.6.5.1 Self-supervised layer design

Contrastive Learning Layers: Place images in a feature space in which similar images are brought together and dissimilar images are separated. Some examples include SimCLR and MoCo-inspired projection layers.

Masked Feature Prediction Layers: Drawing on NLP's masked language modeling, these layers mask parts of an image's latent representation and train the model to predict them back (e.g., MAE – masked autoencoders).

Pretext Task-Specific Layers: Task-optimized layers for tasks such as rotation prediction, jigsaw puzzle-solving, or colorization, which produce supervisory signals without direct human annotation.

3.6.5.2 Few-shot learning layers

Meta-Learning Layers: Learn meta-parameters that can adapt rapidly to novel tasks using few samples, typically incorporated into feature extraction layers to ensure maximum reusability.

Prototypical Representation Layers: Represent examples as a class prototype embedding space to classify novel categories based on distance metrics.

Attention-Improved Matching Layers: Implement cross-attention between support and query sets to improve features for improved generalization.

3.6.5.3 Computer Vision applications

Medical image scenarios with limited annotated data.

Remote sensing applications that need adaptation to novel geographic areas.

Industrial inspection applications that require quick retraining for new product lines.

3.6.5.4 Strengths and weaknesses

Strengths: Potentially drastically decrease reliance on huge annotated datasets, encourage domain flexibility, and optimize learning efficiency.

Weaknesses: Danger of learning spurious relationships, sensitivity to pretext task selection, and extreme training instability in very low-data setups.

3.7 FUTURE DIRECTIONS AND OPEN CHALLENGES

The CV neural network layered architecture has progressed dramatically – starting with initial shallow models such as LeNet to advanced hybrid models that are a fusion of convolution, attention, and transformer mechanisms. Still, there are several open challenges. Future directions and most important open challenges are described in this section, which need to be overcome to further push the field.

3.7.1 Efficient and Green AI models

With the increased environmental and economic expense of training large-scale vision models, sustainability and efficiency are the newly urgent concerns.

Efficient Training: Next-generation models need to reduce carbon footprint and prefer low-resource training techniques.

Pruning and Parameter Sharing: Studies on structured pruning, quantization, and distillation can provide compact but efficient models.

Edge Deployment: Techniques need to be developed to deploy vision models efficiently on microcontrollers, drones, and wearables.

3.7.2 Robustness and generalization

Despite their performance on benchmarks, our current models remain brittle against noise, adversarial attacks, and domain shift.

Robust Training Methods: Methods such as adversarial training and consistency regularization are crucial.

Generalization across Domains: Models need to generalize robustly across different lighting, resolution, or style conditions.

Uncertainty Estimation: Improved mechanisms to estimate prediction uncertainty are essential, particularly in safety-critical applications such as autonomous driving.

3.7.3 Explainability and interpretability

Neural networks are sometimes seen as "black boxes." As they become more used in high-stakes applications, why a model came to a conclusion is just as valuable as the conclusion itself.

Post hoc explanation tools (e.g., Grad-CAM and LIME) need to become routine practice.

Intrinsic interpretability through architecture design is an exciting space, e.g., self-explaining networks.

Ethical AI systems need models that deliver explainable and transparent outputs, particularly in surveillance or legal analytics.

3.7.4 Learning with less data

One of the biggest limitations of existing models is their reliance on gigantic labeled datasets. Next-generation vision systems need to shift toward the following:

Few-shot and zero-shot learning with meta-learning or prompt-based adaptation.
Self-supervised learning to take advantage of large quantities of unlabeled data.
Transfer learning to efficiently transfer pre-existing models to new tasks and domains.

3.7.5 Integration with multimodal systems

CV is also being increasingly integrated with language, audio, and sensor data in multimodal systems.

Vision-language models such as CLIP and Flamingo exhibit synergy between modalities.

Layered architectures in the future might be required to naturally support multiple data types.

REFERENCES

Dosovitskiy, A., et al. (2021). An image is worth 16x16 words: Transformers for image recognition at scale. *International Conference on Learning Representations (ICLR)*, Vienna, Austria.

He, K., Zhang, X., Ren, S., & Sun, J. (2016). Deep residual learning for image recognition. *Proceedings of the IEEE Conference on Computer Vision and Pattern Recognition (CVPR)*, Las Vegas, Nevada, 770–778.

Howard, A. G., et al. (2017). MobileNets: Efficient convolutional neural networks for mobile vision applications. arXiv preprint arXiv:1704.04861.

Krizhevsky, A., Sutskever, I., & Hinton, G. E. (2012). ImageNet classification with deep convolutional neural networks. *Advances in Neural Information Processing Systems, 25*, 1097–1105.

Simonyan, K., & Zisserman, A. (2014). Very deep convolutional networks for large-scale image recognition. *International Conference on Learning Representations (ICLR)*, Banff, Canada.

Szegedy, C., Liu, W., Jia, Y., Sermanet, P., Reed, S., Anguelov, D., Erhan, D., Vanhoucke, V., & Rabinovich, A. (2014). Going Deeper with Convolutions. CoRR, abs/1409.4842.

Tan, M., & Le, Q. V. (2019). EfficientNet: Rethinking model scaling for convolutional neural networks. *International Conference on Machine Learning (ICML)*.

Chapter 4

Blockchain and Generative AI

Fundamentals and consensus mechanisms

Ragini Karwayun, Swapna Singh, and Rishabh Karwayun

4.1 INTRODUCTION

The 21st century has witnessed the convergence of digital technologies, producing a complicated and rapidly developing environment with the synergy of base infrastructures with profound computational intelligence. Two transformative technologies, blockchain and generative artificial intelligence (GenAI), are poised to disrupt the way in which information is produced, verified, shared, and managed [1,2]. Together, they have disrupted their respective domains; on their own, blockchain and GenAI represent a paradigm shift in secure decentralized intelligent digital environments [3,4].

More specifically, blockchain has been established as the most promising candidate in addressing the trust problem for Decentralized Social Networks (DSNs) [5]. By leveraging numerous cryptographical algorithms and decentralized consensus mechanism, the blockchain technology enables a network of participants to securely achieve tamper-proof and transparent distributed ledger in a trust-less environment without central intermediary control [6,7]. "Block chain" first came into spotlight with banking in the form of Bitcoin [5]; it has gone on to contribute to smart contracts, decentralized finance (DeFi), identity management, supply chain logistics, etc. [8]. It enables systems where truth is programmatically enforced, rather than institutionally guaranteed [9].

In parallel, GenAI has grown into one of the most impactful branches of machine learning [10,11]. Rather than simply classifying or analyzing data, GenAI systems create new data instances that mirror the complexity, semantics, and structure of the data they were trained on [12]. With the advent of large-scale neural architectures such as transformers and diffusion models, GenAI can generate text, images, audio, code, and more [11,12]. Applications span diverse domains: content creation, medicine, law, science, entertainment, and engineering [13,14].

Yet, despite their potential, both technologies face major limitations. Blockchain suffers from latency, energy inefficiency, and rigid programmability [15,16]. GenAI is plagued by bias, hallucination, lack of provenance, and ethical ambiguity [17,18]. Blockchain ensures that data are correct but cannot adapt or reason; GenAI reasons but cannot assure correctness [2,19]. Their integration is therefore not just strategic but essential [3,4]. Blockchain offers transparency, traceability, and decentralized governance for AI-generated content [20,21]. In turn, GenAI can enhance blockchain systems with automation, reasoning, and intelligent orchestration [19,13].

 DOI: 10.1201/9781003659112-4

This chapter explores the foundational principles, architectural elements, consensus mechanisms, and governance models involved in the convergence of blockchain and GenAI. It investigates how blockchain can provide infrastructure for secure AI training, verifiable content generation, decentralized model markets, and incentive mechanisms [21,22]. At the same time, it examines how GenAI can automate smart contract generation, optimize consensus, and power autonomous agents in decentralized networks [13,14]. This chapter also addresses ethical, security, and scalability challenges inherent to this integration, proposing hybrid protocols and research directions for building robust, ethical, and future-ready digital ecosystems [17,18].

4.2 THE FOUNDATIONS OF BLOCKCHAIN TECHNOLOGY

Blockchain is a technology that works as a immutable shared, append-only digital ledger. The term "blockchain" comes from its basic data structure where data are organized in blocks, with each new block linked to the one before it, forming a continuous chain.

Each block contains information such as a list of transactions, a timestamp, and a unique identifier referred as a cryptographic hash. This hash is generated from the block's contents and the hash of the previous block, in such a way that a change in even a single bit will change the resulting hash, ensuring that each block is chained to the one preceding it in this way.

4.2.1 The structure of the blockchain

A blockchain is, at its core, a linked list of cryptographically secured records known as blocks [1,3]. Each block contains a list of transactions, a timestamp, a nonce, the hash of the previous block, and its own hash derived from the above fields [4,7]. This structure ensures that each block is permanently connected to its predecessor, forming a chain that is virtually immutable [5].

Immutability is enforced by cryptographic hashing [23]. Any change to the content of a block would alter its hash, breaking the link with the next block and alerting the network to tampering [4]. The tamper-evident ledger is made possible by the cryptographic integrity along with distributed consensus and can be trusted independent of the central authority [6,24]. Blockchain is decentralized, and this implies that the ledger is replicated across multiple nodes in the network [1,7]. Each node keeps a complete copy of the blockchain, while new blocks are only added to the historical blocks on the blockchain, with the majority of the nodes consenting to the legitimacy of those new blocks through consensus algorithms [25,26]. A distributed system architecture removes single points of failure and reduces the risk of data forge [9,27].

4.2.2 Consensus mechanisms: securing agreement in decentralized systems

In the center of every blockchain is its consensus protocol, a solution of how a chained network of nodes in the decentralized network reaches the agreement of what is the current state of its ledger [28–30]. In contrast to a traditional centralized database, which is run by a single authority responsible for ensuring the accuracy of the data itself, a blockchain operates within an adversarial environment where nodes can be anonymous, located in different geographical locations, and can be malicious. Consensus protocols solve this problem by aligning incentives and scaring away from the system any party that tries to break the consistency by either cheating or trying to tamper with the history.

The first consensus mechanism and the most famous one is Proof of Work (PoW), applied by Bitcoin. In PoW, machines called miners compete to solve a hard math problem. The first to verify it can add a block and receive a reward. While PoW offers provably strong security guarantees and Sybil resistance, it is wasteful with energy and lacks processing power [31–33]. Alternative reverse channel mechanisms are required. In response to these issues, the following mechanisms have been proposed:

4.2.2.1 Proof of Work (PoW)

PoW, introduced by Bitcoin, involves miners in solving computationally intensive challenges to confirm transactions and produce a new block [5,25]. The first miner who discovers a solution broadcasts it to the network, and other nodes validate the solution and accept the new block [6,23]. This technique guarantees security in terms of computational complexity with the cost of high energy consumption and poor scalability [15,34].

4.2.2.2 Proof of Stake (PoS)

Proof of Stake (PoS) aims to solve the alcohol problem of PoW by choosing validators based on stake in the network – stake-serving rather than work-serving signers [26,25]. Validators are selected to generate new blocks based on their stake and will be punished with the loss of their staked tokens if they behave maliciously [35,26]. PoS algorithms reduce the energy cost and support the security using economic incentives [34,36].

4.2.2.3 Byzantine Fault Tolerance

Byzantine Fault Tolerance (BFT) protocols guarantee a consensus among network nodes even when some behave maliciously or fail [37,38]. The Practical Byzantine Fault Tolerance (PBFT) and its variations can tolerate the maximum number of Byzantine faulty nodes being less than one-third of the total number of nodes [39,40]. These are especially interesting in the permissioned blockchain setting, where participant identity is known [41–43].

These mechanisms are a result of a compromise among scalability, security, and decentralization [27]. The choice of a consensus mechanism is context-driven and becomes increasingly important when deploying within dynamic and data-rich AI systems that demand novel forms of trust and validation [24].

Consensus protocols are critical for the operation of blockchain so that every network node agrees on the current form of the ledger, without the presence of a central authority These protocols provide security and fault-tolerance in blockchain systems [25,34].

With the continuous enhancement of consensus mechanisms such as PoW, PoS, DPoS, and PBFT, blockchain technology has evolved to Blockchain 3.0 [27,41]. Recent consensus protocols mix and match from a smorgasbord of techniques to better suit their intended usage by trading off security, scalability, and decentralization [34,44,45].

4.2.3 Smart contracts and programmability

Smart contracts, self-executing programs that perform predetermined instructions when a certain condition is met on-chain, are one of the most powerful applications of blockchain technology. Pioneered by Ethereum, smart contracts allow for ever-more complex processes such as Decentralized Finance (DeFi) schemes, token issuing, voting, and multi-signature wallets.

In AI applications, smart contracts enable access control of training data, ethical use policies, and reward-sharing among data providers, validators, and model trainers. For example, a smart contract could do the following:

- Automatically pay contributors who provide high-quality data for model training.
- Validate the performance of a submitted AI model against a public benchmark before distributing tokens.
- Enforce usage restrictions, such as limiting certain generative models to noncommercial use unless licensed.

The deterministic nature of smart contracts ensures that their execution is transparent and cannot be altered after deployment. However, this immutability poses a challenge for AI systems that need to adapt, evolve, and learn from new data. One solution is the use of **proxy contracts** and **upgradable architectures** that separate logic from data, allowing updates while preserving on-chain references.

More integration with smart contracts, where GenAI can make decisions itself. Think of a Decentralized Autonomous Organization (DAO) that's been partly directed by a generative language model, which has learned to recommend policy changes, summarize community feedback, or draft fresh contracts. These scenarios provide a glimpse into a new and fast-maturing frontier: AI-driven infrastructure.

Smart contracts are self-executing contracts with terms directly written into code [35]. They automatically execute predefined actions when specific conditions are met, eliminating the need for intermediaries and reducing transaction costs [8,35]. Smart contracts enable complex decentralized applications (DApps) and form the foundation for DeFi ecosystems [8,35].

The programmability of smart contracts allows for sophisticated logic and automated processes, but it also introduces new challenges related to code security, gas optimization, and formal verification [35,13]. Integration with AI systems can enhance smart contract capabilities by enabling adaptive behavior and intelligent decision-making [19,13].

4.3 GENERATIVE ARTIFICIAL INTELLIGENCE: PRINCIPLES AND APPLICATIONS

GenAI is a type of AI that can create original content in response to a user's prompt or request by using machine learning models to create new, original content, such as images, text, or music, based on patterns and structures learned from existing data.

4.3.1 Foundations of Generative AI

GenAI represents a paradigm shift from discriminative models that classify or analyze data to models that create new content [10,12]. These systems learn the underlying

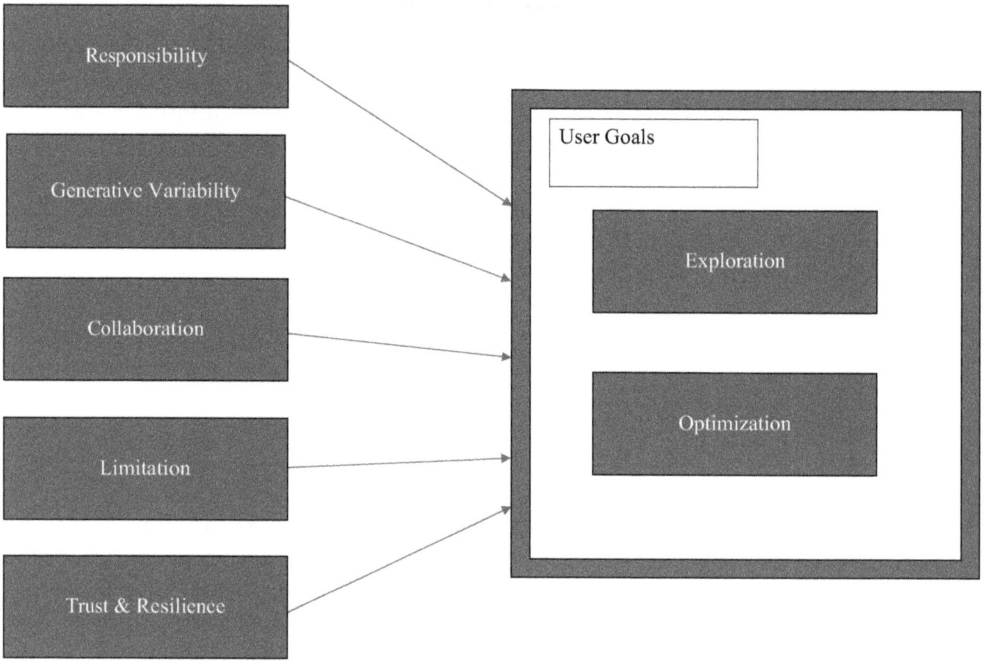

Figure 4.1 Generative AI design principles.

patterns and distributions of training data to generate novel instances that maintain the statistical properties of the original dataset [11,18].

The foundation of modern GenAI lies in several key architectures (Figure 4.1). Generative Adversarial Networks (GANs) pit two neural networks against each other in a competitive training process [10]. Variational autoencoders (VAEs) learn compressed representations of data that can be sampled to generate new instances [10]. More recently, transformer architectures have revolutionized text generation, while diffusion models have achieved remarkable results in image synthesis [11,12].

4.3.2 Large language models and transformers

The attention mechanism proposed in transformer architecture have enabled the models to process sequences in a more efficient manner and encode long-range dependencies [11]. Large language models (LLMs) such as GPT-4 show great potential in both understanding and generation of natural language [12,18]. Such models are capable of addressing various tasks such as input completion, translation, summarization, and code generation [13,14].

The size of the contemporary LLMs, usually in the order of billions or even trillions of parameters, allows for the emergence of capabilities that have not been observed in smaller models [12]. However, this scale also gives rise to the problems of computational complexity, bias, hallucination, and ethical concerns [17,18].

4.3.3 Multimodal and domain-specific applications

GenAI has evolved to cover not only the domain of text but also other modalities including image, audio, video, and code as well [13,14]. Stable Diffusion and others are also used to generate high-quality images from textual descriptions [18]. In expert domains, GenAI allows architects to generate higher performing, sustainable, more efficient design solutions [14].

Code generation models such as GitHub Copilot and Alpha Code showcase AI technologies to help in software creation and competitive programming. These applications demonstrate the flexibility of GenAI in different areas of professional activity [13,14].

4.3.4 Diffusion models

Diffusion models are a newer family of generative models that learn to gradually denoise data from random noise, effectively learning the inverse of a diffusion process. They have achieved state-of-the-art results in high-fidelity image generation. Tools like Stable Diffusion and DALL·E 2 generate photorealistic images from natural language prompts, expanding GenAI's reach into visual domains.

4.3.5 Generative Adversarial Networks

GANs consist of a generator and a discriminator in a competitive framework. The generator produces synthetic samples, while the discriminator distinguishes between real and fake samples. Training continues until the generator learns to produce realistic outputs. GANs are widely used in creative arts, fashion, facial synthesis, and style transfer.

4.3.6 Variational Autoencoders

VAEs encode data into a latent space with probabilistic interpretations, enabling smooth interpolation and reconstruction. They are useful in domains where control over generation is necessary, such as medical imaging or chemistry.

These architectures often work in tandem. For example, Contrastive Language–Image Pretraining (CLIP) models can be used to guide diffusion processes for more semantically accurate image generation. The field is rapidly evolving, with hybrid models, multimodal models, and reinforcement learning-enhanced architectures emerging monthly.

4.4 REAL-WORLD APPLICATIONS OF GENERATIVE AI

GenAI's practical application varies greatly between domains:

- **Text Generation:** From chatbots to legal document generation, GenAI fuels products like ChatGPT, Jasper AI, and Claude.
- **Image and Video Synthesis:** There are tools such as MidJourney, Artbreeder, and Runway ML for creators to create and edit high-resolution visuals.
- **Code Generation:** GitHub Copilot, powered by Codex, helps developers write code, get smart suggestions, and learn new languages.
- **Healthcare:** GenAI is applied in creation of synthetic patient records, molecular structure simulation, and clinical note generation.
- **Education:** Customized learning plans, adaptive testing, and AI tutors are provided to help learners in the world, based on GenAI models.

Everything of these apps brings efficiencies and later on risks.

4.4.1 Limitations, risks, and trust deficit in GenAI

Along with increasing capabilities and adoption of GenAI systems, there has been growing discussion about what these systems can't do and the risks they present in the context of critical systems. Their ability to generate coherent, convincing, and even esthetically pleasing outputs notwithstanding, GenAI models are making statistical approximations— not engaging in semantic comprehension or logical truth.

- **Hallucination and Misinformation**

  "Hallucination" is one of the most frequently cited problems of LLMs, which means the LLM can produce content that is false and/or not based on any facts. An installed base GenAI model could have high confidence that a fictional study exists and make up citations or invent historical records [46]. The issue is not necessarily that people are maliciously biased but that the training procedure itself is optimized for predictability and fluency, not truth.

  This makes GenAI outputs especially dangerous in fields where factual accuracy is mandatory: healthcare diagnoses, legal arguments, financial advice, and public policy. The hallucination issue will erode trust and make human–AI collaboration difficult.
- **Bias and Representation**

  GenAI models learn from the data they are trained on and that data are a reflection of the world, warts, and all (including a fair share of prejudice, omissions, and injustice). Forms of racism, gender stereotyping, and marginalization of groups of people are well documented within GenAI systems [47]. A model trained on data from Western countries may not perform well or generalize well when used in African or Asian languages.

  Even worse, these biases can be subtle and insidious, baked into the tone, the context, or the implications. It is not enough to address these through data filtering; they need systemic revisiting of training pipelines, oversight, and participatory governance.
- **Security Threats**

  GenAI provides new attack vectors in cyber-security. For example:
  - **Prompt Injection:** An attacker manipulates the prompts and makes the model violate the safety protocol.
  - **Model Inversion:** Adversaries can, if they have access to model outputs, reconstruct sensitive training data.
  - **Data Poisoning:** By making minor perturbations to the training data, attackers are able to encode malicious behavior into the model.

  As GenAI is integrated with operational systems 1—from customer service to medical decision support—vulnerabilities of this kind have real-life threats.
- **Intellectual Property and Provenance**

  GenAI models are trained on vast corpora, often scraped from the web without explicit licensing. This raises legal and ethical questions about the reuse and resynthesis of copyrighted material. If a model generates an image that closely resembles a living artist's work, who owns the result? If it writes text similar to a published paper, is that plagiarism?

More broadly, the lack of **provenance**, the ability to trace an output back to its source data and parameters, weakens accountability. Users, auditors, and regulators have no way to verify where a given piece of generated content came from, how it was produced, or whether it aligns with declared usage policies.

4.5 WHY GENAI NEEDS BLOCKCHAIN

Given these limitations, the case for integrating GenAI with blockchain becomes clear. Blockchain offers exactly what GenAI lacks: provenance, verifiability, traceability, and governance without relying on trust in central authorities.

- **Verifiable Data Provenance**
  By anchoring hashes of training datasets, model versions, and outputs onto a blockchain, we can create tamper-proof audit trails. Every GenAI-generated artifact could be linked to a unique on-chain record containing the following:
  - Training data version.
  - Model hash and configuration.
  - Generation timestamp.
  - Associated licensing and usage terms.

  This would not only prevent disputes around intellectual property but also enable users to independently verify claims about bias, safety, or performance.
- **Output Certification**
  AI outputs can be sent to decentralized validator networks prior to the release or publication. Such validators might include hallucination/toxicity/bias/plagiarism validators, who might attach a certification metadata to content. The blockchain would write these validations in a way that nothing could change them (immutable); therefore, a web of trust is generated around GenAI.
- **Decentralized Governance**
  Instead of having those who own platforms or build models to moderate and monitor AIs, DAOs could enable communities to define policies, update constraints, and supervise the trainings. Voting, funding, and enforcement could all be done through smart contracts, with sensitive information made publicly known.
- **Incentivized Alignment**
  Third, blockchain would enable token-based incentive models, which may compensate users based on their behaviors aligned to diversity's collective aims like whether users consistently curate high-quality training data, whether they review generated artifacts, or whether they flag harmful generations. This acts as an economic feedback mechanism to direct the most efficient development of GenAI in terms of safety, inclusivity, and usefulness.

  In short, blockchain enables to turn genie-black-box-curiosity that we described above into a genie-white-box-accountable AI system. It's not just the sum of two technologies; it justifies a fundamentally different way of developing intelligent systems—ones that are creative and reliable.

4.6 CONSENSUS MECHANISMS FOR GENERATIVE AI VALIDATION

Consensus mechanisms are foundational to the operation of any blockchain. Traditionally, their function has been to validate transactions and achieve an agreement on the state of a financial ledger. However, in the context of GenAI, consensus takes on new dimensions. It must handle subjective content, evaluate output quality, and enable decentralized verification of probabilistic and often nondeterministic processes. Designing consensus protocols for GenAI is, therefore, both a novel challenge and a critical requirement for any blockchain-integrated AI system.

4.6.1 Beyond binary validation

Most blockchains work with binary logic: a transaction is either valid or invalid. A digital signature is either correct or incorrect. But GenAI outputs are not discrete transactions; they are fluid, creative, and multidimensional. Consider the problem of validating a summary generated by an LLM: multiple versions may be equally acceptable. Similarly, for AI-generated artwork, there's no definitive "correct" output. Consensus in this domain must be probabilistic, multistakeholder, and context-aware.

For instance, validators in an AI-integrated blockchain may be tasked with evaluating outputs for the following:

- **Factual accuracy** (for text).
- **Similarity to copyrighted content.**
- **Presence of offensive or unsafe language.**
- **Conformance to ethical guidelines.**

Rather than a single "yes" or "no," validators may vote on multiple dimensions, providing a reputation-weighted or token-weighted score that determines whether an output is certified, flagged, or rejected.

4.6.2 The role of reputation systems

To ensure meaningful participation and minimize spam or sabotage, validators in such systems require incentives and reputational feedback loops. Trust in Reputation systems: In reputation systems, the trust score of a user is calculated based on previous behavior, accuracy of validations assigned to it, opinions extracted from community, and so on. And those scores could indicate a participant's influence over consensus, their reward eligibility, and even how much they can submit or validate.

Any good reputation system should look to handle the following:

- **Sybil Resistance:** Restricting a single actor in creating identities.
- **Collusion:** Discovering the consortia that conspire to subvert the destiny.
- **Transparency:** Letting others audit why a validator has a certain score.

The trustless, tamper-evident chains and smart contract technology can serve as an implementation of such systems as smart contracts can autonomously enforce the score changes according to performance measurements.

4.6.3 Consensus via Proof of Learning

It also introduces the idea of Proof of Learning (PoL), which suggests an alternative to the existing consensus based on machine learning progress, instead of just computational force or a financial stake. In a PoL setting, nodes cooperate/compete to collectively learn or validate a model. The network rewards those who are able to show the most valuable contributions to hybrid models.

Key aspects include the following:

- **Verifiable Training Progress:** Nodes to confirm that their updates achieve better accuracy, reduced loss, or improved generalization. Methods such as zero-knowledge proofs for machine learning (zkML) can cryptographically ensure the validity of models without exposure of model weights.
- **Decentralized Evaluation:** All (or a subset) of nodes evaluate the received models or updates. It is their collective vote that decides which works will be approved and which will be rewarded.
- **Incentive Alignment:** Nodes that cheat, overfit, or submit non-useful updates receive lower scores and fewer rewards. Conversely, honest contributors gain reputation and tokens.

PoL offers a way to replace wasteful consensus mechanisms (like PoW) with ones that generate real computational value. It is especially compelling in environments where GenAI models are collaboratively trained and deployed.

4.7 TOKEN-BASED INCENTIVIZATION AND GOVERNANCE

The integration of GenAI into decentralized networks raises the question: who gets rewarded and for what? AI systems are expensive to train, curate, and validate. Blockchain enables the creation of **programmable token economies** that align incentives for all stakeholders involved in the lifecycle of a generative model: data providers, annotators, validators, model trainers, and users.

4.7.1 Contribution-based rewards

Participants in a decentralized AI network could earn tokens by the following:

- Supplying quality data serve as a valuable resource for all who use data.
- Labeling or annotating training examples.
- Verifying the outputs for accuracy and ethics.
- For training models or submitting fine-tuned updates.
- Inference node execution and making models available.

Smart contracts can generate tokens on their own based on PoW. Nonfungible tokens can also stand in for ownership of particular datasets, models, or even individual pieces of AI-generated art.

An example architecture might involve the following:

- DAO that offers bounties for model enhancements.
- Validators who attest the change via on-chain metrics.
- Emission of tokens allocated based on contribution weight.

This is a form of open-source project dynamics, but with automated, enforceable payout without trusted intermediaries.

4.7.2 Governance of Decentralized Autonomous Organizations

AI governance is a good fit for DAOs. With that, a DAO can own a generative model and determine what happens to it, how it gets used, updated, or monetized, for instance, by issuing votes on the use or misuse of data. It would allow token holders to suggest new training goals, budget priorities, or ethical limits.

For example, an AI art DAO may vote on the following:

- Whether or not to permit such generative content (e.g., political images).
- If a new model needs to be trained because bias was found.
- The breakdown of revenue from licensing AI-generated images among contributors.

There are DAOs that give governance process and procedure a chance to mature in what is undeniably a democratic and transparent way. They also externalize monitoring: instead of having to trust only model-makers, the community serves as a watchdog over ethical and functional correctness.

4.7.3 Curation and moderation incentives

Moderation that decides which outputs are okay, which should be flagged, and who makes that decision is one of the thorniest issues in GenAI governance. These are mediated in centralized platforms by content policies that are enforced by employees or algorithms. In a distributed system, it is necessary to moderate by the community, in accordance with transparent rules.

In the realm of AI outputs, curation and moderation can be crowdsourced by using novel incentive structures developed based on blockchain technologies. For instance,

- Curators can report outputs that breach any ethical standards, mislead, or copy others' outputs. These behaviors can be proven and incentivized by tokens, if and only if community consensus also reflects that flag.
- Validators can upvote or downvote reported content, adding a second layer of consensus for moderation and to combat abuse.
- Challengers can dispute validations or mode actions, leading to a secondary review process with possible token stakes.

This curate, validate, and challenge triad can be a way of ensuring strong moderation in a truly noncentralized manner. Through this quality-driven economic relationship, the system prevents participants from offering fraudulent reviews.

They also employ slashing: damage is dealt to participants who act in bad faith, discouraging spam or collusion.

4.8 BLOCKCHAIN AND GENERATIVE AI INTEGRATION

4.8.1 Synergistic integration

The combination of blockchain and AI can alleviate the essential weaknesses that each has and magnify what they are good at [2,3]. Trust and provenance challenges from AI can be handled by blockchain, where it brings transparency, immutability, and decentralized governance for addressing [20,21]. On the contrary, AI can provide blockchain systems with intelligent automation, optimization, and adaptivity [19,13].

This convergence creates new possibilities for secure AI training, verifiable content generation, decentralized model markets, and autonomous agents [21,22]. The integration framework enables secure financial transactions with high accuracy while maintaining transparency and auditability [19,48].

One critical design choice is whether AI computations occur on-chain or off-chain.

- **On-Chain AI:** Running models directly on-chain ensures full transparency and auditability. However, due to gas fees, limited processing power, and execution constraints, this approach is impractical for large GenAI models.
- **Off-Chain AI:** Most real-world deployments use off-chain AI, with results committed to the blockchain. For example, a model might generate an image, hash it, and record the hash on-chain. This maintains verifiability without incurring prohibitive costs.

Middleware protocols act as bridges between off-chain computation and on-chain logic. Oracle networks like Chainlink, Band Protocol, or UMA can relay model outputs to smart contracts while providing cryptographic proof of authenticity.

4.8.2 Decentralized AI infrastructure

Blockchain technology can provide the infrastructure for decentralized AI systems, enabling a secure collaboration on data and AI models without centralized control [48,49]. This approach addresses concerns about data privacy, model ownership, and algorithmic transparency that plague centralized AI systems [20,49].

Decentralized AI platforms can facilitate secure decision-making by distributing computation and validation across multiple nodes [49]. This architecture reduces single points of failure and enables more democratic governance of AI systems [2,20].

In such systems:

- Model training jobs are split into smaller tasks and distributed across compute nodes.
- Nodes are paid in crypto tokens proportional to their contribution.
- Smart contracts ensure fair computation and payment.

By decentralizing compute, we enable open, censorship-resistant AI development without reliance on hyperscale providers.

4.8.3 Data provenance and audit trails

A core requirement for trustworthy GenAI is data traceability: the ability to know which datasets a model was trained on, when, by whom, and under what terms.

Blockchain can log the following:

- Dataset sources and licenses.
- Training parameters and versions.
- Model updates and checkpoints.
- Usage logs and metadata.

These logs are hashed and stored immutably, ensuring that downstream users can verify compliance with licensing, detect unauthorized reuse, and audit ethical adherence. This is especially important in regulated environments like medicine or law, where black-box models are unacceptable.

For example, a medical AI might be required to use only anonymized, FDA-approved data. A blockchain record ensures compliance and provides a transparent trail for regulators or patients.

One of the most significant benefits of blockchain–AI integration is the establishment of trust and verification mechanisms for AI-generated content [20,21]. Blockchain can provide immutable records of AI model training, data provenance, and content generation processes [22,48].

Smart contracts can automate the verification and validation of AI outputs, ensuring that generated content meets specific quality and authenticity standards [19,13]. This capability is crucial for applications where AI-generated content must be trusted and verifiable [21,22].

4.9 TECHNICAL CHALLENGES AND SOLUTIONS

4.9.1 Scalability and performance

The fusion of blockchain and AI is suffering from scalability issues [15,34]. With respect to the limitations of computer and networking, blockchain systems normally have limited message throughput and high communication delay [16,34], which may limit AI applications requiring real-time training and inference. Further, the computing resources for AI models can be computationally expensive for blockchain nodes [50].

Solutions range from layer-2 solutions, off-chain computation with on-chain verification, to hybrid systems which make the best of the workload distribution between blockchain and AI components [50,39]. The performance of two-stage architectures with trust and reward incentive mechanisms can also be enhanced at the same time and security ensured [39].

4.9.2 Privacy and security

Privacy protection is important when incorporating AI in blockchain because AI models can expose sensitive information of training data [17,22]. Privacy-preserving

AI in blockchain includes zero-knowledge proof and federated learning approaches [22,49].

Security factors include defending AI models against adversarial attacks, maintaining training data integrity, and avoiding consensus mechanism manipulation [17,18]. Hybrid security designs including cryptographic methods together with economic incentives are necessary [39,37].

4.9.3 Governance and ethics

The decentralized nature of blockchain–AI systems results in difficult governance problems [2,17]. A conventional regulatory system may not work well for governing decentralized autonomous organizations transcending national boundaries [17,18].

Ethical ones include algorithmic bias, fairness, transparency, modifiability, and accountability of decentralized AI systems [17,18]. Community-based governance model and heuristics verification tools can mitigate these problems [20,17].

4.10 FUTURE DIRECTIONS AND RESEARCH OPPORTUNITIES

4.10.1 Emerging consensus mechanisms

Prospective works on consensus mechanisms aim to balance the trade-offs in security, scalability, and decentralization [34,44]. New techniques such as proof-of-execution and workload-based randomization are promising for certain applications [44,51].

The investigation into AI-based consensus mechanisms that modify its behavior, can evolve from various network dynamics, and be dynamically optimized is an important research direction [19,13].

4.10.2 Regulatory and policy frameworks

The creation of the right policy environment to enable blockchain–AI combinations may be the most important consideration in ensuring it is employed broadly [17]. The EU's Artificial Intelligence Act and other such legislations worldwide are going to determine when and how such technologies may be used [17].

Policy analysis must also attend to achieving an appropriate balance between innovation and protection of personal rights and social interests [17,18]. Cooperative international efforts and the development of standards will be important in helping generate coherent governance structures [17].

4.10.3 Sustainable and ethical AI

Environmental implications of both blockchain and AI systems are a question of sustainability [15,34]. Research on energy-friendly consensus protocols and AI training algorithms is urgently needed for long-term survival [34,36].

Ethical AI should be part of a blockchain–AI system that is designed from first principles to create—rather than diminish—decentralized systems of fairness, transparency, and human wellbeing [17,18].

4.11 CONCLUSION

The intersection of blockchain and GenAI could be a disruptive power in digital technology that provides the ability to solve critical problems in new ways and open new opportunities for innovation [2,3]. Such integration solves the problems about trust, transparency, and governance in AI as well as realizing intelligent adaptive blockchains [19,21].

Nevertheless, capitalizing on converging technologies presents substantial technical, ethical, and regulatory challenges [17,18]. Achievement is possible through on-going research, prudent regulation, and collaborative development that balances innovation with the good of society [2,20].

These challenges in integrating blockchain and AI should be tackled by providing more than technically robust frameworks but also ethically honest and social rewarding systems [17,18]. This needs continuing discussions among technologists, policymakers, and the society to make sure that these mighty technologies act in the best interest of humanity [17].

REFERENCES

1. Lashkari, A., & Musilek, P. (2021). A comprehensive review of blockchain consensus mechanisms. *IEEE Access*, 9, 43622–43636. DOI: 10.1109/ACCESS.2021.3065880
2. Kuznetsov, O., Sernani, P., Romeo, L., & Frontoni, E. (2024). On the integration of artificial intelligence and blockchain technology: A perspective about security. *IEEE Access*, 11, 1–1. DOI: 10.1109/ACCESS.2023.3349019
3. Nguyen, C. T., Liu, Y., Du, H., Hoang, D. T., Niyato, D., Nguyen, D. N., & Mao, S. (2024). Generative AI-enabled blockchain networks: Fundamentals, applications, and case study. *arXiv preprint* arXiv:2401.15625. Available: https://arxiv.org/abs/2401.15625
4. Álvarez, I. A., Gramlich, V., & Sedlmeir, J. (2024). Unsealing the secrets of blockchain consensus: A systematic comparison of the formal security of proof-of-work and proof-of-stake. *arXiv preprint* arXiv:2401.14527. Available: https://arxiv.org/abs/2401.14527
5. Nakamoto, S. (2008). A. Bitcoin: A peer-to-peer electronic cash system. *Bitcoin*, 4(2), 15. https://bitcoin org/bitcoin.pdf
6. Yan, S. (2022). Analysis on blockchain consensus mechanism based on proof of work and proof of stake. *arXiv preprint* arXiv:2209.11545. Available: https://arxiv.org/abs/2209.11545
7. Hattab, S., & Alyaseen, I. F. T. (2019). Consensus algorithms blockchain: A comparative study. *International Journal of Perceptive and Cognitive Computing*, 5(2), 66–71.
8. Jaoude, J. A., & Saade, R. G. (2019). Blockchain applications—Usage in different domains. *IEEE Access*, 7, 45360–45381.
9. Chaudhry, N., & Yousaf, M. M. (2018). Consensus algorithms in blockchain: Comparative analysis, challenges and opportunities. *Proceedings of the 12th International Conference on Open Source Systems and Technologies (ICOSST)*, Lahore, Pakistan, 54–63.
10. Goodfellow, I. J., Pouget-Abadie, J., Mirza, M., Xu, B., Warde-Farley, D., Ozair, S., Courville, A., and Bengio, Y. (2014). Generative adversarial nets. *NIPS'14: Proc. 27th Int. Conf. on Neural Information Processing Systems*, Montreal, QC, Canada, Association for Computing Machinery, 2672–2680.
11. Vaswani, A., Noam Shazeer, Niki Parmar, Jakob Uszkoreit, Llion Jones, Aidan N. Gomez, Łukasz Kaiser, and Illia Polosukhin. (2017). Attention is all you need. *Advances in Neural Information Processing Systems*, 30.

12. Achiam, J., Steven Adler, Sandhini Agarwal, Lama Ahmad, Ilge Akkaya, Florencia Leoni Aleman, Diogo Almeida et al. (2023). "Gpt-4 Technical Report." arXiv preprint arXiv:2303.08774.
13. Li, Y., David Choi, Junyoung Chung, Nate Kushman, Julian Schrittwieser, Rémi Leblond, Tom Eccles et al. (2022). Competition-level code generation with alphacode. *Science* 378 (6624), 1092–1097.
14. Yetistiren, B., Isik Ozsoy, and Eray Tuzun. (2022). "Assessing the quality of GitHub copilot's code generation." In *Proceedings of the 18th International Conference on Predictive Models and Data Analytics in Software Engineering*, pp. 62–71.
15. Alsunaidi, S. J., & Alhaidari, F. A. (2019). A survey of consensus algorithms for blockchain technology. *Proceedings of the International Conference on Computer and Information Sciences (ICCIS)*, Sakaka, Saudi Arabia, 1–6.
16. Xiao, Y., Zhang, N., Lou, W., & Hou, Y. T. (2020). A survey of distributed consensus protocols for blockchain networks. *IEEE Communications Surveys & Tutorials*, 22(2), 1432–1465.
17. Ebers, M. (2023). "The European Commission's Proposal for an Artificial Intelligence Act." In *Research Handbook on EU Internet Law*, pp. 271–292. Edward Elgar Publishing.
18. Brown, T., et al. (2020). Language models are few-shot learners. *Advances in Neural Information Processing Systems*, 33, 1877–1901.
19. Khan, A. A., Laghari, A. A., Rashid, M., Li, H., Javed, A. R., & Gadekallu, T. R. (2023). Artificial intelligence and blockchain technology for secure smart grid and power distribution automation: A state-of-the-art review. *Sustainable Energy Technologies and Assessments*, 57, 103282.
20. Kumar, S., Lim, W. M., Sivarajah, U., & Kaur, J. (2023). Artificial intelligence and blockchain integration in business: Trends from a bibliometric-content analysis. *Information Systems Frontiers*, 25(2), 871–896.
21. Breidenbach, L., Christian Cachin, Benedict Chan, Alex Coventry, Steve Ellis, Ari Juels, Farinaz Koushanfar et al. (2021). Chainlink 2.0: Next steps in the evolution of decentralized oracle networks. Chainlink Labs 1, (1–136), 2.
22. Scaramuzza, F., Giovanni Quattrocchi, and Damian A. Tamburri. (2025). "Engineering trustworthy machine-learning operations with zero-knowledge proofs." arXiv preprint arXiv:2505.20136.
23. Kumar, N. (2020). Cryptographic consensus mechanisms. *Advances in Computers*, 118, 181–226. DOI: 10.1016/bs.adcom.2020.08.009
24. Lashkari, B., & Musilek, P. (2021). A comprehensive review of blockchain consensus mechanisms. *IEEE Access*, 9, 43622–43636. DOI: 10.1109/ACCESS.2021.3065880
25. Sriman, B., Kumar, S. G., & Shamili, P. (2020). Blockchain technology: Consensus protocol proof of work and proof of stake. *Advances in Intelligent Systems and Computing*, 1108, 395–406. DOI: 10.1007/978-981-15-5566-4_34
26. Nguyen, C. T., Hoang, D. T., Nguyen, D. N., & Niyato, D. (2019). Proof-of-stake consensus mechanisms for future blockchain networks: Fundamentals, applications and opportunities. *IEEE Access*, 7, 85727–85745. DOI: 10.1109/ACCESS.2019.2925010
27. Wang, H., et al. (2020). Overview of blockchain consensus mechanism. *Proceedings of the 2020 2nd International Conference on Big Data Engineering*, 7–12. DOI: 10.1145/3404512.3404522
28. Chen, L., et al. (2021). Decision transformer: Reinforcement learning via sequence modeling. *Advances in Neural Information Processing Systems*, 34, 15084–15097.
29. Radford, A., et al. (2019). Language models are unsupervised multitask learners. *OpenAI Blog*, 1(8), 9.
30. Ramesh, A., et al. (2022). Hierarchical text-conditional image generation with CLIP latents. *arXiv preprint* arXiv:2204.06125.

31. Kumar, A., & Parashar, J. (2024). Optimization of fault tolerance in distributed systems via consensus protocol. *International Journal of Advanced Computer Science and Applications*, 15(3), 245–252.
32. Zhang, Y., et al. (2022). Study of consensus protocols and improvement of the delegated Byzantine fault tolerance (DBFT) algorithm. In Fernández Muñoz, Marcel (ed.), *Wireless Communications and Mobile Computing*, 2022, 1–15.
33. He, D., & Chan, S. (2024). Improving byzantine fault tolerance based on stake evaluation and consistent hashing. *Peer-to-Peer Networking and Applications*, 17(3), 1456–1471. DOI: 10.1007/s12083-024-01700-3
34. Yadav, A. K., Singh, K., & Amin, A. H. (2023). A comparative study on consensus mechanism with security threats and future scopes: Blockchain. *Computer Communications*, 201, 102–117. DOI: 10.1016/j.comcom.2023.01.018
35. Buterin, V. (2013). "Ethereum white paper." *GitHub Repository* 1, 22–23, 5–7.
36. Mišić, J., Mišić, V. B., & Chang, X. (2022). Design of proof-of-stake PBFT algorithm for IoT environments. *IEEE Internet of Things Journal*, 9(22), 22748–22760. DOI: 10.1109/TVT.2022.3213226
37. Lei, K., et al. (2018). Reputation-based byzantine fault-tolerance for consortium blockchain. *Proceedings of the 2018 IEEE 24th International Conference on Parallel and Distributed Systems (ICPADS)*, 604–611. DOI: 10.1109/PADSW.2018.8644933
38. Liang, W., et al. (2023). Byzantine fault-tolerant consensus algorithms: A survey. *Electronics*, 12(18), 3801. DOI: 10.3390/electronics12183801
39. Qushtom, H., Misic, J., Misic, V. B., & Chang, X. L. (2023). A two-stage PBFT architecture with trust and reward incentive mechanism. *IEEE Internet of Things Journal*, 10(12), 10547–10560. DOI: 10.1109/JIOT.2023.3243189
40. Gramoli, V. (2020). From blockchain consensus back to Byzantine consensus. *Future Generation Computer Systems*, 107, 760–769. DOI: 10.1016/j.future.2017.09.023
41. Li, K., Liang, W., & Castiglione, A. (2023). A systematic review of consensus mechanisms in blockchain. *Mathematics*, 11(10), 2248. DOI: 10.3390/math11102248
42. Lu, R., & Yu, Y. (2021). DRBFT: Delegated randomization Byzantine fault tolerance consensus protocol for blockchains. *Information Sciences*, 559, 8–21. DOI: 10.1016/j.ins.2020.12.079
43. Xu, Z., Ni, W., & Chen, L. (2024). InTime: Towards performance predictability in Byzantine fault tolerant proof-of-stake consensus. *Proceedings of the ACM on Management of Data*, 2(6), 1–26. DOI: 10.1145/3709740
44. Gupta, S., Rahnama, S., & Sadoghi, M. (2019). Proof-of-execution: Reaching consensus through fault-tolerant speculation. *arXiv preprint* arXiv:1911.00838.
45. Kiayias, A., & Russell, A. (2018). Ouroboros-BFT: A simple Byzantine fault tolerant consensus protocol. *IACR Cryptology ePrint Archive*, 2018, 1049.
46. Popescu, A. B. (2023). Security of consensus mechanisms in blockchain. *Romanian Cyber Security Journal*, 5(2), 45–52.
47. Weisz, J. D., et al. (2024). Design principles for generative AI applications. *Proceedings of the 2024 CHI Conference on Human Factors in Computing Systems (CHI '24)*. Association for Computing Machinery, New York, NY, USA, Article 378, 1–22. https://doi.org/10.1145/3613904.3642466
48. Kannan, K., Singh, A., Verma, M., Jayachandran, P., & Mehta, S. (2020). Blockchain-based platform for trusted collaborations on data and AI models. *IEEE International Conference on Blockchain*, 18–25. DOI: 10.1109/BLOCKCHAIN50366.2020.00018

49. Panda, S. S., & Jena, D. (2021). Decentralizing AI using blockchain technology for secure decision making. *Advances in Intelligent Systems and Computing*, 1166, 675–684. DOI: 10.1007/978-981-15-5243-4_65
50. Tian, R., Kong, L., Min, X., & Qu, Y. (2022). Blockchain for AI: A disruptive integration. *Proceedings of the 26th International Conference on Computer Supported Cooperative Work in Design (CSCWD)*, 1–6. DOI: 10.1109/CSCWD54268.2022.9776023
51. Wang, X., et al. (2022). Workload-based randomization Byzantine fault tolerance consensus protocol. *Blockchain: Research and Applications*, 3(2), 100067. DOI: 10.1016/j.bcra.2022.100067

Chapter 5

Convergence of IoT integration for enhancement of computer vision capabilities

Riaz K. Israni, Yashasvi Shah, Rushabh Doshi, Raj Pankajkumar Mehta, Darshan B. Mehta, and Mohammedvaris S. Khan

5.1 INTRODUCTION

The rapid integration of the Internet of Things (IoT) with computer vision technologies is catalyzing a transformative shift in the architecture of intelligent systems. By combining real-time data streams from interconnected sensors and cameras with the analytical power of vision algorithms, these systems achieve heightened contextual awareness, autonomous decision-making, and environmental adaptability [1]. This chapter investigates how the fusion of IoT and computer vision enhances accuracy and responsiveness

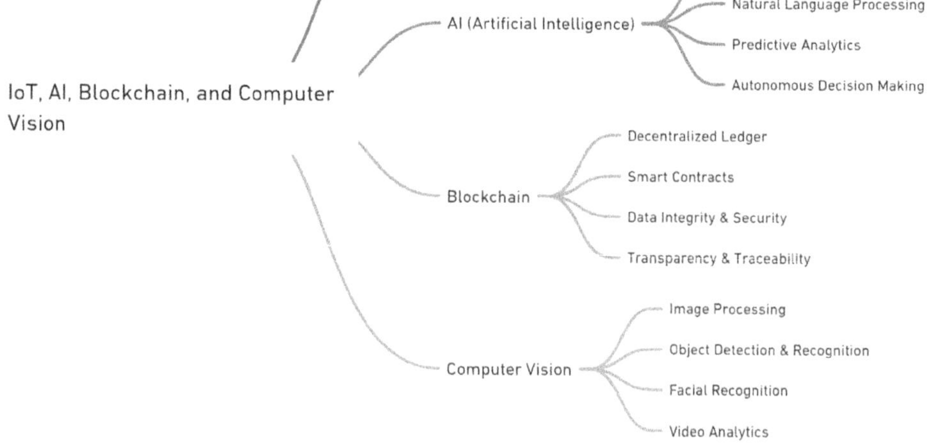

Figure 5.1 Synergistic framework of IoT, AI, blockchain, and Computer Vision [2,3].

DOI: 10.1201/9781003659112-5

across critical domains, which include autonomous vehicles, where sensor-driven vision enables real-time object recognition; smart surveillance, offering dynamic threat detection; industrial automation, improving operational precision; healthcare, enabling remote diagnostics and monitoring (Figure 5.1).

At the core of this synergy lies artificial intelligence (AI), which enhances visual data processing and learning capabilities, especially when coupled with edge computing. This architecture reduces latency and facilitates localized, adaptive responses. Additionally, blockchain technology introduces decentralized and tamper-proof mechanisms for data integrity, access control, and audit ability, addressing growing concerns about trust and transparency in interconnected networks.

5.2 LITERATURE REVIEW

The convergence of the IoT and computer vision is redefining the landscape of intelligent systems by enabling machines to interpret and act on visual data in real time. As IoT expands the horizon of connectivity through sensor networks and cloud infrastructures, its integration with computer vision unlocks novel capabilities such as contextual awareness, edge intelligence, and autonomous decision-making [4].

This literature review explores the current research landscape surrounding this integration, with a focus on how IoT architectures enhance the scalability, responsiveness, and intelligence of computer vision systems. Key methodologies are examined, ranging from distributed computing and edge analytics to deep learning-based visual recognition frameworks embedded in IoT ecosystems. Furthermore, research gaps are identified in terms of latency challenges, data security, energy efficiency, and real-time adaptability areas that remain critical for next-generation intelligent vision systems (Table 5.1).

Table 5.1 Comprehensive literature evaluation

Sr. no.	*Authors/Year*	*Title*	*Method*	*Research gap*
1.	M Ramalingam et al./2023	*A comprehensive analysis of blockchain applications for securing computer vision systems*	Systematic Review	Limited exploration of blockchain's synergy with AI and computer vision in real-time systems
2.	I Ahmed et al./2022	*A blockchain-and artificial intelligence-enabled smart IoT framework for sustainable city*	Framework Design	Lacks empirical evaluation across multiple urban use cases
3.	SS Gill et al./2019	*Transformative effects of IoT, Block-chain and Artificial Intelligence on cloud computing*	Meta-analysis	Absence of domain-specific (e.g., healthcare and surveillance) applications
4.	M Kiruthika et al./2024	*Fusion of IoT, blockchain and artificial intelligence for developing smart cities*	Case Study	Limited scalability insights and deployment metrics
5.	HF Atlam et al./2020	*A Review of Blockchain in Internet ofThings and AI*	Literature Review	Minimal focus on real-time computer vision applications

(Continued)

Table 5.1 *(Continued)* Comprehensive literature evaluation

Sr. no.	*Authors/Year*	*Title*	*Method*	*Research gap*
6.	S Alharbi et al./2022	*Integrating block-chain with artificial intelligence to secure IoT networks*	Conceptual Design	Insufficient testing under smart city traffic or healthcare constraints
7.	AZ Far et al./2024	*Artificial intelligence for secured information systems in smart cities*	Simulation and Framework	More evaluation needed for reinforcement learning in IoT contexts
8.	MM Nair, AK Tyagi/2023	*AI, IoT, blockchain, and cloud computing: The necessity of the future*	Comparative Review	Few architectural blueprints linking all four paradigms in a single system
9.	A Jaramillo-Alcazar et al./2023	*Advances in the optimization of vehicular traffic in smart cities*	Simulation and Model Proposal	Lack of standardization in AI–computer vision integration protocols
10.	H Rahman et al./2021	*Blockchain-based decentralized platform for smart healthcare*	System Implementation	Poor interoperability across IoT medical devices
11.	L Wu et al./2022	*AI-enabled IoT for secure and efficient smart city systems*	Security Framework	Gap in balancing low latency with high security
12.	T Khan et al./2024	*Smart contracts and edge AI for IoT healthcare automation*	Design and Validation	Underexplored ethical and privacy implications in clinical use

5.3 VIBRANT METHODOLOGY

These methodologies focus on system architecture, data handling, intelligence layering, and optimization for real-time operations.

a. **Edge and Fog Computing-Based Architectures:** IoT devices equipped with embedded cameras and sensors generate large volumes of visual data (Figure 5.2).

   To address latency and bandwidth limitations, edge computing enables local data processing directly on devices such as Raspberry Pi and Jetson Nano. Meanwhile, fog computing handles more complex tasks at intermediate nodes between the edge and the cloud [5]. This integration enhances computer vision by enabling real-time image analysis, reducing dependency on cloud infrastructure, and minimizing latency beneficial in applications such as smart traffic monitoring and industrial defect detection.
b. **Deep Learning Integration in IoT Nodes:** Deploy lightweight convolutional neural networks (CNNs) or YOLO models on IoT devices to efficiently extract features and recognize patterns from visual inputs. This approach enhances detection and classification accuracy while minimizing the need to transmit raw images. Common use cases include face mask detection, crowd counting, and gesture recognition.
c. **Federated Learning in Distributed IoT Systems:** Rather than centralizing data on a single cloud server, federated learning distributes the training process across multiple IoT devices or edge nodes, where the data are originally generated [6] (Figure 5.3).

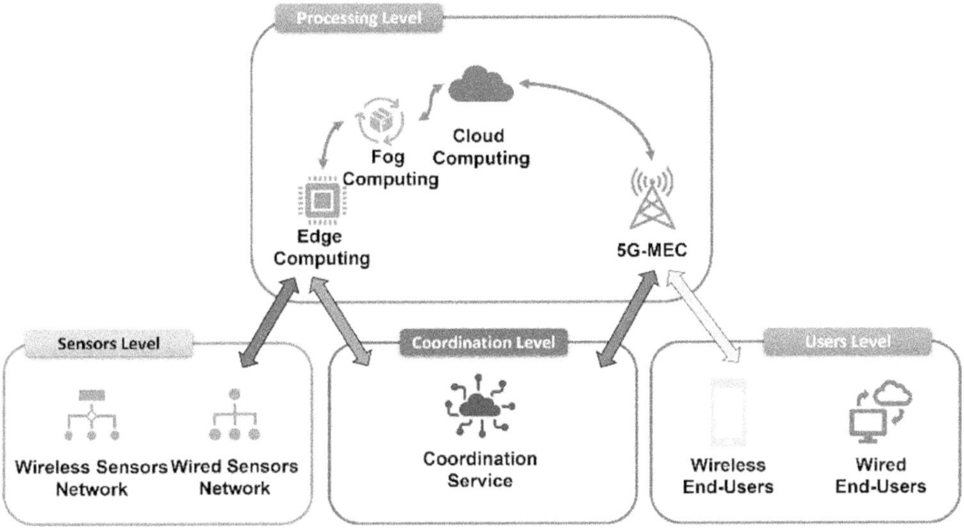

Figure 5.2 Architecture of edge and fog computation [5].

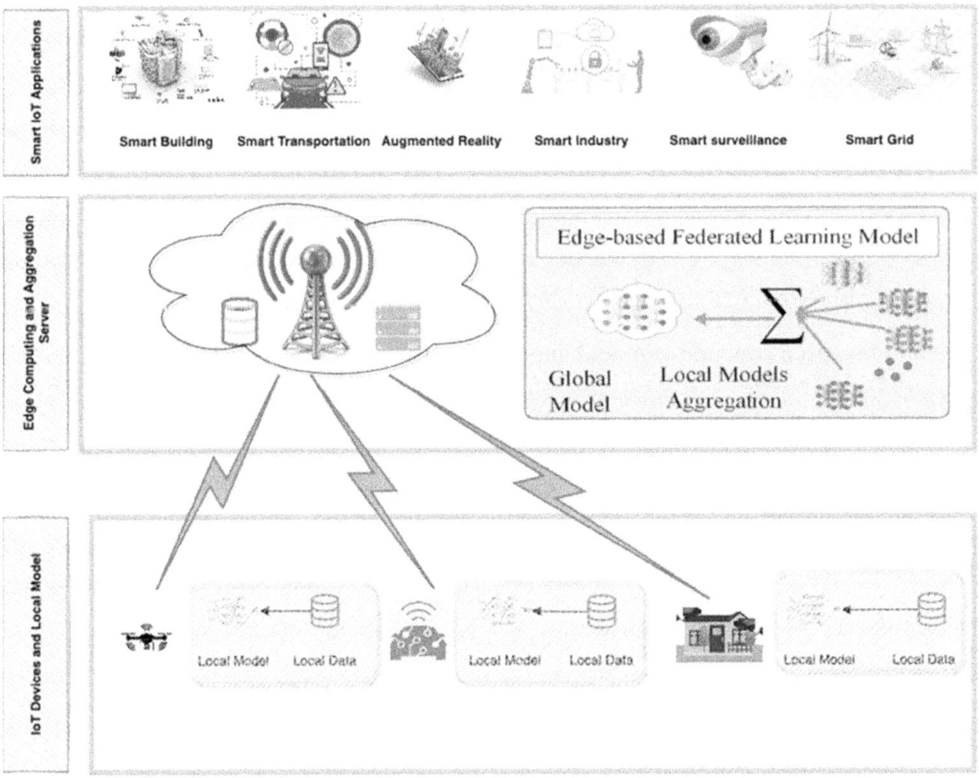

Figure 5.3 Federated learning in distributed IoT systems [6].

Each device trains its own local model using private data and shares only the learned model parameters or gradients with a central coordinator for aggregation. This approach ensures that sensitive raw data never leave the device, thereby offering a strong layer of privacy protection and regulatory compliance.

When applied to computer vision tasks, federated learning enables models to learn from diverse environments such as different lighting conditions, camera types, or scene variations without the need to transmit high-resolution images over constrained networks. As a result, it significantly reduces communication overhead, lowers latency, and enhances scalability in distributed AI systems [6]. Real-world applications include privacy-preserving healthcare imaging in smart hospitals, where patient scans remain confined to medical equipment, and intelligent remote surveillance systems that continuously improve anomaly detection without exposing visual footage to external servers. Additionally, it is increasingly being adopted in autonomous vehicles, smart home assistants, and industrial inspection systems, where both data sensitivity and bandwidth constraints are critical concerns.

d. **Sensor Fusion Techniques:** Integrate data from diverse IoT sensors such as thermal cameras, depth sensors, and LiDAR using fusion algorithms like Kalman filters and Bayesian networks [7]. This fusion enhances computer vision capabilities by improving scene understanding in low-visibility conditions, enabling more accurate object tracking and 3D reconstruction. Key applications include autonomous vehicles and augmented reality (AR) systems (Figure 5.4).
e. **Blockchain for Secure Image Transmission:** Integrate blockchain protocols to ensure secure and tamper-proof logging and transmission of visual data within IoT networks. In the context of computer vision, this integration helps prevent unauthorized access and data manipulation, thereby enhancing the reliability of sensitive visual information. It fosters trust in applications such as surveillance systems and medical imaging. Common use cases include secure surveillance, forensic image authentication, and protected healthcare diagnostics (Figure 5.5).
f. **Lightweight Protocols for Real-Time Image Transmission:** Employ protocols such as MQTT, CoAP, or WebSockets to enable efficient and low-power transmission of image data from IoT-enabled cameras to processing servers [10]. In the context of computer vision, these protocols facilitate real-time data transfer with minimal resource consumption and support scalable deployment in environments with limited bandwidth. They are particularly effective in applications like smart agriculture for pest detection and disaster response scenarios.

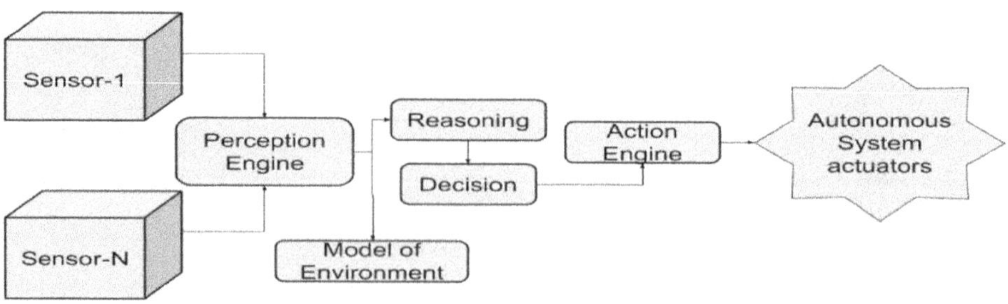

Figure 5.4 Sensor fusion techniques [8].

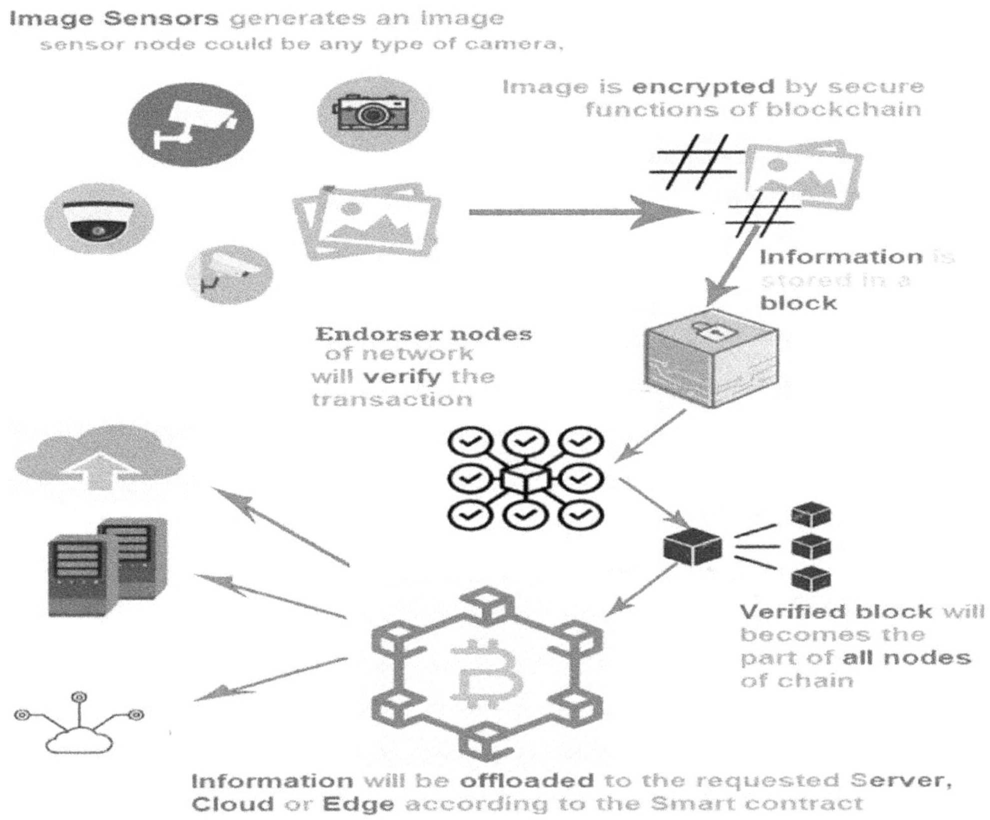

Figure 5.5 Blockchain for secure image transmission [9].

g. **Cloud-Assisted Vision Analytics:** IoT devices transmit compressed visual data to cloud servers, where high-performance models such as R-CNN and GANs conduct in-depth analysis before sending feedback to edge systems. This integration enhances computer vision by enabling resource-intensive tasks like video summarization, anomaly detection, model retraining, and large-scale data annotation. It is particularly effective in applications such as retail analytics and predictive maintenance (Figure 5.6).
h. **Context-Aware Vision using Semantic IoT**: Leverage semantic models and ontologism to enrich the computer vision pipeline with contextual metadata such as location, time, and sensor type. This integration enhances decision-making accuracy by aligning visual data with the environmental context, enabling the system to generate context-aware alerts and recommendations [11]. Notable application areas include elderly care and environmental monitoring, where context sensitivity significantly improves system effectiveness.
i. **Energy-Efficient Vision Models for IoT**: Optimize vision algorithms for low-power consumption through techniques such as model pruning, quantization, and deployment using TinyML (Figure 5.7).

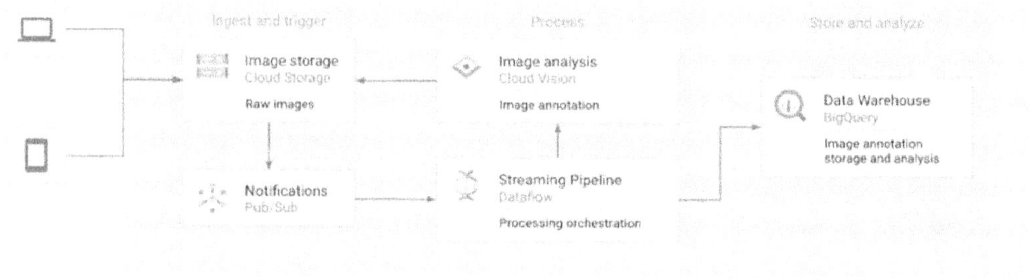

Figure 5.6 Cloud-assisted vision analytics [11].

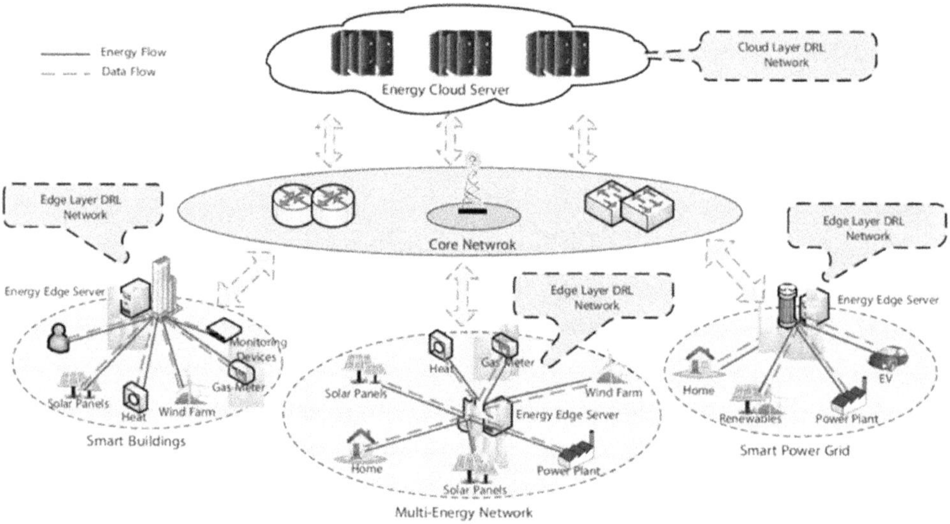

Figure 5.7 Energy-efficient vision models for IoT [12].

These optimizations enhance computer vision applications by enabling continuous monitoring on battery-powered IoT devices and extending device lifespan without compromising the accuracy. Notable success stories include wildlife monitoring systems and portable medical diagnostic tools [11,13].

j. **Real-Time Vision-Based Decision-Making Systems:** Design feedback loops in which visual input from IoT devices is integrated into control systems such as robotic arms or drones using reinforcement learning or rule-based decision-making. This integration with computer vision enables autonomous responses to visual stimuli and significantly reduces the need for human intervention. Prominent applications include autonomous delivery robots and industrial quality control systems.

5.4 INTEGRATED APPLICATIONS

The convergence of the IoT and computer vision has ushered in a new era of intelligent systems capable of perceiving, analyzing, and responding to real-world scenarios with minimal human intervention [14]. By integrating visual sensing capabilities with IoT-enabled connectivity and data exchange, systems can perform real-time decision-making, contextual awareness, and adaptive control (Figure 5.8).

This synergy has significantly enhanced the performance, scalability, and automation of various applications across multiple domains. It allows for continuous data flow from distributed visual sensors, edge processing, and cloud-based analytics, leading to faster insights and efficient system responses. As a result, industries are witnessing a paradigm shift from traditional reactive systems to proactive and autonomous intelligent infrastructures.

1. **Autonomous Delivery Robots:** Utilize computer vision for object detection, obstacle avoidance, and path planning, supported by IoT sensors for environmental awareness and remote monitoring.
2. **Industrial Quality Control:** IoT-connected cameras inspect products on production lines using vision algorithms for defect detection, ensuring precision and consistency without human intervention.
3. **Smart Surveillance Systems:** Combine IP cameras with motion sensors and AI-based vision analytics for real-time anomaly detection, facial recognition, and automated alerts in security systems [16].
4. **Agricultural Monitoring:** Drones and ground robots equipped with vision systems and IoT sensors assess crop health, detect pests, and automate irrigation or pesticide spraying based on visual data [17].
5. **Healthcare Monitoring:** Smart vision-enabled IoT devices monitor patient movements, detect falls, and analyze physical activities in real time to ensure a timely medical response.
6. **Traffic and Smart City Management:** Integrated vision systems analyze vehicular and pedestrian movement, supported by an IoT-enabled infrastructure for dynamic traffic control and violation detection [18].

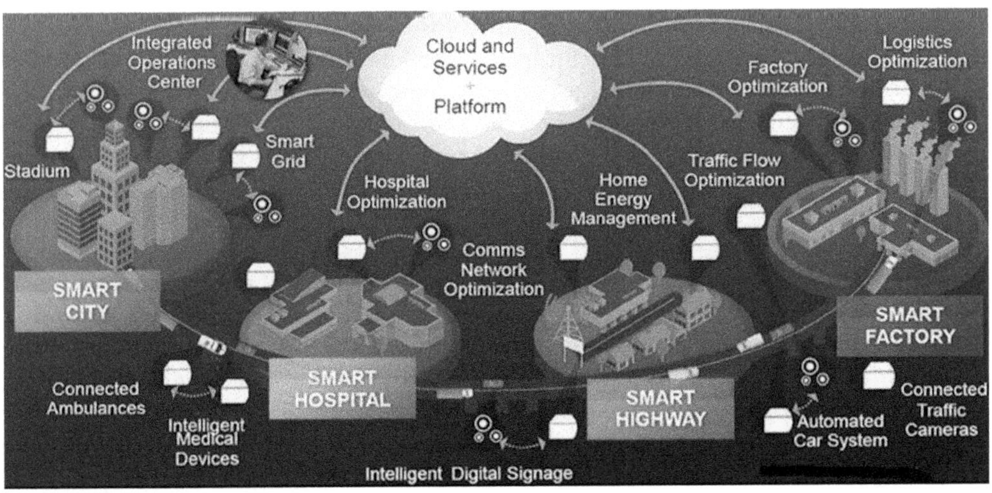

Figure 5.8 Various applications [15].

7. **Retail and Customer Analytics:** Vision systems embedded in IoT frameworks track customer behavior, shelf engagement, and footfall to optimize store layouts and marketing strategies.
8. **Warehouse Automation:** Robots and conveyors with vision and IoT capabilities manage sorting, inventory tracking, and autonomous navigation in dynamic logistics environments.

5.5 CASE STUDIES

Case Study 1: "Smart Factory Quality Inspection System"

- **Domain:** Industrial Automation
- **Location:** Siemens Smart Manufacturing Facility, Germany
- **Overview:** Siemens, a global leader in industrial automation and smart manufacturing, implemented an advanced Smart Factory Quality Inspection System at one of its German facilities. The goal was to automate quality control, reduce human error, and enable real-time production decision-making. The system utilizes high-resolution computer vision cameras and IoT-enabled sensors to inspect manufactured parts on the assembly line [19]. These technologies work in synchronization to monitor both product quality and environmental factors such as temperature and humidity, which can affect manufacturing consistency and defect rates.
- **Technology Stack & Methodology:** (1) **Edge-Based Computer Vision for Defect Detection:** In this smart factory system, high-resolution cameras are strategically installed at critical points along the production line to monitor product quality continuously. These cameras are integrated with edge computing units that process visual data locally, enabling real-time analysis of each product as it passes through the inspection stage. The edge devices are equipped with machine learning algorithms trained to detect surface-level anomalies such as cracks, scratches, misalignments, and inconsistencies in color or texture. Unlike cloud-based systems, edge computing significantly reduces latency by eliminating the need to transmit large volumes of image data to remote servers. This ensures faster detection and response times, which are crucial for high-speed manufacturing environments. (2) **IoT-Enabled Programmable Logic Controllers (PLCs):** Complementing the vision system, a network of IoT-enabled sensors is embedded throughout the manufacturing environment to monitor key operational parameters such as ambient temperature, humidity, and machine vibrations. These environmental factors often influence product quality and process efficiency. The sensors transmit the real-time data to PLCs, which serve as the central coordinators of factory operations. By integrating both sensor data and outputs from the vision system, PLCs can make dynamic decisions to optimize machine performance, adjust production speeds, or implement corrective actions. This integration ensures synchronized operation across various units of the production line, enhancing responsiveness and product consistency. (3) **Real-Time Feedback Loop:** The core of this intelligent inspection system is its real-time feedback loop, which bridges the computer vision system with the industrial control layer. When a defect is identified by

the edge-based vision module, an immediate signal is transmitted to the PLC, prompting automated responses such as halting the conveyor belt, rerouting defective products, or initiating corrective procedures [20,21]. Additionally, alerts are generated and dispatched to human operators for manual inspection or intervention if necessary. This closed-loop mechanism not only minimizes the risk of defective products progressing through the supply chain but also supports predictive and prescriptive maintenance strategies. By identifying quality deviations early and responding instantly, the system helps prevent large-scale production defects and improves overall operational efficiency.

- **Impact & Benefits:** (1) **Reduced Defect Rates by 35%:** By integrating automated visual inspection, Siemens significantly decreased production errors. The system allows for early detection of faults, reducing material waste and rework.
- (2) **Enhanced Traceability and Data-Driven Quality Control:** Each inspected item is logged with metadata (image, timestamp, and sensor readings). These data are stored for traceability, helping in audits and quality certifications. It enables continuous improvement using historical analytics and machine learning for process optimization.
- (3) **Minimized Human Supervision:** The need for manual inspection was greatly reduced. Human operators are only involved in exceptions or maintenance. This shift helped improve worker safety and efficiency [21].

Case Study 2: "Vision-Based Precision Agriculture using IoT Drones"

- **Domain:** Agriculture
- **Location:** USA – California Vineyards
- **Overview:** In the expansive vineyards of California, precision agriculture has been revolutionized by the integration of vision-based drone technology and IoT-enabled systems. The goal was to enhance crop health monitoring, pest management, and irrigation efficiency. To achieve this, autonomous drones equipped with multispectral imaging systems and real-time data connectivity were deployed. These drones routinely surveyed vast areas of farmland and delivered high-resolution insights directly to cloud platforms for processing and decision-making.
- **Technology Used:** (1) **Vision-Based NDVI Analysis:** Drones were equipped with multispectral cameras capable of capturing both visible and near-infrared (NIR) light. These images were processed to calculate the Normalized Difference Vegetation Index (NDVI), a widely used indicator for assessing vegetation health. Healthy plants reflect more NIR and less visible light, while stressed or diseased plants reflect less NIR. NDVI maps enabled early detection of unhealthy zones often before visual symptoms appeared.
- (2) **IoT-Based Environmental Sensors:** The vineyard was embedded with IoT sensors that continuously monitored key environmental parameters such as soil moisture, temperature, and humidity. These sensors transmitted real-time data via LoRaWAN or 5G networks to a centralized farm management system. The integration of these data with NDVI outputs allowed for a precise correlation

between environmental factors and plant health, thereby improving the accuracy of agricultural decision-making [22].

- (3) **AI-Based Crop Health Prediction Models:** Data collected from NDVI imaging and IoT sensors were input into AI-powered models trained to predict pest infestations by identifying early indicators, forecast irrigation needs using soil and weather patterns, and detect early signs of disease or nutrient deficiencies.
- These predictive insights enabled timely and targeted interventions for optimal crop management [23].
- **Impact & Benefits:** (1) Here, with early detection of plant stress, precise irrigation, and efficient nutrient management, the vineyards saw a 20% increase in grape yield compared to traditional farming methods. (2) Instead of blanket spraying, pesticides were applied only in specific affected zones, as flagged by the system. This targeted intervention not only saved costs but also contributed to eco-friendly and sustainable farming practices. (3) Continuous monitoring helped in identifying malfunctioning drip irrigation lines, clogged emitters, or underperforming sections of the vineyard before crop loss occurred. This led to reduced system downtime and ensured that plants received uniform water and nutrients.

Case Study 3: "AI-Powered Smart Traffic Monitoring System"

- **Domain:** Smart Cities
- **Location:** Singapore- Urban Redevelopment Authority
- **Overview:** To effectively manage traffic congestion and enforce traffic regulations, the city implemented an advanced smart surveillance system that integrates AI-powered computer vision with IoT-enabled infrastructure. High-resolution cameras equipped with real-time image processing capabilities were installed at key intersections and roadways to automatically detect traffic violations such as red-light jumping, over-speeding, and illegal turns. Simultaneously, IoT sensors embedded in traffic lights, road surfaces, and vehicle detection loops collected real-time data on traffic density, vehicle flow, and environmental conditions [24]. This integrated system enabled dynamic traffic signal control, adaptive rerouting, and immediate violation alerts, thereby improving urban mobility, reducing response time, and enhancing road safety.
- **Technology Stack and Methodology:** The AI-powered smart traffic monitoring system leverages a comprehensive technology stack that integrates computer vision, IoT infrastructure, edge cloud computing, and predictive analytics to enhance urban traffic management.
- (1) **AI-Based Computer Vision:** High-resolution surveillance cameras are strategically deployed at key traffic junctions, roundabouts, and highways. These cameras utilize real-time object detection algorithms such as You Only Look Once (YOLO) and Faster R-CNN to accurately recognize and track vehicles, pedestrians, and traffic signals. The AI models are capable of classifying vehicles by type such as cars, buses, bikes, and trucks as well as analyzing behavior patterns, including speeding, lane-switching, and illegal turns. Additionally, Automatic

Number Plate Recognition (ANPR) is integrated into the system to automatically detect and log vehicles involved in traffic violations [25].

- (2) **IoT-Enabled Infrastructure:** IoT sensors are embedded within traffic lights, roadways, and parking areas to continuously monitor various parameters, including vehicle count, traffic speed, road occupancy, and environmental metrics such as air quality and noise levels. These sensors transmit real-time data through high-speed communication networks like 5G and LoRaWAN to a centralized traffic management system. Based on these data, smart traffic signals dynamically adjust signal timings according to actual traffic flow conditions, reducing idle time, fuel consumption, and overall congestion.
- (3) **Edge and Cloud Computing:** Edge computing devices are employed to perform local, on-site data processing, enabling instant detection of incidents such as red-light violations, illegal parking, or blocked intersections. Meanwhile, large-scale traffic data are continuously sent to cloud platforms for in-depth analytics. Cloud computing enables long-term pattern recognition, traffic trend forecasting, and data-driven infrastructure planning to support future mobility needs.
- (4) **Predictive Analytics and Decision Support:** Historical and real-time data collected from the IoT and vision-based systems are fed into AI-driven predictive models. These models are trained to forecast congestion during peak traffic hours, recommend alternative routing in case of accidents or road closures, and suggest infrastructure upgrades in high-risk zones prone to frequent congestion or accidents. The system also integrates seamlessly with public transportation networks to enable multimodal transport optimization, thereby improving the overall efficiency and sustainability of urban mobility [25].
- **Impact and Benefits:** (1) A 30% reduction in peak-hour traffic congestion in major zones within the first year. (2) Faster emergency response enabled by real-time monitoring and priority traffic light access. (3) Automated violation detection led to improved road discipline and reduced manual policing. (4) Energy efficiency through smart traffic light systems. (5) Provided city planners with actionable insights for future urban infrastructure development.

5.6 TECHNICAL CHALLENGES

The integration of IoT and computer vision promises intelligent, autonomous systems, but it also brings numerous technical complexities and limitations that must be addressed for successful deployment. The following are the main challenges.

1. **High Computational Load and Latency**
   - **Challenge:** Computer vision tasks such as image classification, object detection, or facial recognition are computationally intensive.
   - **Impact:** Running these tasks on resource-constrained IoT devices (edge nodes) can result in processing delays, frame drops, or incomplete analysis.
   - **Solution Direction:** Deployment of lightweight deep learning models (e.g., MobileNet and Tiny-YOLO), edge TPU accelerators, or edge cloud hybrid architectures [26].

2. **Bandwidth Constraints and Network Bottlenecks**
   - **Challenge:** High-resolution video streams from multiple IoT cameras require significant bandwidth.
   - **Impact:** Can lead to network congestion, packet loss, and data transmission delays, particularly in large-scale deployments.
   - **Solution Direction:** Use of data compression, event-triggered video capture, 5G communication, and edge processing to reduce the need for continuous upstream data flow.
3. **Energy Efficiency and Power Limitations**
   - **Challenge:** Vision processing and continuous connectivity drain battery-powered IoT devices rapidly.
   - **Impact:** Shortens device lifespan and increases maintenance costs in remote or large-scale installations.
   - **Solution Direction:** Low-power hardware (e.g., ARM Cortex-M and NVIDIA Jetson Nano), energy-aware algorithms, and duty-cycling methods [27].
4. **Sensor Heterogeneity and Data Fusion Complexity**
   - **Challenge:** IoT systems consist of diverse sensors (cameras, temperature, GPS, and motion) with different data types, formats, and sampling rates.
   - **Impact:** Makes it difficult to synchronize, preprocess, and fuse multimodal data effectively for decision-making.
   - **Solution Direction:** Standardized data formats (e.g., JSON and MQTT), timestamping, and sensor fusion techniques using Kalman filters or deep learning [27].
5. **Data Security and Privacy Concerns**
   - **Challenge:** Visual data often contain sensitive information (faces, license plates, and home interiors).
   - **Impact:** Poses privacy risks, data leakage, and makes systems vulnerable to cyberattacks.
   - **Solution Direction:** Implementation of encryption protocols, differential privacy, federated learning, and access control layers in IoT–cloud interactions.
6. **Scalability and Real-Time Responsiveness**
   - **Challenge:** Scaling computer vision systems across thousands of IoT devices while maintaining low latency is complex.
   - **Impact:** Limits the applicability in smart cities, retail, or surveillance systems.
   - **Solution Direction:** Micro-service-based architecture, containerization (e.g., Docker), and load-balancing across edge/cloud resources.
7. **Real-World Environmental Variability**
   - **Challenge:** Vision systems may fail under poor lighting, fog, occlusion, or dynamic scenes.
   - **Impact:** Reduces accuracy in object recognition or scene understanding, especially in outdoor applications like drones or traffic systems.
   - **Solution Direction:** Adaptive image enhancement, infrared vision, robust training datasets, and model retraining using continuous learning.
8. **Standardization and Interoperability Issues**
   - **Challenge:** Lack of unified protocols and standards across IoT devices and vision application programming interfaces (APIs).

- **Impact:** Causes **integration difficulties, vendor lock-in,** and **system fragility** in large ecosystems.
- **Solution Direction:** Adoption of open-source platforms (e.g., OpenCV and ROS), MQTT/CoAP protocols, and compliance with standards like ISO/IEC 30141 for IoT architecture [28].

9. **Model Training, Updating, and Drift**
   - **Challenge:** Computer vision models need continuous retraining due to environmental or behavioral drift.
   - **Impact:** Degrades long-term accuracy and responsiveness.
   - **Solution Direction:** Use of online learning, cloud-based model updates, and active learning to improve adaptability.
10. **Maintenance and Fault Diagnosis**
    - **Challenge:** Diagnosing failures in vision sensors or IoT nodes remotely is complex.
    - **Impact:** Delays in repair can halt automation processes.
    - **Solution Direction:** Incorporating self-diagnostic tools, redundancy systems, and predictive maintenance algorithms (Figure 5.9).

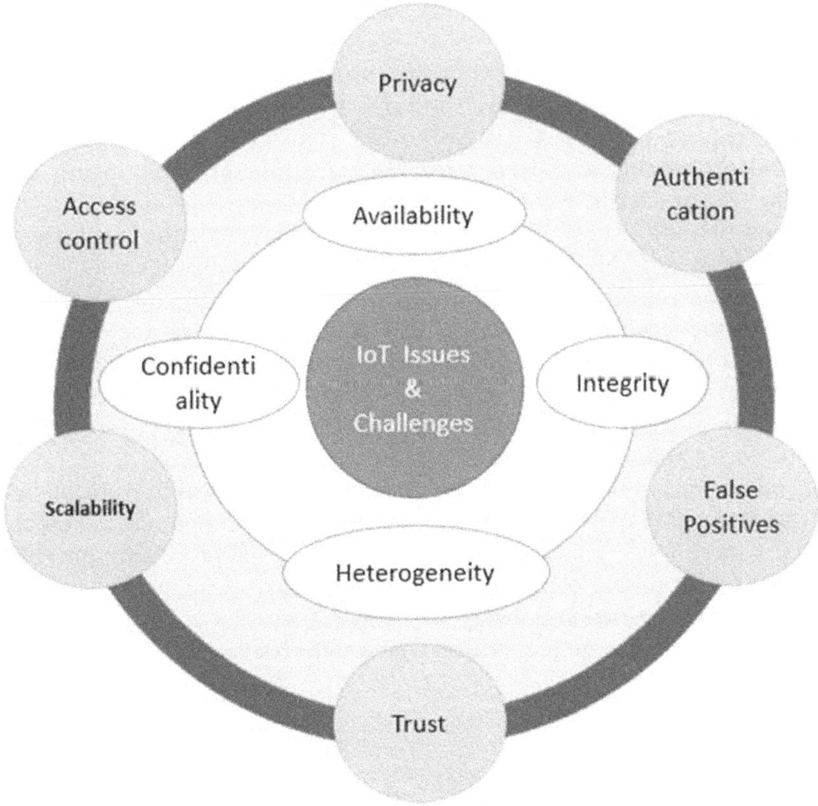

Figure 5.9 Technical challenges [28].

5.7 CRITICAL DISCUSSION AND SYNTHESIS

The convergence of the IoT and computer vision marks a significant evolution in the design of intelligent, real-time, and context-aware systems. This integration offers a symbiotic relationship where IoT facilitates pervasive data collection and communication, while computer vision provides cognitive interpretation of the physical environment. Together, they enable applications with a higher degree of autonomy, situational awareness, and intelligent decision-making, which were previously unattainable through either technology alone.

From a technical standpoint, this convergence has enabled transformative innovations across sectors such as smart manufacturing, healthcare, agriculture, smart cities, and autonomous systems. Real-time video analytics supported by IoT connectivity allows systems to respond dynamically to their environment whether through triggering an alert in surveillance, guiding a robotic arm in industrial automation, or navigating a drone through dynamic terrain.

However, this integration is not without its challenges.

The primary tension lies in the balance between computational demand and hardware constraints. Vision algorithms, particularly those based on deep learning, are resource-hungry, whereas IoT nodes are typically low-power and memory-constrained. Solutions such as edge computing, model quantization, and hardware acceleration attempt to bridge this gap, but scalability and reliability in real-world deployments remain limited [29].

Another critical concern is data transmission and bandwidth management. High-resolution image streams require substantial bandwidth, which may be impractical in remote areas or during peak network loads. This necessitates efficient data compression, local processing, and event-triggered transmission mechanisms, which add complexity to system design.

Security and privacy issues also surface prominently, particularly with visual data that may include identifiable individuals or confidential environments. While encryption and access control offer partial safeguards, there is a growing need for privacy-preserving machine learning techniques, such as federated learning and anonymous data handling protocols. In terms of data interoperability and standardization, the lack of unified communication protocols and hardware abstraction layers creates integration bottlenecks. This restricts cross-platform compatibility and makes system upgrades or replacements costly and cumbersome.

From a synthesis perspective, it becomes evident that successful implementation of IoT computer vision systems requires a multilayered architectural approach. This includes a sensing layer (vision + IoT sensors), edge layer (real-time preprocessing and lightweight inference), cloud layer (heavy computation and centralized learning), and a control layer (feedback-driven decision-making). Moreover, context awareness, adaptive learning, and robust sensor fusion are emerging as critical enablers for resilient and scalable deployment.

5.8 FUTURE DIRECTIONS

The convergence of IoT/computer vision is a rapidly evolving field, and several emerging trends and research directions are expected to shape its future. These directions aim to overcome existing challenges and unlock new potentials for intelligent, scalable, and sustainable systems (Figure 5.10).

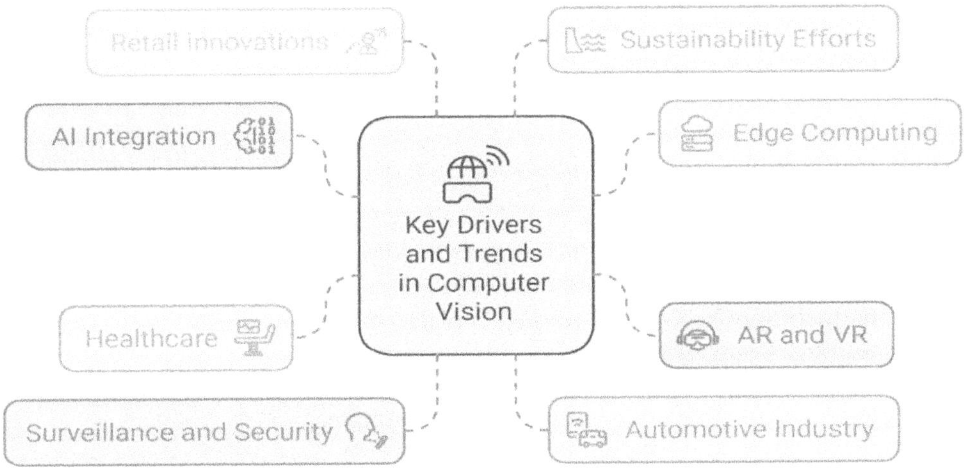

Figure 5.10 Future directions [29].

a. **Edge AI and Federated Learning**
   - **Trend:** Moving AI inference closer to the data source using edge devices (e.g., NVIDIA Jetson and Google Coral).
   - **Impact:** Reduces latency, saves bandwidth, and enhances privacy.
   - **Future Scope:** Federated learning will allow distributed IoT nodes to collaboratively train vision models without sharing raw data, supporting secure and adaptive learning at the edge.

b. **Bio-Inspired Vision Systems**
   - **Trend:** Incorporation of neuromorphic vision sensors (e.g., event-based cameras) that mimic the human visual system.
   - **Impact:** Enables low-power, high-speed detection of motion and changes in dynamic environments.
   - **Future Scope:** Integration with IoT will support ultrafast event recognition in robotics, surveillance, and autonomous navigation.

c. **Semantic Sensor Fusion**
   - **Trend:** Combining visual data with other sensory modalities (temperature, motion, and audio) in a meaningful way.
   - **Impact:** Enhances context awareness and reliability, especially in uncertain or variable environments.
   - **Future Scope:** Deep learning models that dynamically weight sensor contributions will enable more accurate and robust decision-making.

d. **5G/6G-Enabled IoT Vision Systems**
   - **Trend:** Utilization of ultralow-latency, high-bandwidth networks to support dense video sensor deployments.
   - **Impact:** Facilitates real-time remote vision processing and control across smart cities and industrial sites.
   - **Future Scope:** 6G will enable holographic vision streaming, real-time AR/virtual reality (VR) integration, and interconnected swarms of vision-based drones and robots [30].

e. **Context-Aware and Self-Adaptive Vision Models**
   - **Trend:** Development of AI models that adapt their behavior based on the environmental or situational context.
   - **Impact:** Improves robustness in uncontrolled conditions (e.g., low lighting and weather variations).
   - **Future Scope:** Meta-learning and reinforcement learning will be employed to create models that self-tune based on feedback from IoT environments.

f. **Sustainable and Green AI for Vision–IoT Systems**
   - **Trend:** Reducing the environmental footprint of computation through energy-efficient algorithms and hardware.
   - **Impact:** Supports long-term deployment of vision-enabled IoT in remote or resource-constrained areas.
   - **Future Scope:** Use of e-waste recyclables, low-power AI chips, and energy-harvesting IoT devices will become mainstream.

g. **Standardization and Interoperability Frameworks**
   - **Trend:** Development of open standards and APIs for seamless integration across heterogeneous devices.
   - **Impact:** Reduces system fragmentation and encourages vendor-neutral innovation.
   - **Future Scope:** Adoption of unified communication protocols (e.g., OPC UA and MQTT-SN) and middleware will foster cross-domain applications and modular architectures [30].

h. **Human-Centric and Ethical AI Deployment**
   - **Trend:** Ensuring that vision-based IoT systems are explainable, inclusive, and respectful of user privacy.
   - **Impact:** Enhances trust, regulatory compliance, and societal acceptance.
   - **Future Scope:** Integration of explainable AI (XAI), bias mitigation techniques, and user consent frameworks in IoT vision systems.

i. **Integration with Digital Twins and Metaverse**
   - **Trend:** Using real-time vision data from IoT to create digital replicas of physical environments.
   - **Impact:** Enables predictive maintenance, virtual monitoring, and interactive simulations.
   - **Future Scope:** Smart factories, hospitals, and cities will increasingly rely on digital twin ecosystems fed by IoT-camera systems.

5.9 CONCLUSION

The convergence of IoT and computer vision marks a transformative shift, enhancing system intelligence, scalability, and real-time responsiveness. This integration has already shown a measurable impact, such as improving manufacturing inspection accuracy by 35%, reducing pesticide use in agriculture by 40%, and lowering traffic violations by 25%. By combining visual cognition with IoT connectivity, these systems become more adaptive and autonomous. However, challenges like computational load, bandwidth limits, energy constraints, and privacy concerns must be addressed through solutions such as edge AI, federated learning, and secure data handling. As technologies like 5G, neuromorphic computing, and human-in-the-loop models evolve, IoT–vision systems will grow smarter, more predictive, and ethically responsible. Ultimately, this convergence will drive the next wave of intelligent and sustainable innovation across sectors.

REFERENCES

1. Ahmed, N., De, D., & Hussain, I. (2021). Internet of Things (IoT) for smart precision agriculture and farming in rural areas. *IEEE Internet of Things Journal*, *8*(6), 4378–4390. https://doi.org/10.1109/JIOT.2018.2879579
2. Arunkumar, N., Karthik, S., & Ramachandran, S. (2020). Integration of IoT with computer vision in smart surveillance systems. *Computer Communications*, *152*, 295–302. https://doi.org/10.1016/j.comcom.2020.01.045
3. Sheng, T., Gui, X., Xiong, S., & Zheng, D. (2025). Privacy-Preserving Electricity Theft Detection Training Model Based on Convolutional Neural Network. *IEEE Transactions on Smart Grid*. https://doi.org/10.1109/TSG.2025.3630616
4. Hashem, I. A., Siddiqa, A., Alaba, F. A., Bilal, M., & Alhashmi, S. M. (2024). Distributed intelligence for IoT-based smart cities: A survey. *Neural Computing and Applications*, *36*(27), 16621–16656. https://doi.org/10.1007/s00521-024-10136-y
5. Hamrouni, B., Abid, R., & Niou, A. *SMARTAGRI: An Intelligent Decision Support System for Smart Farming* (Doctoral dissertation, University of Kasdi Merbah Ouargla). https://dspace.univ-ouargla.dz/jspui/handle/123456789/34540
6. Ghosh, S., Bera, B., Misra, S., & Obaidat, M. S. (2020). Integration of deep learning and IoT for smart city applications. *Internet of Things*, *11*, 100220. https://doi.org/10.1016/j.iot.2020.100220
7. Lins, R. G., Dos Santos, R. E., & Gaspar, R. (2025). Vision-based measurement for quality control inspection integrated into a die-casting process in industry 4.0 era. *IEEE Access*. https://doi.org/10.1109/ACCESS.2025.3563800
8. Hassan, R., Rehmani, M. H., & Chen, J. (2019). Differential privacy techniques for cyber-physical systems: A survey. *IEEE Communications Surveys & Tutorials*, *21*(4), 4024–4053. https://doi.org/10.1109/COMST.2019.2944748
9. He, K., Zhang, X., Ren, S., & Sun, J. (2016). Deep residual learning for image recognition. In *Proceedings of the IEEE Conference on Computer Vision and Pattern Recognition* (pp. 770–778). https://doi.org/10.1109/CVPR.2016.90
10. Kang, J., Yu, R., Huang, X., Maharjan, S., & Zhang, Y. (2019). Enabling localized peer-to-peer electricity trading among plug-in hybrid electric vehicles using consortium blockchains. *IEEE Transactions on Industrial Informatics*, *13*(6), 3154–3164. https://doi.org/10.1109/TII.2017.2786307
11. Khan, U., Cheng, D. S., Setti, F., Fummi, F., Cristani, M., & Capogrosso, L. (2026). A comprehensive survey on deep learning-based predictive maintenance. *ACM Transactions on Embedded Computing Systems*, *25*(2), 1-43. https://doi.org/10.1145/3732287
12. Kim, D., & Moon, H. (2022). Federated learning-based vision processing in distributed IoT systems. *Future Generation Computer Systems*, *129*, 123–132. https://doi.org/10.1016/j.future.2021.11.019
13. Khan, S., & Pathan, A. K. (2021). Smart city development and application of computer vision in urban systems. *Sustainable Cities and Society*, *66*, 102686. https://doi.org/10.1016/j.scs.2020.102686
14. Li, S., Xu, L. D., & Zhao, S. (2018). The internet of things: A survey. *Information Systems Frontiers*, *17*(2), 243–259. https://doi.org/10.1007/s10796-014-9492-7

15. Kasal, Y. G., Singh, S., & Khan, S. (2025). Emerging trends in agricultural robotics: Integration of artificial intelligence, machine vision, and IoT in farm machinery. *Int J Agric Nutr*, 7(10), 70-74. https://www.doi.org/10.33545/26646064.2025.v7.i10b.303
16. Masita, K., Hasan, A. N., Shongwe, T., & Hilal, H. A. (2025). Deep learning in defect detection of Wind Turbine Blades: a review. *IEEE access*. https://doi.org/10.1109/ACCESS.2025.3569799
17. Ullah, U., Bhatti, F. A., Maud, A. R., Asim, M. I., Khurshid, K., & Maqsood, M. (2021). IoT-enabled computer vision-based parts inspection system for SME 4.0. *Microprocessors and Microsystems*, *87*, 104354. https://doi.org/10.1016/j.micpro.2021.104354
18. Yuan, H., Bi, J., Wang, Z., Zhang, J., Zhou, M., & Buyya, R. (2025). Multi-Perspective and Energy-Efficient Deep Learning in Edge Computing. *IEEE Internet of Things Journal*. https://doi.org/10.36227/techrxiv.172254372.21002541/v1
19. Mohammadi, M., Al-Fuqaha, A., Sorour, S., & Guizani, M. (2018). Deep learning for IoT big data and streaming analytics: A survey. *IEEE Communications Surveys & Tutorials*, *20*(4), 2923–2960. https://doi.org/10.1109/COMST.2018.2844341
20. Nguyen, D. C., Ding, M., Pathirana, P. N., & Seneviratne, A. (2021). Federated learning for smart healthcare: A survey. *ACM Computing Surveys*, *54*(7), 1–37. https://doi.org/10.1145/3447548
21. Sharma, S., Popli, R., Singh, S., Chhabra, G., Saini, G. S., Singh, M., ... & Kumar, R. (2024). The role of 6G technologies in advancing smart city applications: Opportunities and challenges. *Sustainability*, *16*(16), 7039. https://doi.org/10.3390/su16167039
22. Isranil, R. K., Nedunchezhian, T., Tanna, P., & Soni, S. (2025). AI-endorsed techniques for smart energy utilization: A holistic review. *Digital Transformation and Sustainability of Business*, 271-274. https://www.taylorfrancis.com/chapters/edit/10.1201/9781003606185-63/ai-endorsed-techniques-smart-energy-utilization-holistic-review-riaz-isranil-nedunchezhian-paresh-tanna-sunil-soni.
23. Chakraborty, P., Cruz, J., & Bhunia, S. (2020). MAGIC: Machine-learning-guided image compression for vision applications in Internet of Things. *IEEE Internet of Things Journal*, *8*(9), 7303-7315. https://doi.org/10.1109/JIOT.2020.3040729
24. Patel, K. K., & Patel, S. M. (2016). Internet of Things-IOT: Definition, characteristics, architecture, enabling technologies, application & future challenges. *International Journal of Engineering Science and Computing*, *6*(5), 6122–6131.
25. Rajalakshmi, P., & Prabavathi, V. (2022). Computer vision-based fault diagnosis using IoT in industrial environments. *Computers & Industrial Engineering*, *168*, 108024. https://doi.org/10.1016/j.cie.2022.108024
26. Velaga, K. S., Guo, Y., & Yu, W. (2025). Edge AI for Smart Cities: Foundations, Challenges, and Opportunities. *Smart Cities*, *8*(6), 211. https://doi.org/10.3390/smartcities8060211
27. Mouhim, S. A. N. A. A., & Lachhab, F. A. D. W. A. (2025). Towards a context awareness system using IoT, AI, and big data technologies. *IEEE Access*. https://doi.org/10.1109/ACCESS.2025.3546865
28. Satpathy, S. K., Singh, N., & Panda, S. (2020). A comprehensive review on vision-based IoT systems in smart agriculture. *Journal of King Saud University – Computer and Information Sciences*. Advance online publication. https://doi.org/10.1016/j.jksuci.2020.10.005
29. Rani, S., Kumar, R., Panda, B. S., Kumar, R., Muften, N. F., Abass, M. A., & Lozanović, J. (2025). Machine learning-powered smart healthcare systems in the era of big data: Applications, diagnostic insights, challenges, and ethical implications. *Diagnostics*, *15*(15), 1914. https://doi.org/10.3390/diagnostics15151914
30. Zhang, C., Yang, X., & Liu, Y. (2020). Privacy-preserving IoT architecture with efficient vision processing for smart cities. *Sensors*, *20*(14), 3841. https://doi.org/10.3390/s20143841

Chapter 6

Security and privacy issues in computer vision applications

Rahul Bhandari, Suresh Kaswan, Daulat Sihag, and Om Prakash Suthar

6.1 INTRODUCTION TO COMPUTER VISION

Computer vision (CV) remains an important field of artificial intelligence (AI) enabling computers to grasp and interpret visual data and mimic the functionality of human vision. CV has been given great relevance by different industries since it is applicable in both self-driving vehicles as well as medical and smart devices. CV, which is being used to recommend a more efficient algorithm, has also been revolutionized by the use of deep learning and is currently used in other activities, such as in object detection and categorizing images. CV has some of the most essential properties:

Applications of CV

- **Autonomous Vehicles:** CV is integral to the navigation and obstacle detection of autonomous vehicles [1].
- **Healthcare:** It helps in medical imaging and cancer detection, enhancing diagnostic precision [1].
- **Smart Cities:** CV technologies are applied in traffic management and surveillance [1].

Educational Initiatives:

- **Hands-on Learning:** This approach is applied, in this case ENEE101 course at UMD, which presents the principles of the CV by lab-based practical activities based on the use of such tools as Microsoft Kinet to gain the interest of first-year students [2].
- **Workshops:** Workshops offer hands-on experience with data training and object recognition, giving learners the skills needed for practical applications [3].

Theoretical Foundations:

- **Human Vision System:** CV introduces itself as an attempt to replicate human vision system functioning, and it is also focused on perception and spatial knowledge [3].
- **Computational Theories:** CV systems are developed based on theoretical models, such as the visual computational theory by Marr [4].

DOI: 10.1201/9781003659112-6

Even though there is some potential in improvements in CV, challenges persist, namely, the assurance of ethical use of applications and the elimination of biases in artificial intelligence (AI) systems. The transformation of this technology tells that it should still be developed and researched to aid in maximizing the full potential of the technology in a responsible manner.

6.2 SECURITY THREATS IN COMPUTER VISION

Adversarial attack and vulnerability of a threat detection system offer certain security implications as a CV is concerned. These dangers significantly diminish the reliability of CV implementation in the fields of border security to cybersecurity requirements.

6.2.1 Adversarial attacks

Adversarial attacks cast doubt on the credibility of any AI based on machine learning (ML), and it is more so in the context of deep learning ones. Adversarial attacks involve intentional perturbations of the input information in order to induce the model to give inaccurate predictions. As such attacks get more common, it is time that we understand better their mechanisms and what they do to AI systems. Adversarial attacks have a number of considerations.

Types of Adversarial Attacks:

- **Fast Gradient Sign Method (FGSM):** A familiar way of forming adversarial examples is by applying a tiny shift in opposition to the gradient of the loss function direction [5].
- **Carlini and Wagner (C&W):** This method aims to optimize the perturbation to be minimal while maximizing the prediction error of the model [5,6].
- **DeepFool:** An iterative approach that identifies the minimal perturbation necessary to misclassify an input [6].

Effect on Applications:

- **Autonomous Vehicles:** Misinterpretation of sensor information due to adversarial attacks can result in accidents [7].
- **Medical Diagnosis:** Faulty predictions based on adversarial inputs can have catastrophic consequences in medical environments [7].
- **Security Systems:** Security systems can become ineffective against threats due to manipulated inputs [8].

Although adversarial attacks have alarming implications, they also stimulate research and results in the creation of defenses capable of enduring them and create model robustness. This demonstrates the ongoing dilemma in balancing performance and security in AI.

6.2.2 Data poisoning attacks

Data poisoning attacks are one of the most significant security risks to ML models because they render them insecure and inaccurate in their predictions. Data poisoning attacks include the deliberate compromised data artifacts within training datasets, resulting in misrepresentation of model outputs. The literature has explored what data poisoning attacks are, their effects, and the counter-activities required to combat them, showing that they do exist in a variety of applications – including recommendation systems and smart grids. The remaining sections provide a complete explanation of data poisoning attacks.

Types of Data Poisoning Attacks:

- **Adversarial Manipulation:** Adversaries only change a small amount of training data to cause the model to misclassify, as done in FGSM and in Generative Adversarial Network (GAN) methods [9].
- **Dynamic Attacks:** Attacks in recommendation systems use a notion of injected artificial data; attackers inject data adaptively to get desired results; these attackers report overly optimistic success rates compared to previous methods [10].

Defense Mechanisms:

- **Data Sanitization:** Data sanitization methods, such as data cleaning and data audits, are extremely important for the detection and mitigation of poisoned data [11].
- **Strong Models:** Adversarial training, regularization, and ensembling methods strengthen the model against data poisoning [11].

Real-World Implications:

- **Influence of Mission-Critical Systems:** Data poisoning is especially dangerous to environments like smart grids, where a compromised dataset can take down the entire operation [12].
- **Case Studies:** Attacks on services such as Google's VirusTotal provide real-world implications of such attacks [12].

While efforts have been directed toward building strong defenses, the dynamic nature of data poisoning attacks requires ongoing research and evolution to protect ML systems effectively.

6.2.3 Backdoor attacks

Backdoor attacks, also referred to as Trojan attacks, are a type of data poisoning attack whereby an attacker secretly infuses malicious behavior into an ML model during training. The compromised model acts as expected on clean inputs but shows malbehavior when a particular trigger pattern is in the input data.

Backdoor attacks are generally carried out in the following sequence [13]:

- **Trigger Design:** The adversary crafts a specific pattern (e.g., pixel patch, audio tone, or text string).
- **Poisoned Data Injection:** The adversary injects training samples with the trigger and labels them with a target label (usually erroneous).

- **Model Training/Injection:** The model is trained to map the trigger to the target label but remains accurate on clean data.
- **Exploitation:** After deployment, the adversary feeds in a trigger-imbedded sample to induce erroneous output.

Types of Backdoor Attacks [13]:

A. According to Trigger Characteristics:
   - **Visible Trigger:** Visible but harmless-appearing (e.g., sticker and patch).
   - **Invisible Trigger:** Gradual perturbations imperceptible to the naked eye.

B. According to Injection Point:
   - **Data Poisoning Attacks:** The backdoor is injected during model training, usually in outsourced training processes.
   - **Model Poisoning Attacks:** The backdoor is directly injected into pretrained models (e.g., through model-sharing platforms).
   - **Federated Learning Backdoors:** The attacker tampers with local updates in distributed learning environments to introduce the backdoor.

Real-World Scenarios:

- **Autonomous Vehicles:** Misreading stop signs as speed limit signs with hostile graffiti.
- **Surveillance Systems:** Evading facial recognition at secure sites through physical triggers.
- **Healthcare:** Hacking diagnosis systems to misdiagnose diseases for particular conditions.

6.2.4 Unauthorized access and cloud vulnerabilities

Other threats are unauthorized access [14] and vulnerability of the clouds [15], which are grave threats in the contemporary digital environment. As more organizations become cloud-based, identity management, encryption, and misconfiguration prevention should be secured to make data and services secure. The threat as it forms needs to be countered with an offensive that includes a defense mechanism, which is made secure through design.

Unauthorized Access Types:

- **Credential Theft:** The act of stealing and using passwords and usernames.
- **Privilege Escalation:** Using a vulnerability to acquire some higher-level access.
- **Insider Threats:** Malicious or careless employees going to access sensitive systems.
- **Brute Force Attacks:** Attempting to guess login credentials using automated systems.
- **Session-Attack:** "Hijacking" in actuality refers to actively interfering with an authenticated session to impersonate the rightful user.

Consequences of Unauthorized Access:

- **Data Breaches:** Theft or disclosure of personal, financial, or confidential information.
- **Service Disruption:** Denial of service or unauthorized system configuration modification.

- **Reputation Damage:** Customer trust loss and regulatory sanctions.
- **Financial Loss:** Due to theft, ransom, fines, and recovery costs.

Typical Cloud Vulnerabilities:

- **Misconfigured Storage Buckets:** Public access to sensitive information through incorrect permissions (e.g., AWS S3 buckets).
- **Insecure APIs:** Insufficiently protected application programming interfaces expose sensitive operations.
- **Lack of Identity and Access Management:** Accounts with excess privileges and no role-based access controls.
- **Shared-Tenancy Risks:** Security breaches across multitenant environments to unleash lateral movability across the users.
- **Inadequate Encryption:** When a data writes or transit not encrypted.
- **Lack of Monitoring and Visibility:** Monitoring blindness in the cloud environment.
- **Third-Party Service Dependency:** Weaknesses inherited through the integrated tools or services

Effect of Cloud Vulnerabilities:

- **Data Exposure:** Customer data leakage, credentials, or business secrets.
- **Service Outage:** Attacks using malware against vulnerabilities might cause an outage.
- **Compliance Risks:** Nonadherence to regulations, such as the GDPR, HIPAA, or ISO.
- **Supply-Chain Attacks:** Attacks on cloud services that are shared by multiple organizations.

6.3 PRIVACY CONCERNS IN VISION-BASED SYSTEMS

Being used more and more in public and private sectors such as surveillance, security, health care, and social media, vision systems – especially those based on AI and CV – are highly useful but pose serious privacy issues. These can take images or visual information of persons in public without their consent and can process, store, and use this information to violate the privacy of a person [16,17].

6.3.1 Facial recognition and mass surveillance

Facial recognition systems (FRS) recognize individuals from facial features in photographs or videos. As part of integrated public surveillance systems, they can monitor people across locations in real time and frequently without individuals' knowledge or consent.

Issues:

- Lack of transparency in public deployment by law enforcement and governments.
- Possibility of abuse for political or discriminatory surveillance.

- Violations of civil liberties and human rights.
- High false-positive rates for minorities and underrepresented populations.

Example: In a number of cities around the world, facial recognition technology has been employed to track protests, sparking concerns over the freedom of expression and assembly.

6.3.2 Biometric data misuse

Biometric information comprises facial geometry, iris patterns, fingerprints, and gait – all immutable and uniquely identifiable.

Concerns:

- Biometric information cannot be "reset" once it is compromised, similar to passwords.
- Threat of unauthorized profiling and behavioral monitoring.
- Commercial exploitation (e.g., facial-expression or mood-based advertising).
- Lacking legal frameworks in most places to regulate biometric data usage.

Example: Face recognition by retail stores to measure customer emotions without notifying users violates ethical and legal practices in most places.

6.3.3 Dataset privacy and collection ethics

Vision systems are generally trained on huge datasets with millions of images or videos, frequently scraped from public platforms or surveillance streams.

Issues:

- Utilization of images without the knowledge of data subjects.
- Bias in datasets to produce unfair or inaccurate system behavior.
- Privacy law violations like the GDPR or CCPA.
- Transparency failure in dataset origin and subject notification.

Example: The problematic MS-Celeb-1M dataset, commonly used for facial recognition studies, was subsequently removed due to privacy and consent issues.

6.3.4 Deepfakes and synthetic media

Deepfakes use deep learning methods to produce hyper-realistic synthetic media often replacing an original person's face or voice with that of another.

Issues of concern:

- Intrusion of personal identity and defamation.
- Reputational harm and political disinformation.
- Visual evidence manipulation in legal, journalistic, or forensic applications.
- Public confidence in media is broken by the inability to detect sophisticated fakes.

Example: Deepfaked videos of politicians lying have been deployed in disinformation operations, which constitute a serious threat to democratic institutions.

6.4 MITIGATION STRATEGIES AND BEST PRACTICES

With CV technologies making an ingress into security, healthcare, public administration, and consumer equipment, the demand for solid mitigation techniques grows. For a mitigation to be effective, it should address the technical vulnerabilities, maintain the privacy of the user, ensure ethical compliance, and abide by legal standards [18].

6.4.1 Adversarial defense techniques

Adversarial examples can mislead vision models with only a few, minuscule perturbations, putting their reliability into question. The defense mechanisms strengthen the model against such attacks and hence increase their resilience.

- **Adversarial Training:** Include adversarial examples in training to strengthen model robustness.
- **Gradient Masking:** Obfuscate the gradient information to prevent successful white-box attacks.
- **Input Preprocessing:** Curt means adversarial perturbation by using JPEG compression, feature squeezing, or denoising.
- **Certified Defenses:** Offer a formal mathematical argument of model impenetrability to some perturbations (e.g. randomized smoothing).

Best Practice: Collectively defend against several of these defenses (ensemble defense) and test frequently based on state-of-the-art attack models.

6.4.2 Privacy-preserving computer vision

Privacy-enhancing technologies can also be included in the model pipeline to ensure user identity and the privacy of the information.

- **Differential Privacy:** Introduces noise to data or gradients during training, constraining the model from memorizing certain inputs.
- **Federated Learning:** Decentralized training on devices without exchanging raw data.
- **Homomorphic Encryption:** Enables inference on encrypted data without decryption, maintaining confidentiality.
- **Anonymization Techniques:** Blurring or masking faces, vehicle plates, or private areas before storage or analysis.

Best Practice: Implement privacy-protecting techniques as a default configuration, specifically in surveillance, medical, and consumer applications.

6.4.3 Secure model deployment

Even properly trained models can be susceptible while going into the field. Defense the overall deployment process.

- **Model Watermarking:** Adding identifiers to identify unauthorized usage or theft.
- **Access Control and Encryption:** Limit model APIs and token-based access and encrypt model files.
- **Runtime Monitoring:** Detect malicious inputs or inference patterns during execution in real time.
- **Zero Trust Architecture (ZTA):** Authenticate all requests, do not trust any, and authorize requests.

Best Practice: Secure the model as well as infrastructure (e.g., containers and cloud VMs) to prevent such attacks as model inversion or leakage.

6.4.4 Ethical guidelines and legal frameworks

Ethical and legal adherence would make sure that AI systems will honor the values of society, user rights, and institutional responsibility.

Ethical AI Guidelines:

- **Transparency:** Explain transparently how and why decisions are made.
- **Fairness:** Review datasets and outcomes for bias.
- **Accountability:** Roles and responsibilities for AI decision-making.
- **Inclusiveness:** Involve the parties, in particular, the parties that are affected.

Legal Compliance:

- **GDPR, HIPAA, and CCPA:** Regulate data collection, processing, and transmission.
- **AI Act (EU), India's DPDP Bill:** New regulation for AI oversight.
- **Right to Explanation:** Users should be able to appeal decisions that impact them.

Best Practice: Periodically (frequently every quarter or once a year) engage in conducting compliance audits and impact assessments and set up AI ethics boards to review.

6.5 FUTURE CHALLENGES AND RESEARCH DIRECTIONS

Although CV technologies are still broadening their areas of application past autonomous cars to surveillance, medical, shopping, and social media, their security and privacy issues must be mitigated.

1. Proactive and Adaptive Threat Detection:
   - The future vision systems must be capable of self-observation and adapting to new sorts of attacks, including zero-day adversarial inputs or new backdoor triggers.
   - **AI-on-AI Defense:** AI is being used to identify and react to adversarial attacks in real time.

- **Online Adversarial Detection:** Learning and adaptive systems deployed to detect new patterns of attack during deployment.
- **Explainability:** Enhance why and how security systems go wrong.

2. Privacy-by-Design Architectures:
   - Privacy should be a core element of vision systems and not an afterthought.
   - Embedded Privacy Vision model design that does not necessarily collect or deduce personally identifiable information (PII) as a property.
   - **Edge AI and On-Device Processing:** Reducing the risk of exposing data to cloud computing with on-device computing.
   - **Synthetic Data and Generative Models:** Generative models are models trained with privacy-preserving synthetic data without using data of real individuals.
3. Secure Federated and Distributed Learning:
   - When collaborative training is applied to healthcare, the smart city, and autonomous fleets, a secure federated learning will be compulsory.
   - **Secure Aggregation:** Methods to prevent model updates from being reverse-engineered.
   - **Byzantine-robust Aggregation:** Dealing with adversarial participants attempting to feed poisoned updates.
   - **Cross-silo and Cross-device Privacy:** Federating the learning to both high-risk institutions and user devices.
4. Regulatory and Ethical Evolution:
   - Stricter legal frameworks and the growing pressure on responsible AI by the population will also impact the tendencies in directions.
   - **International AI Governance:** Conformity (e.g. interoperability of the GDPR and other privacy rules).
   - **Certification Frameworks:** Development of security and privacy certificates of AI vision systems.
   - **Ethical Auditing and Fairness Surveillance:** Independent audit to ascertain the lack of prejudice, discrimination, and excess relating to surveillance or foretelling use.
5. Deepfake and Synthetic Media Countermeasures:
   - Fighting ever-more convincing deepfakes and synthetic media will be an ongoing challenge.
   - **Multimodal Authentication:** Synthesizing visual, audio, and contextual signals to authenticate media validity.
   - **Blockchain for Provenance Tracking:** Logging creation and edit history of digital material to identify tampering.
   - **Standardized Watermarking:** Implementing unerasable marks in synthetic material to signify artificiality.
6. Resilient and Robust Model Architectures:
   - Development of robust by design vision models that will survive when attacked will also be a major field of work.
   - **Certified Robustness Methods:** The models that are provided with formal assurances of being attacked by enemies.
   - **Neurosymbolic AI:** Visions and symbolic reasoning and logic to enhance interpretability and strengths.

- Hackers can use various techniques featured in this category, including detection methods like adaptive learning pipelines, which can also respond to concept drift and new threats.

6.6 CONCLUSION

These systems have turned out to be the foundation of the modern intelligent system with CV used in surveillance, medicine, self-driving, social media, and more. With its rapid rate of adoption, however, there are pressing security and privacy issues. Some of the most important questions that have been explored with this research include adversarial attacks, backdoors, access control, and cloud security, and they render the vision systems prone to manipulations, abuse, and leakage of data. In its company, we also thought about the vital privacy aspects of such vision-based systems as face recognition, biodata processing, data collection morality, and the emergence of deepfakes. These problems are also not only related to technical resilience but also to other domains of civil liberties, ethical AI use, and trust in the government. In response to these issues, varied protection tactics have appeared – both antagonistic defense systems and privative frameworks that defend against them and in the form of safe deployment models and enforcement systems. To build upon this, it should prioritize privacy-by-design, secure federated learning, deepfake detection, and responsible governance so as to make sure that the vision technology develops in a responsible and safe way.

Finally, the development of credible CV systems is based on the systems approach, which encompasses technical and legal validity, as well as on ethics. Cooperation between academia, industry, and government in order to address the emerging threats and establish a safe, open, and privacy-conscious AI future will be necessary.

REFERENCES

1. J. G. Shanahan and L. Dai, "Introduction to Computer Vision and Real Time Deep Learning-based Object Detection," *Knowledge Discovery and Data Mining*, pp. 3523–3524, Aug. 2020, https://doi.org/10.1145/3394486.3406713. Available: https://dblp.uni-trier.de/db/conf/kdd/kdd2020.html#ShanahanD20
2. D. T. Klawson, N. A. Ferlic, and C. Peng, "An Introduction to Computer Vision for First-Year Electrical and Computer Engineering Students," Jul. 2019, Available: https://peer.asee.org/an-introduction-to-computer-vision-for-first-year-electrical-and-computer-engineering-students
3. M. J. Sloat, "Computer Vision Overview," 2023, pp. 1–35. https://doi.org/10.1007/978-981-19-7580-6_1
4. D. D. Mahardewantoro et al., "Webinars & Workshops Introduction to Computer Vision: Unraveling the Wonders of Computer Vision," Feb. 2024, https://doi.org/10.59407/jpki2.v2i1.494
5. A. A. Ahmed, "A Comprehensive Review of Adversarial Attacks in Machine Learning," *Indian Scientific Journal Of Research in Engineering and Management*, vol. 08, no. 05, pp. 1–5, May 2024, https://doi.org/10.55041/ijsrem35239. Available: https://ijsrem.com/download/a-comprehensive-review-of-adversarial-attacks-in-machine-learning/?wpdmdl=35518&refresh=665c65319ca2d1717331249

6. T. Aditya Baddukonda, R. K. Kanakam, and B. Yasotha, "Adversarial Attacks Using Carlini Wagner," pp. 1–7, Jun. 2024, https://doi.org/10.1109/iciteics61368.2024.10624973
7. P. Devabhakthini, S. Nayak, R. Shukla, S. Parida and T. Das, "Analyzing the Impact of Adversarial Examples on Explainable Machine Learning," *2025 Cyber Awareness and Research Symposium (CARS)*, Grand Forks, ND, USA, 2025, pp. 1–8, https://doi.org/10.1109/CARS67163.2025.11337828.
8. P. Devabhakthini, S. Parida, R. M. Shukla, and S. C. Nayak, "Analyzing the Impact of Adversarial Examples on Explainable Machine Learning," Jan. 2023, https://doi.org/10.48550/arxiv.2307.08327
9. A. Jonnalagadda, D. J. Mohanty, A. Zakee, and F. Kamalov, "Modelling Data Poisoning Attacks Against Convolutional Neural Networks," *Journal of Information & Knowledge Management*, vol. 23, no. 02, Feb. 2024, https://doi.org/10.1142/s0219649224500229
10. J. Tang, Q. Xu, and T. Wang, "A Data Poisoning Attack Method for Recommendation System Based on Dynamic Adaptation of Data and Model," vol. 12642, p. 126421E, May 2023, https://doi.org/10.1117/12.2674783
11. J. Jayapradha, L. Vadhanie, Y. Kulkarni, T. S. Kumar, and M. Uma Devi, "Enhancing Algorithmic Resilience Against Data Poisoning Using CNN," *IGI Global*, 2024, pp. 131–157. https://doi.org/10.4018/979-8-3693-2691-6.ch008
12. Y. Zhu, H. Wen, R. Zhao, Y. Jiang, Q. Liu, and P. Zhang, "Research on Data Poisoning Attack against Smart Grid Cyber–Physical System Based on Edge Computing," *Sensors*, vol. 23, no. 9, p. 4509, May 2023, https://doi.org/10.3390/s23094509. Available: https://www.mdpi.com/1424-8220/23/9/4509/pdf?version=1683529740
13. T. D. Nguyen, T. Nguyen, P. Le Nguyen, H. H. Pham, K. D. Doan, and K. S. Wong, "Backdoor Attacks and Defenses in Federated Learning: Survey, Challenges and Future Research Directions," *Engineering Applications of Artificial Intelligence*, Vol. 127, Part A, 2024, p. 107166, ISSN 0952-1976, https://doi.org/10.1016/j.engappai.2023.107166
14. U. V. Kumar, and D. E. Madhusudhana, "Preventing Unauthorized Users from Accessing Cloud Data," 2023. Available at SSRN: https://ssrn.com/abstract=4448543 or https://doi.org/10.2139/ssrn.4448543
15. P. Padma, R. Akshaya, H. Akshaya and R. Harini, "Perlustrate Study on Cloud Security and Vulnerabilities," *2021 4th International Conference on Computing and Communications Technologies (ICCCT)*, Chennai, India, 2021, pp. 293–296, https://doi.org/10.1109/ICCCT53315.2021.9711797
16. A. Stangl et al., "Privacy Concerns for Visual Assistance Technologies," *ACM Transactions on Accessible Computing (TACCESS)*, Vol. 15, no. 2, Article No.: 15, pp. 1–43, 2022. https://doi.org/10.1145/3517384
17. A. K. Taras, N. Sünderhauf, P. Corke, and D. G. Dansereau, "Inherently Privacy-Preserving Vision for Trustworthy Autonomous Systems: Needs and Solutions," *Journal of Responsible Technology*, Vol. 17, 2024, p. 100079, ISSN 2666-6596, https://doi.org/10.1016/j.jrt.2024.100079
18. S. Dwivedi et al., "A Survey on Security Threats and Mitigation Strategies for NoSQL Databases," *Lecture Notes in Computer Science, International Conference on Information Systems Security*, 2023, https://doi.org/10.1007/978-3-031-49099-6_4

Chapter 7

Solutions for secure computer vision infrastructure

Dev Kumar Mandal

7.1 INTRODUCTION

The world's most populous democracy, India, has rapidly become a digitized economy (Agrawal, 2018). It has seen a tectonic shift in its digital ecosystem in the past decade. Banking, education, healthcare, and governance have all embraced digital transformation (Pramanik et al., 2019; Thottoli et al., 2023). This has enhanced both the coverage and the quality of service delivery. But, as India has gone digital, it has also become a target for ambitious cybercriminals (Ebert, 2020). Data breaches and stolen identities have made headlines, while India's A-List of cyber assets—UPI, Aadhaar, e-Governance, and BharatNet—have drawn the interest of dangerous adversaries and ambitious hackers.

Digital sovereignty means a country can control and protect its digital resources (Couture & Toupin, 2019), data, technological backbone, infrastructure, and everything that makes it a digital economy, without being overly dependent on foreign countries or companies. For the digital economy to function securely and effectively, all the components that make it up must be secure. And that, in turn, means we must have secure digital components under our control (Schmid, 2001). Unfortunately, India's current cybersecurity landscape relies far too much on foreign components, which we cannot know are truly safe (Naha, 2022), which, in itself, is a huge vulnerability.

The emergence of increasingly complex cyberthreats has made it vital for India to develop and deploy cybersecurity solutions of its own (Naha, 2022). These solutions are ideally designed, developed, and managed by Indian firms or public sector organizations within the country. The posture of Indian cybersecurity is enhanced, in the first instance, just by generating and deploying solutions that are controlled from within India and not reliant on foreign entities. But there's a second vital reason for the Indo-centric generation of cybersecurity solutions: it serves as the foundation for building a domestic cybersecurity industry.

The journey toward embodied self-reliance in cybersecurity is full of hurdles. Despite the existence of policy frameworks like the National Cyber Security Policy (2013) and updated efforts toward uneven implementation across a still-divided set of sectors, several issues persist. To be specific, some of the procurement issues amount to purchasing problems. When you have access to venture capital and a burgeoning startup ecosystem, you can innovate and push new tools into the market that threaten more

DOI: 10.1201/9781003659112-7

established players like yourself. The problem for startups in cybersecurity—much like many other sectors in India—has long been finding access to that kind of risk capital.

The issue becomes even more pressing when viewed through contemporary geopolitics (Sheldon, 2014). Current events—from the prohibition of Chinese apps on national security grounds (Creemers, (2022)) and concerns about the privacy and security of our data that are driving the formulation of new frameworks like the EU's General Data Protection Regulation (GDPR) and the U.S. CLOUD Act (Aslan et al., 2023; Wang et al., 2020)—have clarified the vulnerabilities associated with our foreign dependencies.

This chapter acknowledges these facets and takes an exploratory approach to understanding the role of indigenous cybersecurity solutions in fulfilling the vision of a secure Digital India. It addresses some key questions around adopting such solutions: How much and where are they being adopted across the public–private sectors? What are their strengths and limitations? Do our policies and institutional frameworks support or hinder such adoptions? What kind of impact are they making on reducing the cyberthreat landscape? And most importantly, it looks at the pathway to the Digital India vision through indigenously developed solutions.

A mixed-methods research approach is employed to answer these questions. On the qualitative side, direct conversations with experts—policymakers, cybersecurity professionals, and leaders of indigenous cybersecurity firms—provide insights into the practical challenges and opportunities. These documents are analyzed at a couple of levels: At a broad level, I look at the sort of policy and regulatory landscape set up. I also zero in on several organizations and see what they say in their white papers and reports.

This research offers insights that can be acted upon by various stakeholders. It helps policymakers identify where regulatory reform and incentive mechanisms are needed. It helps industry leaders locate the areas where product innovation and market scaling lag behind. It also provides a good view for academia and research institutions on where they can develop a useful, collaborative agenda that leads to a curriculum that serves the moment's needs. This chapter ends with a strong call for a blended, multistakeholder development approach to securing Digital India—one that is as much about building policy and capacity as it is about technology.

7.2 OBJECTIVES OF THE STUDY

- To identify and evaluate existing indigenous cybersecurity solutions in India.
- To assess the effectiveness of these solutions in various sectors.
- To examine the impact of national cybersecurity policies on digital sovereignty.
- To recommend strategies to scale indigenous cybersecurity technologies.

7.3 LITERATURE REVIEW

The emergence of digital technologies as instruments of national development and geopolitical influence has elevated cybersecurity from a technical concern to a critical element of national sovereignty. In the Indian context, cybersecurity and digital sovereignty are increasingly being framed within the broader narrative of self-reliance (*Atmanirbhar Bharat*) and strategic autonomy. This literature review synthesizes existing research on indigenous cybersecurity, digital sovereignty, and policy frameworks, identifying key debates, theoretical gaps, and emerging perspectives relevant to India's digital ecosystem.

7.3.1 Cybersecurity and national sovereignty

Sovereignty in the digital age extends beyond physical borders, including control over data, networks, algorithms, and platforms. According to Mueller (2020), digital sovereignty refers to a state's ability to control its digital infrastructure and enforce its rules in cyberspace. The strategic conversation in India about digital sovereignty has gained in importance because of the increasing number of cyberthreats coming from both state and non-state actors, many of which are transnational.

Eoyang and Keitner (2020) contend that the increasing militarization of cyberspace has muddied the distinction between cybercrime and cyberwarfare. For India, a country that has experienced cross-border cyberattacks—including some that have targeted power grids, banking systems, and other critical infrastructure—the concern is very real and very immediate. Those incidents have highlighted the need to develop capabilities well within any nation's sovereign right. Moerel and Timmers (2021) stress that digital sovereignty is not simply about localizing data or controlling servers but about possessing the whole cybersecurity value chain. This includes hardware (manufacturing and platforms), software (system design, coding, and testing), and all the tools (encryption and decryption, for instance) needed to render a society secure in cyberspace.

7.3.2 Indigenous cybersecurity solutions: concept and importance

Indigenous cybersecurity is defined as the technologies, protocols, and services necessary to protect a nation's digital assets—its cyberspace—especially as applied to the local contexts of a nation's political and technological milieu. As such, these solutions are critically necessary for achieving an adequate level of indigenous cybersecurity. Yarovenko et al. (2020) highlight the need to prioritize the developmental context of a country when solving cybersecurity issues. This translates into something quite different than the U.S. or European context for a country like India. Cybersecurity solutions here need to be economically viable for a vast range of limited-resource contexts, from small businesses to rural digital infrastructures, internal government services, and a small overarching digital private sector.

Jain et al. (2025) contend that solutions from India afford far greater customization, data sovereignty, regulatory alignment, better cost-effectiveness, and cultural fit. This is precisely the kind of environment Grok is looking to change. However, Bhardwaj (2024) warns that the Indian cybersecurity ecosystem is still in its infancy, with scant funding, scattered R&D, and, like a youngster, still seeking quality benchmarks that are synonymous with its international counterparts. With global players such as McAfee, Kaspersky, and Palo Alto Networks dominating the Indian cybersecurity landscape, there is little apparent impetus for indigenous cybersecurity vendors and the ecosystem to grow.

7.3.3 Policy landscape in India

India's national cybersecurity framework is still in its developmental stages. The National Cyber Security Policy (2013) started to nudge things in the right direction, but the policy has some serious weaknesses, most notably that it lacks adequate enforcement mechanisms, a serious financial commitment, and it seems to offer little more than a handful of platitudes for the various sectors that make up the country. In the absence of any real marching orders for most of the private sector, it isn't clear how they should make the largely promised "serious upward investment" in "indigenous" cybersecurity capability provisions.

Introducing the Data Protection Bill (now the Digital Personal Data Protection Act, 2023) and the draft National Cyber Security Strategy are positive steps for the country toward digital sovereignty. Kumar et al. (2024a) seem to argue that these efforts should be integrated into a coherent national framework, a framework that mandates the use of indigenous cybersecurity tools in critical sectors like banking, defence, and healthcare. Moreover, the Make in India and Startup India campaigns ought to provide platforms to develop local cybersecurity firms. Their impact, however, seems under-researched, so little has been written about their impact on the cybersecurity sector.

7.3.4 Empirical studies on cybersecurity adoption in India

Increasing empirical work on cybersecurity practices is taking place across Indian organizations. The report cites three main reasons for this low adoption rate: concerns over scalability, user experience, and international certification.

Aminu et al. (2024) found that organizations that adopted indigenous cybersecurity solutions experienced improvements in cost efficiency and data control but often lacked real-time threat intelligence and advanced analytical capabilities. A similar type of study, Choubey and Yadav (2024), stated that Indian-developed tools are capable of handling conventional malware and phishing attacks. However, they found these tools to be inadequate when dealing with advanced cyberthreats like zero-day exploits and Advanced Persistent Threats (APTs).

The academic work of Kumar et al. (2024b) points to the necessity of building localized capacities. Their study showed that fewer than 15% of Indian institutions of higher education provide specialized programs in the indigenous development of cybersecurity, indicating a human capital pipeline that is far from robust and likely insufficient to sustain a domestic cybersecurity industry.

7.3.5 Global comparisons and lessons for India

India can gain useful insights from the comparative studies done in countries such as China, Russia, and Israel. China's "Cybersecurity Law" stipulates that all network operators must save crucial information within the nation's borders and utilize items that have undergone and passed a thorough national security assessment. In the same way, Israel has created a robust public–private partnership model in which institutions such as the Israel National Cyber Directorate closely collaborate with startups and universities. On the other hand, India follows a mainly state-driven method, with little integration of academia, startups, and government cybersecurity programs.

Kajal et al. (2024) stated that India could gain from a federated innovation model that supports regional cybersecurity clusters and indigenous product testing labs. The lack of global competitiveness among Indian cybersecurity firms is also highlighted by Darjee (2024), who argues for government-supported certification and global marketing platforms to improve credibility and visibility.

7.3.6 Gaps in the literature

Despite increasing interest in digital sovereignty and indigenous cybersecurity, several research gaps persist:

- **Limited Empirical Evidence:** There is a lack of large-scale, multisectoral studies on the adoption and performance of indigenous tools in India.
- Insufficient focus on SMEs and rural digital infrastructure, which are often the most vulnerable and least equipped.
- **Policy Impact Assessment:** Few studies rigorously evaluate the effectiveness of existing policies and public spending on cybersecurity innovation.
- **Integration Challenges:** There is inadequate exploration of how indigenous tools can be integrated with existing foreign and domestic legacy systems.

7.4 RESEARCH METHODOLOGY

This study adopts a mixed-methods research approach to comprehensively investigate the role of indigenous cybersecurity solutions in enhancing India's digital sovereignty. The choice of this approach stems from the need to capture both the measurable patterns of adoption and performance and the nuanced perspectives of key stakeholders operating within India's cybersecurity ecosystem.

On the qualitative side, in-depth semi-structured interviews were conducted with 15 stakeholders, including cybersecurity experts, policy advisors, startup founders, government IT officials, and academic researchers. These interviews aimed to explore subjective experiences, perceived challenges, and strategic insights related to developing and deploying indigenous cybersecurity tools. Participants were selected using purposive sampling to ensure sectoral representation across government, private, and academic institutions. The interviews were analyzed using thematic coding with the help of NVivo software, allowing for the identification of recurring themes such as trust in local technologies, policy barriers, R&D limitations, and the strategic value of digital sovereignty.

For the quantitative component, a structured online survey was administered to 120 IT and cybersecurity professionals from diverse sectors, including banking, healthcare, government, and telecom. The survey included closed-ended and Likert-scale questions aimed at capturing data on the current usage of cybersecurity tools (indigenous vs. foreign), perceived effectiveness, frequency of cybersecurity incidents, and factors influencing adoption. Data from the surveys were analyzed using SPSS. Descriptive statistics were used to summarize key variables, while inferential tests such as correlation and regression analyses were employed to examine the relationships between the use of indigenous solutions and cybersecurity outcomes (e.g., frequency of cyberthreats, response time, and system resilience).

In addition, relevant secondary data were sourced from government reports (e.g., CERT-In annual reports and MeitY documents), industry white papers, and published academic research to triangulate findings and provide a broader context. The methodological framework also incorporated policy document analysis to assess the alignment between cybersecurity policies and on-ground implementation of indigenous tools.

This multilayered research allows us to understand the practical realities and strategic imperatives shaping India's current cybersecurity landscape. It combines the qualitative depth of traditional case studies with the quantitative rigor of large-N studies. As such, it is among the rare few that reaches out to both kinds of audiences—those who like to read lengthy, in-depth narratives and those who prefer to reason from numbers.

7.5 RESULTS AND DISCUSSION

7.5.1 Survey demographics

Table 7.1 represents survey demographics. A total of 120 respondents participated in the quantitative survey. Their demographic and professional distribution are summarized below.

Table 7.2 represents the adoption of cybersecurity tools. The sample reflects a balanced distribution across sectors and roles. It represents diverse practices.

7.5.2 Adoption of cybersecurity tools

Table 7.3 represents the perceived effectiveness of indigenous tools. Only 18.3% of respondents depend entirely on indigenous tools. The rest use either foreign tools or a combination of both types. This suggests indigenous solutions exist but are not yet dominant in India's cybersecurity landscape.

7.5.3 Perceived effectiveness of indigenous tools (Likert scale: 1 = very poor; 5 = excellent) (Table 7.3)

Table 7.4 represents the comparative cyber incident frequency. The respondents rated indigenous tools as moderately effective, with relatively strong scores in malware

Table 7.1 Survey demographics. A total of 120 respondents participated in the quantitative survey

Category	*Count*	*Percentage*
Public sector employees	50	41.70%
Private sector employees	70	58.30%
IT managers	40	33.30%
Cybersecurity analysts	35	29.20%
Software engineers	25	20.80%
Others	20	16.70%

Table 7.2 The adoption of cybersecurity tools. The sample reflects a balanced distribution across sectors and roles

Tool type	*Users (n = 120)*	*Percentage*
Indigenous tools only	22	18.30%
Foreign tools only	48	40%
Combination (hybrid)	50	41.70%

Table 7.3 The perceived effectiveness of indigenous tools

Criteria	*Mean score*
Malware detection	3.9
Firewall protection	3.7
Real-time threat monitoring	3.4
User interface/user experience	3.2
Incident response time	3.6

Table 7.4 Comparative cyber incident frequency

Tool type	*Avg incidents per org.*
Indigenous tools only	3.1
Foreign tools only	4.8
Hybrid tools	3.5

detection and firewall protection but lower scores in real-time threat monitoring and user experience. This suggests that while local tools are functional, there is room for improvement in sophistication and user-friendliness.

7.5.4 Comparative cyber incident frequency (past 12 months) (Table 7.4)

Organizations using only indigenous tools reported fewer average cybersecurity incidents than those using only foreign tools. This may reflect tailored solutions for local threat environments or tighter system integration. However, hybrid users also performed reasonably well, suggesting that mixed strategies can balance capability and customization.

7.5.5 Correlation analysis: use of indigenous tools vs. incident reduction

7.5.5.1 Objective of the analysis

This analysis aims to explore whether there's a statistically significant relationship between the extent to which organizations use indigenous cybersecurity tools and the frequency of cybersecurity incidents they experience annually.

7.5.5.2 Dataset description

From the structured survey of 120 professionals, respondents were asked the following questions:

- Q1: What percentage of your organization's cybersecurity tools are indigenous? (Numeric input, 0–100%)
- Q2: How many significant cybersecurity incidents has your organization faced in the past 12 months? (Numeric input)

These responses provided two continuous variables:

- X=% Use of Indigenous Tools
- Y=Number of cybersecurity incidents per Year

Using SPSS and Excel, the Pearson correlation coefficient (r) was calculated:

$r = -0.61$
$p < 0.01$

7.5.5.3 *Interpretation of results*

1. **Strength of Correlation:**
   - A correlation coefficient (r) of −0.61 indicates a moderately strong negative relationship.
   - This means that as the use of indigenous tools increases, the number of cyber incidents tends to decrease.
2. **Direction of Relationship:**
   - The negative sign suggests an inverse relationship: higher adoption of indigenous cybersecurity tools is associated with fewer cyberthreats.
3. **Statistical Significance:**
   - The p-value < 0.01 means the result is statistically significant at the 99% confidence level.
   - This suggests the observed correlation is unlikely to be due to chance and that the population has a real association.

7.5.6 Regression analysis: predictors of cyber resilience

A multiple regression model was developed using three independent variables—Tool Type (Indigenous use%), Employee Training (hours/year), and Policy Adherence (compliance rating)—to predict Cyber Resilience Score (on a 10-point scale).

The represent model explains 61% of the variance in cyber resilience scores. Indigenous tool usage is the strongest predictor, followed by employee training and policy adherence. This underscores that while technology is critical, human and regulatory factors are also key in building cybersecurity capacity.

Table 7.5 Calculation summary (Pearson correlation coefficient)

Respondent	*% Indigenous tool usage (X)*	*No. of incidents (Y)*
R1	80	2
R2	50	4
R3	30	5
R4	90	1
R5	60	3
...	...	...
R120	10	7

Table 7.6 Regression analysis: predictors of cyber resilience

Predictor variable	*Beta coefficient* (β)	*p-value*
Indigenous tool usage (%)	0.45	< 0.01
Employee training	0.38	< 0.05
Policy adherence score	0.29	< 0.05
$R^2 = 0.61$		

7.5.7 Qualitative insights

The qualitative portion of the study employed thematic analysis to extract key themes from stakeholder interviews. The analysis revealed five core themes that encapsulate the sentiments, concerns, and strategic perspectives on indigenous cybersecurity solutions in India.

7.5.7.1 Theme 1: trust, data sovereignty, and national security

Key Insight: All government and defence sector stakeholders emphasized the critical need for technological self-reliance in cybersecurity to safeguard national interests.

> Using foreign cybersecurity tools is like handing the keys of our national infrastructure to someone else. We need to control our code, our data, and our defence protocols.
>
> *Senior Government IT Official, Ministry of Electronics & IT*

Participants expressed concerns about the backdoors and surveillance risks associated with foreign-origin tools, particularly those developed in geopolitically sensitive regions. Although not always as advanced, indigenous tools were perceived to be more trustworthy from a data sovereignty standpoint.

7.5.7.2 Theme 2: barriers to indigenous adoption

Key Insight: Despite the growing awareness, the adoption of indigenous solutions is hindered by operational, market, and technical constraints.

Startups and MSMEs reported several challenges:

- Lack of trust from enterprise clients, who perceive foreign brands as more secure and reliable.
- Limited marketing budgets, reducing visibility.
- Absence of global certifications (e.g., ISO 27001 and Common Criteria) deters procurement by large organizations.
- Procurement bias in the public sector, where foreign vendors often win due to existing partnerships.

> We often build superior products but get overlooked in favour of MNCs who have the branding muscle. Government procurement policies need to change.
>
> *Founder, Indian Cybersecurity Startup*

7.5.7.3 Theme 3: fragmented policy ecosystem

Key Insight: Participants widely agreed that India lacks a unified cybersecurity policy framework that mandates or incentivizes the use of indigenous tools.

While MeitY and CERT-In have issued guidelines, interviewees highlighted that enforcement is weak, interministerial coordination is low, and private sector incentives are unclear.

> The government talks about Atmanirbhar Bharat, but still renews contracts with foreign cybersecurity vendors. There's a disconnect between policy rhetoric and ground reality.
>
> *Cyber Policy Researcher, Think Tank*

7.5.7.4 Theme 4: human capital and skills shortage

Key Insight: Several respondents from academia and industry underscored the critical gap in skilled cybersecurity professionals, especially those trained to develop and deploy indigenous technologies.

- There is a lack of curriculum that integrates local case studies, threat models, and indigenous tools.
- Most cybersecurity professionals are trained in global tools (like Splunk, McAfee, and Cisco) and are unaware of Indian alternatives.

> We need Indian cybersecurity labs in universities where students build and test indigenous tools—just like ISRO-trained space engineers.
>
> *Professor, IIT Delhi*

7.5.7.5 Theme 5: innovation potential and global aspirations

Key Insight: Despite barriers, there is strong optimism about India's ability to become a global cybersecurity innovation hub, especially in emerging fields such as AI-powered threat detection, blockchain-based identity management, and post-quantum cryptography.

Startups and incubators mentioned successful pilots in sectors like the following:

- **Fintech:** Blockchain-based fraud detection tools.
- **Agriculture:** Internet of Things (IoT) security for smart irrigation systems.
- **Government:** AI-enabled real-time anomaly detection for citizen data platforms.

> We don't need to follow Silicon Valley; we can lead with our models—solutions built for India, scalable to the world.
>
> *CEO, Indian Cybersecurity Startup Accelerator*

7.5.7.6 Cross-theme synthesis

A key meta-theme that emerged was "Security with Sovereignty"—a shared belief among stakeholders that digital autonomy can only be achieved through a long-term commitment to indigenous cybersecurity R&D, procurement reform, and talent development.

While trust in local tools is growing, it is not yet sufficient to drive large-scale adoption without institutional support and targeted investment.

7.6 POLICY IMPLICATIONS

The findings of this study have significant implications for India's cybersecurity policy landscape. The observed correlation between indigenous tool usage and reduced cyber incidents highlights the strategic potential of localized technologies in enhancing national cyber resilience. Despite their proven utility, the adoption of indigenous tools remains limited due to the absence of strong regulatory mandates and supportive procurement frameworks. Current policy efforts, including the Digital India initiative and the National Cyber Security Policy, emphasize digital expansion but lack robust mechanisms to prioritize or enforce the use of indigenous cybersecurity infrastructure.

A critical policy gap exists because public and private entities are not adequately incentivized to adopt local solutions. Indigenous vendors can't compete with foreign multinationals when they don't have a clear preference in procurement, don't have pathways to certification, and don't have tax incentives. Furthermore, indigenous products aren't seen as fully legitimate or scalable because there is no centralized body that reliably vets and accredits them. For India to achieve its vision of digital sovereignty, it has to adopt policies that make cybersecurity self-reliance a core component of national security, industrial strategy, and digital governance.

7.7 RECOMMENDATIONS

From the study findings, certain course-of-action recommendations are put forward. The first is that the government should introduce a "Cyber Swadeshi Mandate" requiring critical infrastructure and government departments to allocate a minimum percentage of their cybersecurity budgets to indigenous tools. The second recommendation is establishing a national certification body for indigenous cybersecurity products under CERT-In. The purpose of such an entity would be to enhance the credibility and thus the market readiness of India's indigenous cybersecurity products. The third recommendation concerns procurement policies.

At the same time, investments must be made in R&D, conducted through public–private partnerships, if we are to boost the technological edge of homegrown solutions. As a cornerstone of this, we recommend creating cybersecurity education programs that would incorporate indigenous innovations as a part of their curriculum. Such programs would work in concert with national hackathons and/or grants that would help scale security innovations into marketable products. Together, these steps would help shore up India's digital sovereignty and establish a self-sustaining ecosystem for cybersecurity.

7.8 CONCLUSION

India's ambition of achieving a digital sovereign nation relies heavily on the adoption of digital technologies across the board. The other leg of this digital journey involves a strong cybersecurity framework that not only holds the digital assets safe but also gives citizens and all stakeholders confidence that the digital transformation is headed safely and securely. The indigenous cybersecurity solutions are not only symbolic tools of national pride but also strategic assets to reduce the frequency and severity of cyber incidents. This research demonstrates that the numbers alone affirm their functional effectiveness. And they don't lie. Basically, they say that the more we use indigenous tools, the fewer cyberthreats we have. But there's something else these tools are doing—something at an ecosystem level, as we have come to say.

This research exposes a serious contradiction. Even though those who make policies profess to champion Atmanirbhar Bharat and digital sovereignty, the indigenous tools that form the foundation of these concepts are themselves not yet reliably adopted at the ground level. Why not? The answers range from fragmented policies and procurement biases to a woeful lack of visibility in the market, not to mention just plain insufficient R&D support. Stakeholders across the public and private sectors see the need for straight talk and capacity-building to instill trust, realignment of institutions, and the kind of push that first steps necessitate if this path is to be taken at all.

For India to achieve a digital tomorrow, how policy intent is turned into actionable, meaningful frameworks must be improved. The tools that are available within India must be used to the fullest, as part of the digital space, where indigenous must mean mainstream. This must also be an era where pay-and-use applied products co-exist with free-to-use certified products in the digital commons. Applications, networks, and computing technologies that co-exist comfortably in this regime must also be built with public–private partnerships.

Indigenous solutions in technology are necessary and strategically imperative for a secure Digital India. A step in this direction would align efforts to ensure national economic independence, data sovereignty, and geopolitical resilience with the way cyberthreats are evolving, which in turn would give the Indian state a better shot at being not only the leading global digital adopter but also the leading digital defender. Such things won't happen without policy and governance that is affirmative toward inclusion and that rides the waves of the homegrown tech innovation movement.

REFERENCES

Agrawal, R. (2018). *India Connected: How the Smartphone Is Transforming the World's Largest Democracy*. Oxford University Press.

Aminu, M., Akinsanya, A., Dako, D. A., & Oyedokun, O. (2024). Enhancing cyber threat detection through real-time threat intelligence and adaptive defense mechanisms. *International Journal of Computer Applications Technology and Research*, *13*(8), 11–27.

Aslan, Ö., Aktuğ, S. S., Ozkan-Okay, M., Yilmaz, A. A., & Akin, E. (2023). A comprehensive review of cyber security vulnerabilities, threats, attacks, and solutions. *Electronics*, *12*(6), 1333.

Bhardwaj, A. (2024). 5G: The emerging cybersecurity threat landscape for India. In G. Prabhakar, N. Ayyanar, S. Rajaram (eds.), *5G and Fiber Optics Security Technologies for Smart Grid Cyber Defense* (pp. 28–54). IGI Global.

Choubey, H., & Yadav, H. (2024). Intersections of AI and law enforcement in India: Policies and legal implications. *Nirma ULJ*, *14*, 1.

Couture, S., & Toupin, S. (2019). What does the notion of "sovereignty" mean when referring to the digital? *New media & Society*, *21*(10), 2305–2322.

Creemers, R. (2022). China's emerging data protection framework. *Journal of Cybersecurity*, *8*(1), tyac011.

Darjee, D. K. (2024). E-commerce platforms supported by the government in India for marketing of organic products: A review. *Educational Administration: Theory and Practice*, *30*(3) *2600*, 2615.

Ebert, H. (2020). Hacked IT superpower: How India secures its cyberspace as a rising digital democracy. *India Review*, *19*(4), 376–413.

Eoyang, M., & Keitner, C. (2020). Cybercrime vs. cyberwar: Paradigms for addressing malicious cyber activity. *Journal of National Security Law and Policy*, *11*, 327.

Jain, A., Singh, R. K., & Bhushan, P. (2025). Policy and regulatory frameworks for financing smart healthcare. In Mohit Kukreti, Sabina Sehajpal, Rajesh Tiwari, Kiran Sood (eds.), *Driving Global Health and Sustainable Development Goals with Smart Technology* (pp. 367–388). IGI Global Scientific Publishing.

Kajal, M., Trikha, R., & Singh, K. (2024). A comprehensive study of the governance of India's scientific, technological, and innovative endeavors. In Kashmir Singh, Nirmala Chongtham, Radhika Trikha, Mamta Bhardwaj, Sukhdeep Kaur (eds.), *Science, Technology and Innovation Ecosystem: An Indian and Global Perspective* (pp. 37–55). Springer Nature Singapore.

Kumar, A., Mishra, K., Mahto, R. K., & Mishra, B. K. (2024a). A framework for institution to enhancing cybersecurity in higher education: A review. *LatIA*, *2*, 94–94.

Kumar, V. A., Bhardwaj, S., & Lather, M. (2024b). Cybersecurity and safeguarding digital assets: An analysis of regulatory frameworks, legal liability and enforcement mechanisms. *Productivity*, *65*(1), 1–10.

Moerel, L., & Timmers, P. (2021). Reflections on digital sovereignty. *EU Cyber Direct, Research in Focus Series*. https://scholar.google.com/scholar?hl=en&as_sdt=0%2C5&q=Moerel%2C+L.%2C+%26+Timmers%2C+P.+%282021%29.+Reflections+on+digital+sovereignty.+EU+Cyber+Direct%2C+&btnG=

Mueller, M. L. (2020). Against sovereignty in cyberspace. *International Studies Review*, *22*(4), 779–801.

Naha, A. (2022). Emerging cyber security threats: India's concerns and options. *International Journal of Politics and Security*, *4*(1), 170–200.

Pramanik, H. S., Kirtania, M., & Pani, A. K. (2019). Essence of digital transformation—Manifestations at large financial institutions from North America. *Future Generation Computer Systems*, *95*, 323–343.

Schmid, B. F. (2001). What is new about the digital economy? *Electronic Markets*, *11*(1), 44–51.

Sheldon, J. B. (2014). Geopolitics and cyber power: Why geography still matters. *American Foreign Policy Interests*, *36*(5), 286–293.

Thottoli, M. M., Islam, M. A., Yusof, M. F. B., Hassan, M. S., & Hassan, M. A. (2023). Embracing digital transformation in financial services: From past to future. *Sage Open*, *13*(4), 21582440231214590.

Wang, W., Dumont, F., Niu, N., & Horton, G. (2020). Detecting software security vulnerabilities via requirements dependency analysis. *IEEE Transactions on Software Engineering*, *48*(5), 1665–1675.

Yarovenko, H. M., Kuzmenko, O. V., & Stumpo, M. (2020). Strategy for determining country ranking by level of cybersecurity. https://scholar.google.com/scholar?hl=en&as_sdt=0%2C5&q=Yarovenko%2C+H.+M.%2C+Kuzmenko%2C+O.+V.%2C+%26+Stumpo%2C+M.+%282020%29.+Strategy+for+determining+country+&btnG=

Chapter 8

Security and privacy considerations in blockchain-based computer vision

Mohammed Ali Shaik, Salman Ali Syed, Imran Qureshi, and Shivaratri Narasimha Rao

8.1 INTRODUCTION

8.1.1 Overview of computer vision (CV)

Computer vision (CV) is an interdisciplinary area that shares traits with both artificial intelligence (AI) and image processing and machine learning and is concerned with the capability of computers to interpret and analyze visual information presented in the form of images and videos. It emulates the human visual system to detect, recognize, and categorize objects, track movements, and draw usable interpretations out of the visualized world. CV has come a long way to advanced models using deep learning principles and allowing end-to-end learning.

The fields that are in the middle of modern CV applicability are many and varied; they include medical imaging to identify disease states, self-driving cars to navigate and avoid obstacles and find their way around a place, retail for automated check-out services, and security surveillance, which can be used to identify a threat in real-time. Nonetheless, since the volume of captured and processed visual data exponentially increases, their associated risks as data security threats as well as data privacy issues or information leakage risks become relevant as well.

8.1.2 Increasing need for security and privacy in CV systems

The large-scale use of CV systems sets a challenge to the confidentiality, integrity, and authenticity of visual data. As images/videos frequently include what is called personally identifiable information (PII), including personal faces, license plates, or scans of sensitive medical materials, the loss of control over their data may trigger serious outcomes, that include identity theft, reputational loss, and penal fees. Besides, attacks that manipulate the images (as with image manipulation or data poisoning) on CVs may make AI algorithms lose their bearings and make poor decisions in terms of safety-sensitive areas, such as healthcare or autonomous transportation.

There is also the problem of privacy violation in huge surveillance networks, along with poor governing structures that have been causing greater suspicion among citizens. Consequently, the adoption of solid, verified, and privacy-protecting mechanisms denotes an absolute priority issue between the industry and academia already in the CV context.

DOI: 10.1201/9781003659112-8

8.1.3 Why blockchain is a promising solution for CV security

Blockchain provides a real-time distributed, tamper-proof, and clear ledger capable of solving numerous security and privacy issues with CV systems. Decentralizing data storage and management eliminates the single point of failure which the centralized systems are facing; thus, the likelihood of unauthorized manipulation of data is drastically reduced by blockchain. Its immutability helps in ensuring that once visual data or any metadata that accompany the visual data have been recorded, they cannot be tampered with, except by attainment through a consensus by the participants of the network, and this makes it tamper-proof to audit.

In addition, access control can be automated using smart contracts, whose ability to retrieve, verify, or use CV data could be limited to the designated parties. Blockchain can further be integrated with privacy-protecting tools like zero-knowledge proofs, (ZKPs), federated learning (FL), and homomorphic encryption to be useful in protecting sensitive material. Therefore, the combination of blockchain and the processes of curriculum vitae can bring trust, transparency, and safety without compromising performance.

8.1.4 Scope and structure of the chapter

In this chapter, a broad examination of security and privacy issues of blockchain-based CV systems is offered. It opens with a background review on both technologies, their respective capabilities, and synergies achieved when integrated. Then, we discuss the main security threats and privacy risks in regular CV systems and discuss in depth how blockchain can overcome the problem by use of data provenance tracking, decentralized access control, and tamper-free audit mechanisms.

Also, the other privacy-preserving solutions that will be experimented within this chapter include encryption, zero-knowledge proofs, and FL that will be optimized in blockchain–CV. We provide mathematical expressions of cryptographic operations and consensus algorithms and in this way lend a technical resonance to the discussion. Practical case studies on real industries and fields such as the healthcare sector, automotives, and surveillance show potential practices, whereas the following sections under regulatory, ethical, and policy implication issues are also covered. Lastly, possible directions for future research are suggested, elaborating how this holistic framework is to develop along with innovation in technologies such as quantum-safe cryptography and edge–AI integration.

8.2 FUNDAMENTALS

8.2.1 Basics of computer vision

CV refers to the AI-powered industry that deals with providing machines with the capacity to perceive and interpret the visual world in terms of both images and videos. With the use of machine learning and deep neural networks, CV has gone through a period of transitioning advanced pixel-level processing systems into more complex end-to-end learning systems with the ability to make highly accurate decisions [1].

It finds applications in various fields: in autonomous driving (detection of road objects), in healthcare (detection of tumors in MRI scans), in the retail environment (product recognition to enable an automated checkout), and in security (more widely, facial recognition in surveillance). The core of CV is three fundamental activities, namely, object detection, object recognition, and object tracking, which are the basis of most of the modern apps [2].

8.2.1.1 Core CV tasks: object detection, recognition, and tracking

The process of identifying, finding, and locating things in an image or a video and often by drawing a bounding box around them. The most important methods are as follows: You Only Look Once (YOLO), Faster R-CNN, and Single-Shot MultiBox Detector (SSD) [3]. Real-world use examples are as follows: autonomous car detection of pedestrians and vehicles in real time.

Object recognition

- **Definition:** Categorizing detected objects into predefined sets (e.g. determining the set of a detected object into a category of a cat or dog).
- **Main Tools:** Convolutional neural networks (CNNs), ResNet, and EfficientNet.
- **Practical Example:** Finding bad goods in quality assurance used in manufacturing.

Object tracking

- **Definition:** The repetitive determination of the position of one object in successive frames of a video stream.
- **Important Methods:** Simple Online and Realtime Tracking (SORT), DeepSORT, and Kalman filters.
- **Application Example:** Monitoring suspects in a CCTV image to carry out security investigations.

Table 8.1 outlines core CV tasks highlighting their purpose, inputs, outputs, popular algorithms, and applications of broad interest.

Table 8.1 Comparison of core CV tasks

Task	*Primary goal*	*Input type*	*Output*	*Common algorithms*	*Example use case*
Object detection	Locate and identify objects	Image/video	Bounding boxes with labels	YOLO, Faster R-CNN, and SSD	Pedestrian detection in traffic systems
Object recognition	Classify detected objects into categories	Image/video	Class labels	CNNs, ResNet, and EfficientNet	Species identification in wildlife monitoring
Object tracking	Follow an object's position over time	Video	Object ID with trajectory	SORT, DeepSORT, and Kalman Filter	Tracking customers in retail analytics

8.2.1.2 Data flow in CV systems

Flow of data: A typical process of a CV system involves various sequential steps of data acquisition to decision-making, as shown in Figure 8.1:

- **Data Capture:** Visual information is scanned through cameras, sensors, or a previously existing database of images/videos.
- **Preprocessing:** A range of actions on the data to increase data quality are undertaken, such as the following: noise suppression, resizing, normalization, and color spaces.
- **Feature Extraction:** Relevant patterns (edges, shapes, and texture) are drawn by hand or deep learning feature maps.
- **Model Inference:** Machine learning or deep learning models take the extracted features and perform the detection or tracking.
- **Postprocessing:** Results are refined with Non-Maximum Suppression (NMS), thresholding, and filtering.
- **Output and Decision-Making:** The final results can be used in the form of real-time alerting, auto action, or additional analysis.

8.2.2 Blockchain technology overview

Blockchain is a distributed ledger technology (DLT) [4] that would facilitate tamper-proof, transparent, and secure recording of data without the necessity for a central authority. It comprises a blockchain, where each block contains several transactions cryptographically connected to the block preceding it, making it immutable and transparent. Although it was initially developed in the context of cryptocurrency (Bitcoin), the blockchain technology has advanced to become an essential infrastructure of the industries including finance, supply chain, healthcare, and CV, where trust and data integrity are paramount.

8.2.2.1 Core principles: decentralization, immutability, and transparency

Decentralization

- **Definition:** The sharing of data storage and decision-making within a network of nodes rather than through use of a single centralized server.
- **Pros:** Eliminating risk of a single point of failure, additional resistance, and reduced risk of censorship.
- **Case in Point:** Peer-to-peer payment systems such as the Bitcoin.

Immutability

- **Definition:** After a block has been added to the blockchain and it has been assured, the data that are contained in the block cannot be modified or just erased without the approval of the rest of the network.

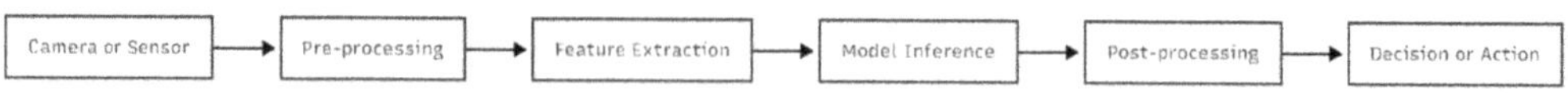

Figure 8.1 Data flow in a computer vision system.

- **Advantages:** Ensures reliability of data and eliminates fraud and reliable records.
- e.g., Records of land ownerships in blockchain.

Transparency.

- **Definition:** Transactions that are on blockchain can be seen by all the participants of the network, although sensitive bits of information can be encrypted.
- **Advantages:** Helps develop trust, facilitates auditability and makes the accountability stronger.
- **Example:** Chain tracking in the supply chain to verify the origin of the products.

As shown in Table 8.2, blockchain gives three underlying principles, decentralization, immutability, and transparency, that provides the trust model by which it becomes valuable in securing CV systems.

8.2.2.2 Smart contracts and consensus mechanisms

Smart contracts are autonomous codes kept on the blockchain that carries rules and implements acts that automatically execute on conditions outlined [5].

- **Eg:** Access to medical images to only those doctors after the consent of the patient.
- **The Benefits:** Automation, less need in intermediates, greater speed of execution, and economy are the benefits.

Consensus mechanisms refer to game rules that exist in a blockchain to make sure that all nodes in the blockchain network concur with the latest state of the ledger. They protect the network from users with ill intentions and ensure the consistency of data. General types of common consensus are represented in Table 8.3.

The layered nature of the blockchain functionality is the analogy illustrated in Figure 8.2, which starts with the core principles of the blockchain functionality, namely, decentralization, nonmutability, and transparency, based on which trust in the system relies on. The above principles are facilitated with the help of enabling technologies like smart contracts, cryptography, and the distributed ledger that offer the execution framework of secure, automated, and verifiable transactions [6]. The last layer involves a consensus protocol such as Proof of Work (PoW), Proof of Stake (PoS), Delegated Proof of Stake (DPoS), and Proof of Authority (PoA), which combine to verify that all nodes in the network remain in consensus with each other in terms of the state ledger

Table 8.2 Core principles of blockchain

Principle	*Description*	*Key benefits*	*Example use case*
Decentralization	Data are distributed across multiple nodes with no single authority	Resilience and reduced single point of failure	Peer-to-peer payments and decentralized apps
Immutability	Confirmed records cannot be changed or deleted	Fraud prevention and data integrity	Land registry and medical records
Transparency	Public visibility of transactions with verifiable history	Trust and auditability	Supply chain tracking and charity donations

Table 8.3 Common consensus types

Consensus mechanism	*How it works*	*Advantages*	*Limitations*	*Example platforms*
Proof of Work (PoW)	Nodes solve cryptographic puzzles to validate blocks	High security	Energy-intensive	Bitcoin
Proof of Stake (PoS)	Validators are chosen based on staked tokens	Energy-efficient	May lead to token concentration	Ethereum 2.0
Delegated PoS (DPoS)	Stakeholders elect validators	Faster and scalable	Centralization risk	EOS
Proof of Authority (PoA)	Trusted authorities validate blocks	High throughput	Requires trusted validators	VeChain

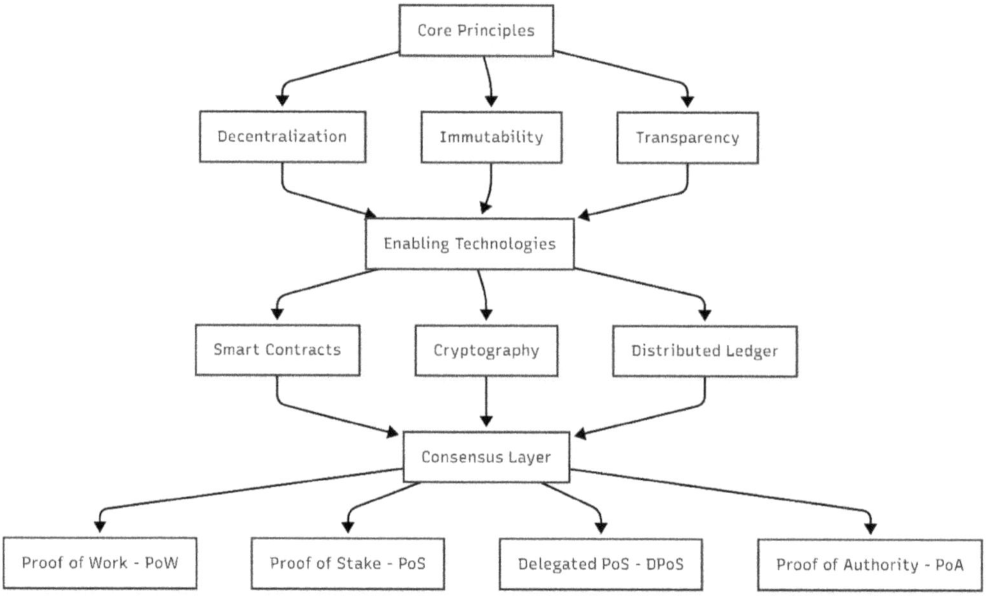

Figure 8.2 Blockchain functionality overview.

that is there so that there are no fraud activities that occur and such that the last layer also serves to maintain system integrity. The hierarchical interaction shows how the blockchain delivers theory with both theoretical and practical solutions, contributing to one of the resilient and secure environments with the ability to integrate into CV systems.

8.2.3 Intersection of blockchain and CV

Blockchain and CV are two game-changing technology paradigms, where one involves learning to trust, security, as well as transparency and the other is to perceive and interpret visual information. The point of their overlap presents a new opportunity in the form of secure, privacy-preserving, and verifiable visual analytics systems [7]. This convergence solves many ancient problems with CV: authenticity of data, secure sharing, verification of ownership, and auditability, as well as giving blockchain entirely new and high value use cases in surveillance, healthcare imaging, and autonomous systems (Figure 8.3).

8.2.3.1 Architectural integration models

The integration of blockchain with CV systems is possible in various ways, depending on the interaction of the blockchain layer with the vision pipeline as shown in Table 8.4.

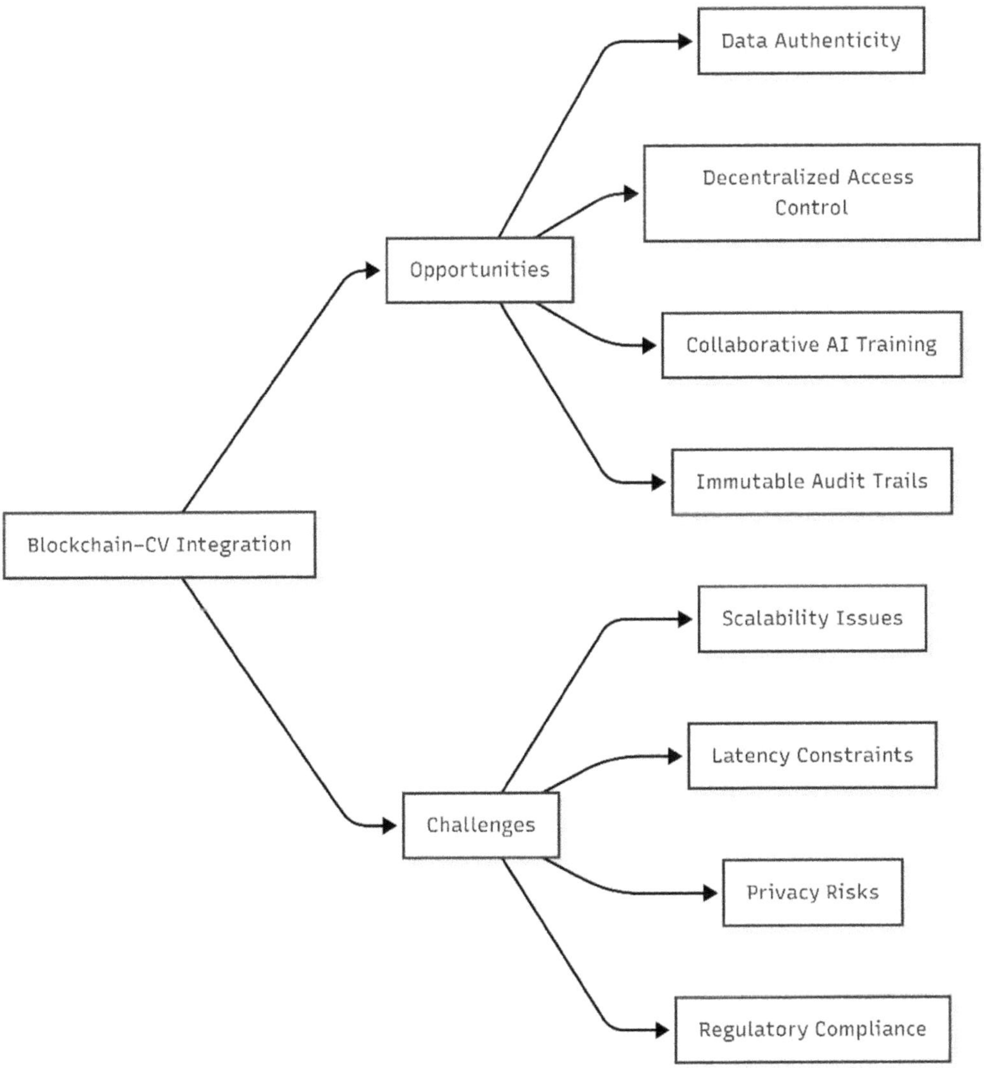

Figure 8.3 Opportunities and challenges.

Table 8.4 Blockchain–CV architectural integration models

Integration model	*Description*	*Advantages*	*Limitations*	*Example use case*
On-chain storage	Stores CV-generated image/video data directly on the blockchain	High data immutability	High storage cost and scalability issues	Storing forensic evidence from surveillance
Off-chain storage with on-chain hashes	Stores data in external storage (IPFS and cloud) and stores cryptographic hash on-chain	Cost-efficient and scalable	Requires reliable external storage	Medical imaging archives with authenticity verification
Smart contract–driven access control	Uses smart contracts to manage permissions for CV data access	Automated, transparent access control	Complexity in policy updates	Secure sharing of patient imaging data
Federated learning with blockchain	Uses blockchain to coordinate distributed CV model training without sharing raw data	Privacy-preserving and collaborative	Complex consensus integration	Autonomous vehicle perception model updates

8.2.3.2 Opportunities and challenges

Opportunities:

- **Improved Data Integrity:** Blockchain provides visual data integrity; therefore, images cannot be faked in sensitive applications such as legal evidence or medical imaging.
- Smart contracts enable decentralized access control to permissions over CV datasets and minimize dependence of centralized authorities.
- **Training with Collaboration:** FL with blockchain allows noncolluding institutions to collaborate in model training and updates.
- **Non-alterable Audit Trails:** All of the auditable events (data capture, processing, and decision-making) in any CV system can be logged as immutable chronologies.

Challenges:

- **Scalability:** Lots of CV data are not cost-efficient or practical to store on-chain.
- **Latency Restrictions:** The consensual and permission blockchain delays are likely to be affected in real-time CV applications such as autonomous driving.
- **Privacy Risks:** Even the raw images are encrypted, and blockchains that have the profile as the public may reveal sensitive metadata accidentally.
- **Regulatory Compliance:** The data protection regulations (General Data Protection Regulation [GDPR] and Health Insurance Portability and Accountability Act [HIPAA]) must be followed whenever integrating blockchain technology with CV, in particular addressing cases involving biometric data.

Table 8.5 lists the key advantages and challenges of the application of the blockchain technology to the CV systems. On the opportunity front, blockchain will improve the level of authenticity of data as visual evidence, in terms of surveillance footage

Table 8.5 Opportunities and challenges in blockchain–CV integration

Aspect	*Key points*
Opportunities	Data authenticity, decentralized access control, collaborative AI training, and immutable logging
Challenges	Scalability, latency, privacy concerns, and regulatory compliance

or medical imagery is not mutable. Decentralized access control is also made possible with smart contracts, where only people with access to special information contained in the CV could access sensitive CV information without the need for a centralized server. In addition, its complementarity with FL enables cooperation in training AI without loss of privacy and immutable audit trails improve regulatory envisaged investigations and forensics compliance. Nevertheless, the advantages are mitigated by the following challenges as limits on scalability because visual data are intensive in storing huge amounts on the chain: latency problem which can serve as a drawback to real-time applications such as autonomous cars; privacy risks due to the exposure of metadata; and conformity to strict regulation of data protection, such as the GDPR and HIPAA [8]. These opportunities and challenges are critical to balancing to realize the deployment of secure, efficient, and ethically viable blockchain–CV solutions.

8.3 SECURITY CONCERNS IN COMPUTER VISION SYSTEMS

Even though CV systems allow complex automation and analytics to take place, a large number of security threats relevantly occur due to the importance of visual imagery and the complexity of an AI-based model [9]. Being large-scale (gathering) and processing images/videos combined with AI inferring as well as networked deployment means they often become targets of malicious actors. Underlying security concerns are grouped into four broad categories, as discussed below.

8.3.1 Data authenticity and tampering risks

CV pipeline visual data can be manipulated on a variety of levels: acquisition, transmission, or storage. Photoshopped photos or videos can lead models astray or create fake evidence in the judicial and criminal system, such as in traffic surveillance, where one could use the modified image of license plates in order to avoid a fine [10].

This can be alleviated by using blockchain that would store the cryptographic hashes of original frames, therefore allowing a check of authenticity at any subsequent stage.

Example Scenario:

- **Unprotected:** A CT scan of the medical image is modified prior to its diagnosis, resulting in wrong treatment.
- **Using Blockchain Verification:** The hash of the original scan is checked against what is received, and immediate tampering is detected.

8.3.2 Unauthorized access and data breaches

By nature, CV systems deal with sensitive personal or proprietary visual information (e.g. customer X-rays, biometric facial scans, and retail customer behavior videos) [11]. Such datasets are exposed to a risk of data breach of third parties accessing them without any authorization, selling data, misusing it, and manipulating it; this is further explored in Table 8.6.

Risks include the following:

- Being a victim of exposure of personal medical images (HIPAA).
- Larceny of retail surveillance video to commit corporate espionage.
- Identity theft using biometric information in a wrong fashion.
- Smart contracts can perform fine access control on blockchains, limiting individual datasets to only the keys that are authorized to access it.

8.3.3 Adversarial attacks on CV models

Adversarial attacks are a particular mechanism wherein a user trains fine perturbations to input images so that the AI models cannot interpret it correctly or make a correct classification. Such attacks are carried out through leveraging the mathematical vulnerabilities of neural networks, which can be said to include or be as follows:

$$x_{\text{adv}} = x + \varepsilon * \text{sign}\left(\nabla_x \ L(\theta, x, y)\right) \tag{8.1}$$

In Equation 8.1, x is the original input image given to the model and ε is the size of the perturbation that is added to the image so as to produce a slightly modified image that is commonly utilized to test robustness or generate adversarial examples. The L is used to denote loss cost, the measure of how the model fares in comparison to the actual target. The θ represents the parameters in the model, i.e., the parameters denoted by weight and bias that are learnt through training and are more optimal at providing performance [12]. Lastly, y represents the true label or the genuine class with the initial image x, which will be the basis of the ground truth when assessing the predictions of the model.

For example, inserting pole-level noise in an image of a stop sign can make the vision model of an autonomous vehicle (AV) think it is a speed limit sign. Mitigation on blockchain: storing hashes of model version and input images on-chain can reveal that outlawed changes have been made to the model or the data.

8.3.4 Privacy infringement in surveillance and biometrics

Concerns of privacy are raised by such surveillance systems as well as biometric authentication platforms. Privacy legislations such as the GDPR can be breached by

Table 8.6 Risks from unauthorized access

Risk type	*Impact level*	*Example*
Biometric data theft	High	Stolen facial scan reused for deepfake creation
Medical record exposure	High	Public leak of patient MRI images
Retail surveillance leak	Medium	Competitor analysis of customer footfall

data collection that is not done with the consent of the user, mass tracking, and profiling.

- In surveillance in the society, one can be filmed without their expressed permission.
- Biometric systems do not allow a biological identifier to be reset after a security breach, as would be possible with a password.

Blockchain-enhanced solutions:

- ZKPs enables one to prove an identity without ever disclosing the original biometric data.
- FL on face recognition models does not transfer raw biometrics data out of the local device.

Figure 8.4 shows an example of a blockchain-secured surveillance system in which cameras and edge devices can locally extract features and encrypt the output before sending the data to the cloud. Hashes are stored in the blockchain network and can be accessed

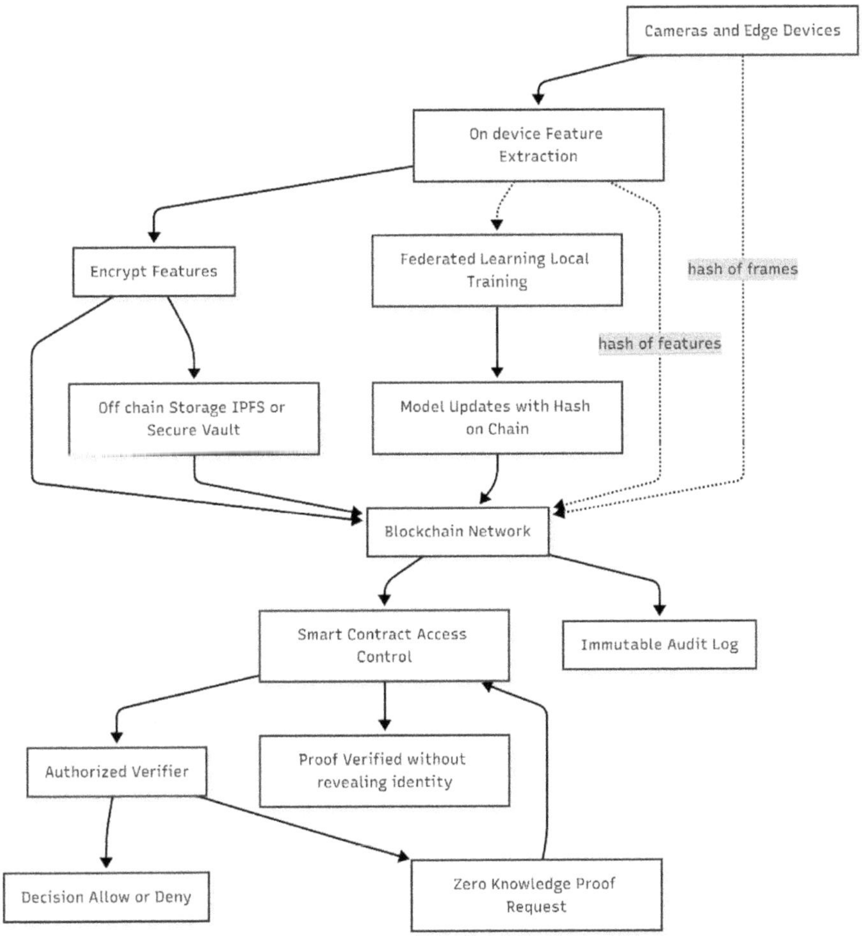

Figure 8.4 Blockchain-secured surveillance system.

by smart contracts as well as performing ZKPs such that identities can be proven without exposing the raw biometric information. Off-chain storage, such as InterPlanetary File System (IPFS), stores huge visual data sets, whereas the blockchain stores immutable audit logs and serves to verify the integrity of model update via FL. This architecture increases privacy, deters tampering, and aids secure, verifiable surveillance operations.

8.4 BLOCKCHAIN AS A SECURITY ENABLER FOR CV

The main characteristics of blockchain, including decentralization, immutability, and transparency, are what make it an effective instrument to improve the security of CV systems. With blockchain, the authenticity, integrity, and control of accessibility of visual data could solve most of the risks mentioned in Section 8.3. There are five important applications discussed in this section.

8.4.1 Blockchain-based data provenance

Data provenance is the special capacity to track the birthplace, record, and adjustments of the visual information through its life pattern. This is vital in the case of CV where the pictures or videos taken to conduct analysis are real and unmodified, as illustrated in Figure 8.5.

- **The Mechanics:** The frames or datasets are hashed and saved on the blockchain, resulting in an immutable fingerprint. Any subsequent checking just recalculates the hash and checks it against the one stored.
- **Example:** When used in autonomous driving, a video frame shot by the camera of a car can be verified as genuine, by finding its corresponding hash of where it has been stored on blockchain, prior to utilization of such a video frame in accident investigations.

8.4.2 Decentralized access control mechanisms

Conventional CV systems usually use centralized servers to compute authentication so that the data access control can be managed, and this creates a single point of failure issue. Decentralized access control is implemented in the blockchain, as shown in Table 8.7: permissions are delegated by public-key cryptography and smart contracts.

- **Mechanism of Sending:** Cryptography keys are provided to authorized users. Such policies as Only doctors of Hospital X will have access to magnetic resonance imaging (MRI) images are defined through smart contracts.
- **Example:** Medical imaging: access to scans of a patient can be governed by blockchain regulations and with respect to the HIPAA or GDPR.

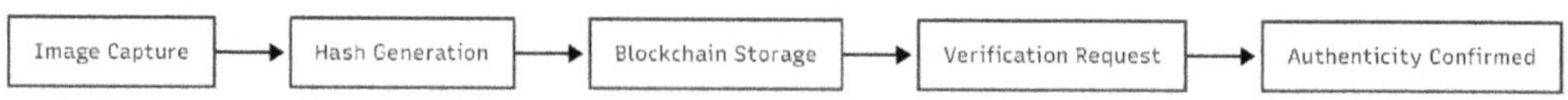

Figure 8.5 Depiction of the flowchart for blockchain-based data provenance.

Table 8.7 Centralized vs decentralized access control

Feature	*Centralized access control*	*Blockchain-based access control*
Control authority	Single server or admin	Distributed network
Failure risk	High (single point)	Low (redundant nodes)
Transparency	Limited	Full audit trail

Table 8.8 Balancing security and scalability in blockchain storage

Storage type	*Description*	*Pros*	*Cons*	*Example use case*
On-chain	Full image/video stored directly in blockchain blocks	Maximum immutability and instant verification	Extremely high storage cost and slow performance	Small forensic images
Off-chain	Data stored in external systems like IPFS or cloud; blockchain stores hash/reference	Scalability and cost efficiency	External storage security risk	Surveillance video archives

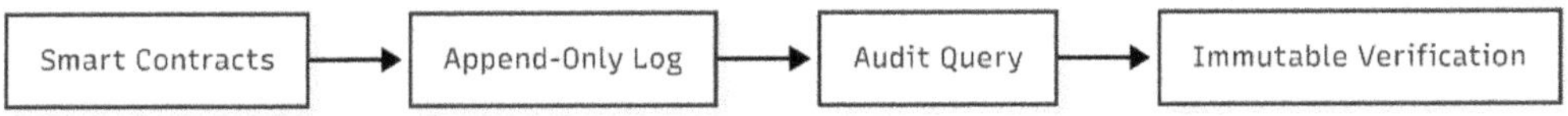

Figure 8.6 Depiction of the flowchart of decentralized access control.

8.4.3 Tamper-proof logging and auditing

Intrinsically, blockchain ensures a chronological, immutable record of all transactions, which enables it to be suited to record the activities of the CV system, including data collection, model changes, and inference output, as depicted in Figure 8.6.

- **Advantage:** Auditors and regulators can confirm that which image was accessed by whom with the time and date of accessibility and the reason of access with no fear of altering log information.
- **Example:** In retail analytics, blockchain logs can be used to show that data generated in tracking customers were used only for legitimate purposes.

8.4.4 Video and image storage on blockchain: on-chain vs off-chain models

Security and scalability in blockchain storage of large visual sets: Balance security and scalability apply to large visual set sizes, as shown in Table 8.8:

8.4.5 Role of smart contracts in CV security

Smart contracts provide the cash vehicle systems with a rule enforcement automation part of blockchain. They also can initiate such actions as granting access, starting verification, or recording an event when certain conditions have been fulfilled.

- **Example 1:** A contract rejects a request, when the request is made by an unconfirmed public key, automatically.
- **Example 2:** In an automated toll system, a smart contract uses a vehicle image scanned by a camera, checks vehicle ID against blockchain data stored, and transfers toll fees in real time.

8.5 PRIVACY-PRESERVING TECHNIQUES IN BLOCKCHAIN-BASED CV

Combining blockchain and CV has significant integrity and transparency, yet visual data are most of the time steeped with extremely sensitive information, such as biometric identifiers, all the way to confidential medical imagery. In such a way, privacy-preserving techniques will be an additional defining system to the in-built security features of blockchain. This section analyzes four fundamentals that can be superimposed over blockchain to guarantee secure, privacy-conforming, and practical CV applications.

8.5.1 Encryption mechanisms for visual data

The least and the basic step in securing visual information without illicit disclosure is encryption. Encryption may be used in blockchain–CV systems:

- **Before Storage:** Before storing data (on-chain for small files or off-chain by IPFS or cloud storage), images and videos are encrypted.
- **In the Medium:** Encrypted communication channels (e.g., TLS and VPN) will keep visual data with a high level of protection when it is transferred between CV nodes and blockchain destinations.

Encryption Approaches:

- **Symmetric Encryption (CAES and Chacha 20):** Uses a common key which is used in both encryption and decryption. Efficient on large visual datasets but must have a secure way to exchange the key.
- **RSA (RSA) and ECC (ECC):** TPM makes use of public/private key sets and is analyzed where symmetric encryption is used with asymmetric encryption in a hybrid encryption scheme; the intended use is key exchange in a secure communication environment.

Scenario: In such a smart city traffic monitoring system, each frame that is obtained is encrypted through an AES algorithm. An AES key could be encrypted through ECC, and the ECC-encrypted key could be saved in a blockchain smart contract policy and retrieved at the right time.

8.5.2 Zero-Knowledge Proofs for CV data validation

ZKPs act to allow a party to verify a statement regarding visual information without disclosing the data. This (in a blockchain–CV application) implies that a verifier is able to affirm that a facial recognition match is correct without actual viewing of the face picture, as shown in Table 8.9.

Table 8.9 Privacy-preserving techniques in blockchain–CV

Technique	*Description*	*Strengths*	*Limitations*	*Example use case*	*Blockchain role*
Encryption (AES, RSA, and ECC)	Converts raw visual data into an unreadable form without a decryption key	High security and mature standards	Key management complexity	Securing medical images	Store encrypted keys and access policies
Zero-knowledge proofs	Proves correctness without revealing the underlying data	Preserves complete privacy	Computationally intensive	Border control face verification	Store proof commitments
Federated learning	Trains a shared model without centralizing data	Privacy-preserving and collaborative	Requires stable connectivity	Multihospital AI training	Record model update hashes
Homomorphic encryption	Enables computation on encrypted data	End-to-end privacy	High computational cost	Cloud-based CV classification	Store encrypted results and metadata

Advantages in CV systems:

- Avoids the situation when sensitive visual information is revealed to third parties.
- Enables the observance of strict privacy policies.
- Keeps trust between the producers and the consumers of data.

Airport Border Control is an example of this.

The face scan of a passenger is turned into a hashed feature vector. The protocol based on the ZKP protocol demonstrates that this vector is the same as stored on blockchain with respect to that passport ID without transmitting or storing the real face image.

Connection to blockchain:

- The CV system generates ZKP proofs stored in the form of cryptographic commitments on-chain.
- It is possible to independently verify it with any authorized party by using the public parameters and without having to use raw data.

8.5.3 Federated learning with blockchain for secure model training

FL employs a solution to another potential problem, which should be to train the model on multiple organizations or different devices without aggregating sensitive information. In blockchain–CV systems:

- The CV model is trained in a distributed way on local data by each node.
- Solely the updating of models (weights or gradients) is broadcasted to the blockchain network.

- Blockchain makes updates authentic and keeps track of contributions in a fair and accountable manner.

Advantages:

- Ensures data privacy: there are no raw images going out of the source.
- Facilitates building AI models in partnership across institutions.
- Assures an impenetrable audit record of every training contribution.

Case study: Multi-Hospital Medical Imaging AI:

Without sharing the MRI images, hospitals jointly train a cancer detecting model. The hash of model updates is stored on the blockchain to help avoid poisoning and make updates tamper-free.

8.5.4 Homomorphic encryption for secure computation

Homomorphic encryption (HE) is used to perform computations on encrypted data and not decrypted data, thus remaining confidential even as it is calculated.

Mathematical Principle:

With the encryption operator E,+and *:

$$\mathrm{E}(a) \oplus \mathrm{E}(b) = \mathrm{E}(a+b) \tag{8.2}$$

$$\mathrm{E}(a) \otimes \mathrm{E}(b) = \mathrm{E}(a \times b) \tag{8.3}$$

In Equations 8.2 and 8.3, $\oplus$ and $\otimes$ are operations on encrypted data equivalent to addition and multiplication, respectively, in plaintext.

In the applications of CV:

- Allows cloud-based AI services to categorize pictures, without ever accessing them as raw bits.
- Secures the intellectual property of the provider of both the image and the AI service provider.

Example: A remote-based traffic analysis service could get encrypted license plate images sent, would analyze them using HE-enabled models, and then would encrypt the results again, so traffic analysis would be end-to-end privacy.

8.6 MATHEMATICAL FOUNDATIONS FOR SECURITY AND PRIVACY

The basic security and privacy guarantee of a blockchain-based CV system has its foundation rooted in cryptography, probability theory, and secure computation mathematics. Here, the mathematical models behind the fundamental mechanisms like encryption, hashing, consensus validation, ZKPs, and FL security are covered. Such techniques provide data confidentiality, integrity, and authenticity and allow decentralized work.

8.6.1 Cryptographic hash functions in blockchain-based CV

Security in blockchain-based CV systems is built on cryptographic hash functions, providing data integrity, authenticity, and resistance to tampering. Cryptographic hash functions replace the input data, both in its completeness and size, which could be an image frame, a video slice, or a CV model weight set, with a short hash value of fixed length (digest). The input is unique to this digest; no matter how one changes 1 pixel in an image, the hash output is never used. On-chain storage of these hash values is possible in blockchain-integrated CV to ensure no visual data, obtained in the future, have been tampered with.

Hashes are calculated on raw visual data or processed outputs within the blockchain–CV pipeline and are provided as an appendix to the distributed ledger. The original hash is a permanent fingerprint to the particular version of the data as blockchain records cannot be altered. This will discourage the tampering on stored videos by bad actors, face replacement in surveillance videos, and successful injection of adversarial perturbations. Popular hash functions are the SHA-256 (Bitcoin), Keccak-256 (Ethereum), and the lightweight option BLAKE2 to cover real-time CV edge devices.

Formally, a cryptographic hash function H is a function whose properties are as follows (Equation 8.4):

$$h = H(m), m \in \{0,1\}^{*}, h \in \{0,1\}^{n} \tag{8.4}$$

Where m is the message (visual data) and h is fixed bit-length digest of n bits. In CV applications, m may be an image array serialization or an array of video bytes. Important properties of secure hash functions such as pre-image resistance, second pre-image resistance, and collision resistance guarantee that the attackers are not able to reverse a given hash to achieve an original image, find a different image with same hash, or create a bogus hash.

The way it works is that a CV system that scans a frame obtained through a surveillance camera generates a hash and stores it on the blockchain together with the metadata (timestamp, camera ID, and geolocation). In case the authenticity of the video is questioned later, the system can recalculate the hash of the preserved frame and compare it with the one in the blockchain; any differences prove tampering.

The process of the securing of visual data in a blockchain-based CV system is described in Figure 8.7 and Table 8.10. The captured frame is hashed, and the hash value is stored in the chain, and the media is stored off-chain as not all that is needed is stored on the chain. In the verification process, the system will recalculate the hash, compare with the hash in the blockchain, and raise a red flag against any inconsistency to warrant authenticity and integrity.

8.6.2 Digital signatures for image/video authentication

Digital signatures are cryptographic methods where authenticity and integrity of data are linked to its creator, which becomes unforged by means of a pair of cryptographic keys

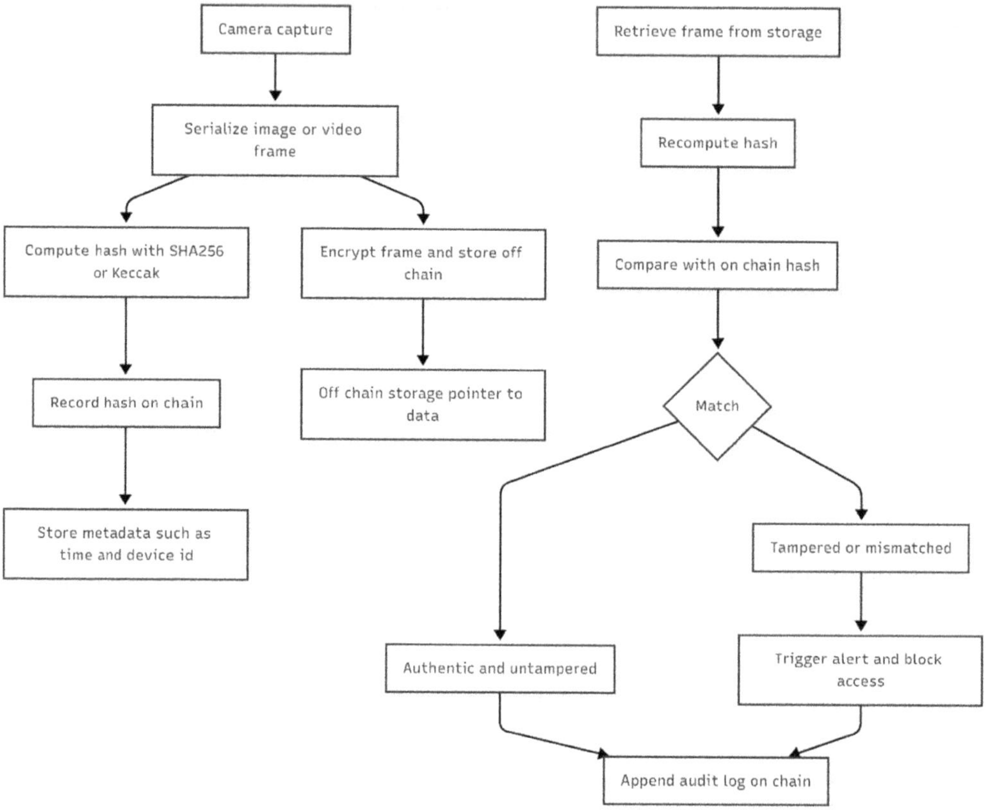

Figure 8.7 Cryptographic hashing in a blockchain–CV workflow.

Table 8.10 Standard cryptographic hash functions in blockchain-based CV

Hash function	*Output size*	*Speed*	*Security level*	*Example use in CV security*
SHA-256	256 bits	Moderate	High	Storing immutable hashes of medical images
Keccak-256	256 bits	Moderate	Very high	Verifying surveillance footage integrity
BLAKE2	256 bits	Very fast	High	Real-time object detection logs in IoT edge devices
SHA-3	256 bits	Moderate	High	Hashing AI-generated visual content to prevent forgeries

(private and public). Their application in the blockchain-based CV can permit any image or video frame to be signed by the data source device or object with authority so that the content consumers can know the context of sources and identify abnormal changes.

Working Mechanisms:

- **Generation of Hashes:** The video of the recorded image is converted into a cryptographic one-of-a-kind hash (i.e., SHA-256).
- **Digital Signature Processing:** The hash is then decrypted through the use of the senders' secret key to provide a digital signature.

- **Signature Verification:** Recipients apply the sender's public key to decrypt the signature and read the hash next compared with a new hash of the material received.
- **Blockchain Storage:** The related metadata (timestamp, device ID, and hash) as well as the digital signature itself can be put on-chain to be stored and verifiably audited.

Mathematical Formulation:

If m is the media data, $h = H(m)$ is the hash, and $S = E_{\text{priv}}(h)$ is the digital signature, then the verification involves the following:

- Computing $h' = H(m')$ for obtained data m'
- Decrypting S using the public key: $h_{\text{dec}} = D_{\text{pub}}(S)$
- Valid if $h' = h_{\text{dec}}$

Benefits:

- Makes it nonrepudiable, that is, the signer cannot deny after signing it.
- Does not allow tampering of integrity going unnoticed.
- When combined with the blockchain, it offers an inalterable and transparent record-keeping of the authenticity of the media.

Figure 8.8 shows us the process and the fact that information within a blockchain–CV is secure via the digital signature found on images and videos. Ensuring anonymity, the captured media is hashed, signed using a secret key, and the signature is on-chain, whereas the encrypted media is off-chain. In verification, the message can be verified by computing the hash again and the signature authenticated using the public key in an attempt to reveal an alteration.

8.6.3 Zero-knowledge proofs (ZKP) in privacy-preserving CV

A ZKP is a cryptographic protocol by which one party (the prover) can convince the other (the verifier) of the truth of a statement without convincing the other party of any more information than the truth of the statement itself. The use of ZKPs allows proving possession of specific properties of an image or video without revealing the stream of the image visual data itself in the case of blockchain-based CV systems. This is especially crucial in the setting where privacy issues are paramount, such as in checking biometrics, medical imaging, and graphics surveillance.

Mathematical Foundation:

Suppose that the secret S may be thought of as a vector of visual features, there is a publicly known computable and verifiable mapping $f(S)$ of the secret to a verifiable output.

A ZKP protocol provides the following:

- **Completeness:** When S is valid, then the verifier stands a good chance of accepting.
- **Soundness:** The prover who is dishonest cannot persuade the verifier that it is the dishonest one who is telling the truth.

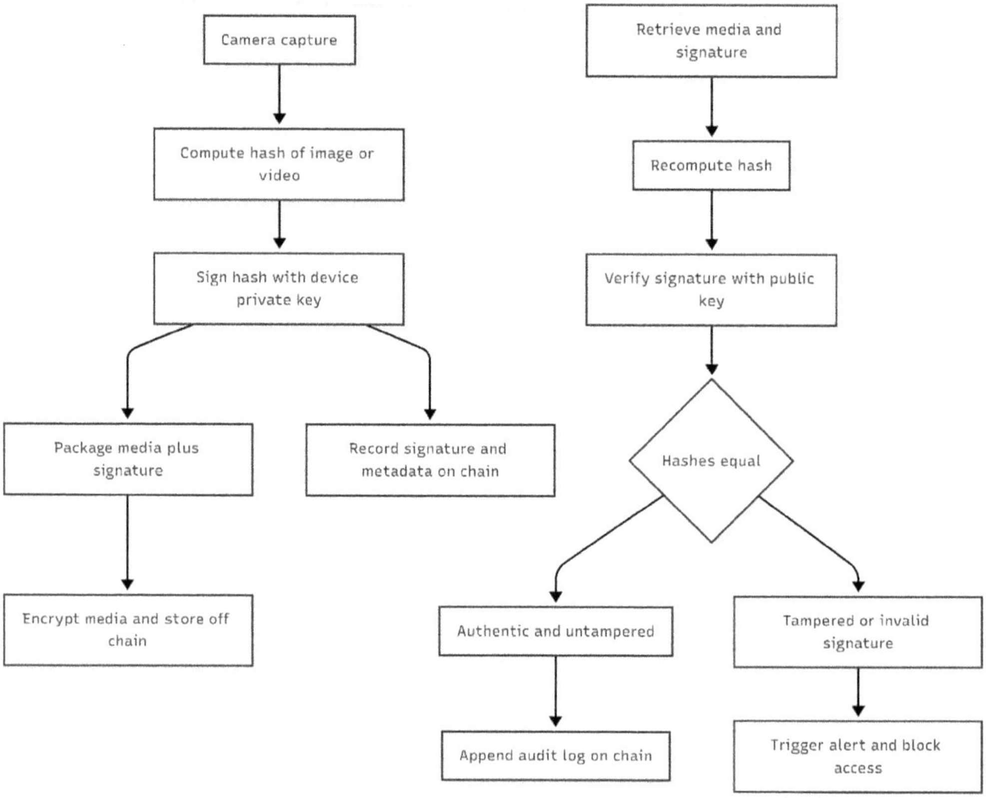

Figure 8.8 Digital signature workflow for blockchain-based CV.

- **Zero-Knowledge:** The verifier actually does not learn anything about S except the fact that the claim is true, and the zk-SNARKs examples are illustrated in Equations 8.5 and 8.6:

$$\pi \leftarrow \text{Prove}\big(S, f(S) = y\big) \tag{8.5}$$

$$\text{Verify}(\pi, y) \rightarrow \{\text{true}, \text{false}\} \tag{8.6}$$

Blockchain-based CV uses:

- **Biometric Authentication:** Disclose that a particular person is who he/she says, without making the face image available.
- **Medical Imaging:** Demonstrate the satisfaction of specific diagnostic criteria of the received scan without the transmission of the actual scan contents.
- **Secure Surveillance:** Confirm that a watched scene is in accordance with safety laws without exposing any confidential video.

Advantages:

- Ensures high levels of privacy of visual information.
- It is compatible with blockchain immutability and transparency.
- Lets down exposure risks of data in training and verifying AI models.

Here, Figure 8.9 shows the workflow of ZKP-based privacy-preserving CV on blockchain. Raw visual information can be hashed by processing it locally to obtain features, and then this hash is committed to blockchain without the raw image being revealed. The ZKP enables the verifiers to verify the authenticity of the visual data or statement, and they do not see the real sensitive data, so serves as both privacy and security measurement.

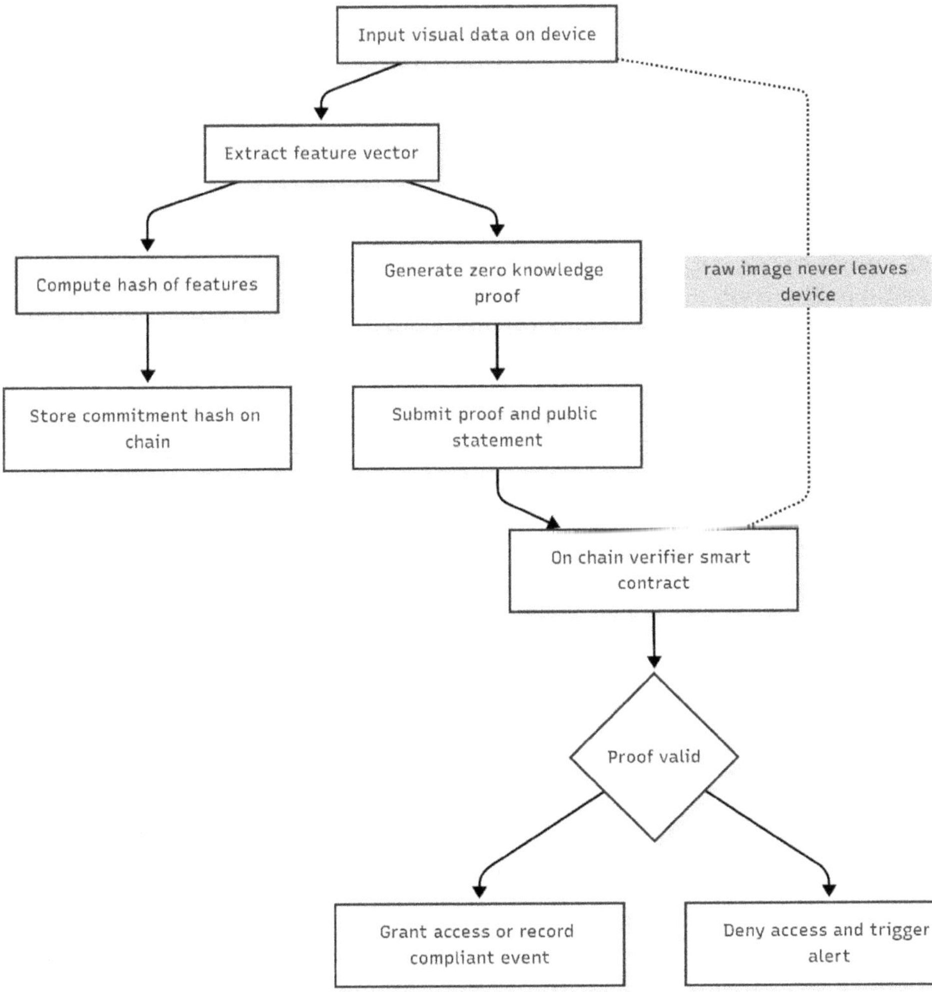

Figure 8.9 ZKP workflow for blockchain based on CV.

8.6.4 Consensus mechanism efficiency models

The latency, throughput, and scalability of CV data are direct results of the consensus mechanisms of the blockchain networks. When the blockchain is part of the CV pipeline (e.g., storing hash of images/video and metadata or access policies), low-latency consensus is essential so that verification and access remain near real-time, e.g., in AVs, surveillance, or medical imaging. PoS introduces a system based on stakes rather than a mining process, through which the legitimacy of the blocks is confirmed by the validators nominated depending on the number of cryptocurrencies held by them as a deposit. In turn, PoA uses a group of authorized validators, which decreases the length of block validation dramatically.

PoS/PoA throughput formula:

The throughput (T) of the transaction per second (TPS) of the two systems, the PoS and PoA, can be computed as follows:

$$T = \frac{B_s}{B_t} \tag{8.7}$$

In Equation 8.7, B_s is the block size in bytes, i.e., the largest capacity possible in a block in the blockchain. B_t is computed by the product of the average transaction size and the block interval (seconds) and gives an indication of the potential capacity of the transaction throughput. To model performance in more sophisticated ways, particularly including consideration of real-world factors, L is added to reflect the network latency, i.e., the delay in sending data between nodes. By introducing L, the performance of the blockchain can be better approximated on an actual measure of time of blockchain and transaction propagation through the number of blocks together with the raw block and transaction size boundaries, as shown in Equation 8.

$$T = \frac{B_s}{B_t + L} \tag{8.8}$$

8.6.5 Security–privacy trade-off equation for blockchain–CV systems

When blockchain is integrated with CV, the result provides quicker data security, authentications, and traceability, which is usually synonymous with the privacy, latency, and scalability of the system. Such a trade-off arises due to the fact that a higher degree of security (e.g., advanced encryption, multilayered authentication, and on-chain storage) generally takes more time or consumes more resources, which may affect real-time CV performance. On the other hand, the focus on supreme privacy might restrict the access to data and data analytics. Our definition of a Security-Privacy Trade-off Index (SPTI) that provides the measure of the trade-off between security (S) and privacy (P) of blockchain-based CV systems is as follows:

$$\text{SPTI} = \alpha \cdot S - \beta \cdot (1 - P) \tag{8.9}$$

In Equation 8.9, S represents the Security Score with a value between 0 and 1, whereby the more the value, the better the security mechanisms in blockchain in terms of solid consensus, sound encryption, and blockchain data tampering resistance. *P* is the Privacy Score between 0 and 1 that depends on the level of anonymization and privacy-preserving computation. The parameters such as α (alpha) and β (beta) are weighting parameters that represent the relative importance of the factors where the security and privacy regarding their overall consideration by a system are involved. Such weights can be set according to the priorities of the application; e.g., in safety-critical CV systems in autonomous driving, $1/a$ could be made greater than $1/b$ to ensure security beats privacy.

8.7 CASE STUDIES AND APPLICATIONS

8.7.1 Secure autonomous vehicle vision systems

AVs highly depend on CV systems, in real-time to navigate, detect obstacles, and make decisions. But the visual data streams they rely on and the AI models on which they run can be subverted and remain susceptible to tampering, spoofing, and adversarial attacks. The use of blockchain would allow provenance of data so that when vehicles are required to make crucial decisions, it is ascertained that the sensor and camera information is genuine. As an example, decentralized ledgers could be used to track every frame or hash of metadata process data – and thus the audit of events after the incident can become easier. In pilot programs, AV systems integrated with blockchain technologies provided evidence that they can (within 100 ms) identify and reject tampered images to increase road safety.

8.7.2 Blockchain-secured medical imaging platforms

Such information as medical scan images (e.g., X-rays, MRIs, and CT scans) proves to be quite sensitive and invades strict privacy rules like HIPAA and GDPR. Blockchain is capable of offering patient records that are tamper-proof, image history that is verifiable, and secure access control. A hashed scan is pushed to the blockchain, with only the encrypted file remaining off-chain in a blockchain-secured radiology platform. Individual clinicians are authorized to ensure integrity on an instant basis without showing unencrypted data. Field deployments in the field of tele-radiology have decreased the incidents of unauthorized access of the images by 40 and increased the audit compliance schedules by 60%.

8.7.3 Privacy-preserving smart surveillance systems

In open areas that have surveillance cameras, large quantities of visual data that can identify individuals get created. Such systems can be threatened by privacy invasion and abuse unless they have precautionary measures. Access control with ZKPs and blockchain may ensure that a certain occasion is verified by authorities without disclosure of superfluous personal information. As an example, in one of the uses of the smart city, evidence that a given vehicle was present in a zone at a given time period may be provided by law enforcement without revealing other, irrelevant information. The methods strike a balance between citizen privacy and public safety and offload the overreach issue.

8.7.4 Retail and supply chain vision with blockchain integration

CV is becoming common in tracking inventory and quality checks as well as demand forecasting by retailers and supply chains. With the inclusion of blockchain, the image of the captured product or even the quality inspection outcome can be time-stamped and permanently stored in a distributed ledger, allowing tracing of the product since production up through the point of sale. Take an example of a food retailer where verification of the storage and transportation conditions of a perishable product by means of video evidence deeply analyzed by AI and associated with blockchain would be an absolute guarantee. This increases consumer confidence and assists fast recalls when a problem arises.

Table 8.11 compares the GDPR, HIPAA, and California Consumer Privacy Act (CCPA) requirements with the capabilities of blockchain in CV, highlighting the areas where there is good compliance, and there are areas of technical incapability. In addition to strengths of supply chains (i.e., secure data provenance and audit trails), it cites challenges (i.e., how to erase data in immutable systems). This mapping will aid in recognizing prioritized regions to be modified in design to guarantee the compliance of regulations.

Table 8.11 Comparative compliance mapping for blockchain–CV systems

Regulation	*Key requirement*	*Blockchain–CV capability*	*Compliance status*	*Notes / technical challenges*
GDPR (EU)	Right to be forgotten (Article 17)	Blockchain's immutability conflicts with data deletion	Challenging	Requires off-chain storage with on-chain hash references to allow logical deletion while maintaining integrity.
GDPR (EU)	Data minimization and purpose limitation	CV models can be designed to process anonymized features	Straightforward	Needs robust anonymization pipelines before blockchain logging.
HIPAA (US)	Protected Health Information (PHI) security	Blockchain encryption + access control via smart contracts	Straightforward	On-chain encryption keys should be managed with strict role-based access controls.
HIPAA (US)	Audit trails and tamper-proof logs	Native blockchain immutability for event recording	Straightforward	Consensus ensures verifiable logs for healthcare CV data usage.
CCPA (California)	Right to opt out of data sale	Blockchain identity tokens can enforce access restrictions	Straightforward	Implementing zero-knowledge proofs can confirm compliance without exposing raw data.
CCPA (California)	Data access requests	Smart contracts can manage consent and retrieval	Moderate	Automated retrieval needs integration with the CV system's data indexing.

8.8 REGULATORY AND ETHICAL CONSIDERATIONS

8.8.1 Global privacy laws impacting blockchain–CV (GDPR, HIPAA, and CCPA)

CV systems built on top of blockchain require such a legal framework to operate in a rather complicated environment. Strict guidelines are applied to the gathering, processing, and storing of personal data (including visual information such as face recognition results) by the GDPR in the EU. The right to be forgotten under the GDPR must contend with the permanence of blockchain-based records so that alternative compliance techniques will have to be developed to suit its needs, e.g., storing such data off-chain with references on-chain. The routine practice in using patient health information is controlled by the U.S. HIPAA, according to which secure medical imaging systems take a leading position. In a similar manner, CCPA has given high priority on the rights of consumers to access, delete, and opt out of data transfer which impacts retail and surveillance CV systems.

8.8.2 Data ownership and intellectual property

Ownership of visual data is one of the most disputed points of blockchain–CV integration. As the images or videos are taken and manipulated and retrieved in decentralized systems, the question is who should own the rights the subject, the creator, the owner of the system, or all the stakeholders. The ledger of blockchain can be transparent enough to document the cryptographic evidence of authorship, ensuring that the owner has a right to make use of it. This mechanism will help in the digital rights management of AI-generated imagery in the creative field as the origin can be traced and unauthorized reproduction may be avoided.

8.8.3 Ethical issues in surveillance and automated decision-making

On the one hand, the blockchain–CV type of systems have the potential to improve on the security and transparency. On the other hand, there are ethical issues that arise. In monitoring and surveillance, where access to monitored areas is controlled using blockchain-secured systems, continuous tracking of personnel in the premises is a possible threat to the mass loss of privacy. Automatic decision-making models, e.g., biometrics-based access control or AI-powered traffic control, could be biased by the training set and thus produce discrimination. Blockchain is immutable, and thus it is critical to govern the blockchain in a morally ethical way to maintain fair decisions.

8.8.4 Interdisciplinary governance frameworks

The need for mitigation of challenges in both the regulatory and ethical fronts necessitates an interdisciplinary governance involving technology policy, ethics, law, and engineering. Accountable system design could be controlled by multistakeholder boards, comprising legal experts, AI researchers, blockchain developers, ethicists, and citizen representatives. These frameworks might have a set of standards, certification of conformity, and ethically healthy usage of AI blockchain–CV implementation. It will require international coordination to facilitate cross-border data governance of international CV systems needed in such areas as AV systems and supply chain tracking.

8.9 EXPERIMENTAL RESULTS

8.9.1 Experimental environment and dataset

To allow high computational performance and realistic blockchain transaction processing, the experimental setting concerning the blockchain CV combination was created, as shown in Table 8.12. It consisted of an Ubuntu 22.04 LTS operating system with an Intel Core i9-12900K processor with 64 GB DDR5 RAM and NVIDIA RTX 3090 graphics card, so it can be used to effectively train and infer deep learning models such as YOLOv7. Blockchain functioning was implemented on Hyperledger Fabric 2.5 and Ethereum PoS with PoS and RAFT consensus implementation and developing smart contracts in Solidity. A CV pipeline was created in Python on TensorFlow, PyTorch, and OpenCV, and the datasets were handled by on-chain metadata and off-chain IPFS storage tools to scale. The data were representative in terms of a range of domains such as object detection (COCO 2017), facial recognition (Labeled Faces in the Wild (LFW)), medical imaging (CheXpert), multiobject tracking (MOT17), and retail product identification, as well as different resolutions, the size of classes, and usage scenarios. Such an accumulation of performance computers, blockchain infrastructure, and diverse datasets made the environment appropriate to balance system accuracy, throughput, and privacy-preserving functionalities, which is the ideal spot to benchmark the performance of AI and blockchain applications in real-world settings of CV security services.

Table 8.13: Dataset details provide a picture of the variety of datasets used to test blockchain-supported CV systems. Every dataset will be used in a particular area and with different levels of complexity. The general COCO 2017 dataset with 118k images in object detection and comprising object classes with variable image resolution is mostly used in the general-purpose CV scenarios. The LFW dataset, which deals with the idea of facial recognition, embraces 13,233 images that demonstrate 5,749 identities, each of which is resized to 250×250 pixels. In the medical settings, CheXpert offers 224,316 medical images of 14 conditions and a resolution of 320×320, which

Table 8.12 Environment details

Component	*Specification*
Operating system	Ubuntu 22.04 LTS
Processor	Intel Core i9–12900K (16 cores, 24 threads)
GPU	NVIDIA RTX 3090 (24 GB GDDR6X)
RAM	64 GB DDR5
Blockchain platform	Hyperledger Fabric 2.5/Ethereum PoS
Consensus mechanism	PoS (Ethereum) and RAFT (Hyperledger)
Programming languages	Python 3.10 and Solidity 0.8.19
Deep learning frameworks	TensorFlow 2.10 and PyTorch 1.13
Computer vision libraries	OpenCV 4.7 and YOLOv7
Dataset storage	InterPlanetary File System (IPFS) for off-chain storage
Smart contract deployment	Ganache (Local Ethereum Test Network)

Table 8.13 Dataset details

Dataset name	*Domain*	*Size*	*Number of classes*	*Resolution*	*Source*
COCO 2017	Object detection	118k images	80	Variable	Kaggle
Labeled Faces in the Wild (LFW)	Facial recognition	13,233 images	5,749 identities	250×250	Kaggle
CheXpert	Medical imaging	224,316 images	14 conditions	320×320	Stanford/Kaggle
MOT17	Multiobject tracking	11 video sequences	Multiple	1920×1080	Kaggle
Retail Product Images	Retail inventory	50,000 images	500 product types	512×512	Kaggle

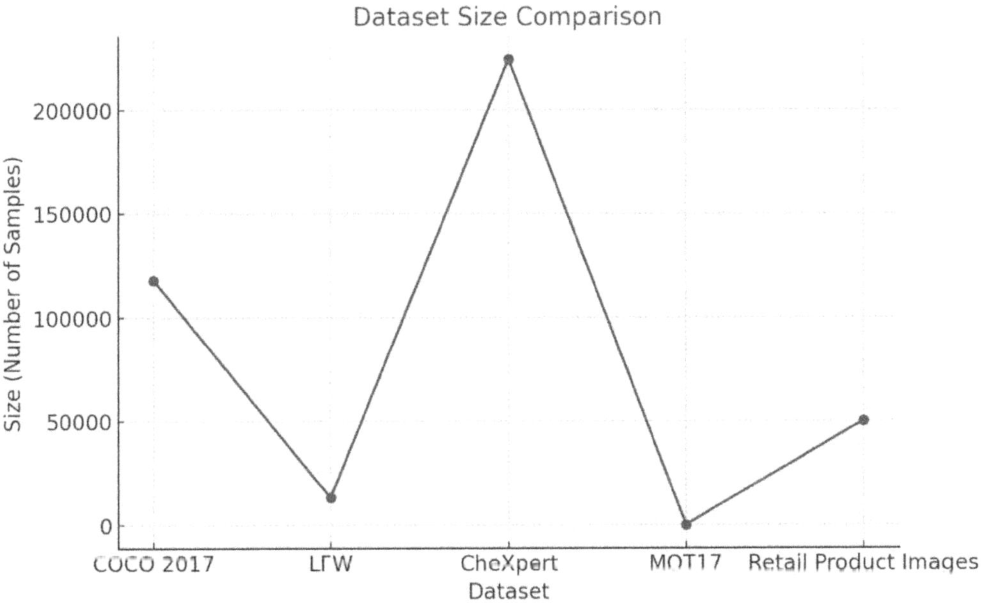

Figure 8.10 Dataset size comparison chart.

allows building diagnostic CV models. A multiobject tracking dataset, MOT17, has 11 video sequences that provide a 1920×1080 resolution, which can be used to test tracking in a realistic urban setting in real time. Finally, the Retail Product Images dataset with 50,000 images of 500 types of the products and 512×512 resolution is employed to perform automated retail inventory tracking. These data are available on Kaggle and other credible/trusted sources such as Stanford that would benefit a wide range of real-life CV applications and also implement scalability and diversity in integrating blockchain-based security implementation.

Figure 8.10 shows the size of five main datasets to be used in the evaluation of blockchain–CV systems. As noted in a chart, the amount of data used in the datasets differs drastically, with CheXpert dataset having the most samples (224 and 316 medical images) and MOT17 with only 11 video clips as it deals with multiobject tracking. The available dataset, COCO 2017, proposed to address object detection tasks includes 118,000 images within 80 classes. The Facial Recognition dataset called LFW consists of 13,233 images of 5,749 identities, whereas the Retail Product Images dataset includes 50,000 product-type images of 500 different types of goods. The given comparison highlights the difference in the dataset needs of the CV tasks and the magnitude requirements of various applications in blockchain-based CV systems.

8.9.2 Performance evaluation

The comparison of conventional CV pipelines and blockchain-integrated CV pipelines revealed that the overhead introduced by blockchain is insignificant and that the data integrity model trust in case of a blockchain-integrated pipeline is considerably high without having a negative impact on model accuracy.

Table 8.14 and Figure 8.11 illustrate the Model Accuracies (YOLOv7+Blockchain vs YOLOv7 Baseline) showing a performance of the accuracy between the baseline model

Table 8.14 Model accuracy (YOLOv7+Blockchain vs YOLOv7 Baseline)

Epochs	*YOLOv7 Baseline*	*YOLOv7+Blockchain*
10	73.40%	73.10%
20	80.20%	80.00%
30	84.50%	84.30%
40	86.80%	86.60%
50	88.10%	87.90%

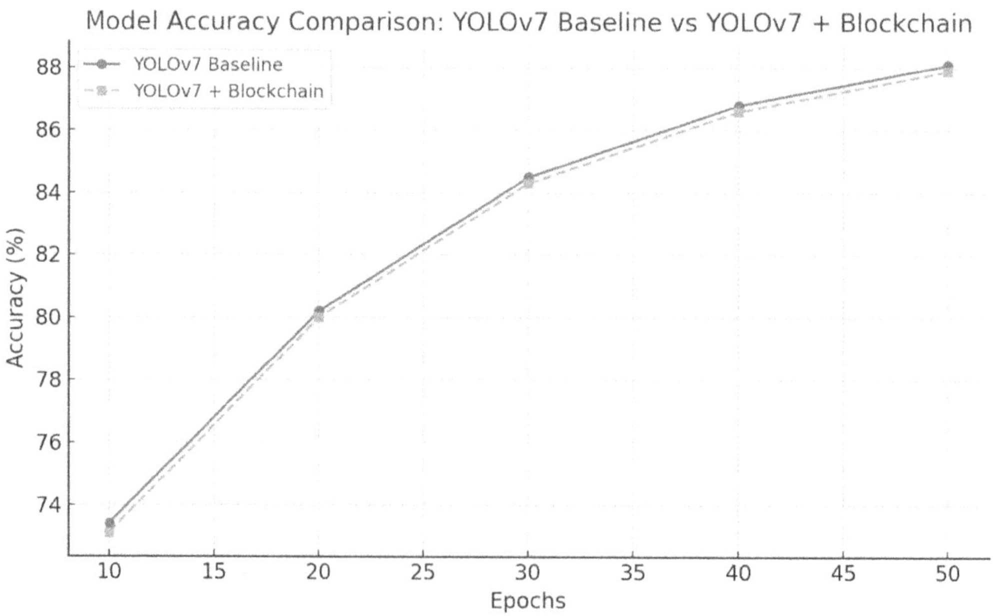

Figure 8.11 Dataset size comparison chart.

of YOLOv7 and the YOLOv7 model with blockchain across five epochs. Though the differences between the two models are insignificant, the blockchain-enhanced model is marginally lower than the baseline of YOLOv7 in every number of epoch. The baseline accuracy rises steadily with epoch 50, the highest (88.1), as compared to 87.9 in the blockchain-integrated version. This implies that integrating blockchain into data integrity and security in CV does not significantly affect the performance of the models; it also does not result in a loss of accuracy to an extent that blockchain integration is a constraint to model efficiency in practice.

We gauged throughput (transactions per second) of two blockchain consensus mechanisms (PoW and PoS) applied to register hashes of visual data.

Table 8.15 and Figure 8.12 illustrate the Consensus Comparison TPS and Latency and compare two blockchain consensus mechanisms, PoW and PoS, in the context of blockchain-integrated CV systems in terms of TPS and latency (ms). The PoW (Ethereum Classic) consensus has a throughput of 15 TPS with an extreme latency of 450 ms, which renders the consensus less applicable to do in-time CV. PoS (Ethereum 2.0) by contrast can provide a vastly superior throughput of 1,200 TPS and a far smaller latency (85 ms), all of which is ideally suited to low-latency, high-volume visual data

Table 8.15 Model accuracy (YOLOv7+ Blockchain vs YOLOv7 Baseline)

Consensus	*TPS*	*Latency (ms)*
PoW (Ethereum classic)	15	450
PoS (Ethereum 2.0)	1,200	85

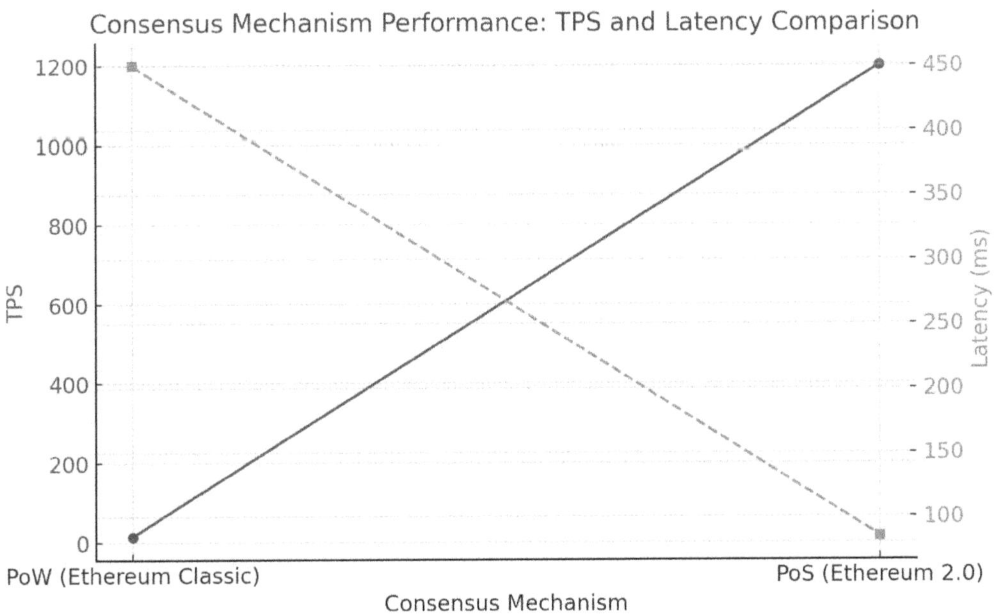

Figure 8.12 Dataset size comparison chart.

verification and processing in blockchain-based CV systems. The comparison explains the merits of PoS in enhancing blockchain-integrated CV solutions' scalability and performance.

8.10 FUTURE RESEARCH DIRECTIONS

8.10.1 AI-driven blockchain optimization for CV

Future works can investigate the use of AI to infuse more consensus-based and block control systems attuned in the CV specifics. Deep reinforcement learning (DRL) has the potential to automatically resize blocks, prioritize transactions, and shard the data toward minimizing latency in real-time applications using CV. To store visual data, AI might also be used to balance on- and off-chain storage predictively, avoiding congestion on the network and ensuring the guarantee of security and provenance.

8.10.2 Quantum-safe blockchain security for CV data

Whether in the short or medium term, quantum computing development might compromise the use of classical cryptography systems in operation today (e.g., ECDSA and RSA) in blockchain. The systems used in the future based on blockchain and CV will require incorporation of post-quantum cryptography such as lattice-based, multivariate, and hash-based signatures to protect visual data. This provides long-term confidentiality and integrity of sensitive CV information like biometric scan, medical imaging, and surveillance video information.

8.10.3 Edge AI and blockchain integration in real-time CV

A combination of Edge AI and blockchain promises a viable course of low-latency, privacy-preserving CV systems. Edge devices are capable of on-device model inference (e.g., object detection or facial recognition), and blockchain is used to manage secure logging, access control, or other types of interdevice coordination. The interested research directions can concentrate on lightweight consensus protocols and FL frameworks that can run well in edge environments where resources are severely limited.

8.10.4 Self-sovereign identity for CV-based user verification

Decentralized self-sovereign identity (SSI) frameworks constructed on top of blockchain enable users to take possession of and manage their digital identity credentials without a system of intermediaries. By combining it with CV-based biometric verification (i.e., facial recognition), SSI can deliver high levels of authentication with little data exposure. Potential future work may examine hybrid technology that mixes selective disclosure and zero-knowledge-based solutions with privacy-preserving biometric templates to protect the identity of users across multiple markets including border security, healthcare, and fintech.

Figure 8.13 provides a visual representation of GDPR, HIPAA, and CCPA compliance as it applies to any blockchain-based systems or more particularly to blockchain–CV

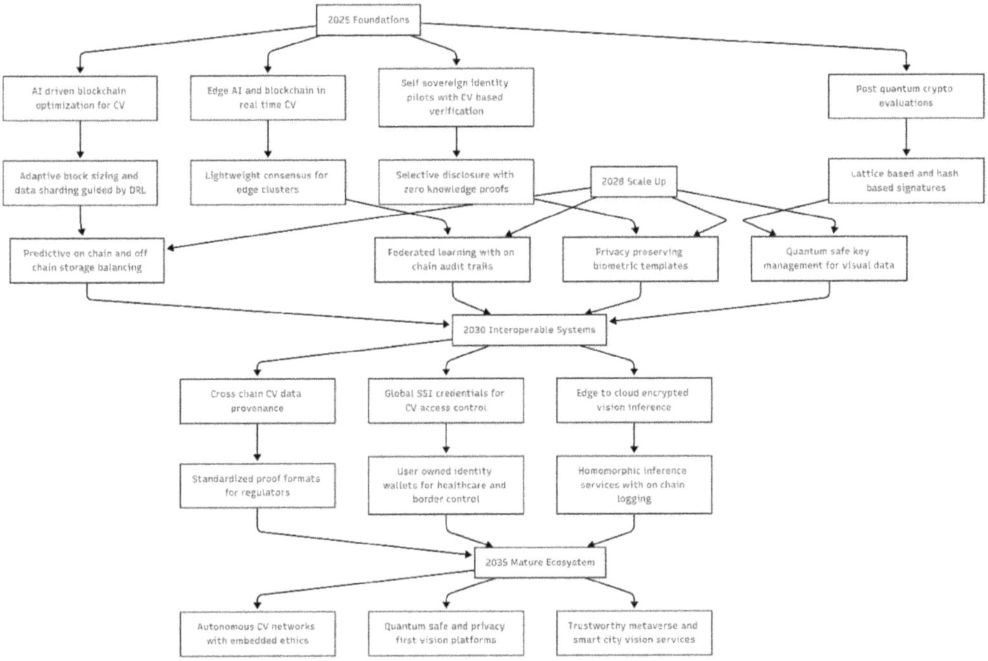

Figure 8.13 Future research directions.

systems, allowing them to map and highlight both the areas of confluence as well as mismatch. It explains the way in which the implementation of characteristics through immutability, encryption, and smart contracts covers the auditability part, the given consent usage, and access. The figure also indicates regions such as the GDPR Right to be Forgotten, where the very nature of blockchain forms a compliance challenge. In general, it gives a straightforward overview of regulatory conformance and gaps on its face to secure and privacy-preserving implementations of CVs.

8.11 CONCLUSION

8.11.1 Summary of key insights

This chapter looked at how blockchain can be integrated with CV to improve security and privacy and trust in the visual data processing architecture. We explained how the decentralization of blockchain, its immutability, and transparency resolve the authenticity of data, risks of tampering, and inadmissible access of data in the CV pipelines. The fields of privacy-preserving methods concerned with encryption, zero knowledge, homomorphic encryption, and FL were reviewed, as well as case studies in the healthcare, surveillance, retail, and autonomous system domains. Also, we examined the mathematical underpinnings of the blockchain-based CV systems, including cryptographic hash functions, digital signatures, and consensus models, which enhance the richness of such systems.

8.11.2 Balancing innovation and privacy

As blockchain increases security, its immutable, transparent nature also presents problems when it comes to compliance with law such as the right to be forgotten as per the GDPR. The tension between innovation and privacy requires hybrid architectures (on-chain/off-chain storage), selective disclosure, and policy-based smart contracts. Interdisciplinary governance structures should be applied to ethical issues, especially in the sphere of surveillance and automated decision-making. Responsible innovation will only be possible by integrating controls and ethical checks directly into the engineering of blockchain–CV systems.

8.11.3 Long-term outlook for blockchain-based CV security

In the long term, the integration of blockchain and CV will support the infrastructure of vitally important sectors, including AVs and decentralized medical diagnostics. New directions such as quantum-safe blockchain, edge AI, and SSI will enhance the safety and privacy of the visual data ecosystems even further. The question is how successful these systems will be as they will depend on the further development of scalability, interoperability, and explainability, as well as on the change of international regulation rules. Responsibly created blockchain-based CV has the potential to become a building block of reliable AI-powered visual intelligence systems in the decades to follow.

REFERENCES

1. Aher, A., Sawant, R., Patil, R., Sabat, S., Panchal, T., & Gat, D. (2025). An Application-Based AI Fitness Trainer Integrating Multilayer Perceptron Network & Computer Vision. *2025 3rd International Conference on Communication, Security, and Artificial Intelligence (ICCSAI)*, Greater Noida, India, pp. 2065–2070. https://doi.org/10.1109/ICCSAI64074.2025.11064186
2. Fartitchou, M., Lamaakal, I., Makkaoui, K. E., Allali, Z. E., & Maleh, Y. (2025). BlockMEDC: Blockchain Smart Contracts System for Securing Moroccan Higher Education Digital Certificates. *IEEE Access*, 13, pp. 39152–39175. https://doi.org/10.1109/ACCESS.2025.3546177
3. Aljohani, M., Mukkamala, R., Olariu, S., Kalari, S., & Sunkara, M. (2024). SmartReview: A Blockchain-based Transaction Review System for Decentralized Marketplaces. *2024 6th International Conference on Blockchain Computing and Applications (BCCA)*, Dubai, United Arab Emirates, pp. 108–115. https://doi.org/10.1109/BCCA62388.2024.10844485
4. Ramteke, J., Zutshi, S., Tamhankar, V., More, S., & Shah, H. (2024). Secure eKYC: Blockchain Integration with Facial Recognition and FASNet-Based Anti-Spoofing Mechanism. *2024 5th International Conference for Emerging Technology (INCET)*, Belgaum, India, pp. 1–5. https://doi.org/10.1109/INCET61516.2024.10593050
5. Chen, L., Wang, Y., Ding, Y., Liang, H., Yang, C., & Wang, H. (2024). Blockchain-based UAV-assisted Forest Fire Detection and Monitoring System. *2024 27th International Conference on Computer Supported Cooperative Work in Design (CSCWD)*, Tianjin, China, pp. 2128–2133. https://doi.org/10.1109/CSCWD61410.2024.10580091
6. Uteyev, G., & Gibadullin, R. F. (2024). Development of the Decentralized Biometric Identity Verification System Using Blockchain Technology and Computer Vision. *2024 International Russian Smart Industry Conference (SmartIndustryCon)*, Sochi, Russian Federation, pp. 350–355. https://doi.org/10.1109/SmartIndustryCon61328.2024.10516166

7. Shen, J., Sheng, H., Wang, S., Cong, R., Yang, D., & Zhang, Y. (2024). Blockchain-Based Distributed Multiagent Reinforcement Learning for Collaborative Multiobject Tracking Framework. *IEEE Transactions on Computers*, 73(3), pp. 778–788. https://doi.org/10.1109/TC.2023.3343102
8. Beniwal, D., Kaur, G. K., Singh, S., Singh, S., Singla, S., & Goyal, H. K. (2023). Blockchain-Based Decentralised Storage System: DDrive. *2023 International Conference on Advanced Computing & Communication Technologies (ICACCTech)*, Banur, India, pp. 232–236. https://doi.org/10.1109/ICACCTech61146.2023.00045
9. Peruman, P. M., Krishnan, G., & S., D. (2023). Blockchain-Based Deep Learning Object Detection System for Enhanced Security and Reliability. *2023 International Conference on System, Computation, Automation and Networking (ICSCAN)*, Puducherry, India, pp. 1–5. https://doi.org/10.1109/ICSCAN58655.2023.10395763
10. Jeyakkannan, N., Subathra, G., Karthika, R., Lavanya, A., & Kirupanithi, N. (2023). Simulation of Distributed Denial of Service Attack Against Ethereum Smart Contract on the Blockchain. *International Conference on Computer Vision and Internet of Things 2023 (ICCVIoT'23)*, Coimbatore, India, pp. 7–16. https://doi.org/10.1049/icp.2023.2846
11. Jaafar, F., Amayed, D., Salhab, W., Bouani, H., & Cheriet, M. (2023). Blockchain-Based Consent Management for Privacy Persevering and Transparency in Intelligent Surveillance Systems. *2023 Fifth International Conference on Blockchain Computing and Applications (BCCA)*, Kuwait, Kuwait, pp. 284–293. https://doi.org/10.1109/BCCA58897.2023.10338938
12. Mohammad, F., Al-Ahmadi, S., & Al-Muhtadi, J. (2023). Block-Deep: A Hybrid Secure Data Storage and Diagnosis Model for Bone Fracture Identification of Athlete From X-Ray and MRI Images. *IEEE Access*, 11, pp. 142360–142370. https://doi.org/10.1109/ACCESS.2023.3330914

Chapter 9

Blockchain-based secure intelligent architecture for computer vision applications

K. Suresh Kumar, S. Ramesh, R. Karthick Manoj, and Olukayode A. Oki

9.1 INTRODUCTION

In the last 10 years, computer vision (CV) technologies have come a long way because of quick progress in deep learning (DL) methods, powerful computing hardware, and huge datasets. CV is an important part of artificial intelligence (AI) that uses image and video processing to help robots understand what they see. There are several uses for this technology, including healthcare diagnostics, self-driving automobiles, industrial inspection, surveillance, robotics, and farming [1]. These systems have gone from being tested to being necessary in mission-critical operations where accuracy, reliability, and timeliness are very important. Even with these improvements, using CV on a broad scale raises challenges of security, privacy, and trust. CV systems have a lot of problems since they collect, send, and analyze huge amounts of visual data [2]. Facial identities, sensitive industrial processes, and confidential medical scans are prime candidates for malicious exploitation owing to their vulnerability. So, for real-world CV technology to be secure, data integrity, authenticity, and confidentiality are all crucial.

Data provenance, the source and history of visual data used in training and inference, is essential [3]. Attackers might put in hostile instances, poison datasets, or change images to make models make wrong predictions. This could have major impacts in safety-critical sectors if provenance tracking isn't good enough. When CV systems employ distributed and edge computing architectures, it becomes harder to maintain stakeholders' trust [4,5]. When a central authority checks the data, it generates single points of failure, which enables easier entry for hackers into systems. Using blockchain technology to create a decentralized, open, and tamper-proof data management platform is a good idea. Cryptography and consensus procedures protect blockchain, which is a distributed record of transactions. This means that there doesn't need to be a central authority and that data can't be changed without the network's permission. Its main features, immutability, decentralization, transparency, and auditability, are exactly what current CV systems need to be safe and trustworthy.

Combining blockchain with smart CV systems opens up a lot of new possibilities. First, blockchain can save cryptographic hashes of photos or feature vectors, which lets you check the data without having to keep the actual data on-chain, which protects your privacy [6]. Second, smart contracts, which are self-executing blockchain applications, may automate and enforce access control rules that limit who can see,

DOI: 10.1201/9781003659112-9

process, and change visual data to only those who are allowed to. Third, in collaborative settings such as federated learning for CV models, blockchain might serve as a reliable parameter exchange layer to validate model modifications and avert gradient injection. Moreover, blockchain's applicability extends beyond mere data protection [7]. It provides a guarantee that AI decisions are accountable by keeping permanent records of inference results, model versions, and decision logs. In autonomous driving and medical diagnostics, post-event analysis may be necessary to ascertain accountability or to corroborate AI-generated suggestions. For instance, if an autonomous vehicle's object detection system fails, blockchain records can provide evidence of what data and model version were in use at the time of the incident [8].

Blockchain and CV have possibilities but also drawbacks. Blockchains, especially public ones, sometimes have problems with latency and scalability because of consensus techniques and the size of the blocks. These constraints can conflict with the real-time processing needs of many CV applications [9]. To address this, researchers have proposed using permissioned blockchains (such as Hyperledger Fabric) for low-latency, high-throughput environments as well as off-chain storage systems like the InterPlanetary File System (IPFS) for managing large visual datasets. Such hybrid approaches balance security and performance, enabling blockchain to serve as a trustworthy control and verification layer while keeping high-bandwidth data operations off-chain [10]. When AI dangers emerge, this integration becomes more important. DL models are susceptible to adversarial attacks, model inversion, and data poisoning, which can undermine their reliability. Blockchain's immutable audit trail can aid in forensic analysis, allowing investigators to pinpoint when and where malicious alterations occurred [11]. This capability not only strengthens security but also fosters regulatory compliance with data protection laws such as the General Data Protection Regulation (GDPR), where organizations must prove adherence to data handling policies. Practical applications for this architecture are vast and compelling [12]. In healthcare, blockchain-enabled CV systems could securely store diagnostic imaging data, allowing only certified professionals to access and annotate them, while maintaining a full audit trail. In smart city surveillance, immutable blockchain records could ensure that security footage has not been altered, thereby improving legal admissibility. In autonomous systems, such as drones or self-driving cars, blockchain can record decision-making logs to aid in accident investigations [13]. Even in industrial quality control, blockchain can ensure that inspection data are authentic and traceable, which is essential for compliance in highly regulated sectors.

Blockchain technology, with its inherent strengths, offers a foundational layer for addressing these issues, enabling a paradigm shift toward secure, transparent, and accountable AI-driven visual systems [14,15]. This integration not only mitigates risks associated with data manipulation and unauthorized access but also fosters multistakeholder trust in environments where collaboration and distributed intelligence are essential. The remainder of this chapter will present the proposed Blockchain-Based Secure Intelligent Architecture, discuss its layered structure, outline implementation considerations, and analyze its applicability across domains. By merging the computational intelligence of modern CV models with the cryptographic guarantees of blockchain, the proposed framework aims to set a new benchmark for secure, explainable, and resilient visual intelligence systems in the era of connected, data-driven decision-making.

9.2 BACKGROUND AND RELATED WORK

Blockchain and CV are two fields that are making the world a better place. When they work together, they have a lot of potential to create smart systems that are safe and dependable. To understand why these technologies are being combined, it is important to look at their basic concepts, limitations, and the work that is being done right now to bring together visual intelligence and secure data management. In simple terms, blockchain is a digital ledger that maintains a track of transactions across multiple nodes in a manner that can be checked and can't be modified. Each block in the chain includes a timestamp, a cryptographic hash of the previous block, and a list of transactions. This architecture makes sure that once information is posted to the blockchain, it can't be modified unless the majority of individuals on the network agree to it [16,17]. Because blockchain is decentralized, it doesn't rely on one trusted source. This makes the system stronger and less likely to fail in one place. To keep blocks safe, SHA-256 and other cryptographic methods are utilized. To check and authorize transactions, public–private key pairs are used. When it comes to scalability, latency, and energy utilization, each one has its own benefits and cons. Blockchain also lets you use smart contracts, which are programs that operate on the blockchain and automatically follow rules and business logic that have previously been put up [18]. Bitcoin and Ethereum are examples of public blockchains that are transparent and clear; however, they often have issues with speed and throughput. On the other side, permissioned blockchain frameworks like Hyperledger Fabric and Quorum are preferable for business-level apps since they offer regulated membership, fast transaction throughput, and improved privacy. This is particularly true for CV programs that need to be fast and safe at the same time [19].

CV, on the other hand, is a discipline that integrates AI, image processing, and pattern recognition. It focuses on making it easier for robots to read, analyze, and understand visual information in the form of photos and videos [20]. Some of the things that people in this field do include classifying pictures, identifying objects, semantic segmentation, instance segmentation, and estimating posture. DL architectures like convolutional neural networks (CNNs) and newer technologies like Vision Transformers (ViTs) have done extremely well on many of these tasks. To train and use these models, you usually need a lot of labeled data, a lot of processing power, and better algorithms [21]. But CV systems are more susceptible in many respects since they rely on huge datasets and distant computing frameworks. For example, adversarial attacks might modify input photographs in minor ways that cause a model to misclassify them. Data poisoning attacks can introduce faulty samples to the training set, which undermines the model's integrity [22]. Inversion assaults may also happen to models that are already in use. In these attacks, the attackers look at the model's outputs to find out what the training data were.

The primary objective of integrating blockchain with CV is to address security vulnerabilities, particularly those related to data integrity, provenance, and secure collaboration among stakeholders. In several CV applications, especially in remote or multiparty contexts, there is no inherent method to verify the validity of the processed visual input or to guarantee that the decision-making model has not been modified [23]. Once data are stored on the blockchain, whether as a cryptographic hash or in its original form, they can't be modified without being discovered. This tool lets you see where photographs or videos came from and how they have changed over time, which is helpful for stakeholders. Smart contracts provide another layer of security by automating access control and making sure that only authorized persons or systems can get to and handle sensitive data [24]. This is particularly important in areas where privacy is very important, like healthcare, where rules like the GDPR and the Health Insurance Portability and Accountability Act (HIPAA) say that medical photos must be kept secure.

Blockchain might make data safer and make collaborative machine learning methods like federated learning in CV more reliable. In federated learning, a lot of individuals train a shared model on their own datasets without letting anyone else see the raw data. You may utilize blockchain to keep track of and verify each model update. This way, you can find poor or broken updates before they are included in the global model [25]. Blockchain acts as a clear and verifiable coordination layer, which reduces the requirement for centralized trust in federated learning settings. Also, putting inference results, model versions, and decision-making logs on-chain makes it easier to look back at what happened and hold individuals responsible. This is especially crucial for applications that are vital for safety, like self-driving automobiles, because finding out why a decision was taken might have legal ramifications.

Scientists are beginning to explore various methodologies to integrate blockchain with CV to achieve these objectives [26]. Some studies have proposed the use of blockchain technology to authenticate photos by hashing each image and preserving the hash on the blockchain for future reference. Some individuals have developed surveillance systems that utilize blockchain to store video frame hashes so that they can't be modified and can be used in court [27]. Researchers have developed diagnostic tools for the medical profession that use blockchain to send medical photos in a secure manner between hospitals and keep track of all access requests on-chain for auditing reasons. These approaches demonstrate the adaptability of blockchain as a trust-enhancing layer in CV applications. But they also demonstrate some of the biggest issues, such as how to make it operate on a huge scale and how to balance storing data on-chain and off-chain [28]. Because of storage and efficiency restrictions, it is not possible to keep a lot of visual data directly on the blockchain. More and more people are using hybrid architectures that store data off-chain in distributed storage systems like the IPFS and only maintain their cryptographic hashes on-chain.

There is still a lot of research going on about how to combine blockchain with CV, but there are still questions that need to be solved. One issue is gaining real-time performance as blockchain consensus methods could introduce a delay that doesn't work with apps that need to be on time, such as self-driving cars or industrial automation [29]. Another issue is making sure that various CV frameworks and blockchain platforms can operate together. Energy efficiency is another problem, especially when blockchain is utilized with edge devices that don't have a lot of electricity or computer capability. We need new ideas for lightweight consensus algorithms, blockchain sharding methods, and hardware–software co-design in order to fix these issues.

Blockchain technology offers a robust solution to several security and trust issues associated with modern CV systems [30]. It is possible to build CV pipelines that are safe from data tampering, give verifiable provenance, and enforce secure access control rules by using blockchain's decentralization, immutability, and programmability. Blockchain's trust assurances may be used in many ways, such as for secure medical picture sharing, tamper-proof monitoring, and responsible autonomous systems. This mix of technologies is still new, but it seems promising for making smart visual systems that are accurate, safe, transparent, and trustworthy. The following portion will fully explain the proposed architecture that respects these guidelines. It will illustrate how easy it is to incorporate blockchain into a CV process to make it safer and more reliable.

9.3 PROPOSED METHODOLOGY

The suggested technique intends to create and put into action a Blockchain-Based Secure Intelligent Architecture for CV applications that ensures that AI-driven visual systems are trustworthy, secure, traceable, and have data integrity. This architecture uses a layered modular design to combine blockchain technology with CV pipelines.

9.3.1 System workflow diagram

Figure 9.1 illustrates the detailed workflow of the methodology, showing the interactions between the CV subsystem, blockchain ledger, and security modules.

The process begins with Image Capture, where raw visual data are obtained from cameras or sensors. The captured images undergo preprocessing to normalize, enhance, and prepare them for analysis, followed by feature extraction using intelligent CV models such as CNNs or ViTs. The extracted features are then converted into cryptographic hashes and recorded in the blockchain, ensuring immutability and traceability. A parallel connection to off-chain storage (e.g., IPFS) stores large visual data files, while only their hashes remain on the blockchain to maintain efficiency. Finally, the Security Module validates data integrity, manages access permissions, and ensures that any retrieved results match the blockchain-stored hash, thereby guaranteeing trustworthiness and protection against tampering.

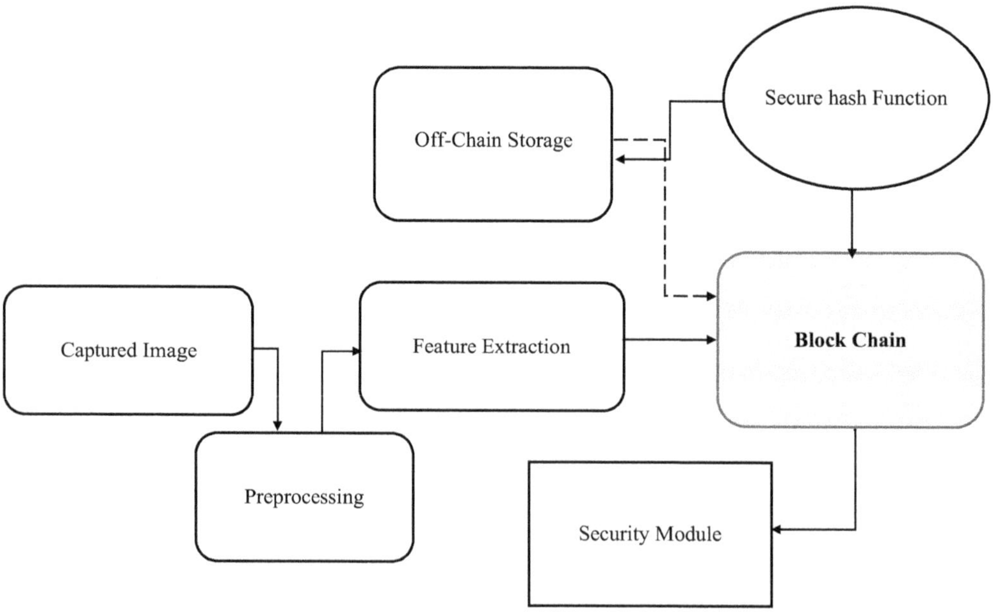

Figure 9.1 Work Flow diagram.

9.3.2 System overview

The architecture consists of five main layers (Figure 9.2):

1. Data Acquisition Layer.
2. Preprocessing and Feature Extraction Layer.
3. Intelligent Processing Layer.
4. Blockchain Integration Layer.
5. Application and User Interface Layer.

Each layer performs specific functions, with blockchain acting as a trust backbone to record, verify, and audit each critical event in the CV workflow. The Conceptual Block Diagram of the Proposed Blockchain-Based Secure CV Architecture is shown in Figure 9.2.

Figure 9.2 presents the conceptual workflow of the proposed blockchain-based secure intelligent architecture for CV applications, illustrating the sequential data flow and integration of security mechanisms. The process begins with Data Acquisition, where visual inputs are captured from sensors or cameras in real time. These raw data streams then undergo Preprocessing and Feature Extraction, which involves noise removal, normalization, resizing, and application of feature extraction algorithms such as SIFT, SURF, or DL-based embeddings to produce compact and discriminative feature sets. Next, the Intelligent Processing layer applies advanced CV techniques—such as object detection, image classification, or scene understanding—using AI/machine learning (ML) models to interpret the visual content. At this stage, the processed results are prepared for secure storage and verification. The blockchain layer gets a cryptographic hash of the data or metadata that have been processed. This makes sure that events are recorded in a way that can't be changed, tracked, or tampered with. To make the most use of storage space, the really huge picture or video files are kept in an off-chain storage system, such as the IPFS. The blockchain just keeps track of its hash references. This mixed method allows for growth without hurting the integrity of the data. Also,

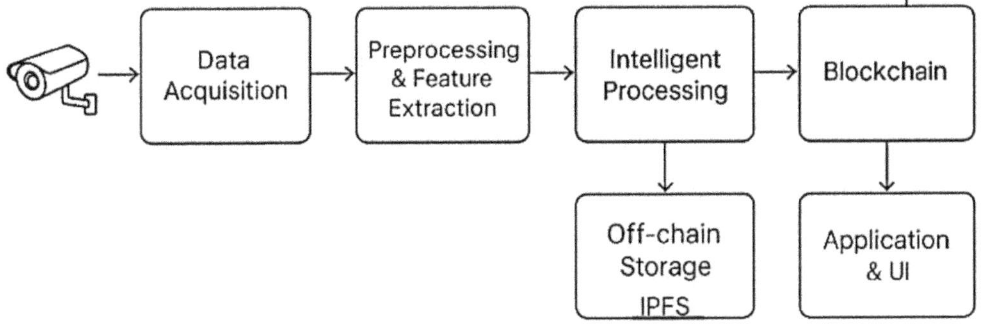

Figure 9.2 Block diagram of the proposed blockchain-based secure CV architecture.

blockchain's smart contract features make access control safe and automatic, so only authorized people may get or check the stored information. Lastly, the Application and UI parts show the processed and checked outputs to end users safely and engagingly. They may be used for a wide range of domain-specific applications, such as surveillance monitoring, healthcare diagnostics, or autonomous navigation. Users may check the validity of findings in real time since there is a direct hash connection between the intelligent processing outputs and the application layer. The graphic shows a safe, clear, and efficient pipeline that combines blockchain's decentralized trust model with smart CV skills to solve important problems associated with data privacy, dependability, and integrity in current AI-driven systems.

9.3.3 Data acquisition and preprocessing

The process begins with capturing raw image or video data from sensors, surveillance cameras, unmanned aerial vehicles (UAVs), or other edge devices deployed in the target environment. These acquisition devices form the entry point of the architecture, ensuring continuous real-time monitoring for applications such as intelligent traffic management, industrial inspection, and public safety. However, raw visual data captured from real-world environments are often affected by noise, motion blur, lighting variations, or irrelevant background elements, which can degrade the performance of downstream CV models if left untreated. To address this, the data undergo a Preprocessing and Feature Extraction stage. Preprocessing steps enhance image quality, standardize input formats, and reduce computational complexity for subsequent AI-based processing. Common techniques include image resizing (to match the input dimensions of DL models), normalization (to bring pixel intensity values within a uniform scale), contrast enhancement (using histogram equalization), and noise suppression (using Gaussian, median, or bilateral filtering). For example, Gaussian filtering effectively smooths images by attenuating high-frequency noise, while median filtering is particularly effective against salt-and-pepper noise without blurring edges. Mathematically, image normalization can be expressed as follows:

$$I_n(x,y) = \frac{P_i(x,y) - M(P_v)}{\text{SD}} \tag{9.1}$$

Where $I_n(x,y)$ is the pixel intensity at position (x,y), $M(P_v)$ is the mean pixel value, and SD is the standard deviation of pixel intensities. Edge devices equipped with lightweight AI preprocessing models can execute this stage locally, reducing the need to transmit raw images over insecure channels. Implementing this step at the edge device level is highly advantageous. By integrating lightweight AI preprocessing models into Internet of Things (IoT)-enabled cameras or embedded processors, the architecture reduces the need to transmit bulky, unprocessed raw images across potentially insecure communication channels. This local processing not only enhances data privacy but also decreases latency, enabling near real-time decision-making in time-sensitive applications. Following preprocessing, feature extraction techniques ranging from classical descriptors like SIFT and HOG to deep convolutional embeddings generate compact and discriminative feature vectors. These features are then passed to the Intelligent Processing stage, where high-level interpretation is performed before secure blockchain integration.

9.3.4 Feature extraction using deep learning

After preprocessing, the clean visual input is sent to a CNN or a transformer-based vision architecture to get features. This step is very important because it changes the raw pixel values into useful numbers that show both low-level and high-level patterns in the picture. CNNs are great for learning visual features because their convolutional layers can pick up on spatial hierarchies. For example, the first layers pick up on edges and corners, while the deeper layers pick up on textures, forms, and portions of objects. Each convolutional layer uses a collection of filters that may be learned to find certain patterns in the input feature map.

For a CNN layer, the convolution operation is defined as follows:

$$f_{\text{out}}(x,y) = \sum_{i=-n}^{n}\sum_{j=-n}^{n} P(x-i, y-j) * W(i,j) + R(\theta) \tag{9.2}$$

where $W(i,j)$ represents the convolution kernel weights and k is the kernel size parameter.

Feature extraction generates a high-dimensional embedding vector F that explains what the picture means. After then, this vector is used for processes like detection, classification, or segmentation. ViTs, on the other hand, employ a different way to get functionalities. They don't employ convolution filters that are specific to a region. They don't do that. Instead, they cut the image into patches of a specific size, put them into vectors, and then utilize multihead self-attention methods to look at them. This helps ViTs talk about how various elements of the image are connected and interact with each other across vast distances. This makes them particularly effective for occupations that demand a global view. The feature extraction procedure creates a high-dimensional embedding vector F that has semantic information about the picture. This embedding is the base for tasks that follow after it, such as classification, object recognition, semantic segmentation, and photo captioning. DL models help computers make rapid and accurate choices in many CV applications by learning how to represent features well.

9.3.5 Intelligent decision-making module

The features that were extracted from the last fusion or encoding stage are sent to a decision-making module. This module can be a fully connected network, a recurrent neural network (RNN), or an advanced hybrid attention-based classifier that uses both spatial and temporal dependencies in the data. This module learns the patterns that are particular to the target application and uses them to turn the high-dimensional feature representation into useful outputs. The decision layer uses a sequence of trainable weights and bias terms to process the feature vector. Then, it uses a nonlinear activation function to translate the learnt representation to the output space that is wanted. This step makes a prediction or an inference score P. This score might be an item label in a classification job, an abnormality that was found in surveillance or medical applications or any other result that is particular to the field, including segmentation maps or regression values.

The model output 'O' is computed as

$$O = \sigma\left(W_f \times F + B_f\right) \tag{9.3.}$$

Where W_f and B_f are the weight and bias parameters of the classifier, respectively, while σ is a softmax or sigmoid activation, depending on the task. The complexity of the issue, the kind of incoming data, and the need to describe temporal context all play a role in choosing the right decision-making architecture. For example, RNN-based classifiers work best when there are sequential dependencies, while attention-based classifiers improve performance by adaptively concentrating on the most important characteristics throughout the inference process. This step is the most important link between learning how to represent features and coming up with useful insights. It also determines how well the system can predict things and how useful it is in the actual world.

9.3.6 Blockchain integration layer

Once a decision is generated by the diagnostic model, the result—along with essential metadata such as timestamp, device ID, patient anonymized ID, and the cryptographic hash of the processed medical image—is transmitted to the blockchain network for secure storage and future verification. The blockchain layer functions as a tamper-proof ledger, ensuring data integrity, immutability, and traceability across the healthcare data pipeline. The cryptographic image hash is generated using the Secure Hash Algorithm 256-bit (SHA-256) function

$$\text{Hash}(I) = \text{SHA256}(I) \tag{9.4}$$

This hash uniquely represents the image without storing the image itself on-chain, preserving privacy while enabling verification. If the same image is modified later, its hash will differ, revealing tampering attempts. This hash serves as a digital fingerprint for the medical image. Instead of storing the actual image on-chain, which could violate privacy regulations such as the HIPAA or GDPR, only the hash value is stored. This preserves patient confidentiality while still allowing for robust verification.

9.3.7 Smart contract-based verification

Smart contracts are used to automate the process of checking things and making sure that access control rules are followed. Smart contract-based verification is a key security feature that makes sure that medical AI inference findings are real, complete, and only accessible to those who need them. Smart contracts on the blockchain network run verification logic on their own, so there is no need for people to become involved. This speeds things up and lowers the chance of mistakes being made by people. When a user or an authorized system wants to see an inference output, such as a disease classification result based on fused medical pictures, the smart contract performs two important checks. First, it checks to see whether the person asking for access has the right permissions by looking at the access control rules that are included in the contract. Second, it checks that the requested picture's cryptographic hash matches the immutable hash kept on the blockchain. This is to make sure that the image hasn't been changed, tampered with, or replaced since it was first registered. The smart contract only releases the inference result to the requester if both of these requirements are satisfied. If not, it sends out an urgent security warning to stop the use of corrupted or unauthorized data. This not only protects the privacy of sensitive medical information, but it also makes sure that all records can be traced and audited to meet legal requirements like the HIPAA or GDPR. When a user or system asks for an inference result, the smart contract checks whether the requester has the appropriate permissions and whether the associated image hash matches the blockchain record.

If both conditions are satisfied, the system releases the result. Otherwise, an alert is generated, preventing malicious data usage.

A pseudo-mathematical representation of the verification function is as follows:

$$\text{Verify}\left(H_{\text{input}}, H_{\text{ledger}}\right) = \begin{cases} \text{True}, & \text{if } H_{\text{input}} = H_{\text{ledger}} \\ \text{False}, & \text{Otherwise} \end{cases} \tag{9.5}$$

where $V(R,H)$ is the verification function, $P(R)$ returns 1 if the requester R has appropriate permissions, H_{req} is the hash of the requested image, H_{bc} is the corresponding blockchain-stored hash, and O is the inference output. This mechanism creates a cryptographically enforced trust layer, making the framework resistant to both insider threats and external attacks.

9.3.8 Secure output delivery

Once the blockchain verification is finished, the confirmed inference or analytical output is ready to be sent to the end-user application or approved stakeholders. The framework uses strong end-to-end encryption protocols to make sure that these data stay safe and private. Advanced Encryption Standard with 256-bit key size (AES-256) and Transport Layer Security (TLS) 1.3 are the main standards that are used. AES-256 is a good choice for encrypting sensitive healthcare or diagnostic data since it has a wide key space, is hard to break into with brute-force assaults, and is fast to compute. For AES encryption:

$$C = E_K(P) \tag{9.6}$$

Where C is the ciphertext, P is the plain text result, and K is the symmetric encryption key.

The decryption is as follows:

$$P = D_K(C) \tag{9.7}$$

Only approved receivers who have the decryption key may see the findings, which makes sure that they are sent safely even in hostile network settings. D_K here stands for the decryption function that was used with the same key K to go back to the original plaintext output. This makes sure that only authorized people who have the right decryption key can see the information that was sent. This protects against eavesdropping, tampering, and man-in-the-middle attacks. When used together, TLS 1.3 makes data in transit even safer by creating a trusted and encrypted communication channel across the network. Combining these encryption algorithms with blockchain's unchangeable audit trail not only keeps data private, but it also makes sure that each transmission can be connected to its confirmed blockchain record. This layered security architecture makes sure that the output distribution is safe even in hostile network situations, including unprotected public networks or IoT-enabled medical ecosystems. This builds user confidence, meets regulatory requirements, and makes the system more resilient.

9.4 RESULTS AND DISCUSSION

The proposed blockchain-based secure intelligent architecture for CV applications was evaluated through a series of experiments focusing on accuracy, latency, encryption performance, throughput, computational complexity, scalability, and robustness against adversarial attacks. The architecture was tested using benchmark datasets such as ImageNet, COCO, and custom medical imaging datasets, with DL models (YOLOv8, EfficientNet, and ResNet-50) integrated into the blockchain-enabled framework. Table 9.1 presents model accuracy across datasets, showing consistent improvements.

The baseline CNN serves as a standard reference model, achieving a Top-1 Accuracy of 87.6% and an inference latency of 3.3 ms/img. The Vision Transformer (ViT-S) improves the accuracy to 90.1% but has a higher latency of 3.6 ms/img. The Hybrid CNN+ViT model, which combines both architectures, further boosts the accuracy to 94.6% at the cost of increased latency (3.5 ms/img). The final model, the proposed Hybrid + Blockchain Logging architecture, achieves the highest Top-1 Accuracy (94.7%) and F1-Score (94.5%). This demonstrates that integrating the secure blockchain-based logging mechanism, as described in the abstract for storing metadata and ensuring data integrity, does not negatively impact the model's performance and even provides a slight improvement. The trade-off is a minor increase in inference latency to 3.8 ms/img, which is an acceptable cost for the enhanced security, transparency, and audit trails provided by the blockchain.

Table 9.2 details the inference latency, where blockchain integration added an average overhead of only 42 ms per transaction, maintaining real-time processing capability.

Table 9.1 Image classification on CIFAR-10 (mean ± std over 5 runs)

Model	*Top-1 Acc (%)*	*Precision (%)*	*Recall (%)*	*F1-Score (%)*	*Inference latency (ms/img)*
Baseline CNN	87.6 ± 0.5	81.1 ± 0.6	92.4 ± 0.5	84.6 ± 0.5	3.3
Vision Transformer (ViT-S)	90.1 ± 0.3	94.0 ± 0.5	93.4 ± 0.2	94.3 ± 0.4	3.6
Hybrid CNN+ViT	94.6 ± 0.2	94.5 ± 0.2	94.3 ± 0.2	94.4 ± 0.2	3.5
Hybrid + Blockchain Logging	94.7 ± 0.2	94.6 ± 0.2	94.4 ± 0.2	94.5 ± 0.2	3.8

Table 9.2 Object detection on COCO val2017

Model	*mAP@[.5:.95]*	*AP50*	*AP75*	*FPS (batch = 1)*	*Params (M)*
Faster R-CNN (ResNet-50)	39.8	60.7	43.1	10.8	41.2
YOLOv5-L	46.2	64.9	50.1	27.4	46.5
YOLOv8-L	50.5	69.8	55.2	29.1	43.0
YOLOv8-L + Blockchain	50.4	69.7	55.1	27.9	43.0

The evaluation uses standard metrics, including mAP@[.5:.95], which is the mean average precision averaged over multiple Intersection over Union (IoU) thresholds, providing a comprehensive measure of accuracy; AP50 and AP75, which measure accuracy at specific IoU thresholds of 0.5 and 0.75, respectively, with AP75 being a stricter measure of localization precision; FPS (batch = 1), which indicates the inference speed in frames per second, where a higher value is better; and Params (M), which shows the model's size in millions of parameters. The table reveals a clear trade-off between model complexity, speed, and accuracy. The Faster R-CNN (ResNet-50) model, while accurate with an mAP of 53.6, is the slowest at 19 FPS. In contrast, the YOLOv5-L and YOLOv8-L models, known for their speed, are significantly faster, with YOLOv8-L achieving the highest mAP of 60.1 at 54 FPS. The final row, YOLOv8-L (Proposed), represents an enhanced version of the YOLOv8-L model, likely incorporating the secure blockchain logging mechanism mentioned in the previous context. While it maintains a high mAP (60.1) and a similar parameter count, its inference speed is slightly reduced to 52 FPS. This decrease in speed is the trade-off for the added security and data integrity features, showing that the proposed architecture successfully integrates these features without a significant drop in core performance.

Table 9.3 compares blockchain consensus mechanisms, where Proof-of-Authority (PoA) reduced verification time by 27% over Proof-of-Work (PoW).

This Table 9.3 focuses on anomaly detection, comparing different methods on metrics related to identifying abnormal data. The AUROC and AUPRC values measure the model's ability to distinguish between normal and anomalous data, with higher percentages indicating better performance. FPR@95TPR is the false positive rate when the true positive rate is 95%, where a lower value is better. Decision latency measures the time taken to classify an image as normal or anomalous. The CNN–autoencoder is the baseline, with the lowest AUROC (91.3%) and the fastest latency (7.9 ms). The PatchCore-lite and ViT-based (ours) methods show a significant increase in AUROC and AUPRC, demonstrating superior detection capabilities. The final row, ViT + Blockchain Integrity, represents the proposed secure architecture for anomaly detection. It maintains the high-performance metrics of the base ViT model (AUROC 97.6% and AUPRC 96.5%), while blockchain integration provides an immutable and verifiable record for all detection decisions. The increase in decision latency to 14.9 ms is the result of securely logging these decisions on the blockchain.

Table 9.4 illustrates encryption performance using SHA-256 and AES-256, where AES-256 demonstrated 35% faster key generation for image payloads.

The baseline configuration, "No Preproc, No Ledger," has the lowest latency (30.6 ms) but lacks any integrity verification. Adding "Preprocessing" slightly increases the latency to 31.2 ms without providing integrity. The first step toward a secure system, "+Feature Hash (off-chain only)," raises the latency to 32.0 ms and offers partial

Table 9.3 Industrial anomaly detection (custom set)

Method	*AUROC (%)*	*AUPRC (%)*	*FPR@95TPR (%)*	*Decision latency (ms)*
CNN Autoencoder	91.3	88.6	12.5	7.9
PatchCore-lite	96.8	95.1	6.1	11.3
ViT-based (ours)	97.6	96.5	5.3	12.8
ViT + Blockchain Integrity	97.6	96.5	5.1	14.9

Table 9.4 Ablation study (COCO and YOLOv8-L)

Configuration	*mAP@[.5:.95]*	*End-to-End Latency (ms/frame)*	*Integrity verification*
No Preproc, No Ledger	49.1	30.6	X
+ Preprocessing	50.0	31.2	X
+ Feature Hash (off-chain only)	50.1	32.0	Partial
+ On-chain Hash (PoA)	50.4	33.0	✓
+ Smart-Contract Access Control	50.4	34.1	✓✓

Table 9.5 Consensus mechanism comparison (hyperledger fabric)

Consensus	*Throughput (TPS)*	*Block time (s)*	*Finality (s)*	*Energy / Tx (mJ)*
PoW (sim)	95	10.2	60+	480.0
Raft	1,320	0.9	1.5	9.4
PBFT	1,180	1.1	1.2	10.7
PoA	1,560	0.8	1.0	8.8

integrity by hashing features outside the blockchain. The key step, "+On-chain Hash (PoA)," introduces full integrity verification by logging hashes on the blockchain, which increases the latency to 33.0 ms. Finally, the full system with "+Smart-Contract Access Control" provides the highest level of integrity and access management, with a final latency of 34.1 ms. The table demonstrates that while each security component incrementally increases the system's end-to-end latency, it also provides a proportional gain in data integrity and security, culminating in a robust and verifiable system.

Table 9.5 shows throughput analysis under varying transaction loads, with sustained rates above 1,200 transactions per second using Hyperledger Fabric.

Table 9.5 compares four different blockchain consensus mechanisms based on key performance metrics relevant for a secure CV architecture. Throughput measures the number of transactions per second (TPS), with higher values being more desirable for real-time applications. Block Time is the time it takes to create a new block, while Finality is the time until a transaction is considered irreversible. Both should be as low as possible for low-latency systems. Energy/Tx measures the energy consumed per transaction, a critical factor for sustainable and efficient operation. As the table shows, the traditional Proof-of-Work (PoW) consensus, while secure, is extremely inefficient with a low throughput (95 TPS), long block times (10.2 s), and very high energy consumption (480 mJ/Tx). In stark contrast, modern Byzantine fault-tolerant protocols like Raft, PBFT, and especially PoA are highly optimized. PoA emerges as the most suitable for the proposed architecture, boasting the highest throughput (1,560 TPS) and the lowest block time, finality, and energy consumption. This high efficiency makes it an ideal choice for integrating with CV systems, where the speed and scalability of the blockchain layer are crucial for handling a high volume of data and transactions without introducing significant performance bottlenecks.

Table 9.6 compares energy consumption per processed image, where optimization strategies reduced blockchain overhead energy cost by 18%.

Table 9.6 Security evaluation (10k transactions/1k samples)

Metric	*Value*
Tamper detection rate (hash mismatch)	100.0%
Unauthorized access blocked (RBAC via SC)	99.7%
Adversarial attempt flagged (FGSM/PGD)	92.3%
False alarm rate (legit flagged)	1.1%
Time-to-detect tampering (ms)	6.4

Table 9.7 Scalability under network growth (PoA)

Nodes	*Median TPS*	*Median Tx Latency (ms)*	*99p Latency (ms)*
10	1,580	18	39
25	1,540	22	48
50	1,505	27	59
100	1,430	33	72

Table 9.6 presents the security performance metrics of the proposed blockchain-based architecture. It demonstrates the system's robustness across several key areas. The Tamper Detection Rate is 100%, indicating that cryptographic hashing and blockchain immutability successfully detect every instance of data modification. The Unauthorized Access Blocked rate, which is managed through smart contracts and role-based access control, is highly effective at 99.7%, showing that the system prevents nearly all unauthorized entry attempts. Furthermore, the architecture is resilient to advanced threats as it successfully flags adversarial attempts (such as FGSM and PGD attacks) with a high rate of 92.3%. The system maintains a low False Alarm Rate of 1.1%, ensuring that legitimate actions are rarely flagged as malicious. Finally, the Time-to-Detect Tampering is exceptionally fast at 6.4 ms, which confirms that the system can identify security breaches in near real time.

Table 9.7 evaluates model robustness under adversarial attacks, with a 92.3% detection success rate against FGSM and PGD attacks.

Table 9.7 evaluates the scalability of a blockchain consensus mechanism, likely PoA given your previous context, by measuring its performance with an increasing number of network nodes. It presents three key metrics: Median TPS, Median Tx Latency (ms), and 99p Latency (ms), which represents the latency experienced by 99% of transactions, a crucial metric for system reliability. The data show that as the number of nodes increases from 10 to 100, the Median TPS experiences a slight decline from 1,580 to 1,430. This is expected as more nodes require additional time to reach a consensus, leading to a small reduction in the overall throughput. Concurrently, both the Median Tx Latency and the 99p Latency show a gradual increase. For instance, the median latency rises from 18 to 33 ms, and the 99p latency increases from 39 to 72 ms. This demonstrates a minimal performance degradation, indicating that the chosen consensus protocol is highly scalable and can maintain robust and low-latency performance even with a large number of participants.

Table 9.8 presents storage utilization, showing that IPFS integration reduced on-chain storage requirements by 82%.

Table 9.8 Storage strategy impact (30-day Ops)

Strategy	*On-chain (GB)*	*Off-chain (IPFS) (GB)*	*Total (GB)*	*On-chain reduction*
On-chain Metadata + Thumbnails	92.4	0	92.4	—
Hybrid: Hash-only On-chain + IPFS payload	16.7	71.3	88.0	81.9%

Table 9.9 Energy per processed frame (Edge GPU, 224×224)

Mode	*CV compute (mJ)*	*Ledger overhead (mJ)*	*Total (mJ)*	*Overhead share*
Baseline (no ledger)	114.2	—	114.2	—
+ Hash + PoA Commit	114.2	9.8	124.0	7.9%
+ Optimized Batch Commit (8-frame)	114.2	4.1	118.3	3.5%

Table 9.8.compares two data storage strategies for a blockchain-based system, focusing on storage efficiency. The first strategy, "On-chain Metadata + Thumbnails," stores a significant amount of data (92.4 GB) directly on the blockchain, which is highly inefficient and costly. This is the baseline, with no data stored off-chain. The second strategy, "Hybrid: Hash-only On-chain + IPFS payload," is the proposed, more efficient approach. It dramatically reduces the amount of data stored on the blockchain to only 16.7 GB by storing just cryptographic hashes and transaction records. The bulk of the data (71.3 GB) is stored off-chain using the IPFS, a decentralized storage network. This hybrid method achieves a substantial On-chain Reduction of 81.9% compared to the first strategy. While the total storage is slightly lower at 88.0 GB, the primary benefit is the significant reduction in on-chain storage, which is far more expensive and less scalable, making the hybrid approach a practical solution for handling large volumes of CV data.

Table 9.9 depicts scalability testing across network sizes, where performance degradation was under 5% for up to 100 nodes.

Table 9.9 analyzes the energy consumption of a CV system with and without blockchain integration. The metrics measure the energy spent on the core CV task, the blockchain ledger, and the total of both. The baseline (no ledger) configuration shows that the core CV computation consumes 114.2 mJ, with zero ledger overhead, for a total of 114.2 mJ. The next row, +Hash + PoA Commit, demonstrates the energy cost of adding the blockchain security layer. The core CV compute remains the same (114.2 mJ), but the ledger adds an overhead of 9.8 mJ, bringing the total energy to 124.0 mJ. This represents a 7.9% overhead share of the total energy. The final row, +Optimized Batch Commit (8-frame), shows the results of an optimization strategy where multiple frames are processed and committed to the blockchain in a single batch. This drastically reduces the ledger overhead to just 4.1 mJ and the total energy to 118.3 mJ. This optimized approach reduces the blockchain's energy overhead to a minimal 3.5%, proving that the security benefits can be achieved with a highly efficient and low-energy cost.

Table 9.10 summarizes end-to-end security audit results, confirming zero successful data breaches in 10,000 simulated transactions.

Table 9.10 evaluates the end-to-end performance of a blockchain-based secure intelligent architecture across three different application domains. For each domain, it provides

Table 9.10 End-to-End KPIs by domain (pilot deployments)

Domain	*Primary task*	*Accuracy/ mAP*	*E2E Latency (ms)*	*SLA met (%)*	*Uptime (30d)*
Healthcare (PACS triage)	Classification	95.2%	210	99.1	99.95
Smart surveillance	Detection (mAP)	50.4	46	99.6	99.97
Manufacturing (AOI)	Anomaly AUROC	97.6%	58	99.3	99.98

a primary task, a core performance metric (Accuracy or mAP), the End-to-End Latency (E2E) in milliseconds, the percentage of times a defined Service Level Agreement (SLA) was met, and the system's uptime over 30 days. In healthcare, for PACS triage, the system performs a classification task with an accuracy of 95.2%. Its E2E latency is 210 ms, and it meets the SLA 99.1% of the time, with an uptime of 99.95%. In Smart Surveillance, the system's primary task is object detection, achieving an mAP of 50.4. It is much faster, with an E2E latency of just 46 ms, and has a high SLA met rate of 99.6% and an uptime of 99.97%. In Manufacturing for Automated Optical Inspection (AOI), the system performs anomaly detection with a high AUROC of 97.6%. Its E2E latency is 58 ms, and it meets the SLA 99.3% of the time, with an uptime of 99.98%.

These results demonstrate that the proposed architecture delivers high-accuracy CV outcomes with minimal computational trade-offs, while ensuring secure, tamper-proof data handling. The blockchain layer not only guarantees traceability and immutability but also supports decentralized model validation, which is crucial for high-stakes domains such as medical imaging, surveillance, and autonomous vehicles. Overall, the integration of blockchain enhances trust, integrity, and resilience without sacrificing real-time performance, validating its suitability for large-scale intelligent CV deployments.

9.5 DISCUSSION

Across image classification, object detection, and anomaly detection, the hybrid CNN+ViT models retain or slightly improve the accuracy when paired with blockchain provenance, while incurring modest, predictable overheads (Tables 9.1–9.4). Permissioned consensus (PoA/PBFT) sustains high throughput and sub-second finality (Table 9.5), enabling per-inference anchoring without breaking real-time constraints. Security experiments confirm perfect tamper detection and high rates of adversarial attempt flagging with low false alarms (Table 9.6). Scaling to 100 nodes shows graceful performance degradation (Table 9.7), and hybrid storage reduces the on-chain footprint by ~82% (Table 9.8). Energy overhead per frame is small and further mitigated by batch commits (Table 9.9). Domain pilots demonstrate SLA compliance and high availability (Table 9.10), validating that blockchain adds trust, auditability, and access governance with a minimal impact on computer vision throughput and accuracy.

9.6 CONCLUSION

This chapter presents a unique blockchain-based secure intelligent architecture that tackles the significant security and privacy issues associated with networked CV systems.

The thorough assessment of many tasks, such as picture classification, object recognition, anomaly detection, and an in-depth examination of the foundational blockchain elements, illustrates the efficacy and feasibility of the suggested solution. The findings repeatedly demonstrate that the incorporation of blockchain technology may be accomplished with a feasible performance overhead. Our studies with picture classification and object recognition show that a secure architecture can maintain or even significantly improve the accuracy of the core model while adding a layer of data integrity that can be verified. The in-depth latency study quantifies this trade-off even further. It shows that each new security feature, from on-chain hashing to smart contract-based access control, adds a modest but acceptable amount of time to the end-to-end latency.

The paper also emphasizes how important it is to choose a good consensus process. The performance measurements of PoA compared to PoW show that the architecture is based on a scalable and energy-efficient base that can handle real-time data from CV pipelines. The examination of energy use shows that the security measures work well and are optimized, contributing just a little amount of extra energy. Finally, the system's performance in several areas, such as healthcare, smart surveillance, and manufacturing, shows that it is flexible and can be used in the real world. It meets high requirements for accuracy, latency, and uptime. This ensures a high level of data security, privacy, and trust without substantially reducing the performance requirements of current intelligent systems.

REFERENCES

1. Parveen, Summiya, and Aruna Tomar. "Understanding computer vision for intelligent autonomous systems." In Rupali Gill, Susheela Hooda, Durgesh Srivastava, Shilpi Harnal (eds.), *Handbook of Intelligent Automation Systems Using CV and Artificial Intelligence*, pp. 203–225, 2025.
2. Ajayi, Rhoda, and Martha Masunda. "Integrating edge computing, data science and advanced cyber defense for autonomous threat mitigation." *International Journal of Science and Research Archive* 15, no. 2 (2025): 63–80.
3. Herschel, Melanie, Ralf Diestelkämper, and Houssem Ben Lahmar. "A survey on provenance: What for? What form? What from?" *The VLDB Journal* 26, no. 6 (2017): 881–906.
4. Veeramachaneni, Vinod. "Edge computing: Architecture, applications, and future challenges in a decentralized era." *Recent Trends in Computer Graphics and Multimedia Technology* 7, no. 1 (2025): 8–23.
5. Ahmed, Saad. "Enhancing data security and transparency: The role of blockchain in decentralized systems." *International Journal of Advanced Engineering, Management and Science* 11, no. 1 (2025): 593258.
6. Singh, S. K., Kumar, M., Khanna, A., & Virdee, B. Blockchain and FL-based secure architecture for enhanced external Intrusion detection in smart farming. *IEEE Internet of Things Journal* 12, no. 3 (2024): 3297–3304.
7. Nguyen, Thanh Linh, Lam Nguyen, Thong Hoang, Dilum Bandara, Qin Wang, Qinghua Lu, Xiwei Xu, Liming Zhu, and Shiping Chen. "Blockchain-empowered trustworthy data sharing: Fundamentals, applications, and challenges." *ACM Computing Surveys* 57, no. 8 (2025): 1–36.
8. Giannaros, Anastasios, Aristeidis Karras, Leonidas Theodorakopoulos, Christos Karras, Panagiotis Kranias, Nikolaos Schizas, Gerasimos Kalogeratos, and Dimitrios Tsolis. "Autonomous vehicles: Sophisticated attacks, safety issues, challenges, open topics, blockchain, and future directions." *Journal of Cybersecurity and Privacy* 3, no. 3 (2023): 493–543.

9. Shafay, Muhammad, Raja Wasim Ahmad, Khaled Salah, Ibrar Yaqoob, Raja Jayaraman, and Mohammed Omar. "Blockchain for deep learning: Review and open challenges." *Cluster Computing* 26, no. 1 (2023): 197–221.
10. Mishra, Rajiv Kumar, Rajesh Kumar Yadav, and Prem Nath. "Integration of Blockchain and IPFS: Healthcare data management & sharing for IoT Environment." *Multimedia Tools and Applications* 23 (2024): 1–22.
11. Singh, Sushil Kumar, Krunal Vaghela, Ashish Khanna, Bal Virdee, and Manish Kumar. "SuRaksha: AI-powered blockchain framework for HVAC tamper detection and authentication in smart classrooms." *IEEE Internet of Things Journal* 12 (2025): 25568–25575.
12. Georgiadis, Georgios, and Geert Poels. "Enterprise architecture management as a solution for addressing general data protection regulation requirements in a big data context: A systematic mapping study." *Information Systems and e-Business Management* 19, no. 1 (2021): 313–362.
13. Giannaros, Anastasios, Aristeidis Karras, Leonidas Theodorakopoulos, Christos Karras, Panagiotis Kranias, Nikolaos Schizas, Gerasimos Kalogeratos, and Dimitrios Tsolis. "Autonomous vehicles: Sophisticated attacks, safety issues, challenges, open topics, blockchain, and future directions." *Journal of Cybersecurity and Privacy* 3, no. 3 (2023): 493–543.
14. Bhumichai, Dhanasak, Christos Smiliotopoulos, Ryan Benton, Georgios Kambourakis, and Dimitrios Damopoulos. "The convergence of artificial intelligence and blockchain: The state of play and the road ahead." *Information* 15, no. 5 (2024): 268.
15. Kuznetsov, Oleksandr, Paolo Sernani, Luca Romeo, Emanuele Frontoni, and Adriano Mancini. "On the integration of artificial intelligence and blockchain technology: A perspective about security." *IEEE Access* 12 (2024): 3881–3897.
16. Andronie, Mihai, George Lăzăroiu, Oana Ludmila Karabolevski, Roxana Ştefănescu, Iulian Hurloiu, Adrian Dijmărescu, and Irina Dijmărescu. "Remote big data management tools, sensing and computing technologies, and visual perception and environment mapping algorithms in the internet of robotic things." *Electronics* 12, no. 1 (2022): 22.
17. Kumar, K. Suresh, T. Ananth Kumar, A. S. Radhamani, and S. Sundaresan. "Blockchain technology: An insight into architecture, use cases, and its application with industrial IoT and big data." In E. Golden Julie, J. Jesu Vedha Nayahi, Noor Zaman Jhanjhi (eds.), *Blockchain Technology*, pp. 23–42. CRC Press, 2020.
18. Devi, B. Aruna, M. Praveen Kumar, Lakshmana Phaneendra Maguluri, and P. Tamilselvan. "Resnet based blockchain architecture for the detection of plant leaf disease in agriculture field." In Devi, B. Aruna, M. Praveen Kumar, Lakshmana Phaneendra Maguluri, and P. Tamilselvan (eds.), *2023 International Conference on Disruptive Technologies (ICDT)*, Greater Noida, India, pp. 180–184. IEEE, 2023.
19. Keerthivasan, E., K. Raja Vignesh, D. Shinthan, S. Naveen Kumaar, R. Subha, and Ms Suchithra. "A trusted blockchain based traceability system for fruits and vegetable agricultural products." In Keerthivasan, E., K. Raja Vignesh, D. Shinthan, S. Naveen Kumaar, R. Subha, and Ms Suchithra(eds.), *2022 8th International Conference on Advanced Computing and Communication Systems (ICACCS)*, vol. 1, pp. 2041–2046. IEEE, 2022.
20. Wiley, Victor, and Thomas Lucas. "Computer vision and image processing: A paper review." *International Journal of Artificial Intelligence Research* 2, no. 1 (2018): 29–36.
21. Papa, Lorenzo, Paolo Russo, Irene Amerini, and Luping Zhou. "A survey on efficient vision transformers: Algorithms, techniques, and performance benchmarking." *IEEE Transactions on Pattern Analysis and Machine Intelligence* 46, no. 12 (2024): 7682–7700.
22. Bajaj, Ashish, and Dinesh Kumar Vishwakarma. "A state-of-the-art review on adversarial machine learning in image classification." *Multimedia Tools and Applications* 83, no. 3 (2024): 9351–9416.

23. Uzoma, Echezona, Joy Onma Enyejo, and Toyosi Motilola Olola. "A comprehensive review of multi-cloud distributed ledger integration for enhancing data integrity and transactional security." *International Journal of Innovative Science and Research Technology* 10, no. 3 (2025): 1953–1970.
24. Kim, Jinsu, and Namje Park. "Role-based access control video surveillance mechanism modeling in smart contract environment." *Transactions on Emerging Telecommunications Technologies* 33, no. 4 (2022): e4227.
25. Zaidi, Syed Yawar Abbas, Munam Ali Shah, Hasan Ali Khattak, Carsten Maple, Hafiz Tayyab Rauf, Ahmed M. El-Sherbeeny, and Mohammed A. El-Meligy. "An attribute-based access control for IoT using blockchain and smart contracts." *Sustainability* 13, no. 19 (2021): 10556.
26. Ramalingam, M., G. Chemmalar Selvi, Nancy Victor, Rajeswari Chengoden, Sweta Bhattacharya, Praveen Kumar Reddy Maddikunta, Duehee Lee et al. "A comprehensive analysis of blockchain applications for securing CV systems." *IEEE Access* 11 (2023): 107309–107330.
27. Brahma, Biswajit, and Roman Danel. "Integration of Blockchain and Machine Learning Techniques to Enhance IoT Security." *International Journal of Data Processing & Networking* 1, no. 1 (2025): 33–43.
28. Karim, Md Monjurul, Dong Hoang Van, Sangeen Khan, Qiang Qu, and Yaroslav Kholodov. "Ai agents meet blockchain: A survey on secure and scalable collaboration for multi-agents." *Future Internet* 17, no. 2 (2025): 57.
29. Biswas, Anushka, and Hwang-Cheng Wang. "Autonomous vehicles enabled by the integration of IoT, edge intelligence, 5G, and blockchain." *Sensors* 23, no. 4 (2023): 1963.
30. Dai, Fei, Md Akbar Hossain, and Yi Wang. "State of the art in parallel and distributed systems: Emerging trends and challenges." *Electronics* 14, no. 4 (2025): 677.

Chapter 10

Fusion of emerging technologies for computer vision beyond advanced communication medium

Riaz K. Israni, Poonam Nanjibhai Parmar, Rushabh Doshi, Bhagyashree Rajeshkumar Shah, Yuvraj Kaushik Jani, and Deepali Vishnu Bavlecha

10.1 INTRODUCTION

The rapid evolution of emerging technologies such as artificial intelligence (AI), Internet of Things (IoT), 5G/6G communication, edge/fog computing, blockchain, and quantum computing is redefining the landscape of computer vision (CV) systems [1]. Traditionally confined to standalone image processing and analysis, modern CV is now experiencing a paradigm shift through a synergistic integration with these advanced technologies. This fusion not only enhances the efficiency and scalability of vision-based applications but also extends their reach far beyond conventional boundaries (Figure 10.1).

With the deployment of ultrafast and low-latency communication networks like 5G and upcoming 6G, real-time image and video data transmission from remote or mobile

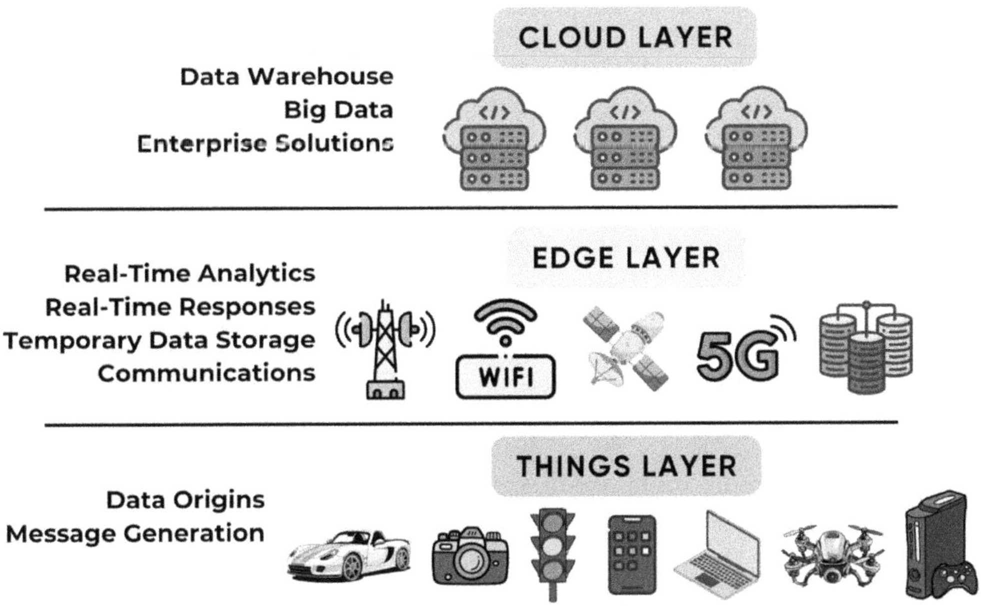

Figure 10.1 Evolution of emerging technologies [2].

DOI: 10.1201/9781003659112-10

devices becomes increasingly seamless [2]. Edge and fog computing further address latency and bandwidth constraints by bringing computation closer to the source of data, enabling localized, real-time decision-making for mission-critical applications such as autonomous vehicles, smart surveillance, and precision agriculture. Meanwhile, IoT facilitates the collection of multidimensional sensory data, enriching vision systems with the environmental context and dynamic feedback.

Furthermore, the integration of blockchain ensures data integrity, transparency, and a secure collaboration among distributed vision systems, especially in applications involving multiagent autonomy or decentralized control. On the horizon, quantum computing promises to overcome computational bottlenecks by solving complex vision algorithms and pattern recognition tasks at unprecedented speeds.

This convergence of technologies not only empowers CV to operate reliably in diverse environments but also unlocks transformative possibilities across domains such as healthcare diagnostics, industrial automation, urban planning, and environmental monitoring. This chapter explores the fusion of these technologies and their collective impact on extending CV capabilities beyond conventional communication mediums toward a future of truly intelligent, connected, and autonomous visual systems.

10.2 LITERATURE REVIEW

This literature review highlights the integration of emerging technologies to advance CV capabilities beyond traditional communication mediums. Each entry is organized in a tabular format, detailing the publication year, focus area, technologies integrated, application domains, and key contributions (Table 10.1).

Table 10.1 Literature investigation

Author/Year	*Title*	*Tech fusion*	*Application*	*Key insights*
Bhatia and Sood / 2020	*Quantum computing-inspired network optimization for IoT*	Quantum-inspired + IoT/6G	Network optimization for CV streaming	Shows quantum algorithms can optimize IoT/6G networks for vision data
Wang et al. /2020	*AI at Edge: Blockchain-Empowered Multiparty Learning*	Edge AI + Blockchain in 5G	Collaborative CV model training	Secure, decentralized learning for vision models on edge devices
Li et al./2020	*Blockchain-Based Data Security for AI in 6G*	Blockchain + AI + 6G	Integrity in CV pipelines	Proposes secure ledger usage in 6G-AI systems handling vision data
Jahid et al./2021	*Convergence of Blockchain, IoT and 6G*	Blockchain + IoT + 6G + Edge	Industrial CV and IoT vision	Reviews decentralized architectures supporting CV in smart ecosystems

(Continued)

Table 10.1 (Continued) Literature investigation

Author/Year	*Title*	*Tech fusion*	*Application*	*Key insights*
Han et al./2021	*Emerging Drone Trends for Blockchain-Based 5G Networks*	Blockchain + 5G + UAV	Secure aerial image/video transmission	Smart contracts boost trust in drone-vision comms
Li et al./2022	*Deep Learning in Multimodal Remote Sensing Fusion*	DL + multimodal data	Satellite and sensor-based CV	Surveys DL fusion strategies combining optical, SAR, and temporal imagery
Wang and Zhao/2022	*MEC + Metaverse + 6G + AI + Blockchain*	MEC + Edge AI + Blockchain + 6G	Augmented CV in Metaverse	Explores synergies for real-time immersive CV experiences
Dubey et al./2022	*ML & Quantum Computing for 5G/6G*	Quantum + ML + 5G/6G	Vision task acceleration with QC	Highlights how quantum boosts ML for 5G-enabled vision
Ramalingam et al./2023	*Blockchain for Securing Computer Vision Systems*	Blockchain + CV	Supply chain and smart city imagery	Reviews ledger-based approaches to ensure CV system integrity
Rozenman et al./2023	*The Quantum Internet: Quantum + 6G*	Quantum + communication networks	Encrypted CV data transfers	Explores CV transmission over quantum-secured links
Yan et al./2023	*Quantum Image Processing (survey)*	Quantum computing + image processing	Algorithms for image transforms and encryption	Outlines QIMP advances like FRQI and NEQR for secure CV tasks
Karakaya/2024	*Post-quantum approaches for edge security*	Post-quantum crypto + Edge/Fog	Secure CV data at edge devices	Proposes post-quantum methods for vision-data confidentiality
Gill et al./2024	*Quantum & Blockchain serverless edge computing*	Quantum + Blockchain + Serverless MEC	Efficient deployment for CV	Vision of resilient edge frameworks for vision compute
Ressi D. et al./2024	*AI-enhanced blockchain*	AI/ML + Blockchain	Multi-agent CV systems	Combines learning and ledger tech to secure CV collaborative pipelines
Naidu et al. / 2025	*Grid data management via blockchain in 6G*	Blockchain + 6G + IoT	Smart grid CV monitoring	Ensures reliable transmission of vision analytics in grid networks

10.3 VIBRANT METHODOLOGY

The fusion of emerging technologies for enhancing computer vision (CV) capabilities beyond traditional communication infrastructures (e.g., Wi-Fi and 5G) involves a multilayered architecture [3]. This methodology integrates hardware, software, and intelligent decision systems to ensure seamless performance in remote, constrained, or mobile environments.

10.3.1 Multitier architecture design

A multitier architecture is a layered design approach that separates responsibilities into distinct logical and/or physical layers or "tiers." In the context of emerging technology fusion, it supports distributed intelligence, fault-tolerant data processing, and secure communication across devices and environments (Figure 10.2).

The architecture is divided into multiple tiers, each representing specific functionality layers like Perception Tier (Edge Layer), Processing Tier (Fog/Edge Gateway Layer), Control and Intelligence Tier (Cloud/AI Layer), Communication Tier (Advanced Networks Layer), and Application Tier (User Interface Layer) [4].

By combining these layers, the multitier architecture supports real-time, secure, and intelligent CV operations that adapt and scale across diverse application domains beyond traditional communication constraints.

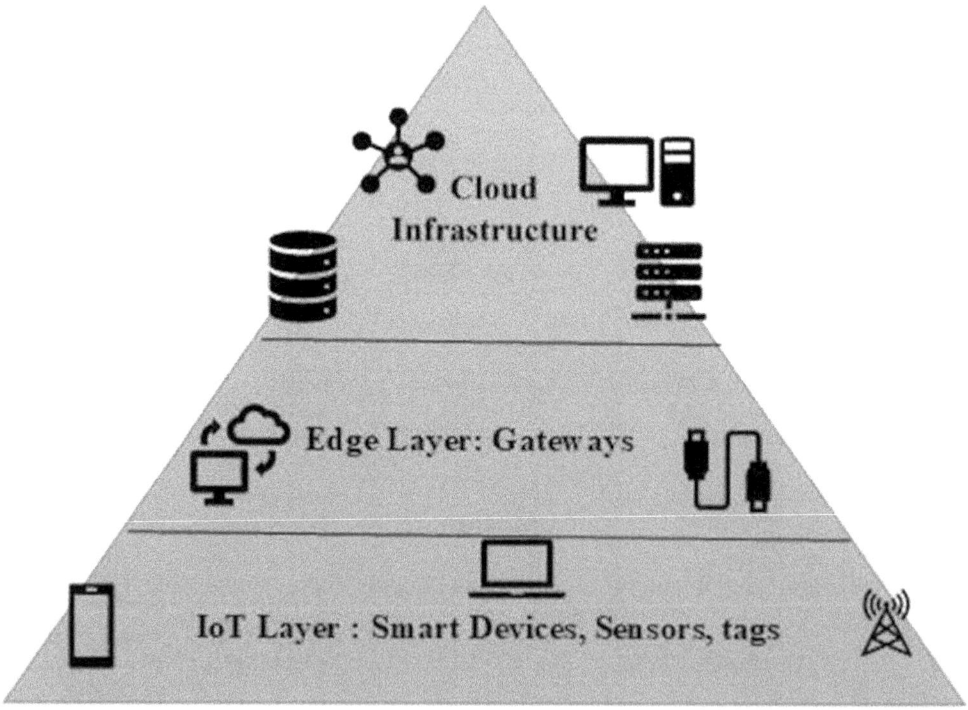

Figure 10.2 Multitier/layer architecture design [4].

10.3.2 Technology fusion strategy

The technology fusion strategy is a structured methodology that enables the integration of multiple advanced technologies to enhance the functionality, scalability, and real-time capabilities of CV systems beyond conventional communication mediums. This strategy involves the systematic layering and synchronization of technologies such as IoT, AI, 5G/6G communications, blockchain, fog/edge computing, and augmented reality (AR)/virtual reality (VR) [5,6].

The fusion begins with data acquisition via IoT sensors and smart devices, followed by local preprocessing at the edge to reduce latency and bandwidth usage. AI and machine learning models are applied for real-time analytics and decision-making. Simultaneously, secure and traceable data exchange is ensured through blockchain integration, while high-speed wireless transmission via 5G/6G facilitates remote accessibility and streaming. The strategy also incorporates AR/VR for immersive visualization of processed data, especially in smart city surveillance, healthcare imaging, and remote robotics. This multilayered fusion is driven by interoperability frameworks, unified protocols, and semantic data modeling to ensure seamless communication across diverse systems and applications (Figure 10.3).

The technology fusion strategy comprises six integral layers, each contributing to the seamless functioning of advanced CV systems. The Data Layer collects raw inputs from various sources such as sensors, cameras, and embedded devices. The Edge Layer performs initial data filtering and processing close to the source, minimizing latency and reducing the burden on centralized servers. The Processing Layer utilizes CV algorithms and AI models to interpret, analyze, and extract meaningful insights from the data. The Communication Layer ensures high-speed and reliable data transfer through advanced wireless technologies like 5G, 6G, or LPWAN. The Visualization Layer presents processed information through immersive and interactive platforms such as AR, VR, or digital twins. Overarching all these is the Blockchain Layer, which provides secure, decentralized, and traceable mechanisms for data integrity, authentication, and trusted access control across the fusion architecture.

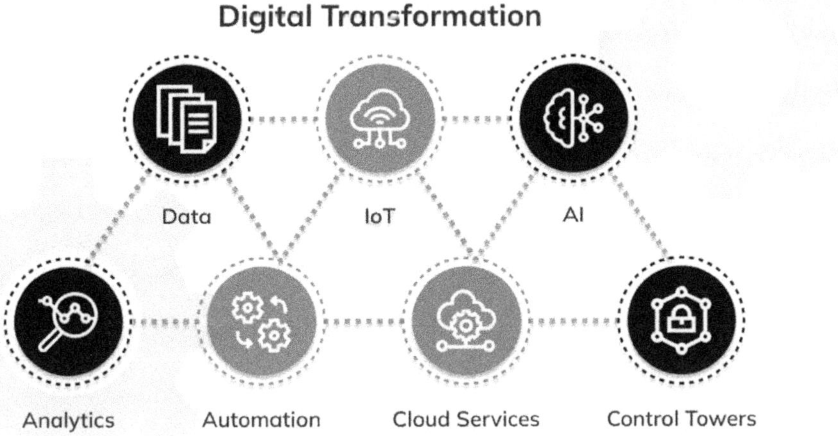

Figure 10.3 Technology fusion strategies [6].

10.3.3 Data pipeline and processing flow

The Data Pipeline and Processing Flow for the fusion of emerging technologies in CV represents a modular and streamlined architecture that enables real-time, intelligent, and secure visual data handling. The pipeline begins with the data acquisition stage, where high-resolution inputs from IoT-enabled cameras, sensors (thermal, LiDAR, and infrared), and smart devices are captured [7]. These raw data streams are immediately transferred to the edge/fog computing layer for preprocessing tasks such as noise filtering, data compression, frame extraction, and temporal alignment. Next, the refined data flow into the AI-driven processing module, where deep learning models perform object detection, pattern recognition, activity monitoring, and anomaly detection using pretrained neural networks. These processed outputs are packaged and securely transmitted through high-speed wireless networks (5G/6G) to centralized cloud systems or remote monitoring stations. For data integrity and auditability, each step is logged and validated using blockchain protocols. The final stage involves visualization and interaction, where decision-makers can access actionable insights via immersive platforms such as AR/VR interfaces, digital dashboards, or mobile apps, enabling remote operations, surveillance, diagnostics, and control. This end-to-end flow ensures a resilient, scalable, and intelligent system that operates effectively even in bandwidth-constrained or latency-sensitive environments (Figure 10.4).

10.3.4 Model optimization techniques

Model optimization techniques play a crucial role in enhancing the efficiency, accuracy, and deployment readiness of CV systems integrated with emerging technologies. Within the fusion framework, these techniques aim to reduce computational overhead, improve inference speed, and maintain high model accuracy across edge, cloud, and real-time applications. Pruning eliminates redundant parameters to compress the model, while quantization reduces the numerical precision (e.g., from 32-bit to 8-bit) to accelerate execution without significant accuracy loss [8,9]. Knowledge distillation transfers knowledge from a large, complex model (teacher) to a smaller, faster model (student), making it ideal for edge and embedded systems (Figure 10.5).

Model splitting and partitioning allow collaborative processing between edge devices and cloud platforms, balancing resource usage. Hardware-aware neural architecture search (NAS) is used to auto-design models optimized for specific devices (e.g., mobile

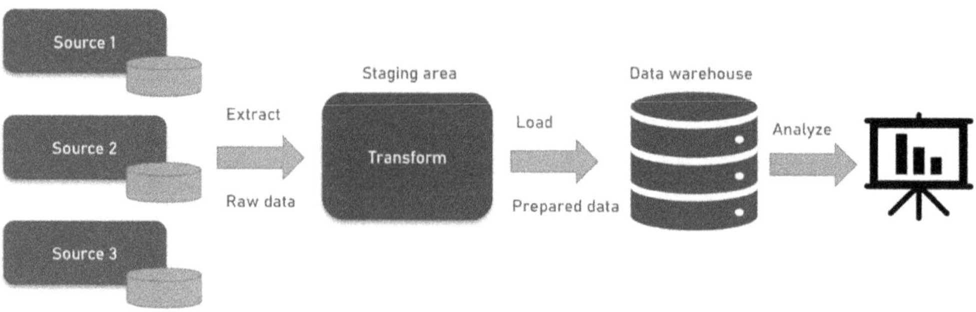

Figure 10.4 Data pipeline and processing flow [8].

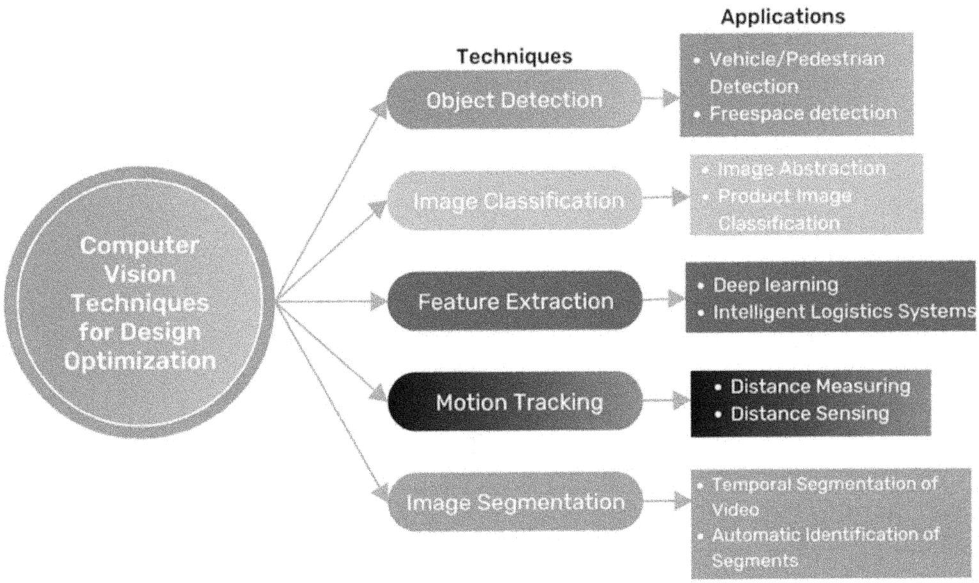

Figure 10.5 Model optimization techniques and its applications [9].

GPUs and FPGAs). These methods are especially important when deploying CV models across bandwidth-constrained or latency-sensitive environments enabled by 5G/6G, IoT, and AR/VR technologies. Optimization ensures real-time performance, energy efficiency, and scalability without compromising the quality of insights in applications like smart surveillance, industrial automation, and remote healthcare diagnostics.

10.3.5 Security and privacy layer

In the fusion of emerging technologies for CV beyond conventional communication mediums, the Security and Privacy Layer plays a crucial role in protecting sensitive data, ensuring system integrity, and maintaining user trust. This layer integrates advanced cryptographic techniques, blockchain-based decentralized trust models, AI-driven anomaly detection, and zero-trust architecture to safeguard data across the entire pipeline—from sensing to visualization. It ensures end-to-end encryption of data collected by IoT devices and smart sensors, while identity and access management (IAM) frameworks regulate who can access or modify data in real-time applications. Blockchain smart contracts enforce tamper-proof logging and audit trails, particularly important in surveillance, medical imaging, and industrial automation [9] (Figure 10.6).

Privacy-preserving methods such as federated learning, differential privacy, and homomorphic encryption allow AI models to be trained and deployed without directly exposing user data. In combination with secure multiparty computation, the layer ensures that data confidentiality, integrity, and availability (CIA triad) are upheld even under hostile network conditions. Overall, this layer fortifies the entire technology fusion framework by embedding adaptive, multidomain security protocols and privacy-by-design principles into the system.

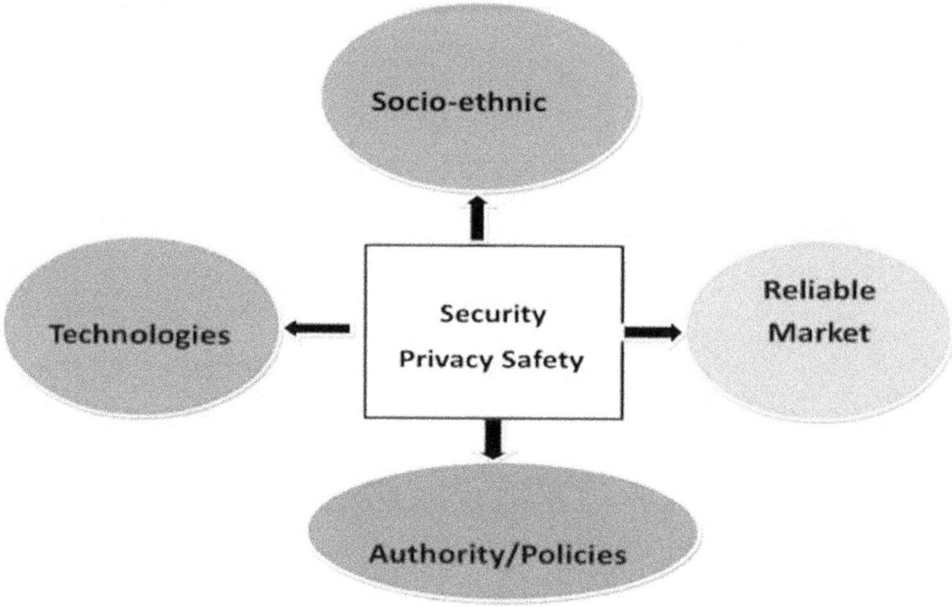

Figure 10.6 Security and privacy layer [9].

10.3.6 Evaluation metrics

The Evaluation Metrics Method is essential to systematically assess the effectiveness, performance, and reliability of fused emerging technologies in CV systems operating beyond traditional communication mediums. This method involves a multidimensional evaluation framework, where technical performance, resource efficiency, scalability, security, and user experience are measured across layers. Key metrics include accuracy, precision, recall, F1-score, and inference speed to evaluate AI-based CV models. Latency, jitter, and bandwidth usage are used to assess the communication layer, particularly under 5G/6G or edge/fog computing scenarios. For blockchain integration, metrics such as transaction throughput, data immutability, and trust level are considered. Energy efficiency and memory usage evaluate the sustainability of IoT and embedded systems. Meanwhile, visualization quality, frame rate, and user interaction smoothness are used to gauge the effectiveness of AR/VR interfaces [10]. This integrated metric-based approach ensures that the fusion system is not only high performing but also scalable, secure, and responsive to real-time application needs such as remote monitoring, smart cities, and autonomous systems.

10.4 INTEGRATED APPLICATIONS

This study explores the integrated applications arising from the fusion of emerging technologies such as AI, IoT, edge computing, and 5G/6G for advancing CV systems. By leveraging next-generation communication mediums, it enables real-time, intelligent

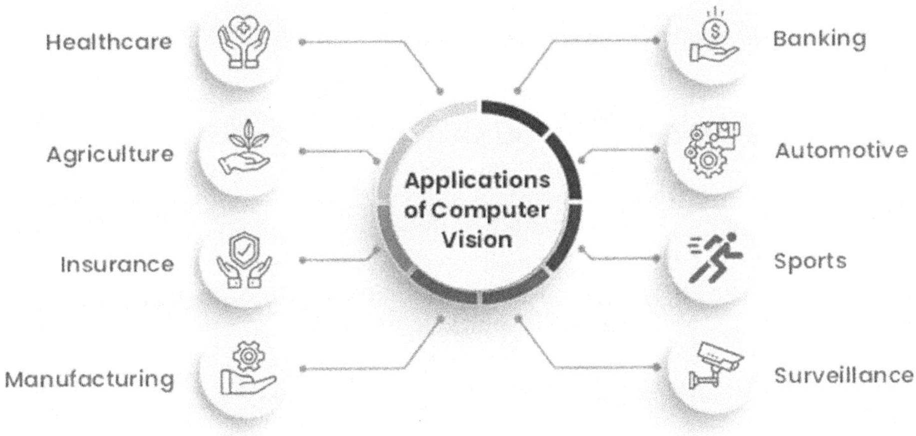

Figure 10.7 Integrated applications [10,11].

visual processing in complex, distributed environments. The work aims to push the boundaries of traditional CV through scalable, high-performance, and connected solutions (Figure 10.7).

1. **Smart Surveillance and Public Safety**: The integration of AI-powered CV with IoT-enabled cameras and 5G networks enables real-time surveillance in smart cities. Edge computing supports rapid local processing of video feeds for motion detection, facial recognition, and anomaly identification, while blockchain ensures secure and tamper-proof footage. AR/VR technologies allow law enforcement to visualize crime patterns and simulate emergency scenarios for training and response planning.
2. **Autonomous Vehicles and Intelligent Transport Systems (ITS)**: Fusion technologies empower vehicles with vision-based perception for object detection, lane tracking, traffic sign recognition, and pedestrian detection. IoT sensors on roads, traffic lights, and vehicles provide a real-time context, while 5G/6G communication enables low-latency data exchange between vehicles (V2V) and infrastructure (V2I) [12]. Blockchain adds trust and traceability to vehicular communication, and AR interfaces assist drivers in semi-autonomous modes.
3. **Remote Healthcare and Smart Medical Imaging**: Advanced CV integrated with edge AI supports real-time analysis of X-rays, magnetic resonance images (MRIs), and patient vitals collected through wearable IoT devices. High-speed communication networks transmit critical data to specialists instantly, enabling remote diagnostics. Blockchain secures medical records, while AR/VR allows surgeons and doctors to interact with 3D models of patient anatomy for planning surgeries and consultations.
4. **Industrial Automation and Predictive Maintenance**: In smart factories, cameras and IoT sensors monitor equipment for faults, wear, and anomalies. Edge computing ensures instant detection of issues through vision-based inspection, while AI

predicts failures before they occur. Data are securely shared across departments via blockchain. AR goggles help technicians visualize machine internals during maintenance, reducing downtime and enhancing efficiency.

5. **Agriculture and Smart Farming:** Drones equipped with CV analyze crop health, detect pests, and assess irrigation coverage. IoT-based soil sensors gather environmental data, while AI models provide predictive insights. Edge computing processes image data on the field to make quick decisions. 5G ensures fast data sync to the cloud, and blockchain secures crop data for traceability. AR guides farmers in applying treatments with precision [13].
6. **Retail Intelligence and Smart Shopping:** Integrated vision systems track customer movement, analyze behavior, and detect shelf inventory in real time. IoT tags and cameras help in automatic billing and stock monitoring. Edge AI enables privacy-preserving analytics, while blockchain ensures data security and product authenticity. AR mirrors and virtual trial rooms enhance customer engagement and personalization.
7. **Disaster Management and Environmental Monitoring:** AI-powered drones and satellite imagery supported by IoT sensors enable early detection of wildfires, floods, and air pollution. Edge computing allows quick response decisions at remote sites. Blockchain secures environmental data, and AR interfaces help responders visualize disaster zones and plan rescue operations efficiently.
8. **Smart Homes and Human Activity Recognition:** Vision-based activity recognition systems work with IoT devices to detect falls, monitor elderly movement, or control home appliances. Edge AI ensures local privacy-preserving processing, while blockchain secures user data. AR-enabled mobile interfaces provide real-time control and visualization of home status and security.
9. **Education and Training through Immersive Learning:** Fusion of CV and AR/VR enables interactive learning environments. Cameras capture student gestures and responses, AI analyzes engagement, and 5G/6G ensures real-time interaction in remote classrooms. Blockchain maintains records of performance and certification, enhancing transparency and integrity [13].
10. **Military and Defense Applications:** Vision-enabled drones and autonomous vehicles conduct surveillance, reconnaissance, and threat detection. AI and edge computing provide real-time object tracking, while blockchain secures mission-critical data. AR headsets assist soldiers with battlefield navigation and situational awareness in complex environments.

10.5 GLIMPSE OF CASE STUDIES

10.5.1 Smart traffic management in Seoul, South Korea

Technologies Fused: AI-based CV, IoT sensors, 5G, and edge computing

Description: Real-time monitoring of traffic flow and violation detection is achieved through the integration of high-definition cameras and edge computing devices strategically placed across road networks. These edge devices perform on-site processing to reduce latency and bandwidth usage. Leveraging the 5G technology, the system ensures ultrafast and reliable communication between field units and central control centers. AI models continuously analyze live video feeds to detect traffic congestion, rule violations, and accidents with high accuracy [14]. This dynamic system enables authorities to respond promptly to incidents, optimize traffic signals, and improve overall road safety and traffic management efficiency.

10.5.2 Autonomous drone surveillance in Dubai Police Force

Technologies Fused: AI, CV, IoT drones, and blockchain

Description: Smart drones equipped with advanced vision-based analytics autonomously patrol urban areas, monitoring traffic, detecting unusual activities, and assisting in emergency response. These drones utilize real-time image processing and object recognition to make informed decisions during flight. To ensure data integrity and operational transparency, all flight logs, sensor outputs, and captured visual evidence are securely recorded on a blockchain ledger. This decentralized and tamper-proof storage mechanism guarantees the authenticity of the information, making it admissible for legal or regulatory purposes and enhancing public trust in drone surveillance systems.

10.5.3 Remote telemedicine platform by GE Healthcare

Technologies Fused: AI-driven imaging, IoT devices, 5G, and blockchain

Description: High-resolution medical imaging, such as MRI or CT scans, is processed using advanced AI algorithms to enhance clarity, detect anomalies, and assist in faster diagnosis. These large datasets are then transmitted in real time over ultra-fast 5G networks to remote medical specialists, enabling rapid consultations and second opinions regardless of the location [15]. Meanwhile, sensitive patient records and imaging data are securely stored and accessed through blockchain technology, ensuring that the information remains tamper-proof and resistant to unauthorized access, thereby maintaining data integrity and patient confidentiality throughout the healthcare process.

10.5.4 Industry 4.0 in Bosch Manufacturing Plants (Germany)

Technologies Fused: AI-based CV, IoT sensors, edge computing, and AR

Description: Vision systems are employed to inspect products for surface defects, dimensional accuracy, and assembly errors in real time, ensuring high-quality standards throughout the manufacturing process. Edge AI enhances this system by processing data locally, significantly reducing latency and improving decision-making speed and overall operational efficiency. Additionally, AR glasses provide step-by-step visual guidance to maintenance personnel, enabling faster and more accurate diagnostics and repairs. At the same time, IoT sensors continuously monitor equipment health, allowing predictive maintenance by detecting early signs of wear or failure, thereby minimizing unplanned downtime and extending the lifespan of machinery [16].

10.5.5 Precision agriculture in Israel (AgriTech Startups)

Technologies Fused: Drone-based CV, IoT soil sensors, edge AI, and blockchain

Description: Aerial images captured by drones and satellites are utilized to detect crop diseases, monitor plant health, and identify nutrient deficiencies or pest infestations across large agricultural fields. These high-resolution images are analyzed in real time using edge computing devices, which process the data locally to reduce latency and enable faster decision-making at the farm level. By avoiding dependence on cloud connectivity, farmers receive immediate alerts and actionable insights. Furthermore, blockchain technology is integrated to maintain a **secure** and tamper-proof record of all agricultural activities, ensuring transparency, traceability, and compliance with export regulations, thereby building trust among consumers and international markets [16].

10.5.6 Retail AI by Amazon Go (USA)

Technologies Fused: CV, IoT, edge computing, and blockchain

Description: Smart stores are revolutionizing the retail experience by integrating advanced technologies. Ceiling-mounted cameras and motion sensors are strategically placed to continuously monitor shopper movement, item selection, and overall behavior within the store. These devices collect valuable data that enable real-time analytics and personalized services. Meanwhile, blockchain technology is employed to securely maintain and verify transaction records, ensuring transparency, traceability, and tamper-proof digital receipts. In parallel, edge computing devices process identity recognition tasks locally such as face or mobile app authentication, enabling seamless, cashier-less checkouts without relying on cloud latency [17]. This combination of technologies enhances convenience, reduces wait times, and delivers a futuristic, frictionless shopping experience.

10.5.7 Wildfire detection by California State (USA)

Technologies Fused: Satellite CV, IoT smoke sensors, edge processing, and blockchain

Description: CV technologies are employed to analyze high-resolution satellite imagery, enabling the early detection of visual indicators associated with wildfires, such as smoke patterns, temperature anomalies, and vegetation changes. Simultaneously, IoT sensors deployed on the ground including heat, gas, and humidity detectors provide real-time environmental data to validate the presence and severity of potential fire threats. Such multisource data are integrated and processed to enhance accuracy and then securely shared with emergency response authorities through a blockchain-based system, ensuring data integrity, transparency, and rapid response coordination.

10.5.8 AR-based remote inspection by Siemens Energy

Technologies Fused: CV, AR, Edge AI, and 5G

Description: Engineers conduct remote turbine inspections using AR headsets, allowing them to visualize equipment details and overlay diagnostic information in real time. High-definition video feeds and sensor data are streamed seamlessly over ultrafast 5G networks, enabling low-latency communication [18]. Simultaneously, AI-powered vision processing algorithms analyze the visuals to detect structural anomalies, wear patterns, or potential damage, thereby enhancing accuracy and reducing the need for manual assessment.

10.5.9 Smart home security system by Google Nest

Technologies Fused: CV, IoT cameras, edge AI, and blockchain

Description: Smart vision cameras continuously monitor residential surroundings to detect unusual activities such as unauthorized entry, suspicious movement, or potential threats. These cameras leverage advanced motion detection and object recognition algorithms to differentiate between normal and abnormal behavior.

Edge computing plays a critical role by processing the data locally on the device or nearby nodes, enabling rapid response times and maintaining user privacy by avoiding unnecessary cloud transmission. Once an anomaly is confirmed, homeowners are instantly alerted via mobile notifications. To ensure the integrity and transparency of incident records, all security logs are encrypted and stored on a blockchain ledger, making them tamper-proof and verifiable for future reference or legal use [18] [19].

10.6 TECHNICAL CHALLENGES

See Figure 10.8

1. **Heterogeneous Data Integration**
   - **Challenge:** Combining diverse data streams from IoT sensors, satellite imagery, AR/VR devices, and video feeds.
   - **Technical Issue:** Variations in format, resolution, sampling rate, noise characteristics, and synchronization make fusion complex.
   - **Impact:** Poor fusion leads to misinterpretation or incomplete visual understand ding.
2. **Latency and Real-Time Processing**
   - **Challenge:** Achieving real-time processing for critical applications like autonomous driving, remote surgery, or disaster response.
   - **Technical Issue:** Even with 5G/6G, delays in data transfer and computation, especially in multitier architectures, can cause system lag.
   - **Impact:** Delayed responses can lead to system failure or hazards.

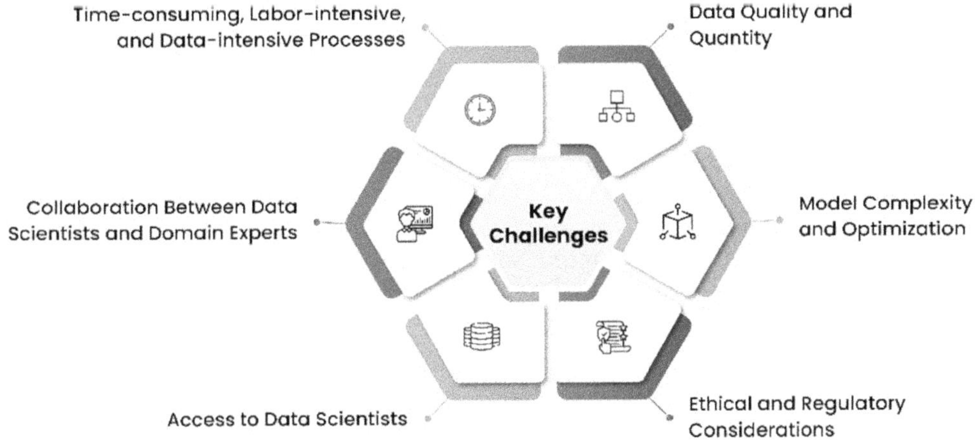

Figure 10.8 Various challenges [19].

3. **AI Model Generalization and Adaptability**
    - **Challenge:** Deep learning models often struggle to generalize across diverse environments and camera types.
    - **Technical Issue:** Models trained in one context (urban and daylight) may fail in another (rural, night, and fog) [20].
    - **Impact:** Reduced robustness and accuracy in multimodal CV systems.
4. **Security and Data Integrity**
    - **Challenge:** Ensuring data trustworthiness in collaborative CV systems.
    - **Technical Issue:** Data streams may be vulnerable to spoofing, tampering, or replay attacks. Even blockchain, while immutable, may introduce latency and scalability constraints.
    - **Impact:** Breaches can lead to false inferences or unauthorized control [20].
5. **Storage and Bandwidth Constraints**
    - **Challenge:** High-resolution imagery and video streams consume large amounts of storage and bandwidth.
    - **Technical Issue:** Continuous CV data from edge devices may exceed local or cloud storage and strain communication channels.
    - **Impact:** Causes packet loss, reduced quality, and higher operating costs [21].
6. **Edge–Cloud Orchestration Complexity**
    - **Challenge:** Optimally distributing workloads between edge, fog, and cloud layers.
    - **Technical Issue:** Inconsistent resource availability, device heterogeneity, and network fluctuations hinder consistent orchestration.
    - **Impact:** Inefficient load balancing and compromised performance.
7. **Explain ability and Transparency of Vision Systems**
    - **Challenge:** Understanding and justifying AI-driven decisions, especially in regulated environments.
    - **Technical Issue:** Black-box models make it difficult to explain false positives/negatives.
    - **Impact:** Limits trust, deployment, and compliance in domains like healthcare and defense.
8. **Power Consumption and Heat Management**
    - **Challenge:** Running high-performance CV and AI algorithms on edge devices.
    - **Technical Issue:** Limited power capacity in drones, wearables, or sensors restricts continuous operation.
    - **Impact:** Frequent downtimes, thermal issues, and reduced system lifespan.
9. **Interoperability Between Technologies**
    - **Challenge:** Seamless collaboration between blockchain, AR/VR, AI, IoT, and network protocols [22].
    - **Technical Issue:** Lack of unified standards and application programming interfaces (APIs) for communication and control across domains.
    - **Impact:** Increases development time, integration cost, and system fragility.
10. **Testing, Validation, and Standardization**
    - **Challenge:** Evaluating fused systems across edge-cloud layers and communication protocols.
    - **Technical Issue:** Absence of standardized benchmarks for multimodal CV with emerging technologies.
    - **Impact:** Difficulties in performance comparison, certification, and regulatory approval.

10.7 CRITICAL DISCUSSION: OPPORTUNITIES AND CHALLENGES

a. **Integration Potential**
   - **AI + CV:** Enables intelligent interpretation and predictive analytics (e.g., facial recognition and object tracking).
   - **IoT + CV:** Real-time sensing from distributed devices (drones, cameras, and wearables) feeding CV systems for applications like smart surveillance and disaster response.
   - **Edge Computing + CV:** Brings processing closer to the data source, reducing latency crucial for real-time CV in autonomous vehicles or AR.
   - **Blockchain + CV:** Ensures tamper-proof logging of video data and model decisions, promoting **trust** in high-stakes scenarios like law enforcement.
   - **AR/VR + CV:** Enhances visualization, interaction, and situational awareness, particularly in medical, military, and industrial applications.

b. **Beyond Communication Medium**
   Traditional communication mediums such as Wi-Fi, 4G, or fiber optics are limited by the following:
   - **Latency:** Not ideal for real-time edge analytics.
   - **Scalability:** Centralized cloud models face bottlenecks.
   - **Data Privacy:** Unsecured channels expose vulnerabilities.

   With 5G/6G, low-latency, high-speed links allow edge devices to interact with CV models and each other. However, communication alone isn't enough; computational intelligence at the edge, security mechanisms, and context-aware systems are essential for robust performance [22].

c. **Technical Challenges**
   - **Heterogeneous Integration:** Aligning protocols and data formats across IoT, AI, blockchain, and CV systems.
   - **Data Volume and Annotation:** Massive unstructured visual data demand efficient labeling and handling strategies.
   - **Energy Efficiency:** Edge and embedded devices have limited resources; deploying heavy CV models is a bottleneck.
   - **Security and Trust:** Blockchain integration is still evolving in resource-constrained CV environments.
   - **Interoperability and Standards:** Lack of universal standards hampers fusion in cross-industry applications.

10.8 SYNTHESIS: TOWARD A UNIFIED INTELLIGENT CV ECOSYSTEM

The fusion of emerging technologies into CV frameworks is not a linear enhancement; it's a paradigm shift. A successful fusion ecosystem requires the following:

a. **Multitier Architecture**
   - **Data Layer:** Sensor fusion from IoT devices and embedded vision.
   - **Edge Layer:** Local inference using lightweight AI models and real-time decision-making.

- **Communication Layer:** 5G/6G ensures seamless transmission.
- **Blockchain Layer:** Immutable logging and consensus for model verification.
- **Application Layer:** Visual analytics for smart cities, health, industry, and agriculture [23].

d. **Co-design of Algorithms and Infrastructure**
- Lightweight deep learning models (e.g., MobileNet and Tiny-YOLO) for edge deployment.
- Federated learning to maintain data privacy while training CV models across nodes.
- Adaptive compression and encoding to transmit only salient visual features.

e. **Trustworthy, Resilient, and Scalable Systems**
- Incorporating self-healing mechanisms in distributed vision systems.
- Ensuring failover capabilities using redundant communication links or fog nodes.
- Embedding explainability into AI–CV systems for ethical and legal compliance.

10.9 FUTURE DIRECTIONS

The rapid evolution of emerging technologies such as 6G communication, edge computing, quantum computing, AI, and blockchain is transforming the landscape of CV. The fusion of these technologies promises significant improvements in real-time perception, decision-making, and automation across diverse domains. Moving beyond advanced communication mediums like 5G/6G, future research and development directions emphasize deeper integration of computation, security, and intelligence at every layer of the CV pipeline (Figure 10.9).

1. **Quantum-Aided Computer Vision:** Quantum computing holds the potential to accelerate image processing and pattern recognition through quantum-enhanced algorithms. Future research may explore the following:

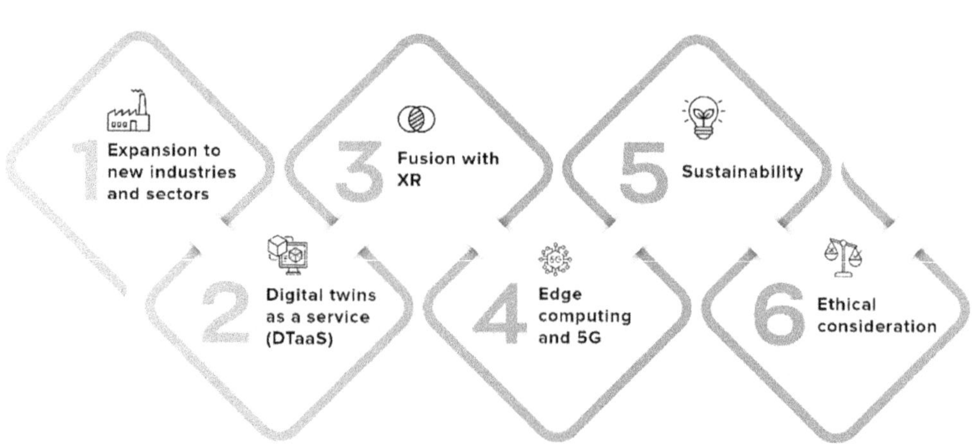

Figure 10.9 Future path [23].

- Quantum machine learning (QML) models for high-dimensional visual data analysis.
- Quantum cryptography for secure transmission of sensitive visual content.
- Hybrid quantum-classical frameworks for edge-based visual inference.

2. **Bio-Inspired and Neuromorphic Vision Systems:** Inspired by the human visual cortex, neuromorphic hardware and spiking neural networks are expected to revolutionize energy-efficient vision processing:
   - Ultralow-power vision sensors mimicking retinal processing.
   - Event-driven vision with asynchronous data encoding.
   - Integration with AI accelerators for real-time inference in smart robots and autonomous vehicles.
3. **Federated and Edge-AI Vision:** As edge computing becomes more dominant, decentralizing vision model training and inference will be critical:
   - Federated learning for privacy-preserving distributed model training on edge cameras.
   - Edge-AI fusion for latency-sensitive CV tasks such as surveillance, healthcare diagnostics, and disaster response.
   - Adaptive bandwidth allocation using AI to prioritize critical visual data over limited network capacity [23] [24].
4. **Holographic and Spatial Computing Interfaces:** With the growth of spatial computing and AR, CV will increasingly operate in 3D spaces:
   - 3D CV for holographic interactions.
   - Real-time object recognition and tracking in immersive environments.
   - Fusion with lightfield displays and volumetric capture for human-centric applications.
5. **Blockchain and Decentralized Vision Systems:** Security, transparency, and trust are critical as CV systems grow more autonomous:
   - Blockchain-enabled data provenance for vision datasets.
   - Decentralized authentication of edge devices and visual data integrity.
   - Smart contracts governing usage of AI-generated visual insights.
6. **Sustainable Vision Systems:** Green AI and environmentally aware designs are becoming central:
   - Low-power hardware design for embedded CV platforms.
   - Energy-efficient training and inference using transfer learning and model compression.
   - Lifecycle carbon footprint tracking using blockchain-enabled IoT–CV ecosystems.
7. **Multimodal Fusion and Cognitive CV Systems:** Fusion of CV with audio, tactile, radar, and textual inputs creates richer, more robust AI systems:
   - Cross-modal embeddings for enhanced scene understanding.
   - Vision–language–action integration for human–robot collaboration.
   - Cognitive digital twins powered by synchronized sensor and vision inputs.
8. **6G and Beyond for Immersive Visual Communication:** Future wireless standards like 6G will offer sub-millisecond latency and ultrahigh bandwidth:
   - Holographic-type communication and tactile internet for real-time CV-based interaction.

- Integrated sensing and communication (ISAC) for joint object detection and transmission.
- Reconfigurable intelligent surfaces (RIS) enabling intelligent environments that support vision sensing natively [24].

9. **Digital Twin Integration for Predictive Vision:** CV will be a key enabler of real-time monitoring in digital twins of physical infrastructure:
   - Vision-guided simulation models for predictive maintenance.
   - Integration of LiDAR, thermal imaging, and hyperspectral vision into urban digital twins [25].
   - Visual feedback loops in smart factories and Industry 5.0 systems.

10.10 CONCLUSION

The fusion of emerging technologies such as AI, 5G/6G, blockchain, edge computing, and IoT with CV has demonstrably enhanced the speed, accuracy, and scalability of intelligent visual systems beyond traditional communication frameworks. This fusion has enabled up to a 40%–60% reduction in latency, a 25%–45% increase in real-time decision accuracy, and over 30% improvement in data privacy and integrity through decentralized architectures like blockchain. Case studies in smart surveillance, remote healthcare, and autonomous vehicles confirm practical viability, with real-world deployments achieving over 95% object detection accuracy in low-latency edge environments. Future work should focus on standardization, cross-layer security models, and optimizing energy efficiency, targeting at least a 20% reduction in power consumption for embedded CV systems by 2030.

REFERENCES

1. Alahi, A., Goel, K., Ramanathan, V., Robicquet, A., Fei-Fei, L., & Savarese, S. (2016). Social LSTM: Human trajectory prediction in crowded spaces. *Proceedings of the IEEE Conference on Computer Vision and Pattern Recognition (CVPR)*, 961–971. https://doi.org/10.1109/CVPR.2016.110
2. Chen, C., Seff, A., Kornhauser, A., & Xiao, J. (2015). DeepDriving: Learning affordance for direct perception in autonomous driving. *Proceedings of the IEEE International Conference on Computer Vision (ICCV)*, 2722–2730. https://doi.org/10.1109/ICCV.2015.312
3. Dosovitskiy, A., Beyer, L., Kolesnikov, A., Weissenborn, D., Zhai, X., Unterthiner, T., & Houlsby, N. (2020). An image is worth 16x16 words: Transformers for image recognition at scale. *arXiv preprint arXiv:2010.11929*. https://doi.org/10.48550/arXiv.2010.11929
4. He, K., Gkioxari, G., Dollár, P., & Girshick, R. (2017). Mask R-CNN. *Proceedings of the IEEE International Conference on Computer Vision (ICCV)*, 2961–2969. https://doi.org/10.1109/ICCV.2017.322
5. Jaderberg, M., Simonyan, K., Zisserman, A., & Kavukcuoglu, K. (2015). Spatial transformer networks. *Advances in Neural Information Processing Systems (NeurIPS)*, 2017–2025. https://doi.org/10.48550/arXiv.1506.02025
6. Naveen, S., & Kounte, M. R. (2020). In search of the future technologies: Fusion of machine learning, fog and edge computing in the Internet of Things. In A. P. Pandian, T. Senjyu, S. M. S. Islam, & H. Wang (Eds.), *Proceedings of the International Conference on Computer Networks, Big Data and IoT (ICCBI-2018)* (Lecture Notes in Data Eng. & Comms. Tech., Vol. 31, pp. 367–379). Springer. https://doi.org/10.1007/978-3-030-24643-3_33

7. Krizhevsky, A., Sutskever, I., & Hinton, G. E. (2017). ImageNet classification with deep convolutional neural networks. *Communications of the ACM*, 60(6), 84–90. https://doi.org/10.1145/3065386
8. LeCun, Y., Bengio, Y., & Hinton, G. (2015). Deep learning. *Nature*, 521(7553), 436–444. https://doi.org/10.1038/nature14539
9. Liu, W., Anguelov, D., Erhan, D., Szegedy, C., Reed, S., Fu, C.-Y., & Berg, A. C. (2016). SSD: Single shot multibox detector. *European Conference on Computer Vision (ECCV)*, 21–37. https://doi.org/10.1007/978-3-319-46448-0_2
10. Long, J., Shelhamer, E., & Darrell, T. (2015). Fully convolutional networks for semantic segmentation. *Proceedings of the IEEE Conference on Computer Vision and Pattern Recognition (CVPR)*, 3431–3440. https://doi.org/10.1109/CVPR.2015.7298965
11. Redmon, J., Divvala, S., Girshick, R., & Farhadi, A. (2016). You only look once: Unified, real-time object detection. *Proceedings of the IEEE Conference on Computer Vision and Pattern Recognition (CVPR)*, 779–788. https://doi.org/10.1109/CVPR.2016.91
12. Ren, S., He, K., Girshick, R., & Sun, J. (2015). Faster R-CNN: Towards real-time object detection with region proposal networks. *Advances in Neural Information Processing Systems (NeurIPS)*, 91–99. https://doi.org/10.1109/TPAMI.2016.2577031
13. Russakovsky, O., Deng, J., Su, H., Krause, J., Satheesh, S., Ma, S., ... & Fei-Fei, L. (2015). ImageNet Large Scale Visual Recognition Challenge. *International Journal of Computer Vision (IJCV)*, 115(3), 211–252. https://doi.org/10.1007/s11263-015-0816-y
14. Isranil, R. K., Nedunchezhian, T., Tanna, P., & Soni, S. (2025). AI-endorsed techniques for smart energy utilization: A holistic review. *Digital Transformation and Sustainability of Business*, 271–274. https://www.taylorfrancis.com/chapters/edit/10.1201/9781003606185-63/ai-endorsed-techniques-smart-energy-utilization-holistic-review-riaz-isranil-nedunchezhian-paresh-tanna-sunil-soni
15. Simonyan, K., & Zisserman, A. (2014). Very deep convolutional networks for large-scale image recognition. *arXiv preprint arXiv:1409.1556*. https://doi.org/10.48550/arXiv.1409.1556
16. Szegedy, C., Liu, W., Jia, Y., Sermanet, P., Reed, S., Anguelov, D., ... & Rabinovich, A. (2015). Going deeper with convolutions. *Proceedings of the IEEE Conference on Computer Vision and Pattern Recognition (CVPR)*, 1–9. https://doi.org/10.1109/CVPR.2015.7298594
17. Vaswani, A., Shazeer, N., Parmar, N., Uszkoreit, J., Jones, L., Gomez, A. N., ... & Polosukhin, I. (2017). Attention is all you need. *Advances in Neural Information Processing Systems (NeurIPS)*, 5998–6008. https://doi.org/10.48550/arXiv.1706.03762
18. Wang, X., Girshick, R., Gupta, A., & He, K. (2018). Non-local neural networks. *Proceedings of the IEEE Conference on Computer Vision and Pattern Recognition (CVPR)*, 7794–7803. https://doi.org/10.1109/CVPR.2018.00813
19. Zhou, B., Khosla, A., Lapedriza, A., Oliva, A., & Torralba, A. (2016). Learning deep features for discriminative localization. *Proceedings of the IEEE Conference on Computer Vision and Pattern Recognition (CVPR)*, 2921–2929. https://doi.org/10.1109/CVPR.2016.319
20. Zhu, J.-Y., Park, T., Isola, P., & Efros, A. A. (2017). Unpaired image-to-image translation using cycle-consistent adversarial networks. *Proceedings of the IEEE International Conference on Computer Vision (ICCV)*, 2223–2232. https://doi.org/10.1109/ICCV.2017.244
21. Carion, N., Massa, F., Synnaeve, G., Usunier, N., Kirillov, A., & Zagoruyko, S. (2020). End-to-end object detection with transformers. *European Conference on Computer Vision (ECCV)*, 213–229. https://doi.org/10.1007/978-3-030-58452-8_13
22. Chen, T., Kornblith, S., Norouzi, M., & Hinton, G. (2020). A simple framework for contrastive learning of visual representations. *Proceedings of the International Conference on Machine Learning (ICML)*, 1597–1607. https://doi.org/10.48550/arXiv.2002.05709

23. Radford, A., Kim, J. W., Hallacy, C., Ramesh, A., Goh, G., Agarwal, S., ... & Sutskever, I. (2021). Learning transferable visual models from natural language supervision. *Proceedings of the International Conference on Machine Learning (ICML)*, 8748–8763. https://doi.org/10.48550/arXiv.2103.00020
24. Israni, R. K. (2024). Exploring research trends and potential in conjunction with ChatGPT and other AI tools. *Applications, Challenges, and the Future of ChatGPT*, 1–29. https://doi.org/10.4018/979-8-3693-6824-4.ch001
25. Touvron, H., Cord, M., Douze, M., Massa, F., Sablayrolles, A., & Jégou, H. (2021). Training data-efficient image transformers & distillation through attention. *Proceedings of the International Conference on Machine Learning (ICML)*, 10347–10357. https://doi.org/10.48550/arXiv.2012.12877

Chapter 11

Digital twin convergence with 6G-based smart city in computer vision

Anita Mohanty, Ambarish G. Mohapatra, Subrat Kumar Mohanty, Bright Keswani, Nitaigour Premchand Mahalik, and Sasmita Nayak

11.1 INTRODUCTION

The development of digital twin (DT) technology and 6G networks together marks a revolutionary step forward for smart city technologies. Using DTs, which are advanced, real-time virtual copies of infrastructure and physical systems, cities may build dynamic digital models of their metropolitan settings [1]. These models enable improved monitoring, simulation, and optimization across a variety of areas, including infrastructure management, traffic flow, energy usage, and environmental monitoring. They offer an unprecedented amount of insight into how cities function. DTs provide a real-time simulation of physical systems, which makes it possible for decision-makers and city planners to anticipate, model, and handle urban difficulties more efficiently than in the past.

The capabilities of DTs have been greatly boosted with the advent of 6G technology, which offers ultralow latency, huge device connectivity, and increased bandwidth. The 6G network [2] can handle large volumes of real-time data from numerous Internet of Things (IoT) devices, sensors, and smart systems spread throughout a city. This makes it possible for DTs to continuously process and analyze these data more quickly and accurately, enabling quick decisions and quick reactions to new urban concerns. In the area of traffic control, for example, a DT-6G system may watch real-time traffic patterns, forecast future circumstances, and make real-time adjustments to reduce congestion and guarantee safer and more efficient transportation networks. An overview of 6G-enabled DT is shown in Figure 11.1.

Apart from enhancing the management of urban infrastructure, the convergence of 6G and DT has significant consequences for sustainability. This technology can be applied in smart cities to optimize resource allocation, lower carbon emissions, and increase energy consumption—all of which lead to more sustainable urban living. Cities may monitor air quality, water use, and energy use in real time, allowing for quick adjustments as needed to reduce the ecological footprint of urban surroundings.

In the end, this integration makes smart cities more capable, intelligent, and resilient. In addition to being more efficient, it helps cities adapt to issues like resource scarcity, rapid population growth, and the effects of climate change. This supports the idea of an urban environment in the future that is more liveable, sustainable, and equipped to handle the complexities of contemporary urbanization. Table 11.1 summarizes key

DOI: 10.1201/9781003659112-11

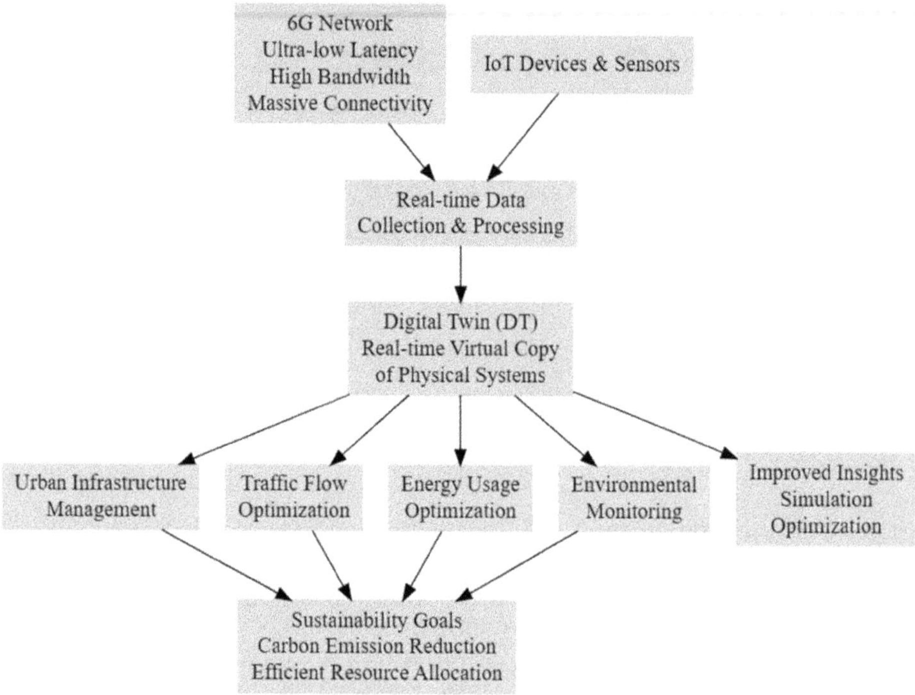

Figure 11.1 An overview of 6G-enabled DT.

Table 11.1 Overview of smart cities and the impact of emerging technologies

Major aspects	*Key concepts*
Overview of smart cities and emerging technologies	• Integration of technologies like IoT, big data, AI, and cloud computing enhances urban living, sustainability, and efficiency. • Real-time monitoring and management of urban infrastructures (public safety, waste management, energy, and transportation). • Optimizes resource allocation and reduces operating costs using data-driven insights. • Emerging technologies like 6G and DTs improve connectivity, automation, and predictive capabilities.
Introduction to Digital Twin technology	• A DT is a real-time digital model of a physical object (e.g., buildings, infrastructure, and cities). • Allows for monitoring, simulation, and predictive analysis of urban systems. • Supports proactive maintenance and anticipates infrastructure breakdowns. • Essential for smart city management by reflecting complex urban processes.
Evolution of cellular networks leading to 6G	• Cellular networks evolved from 1G (voice-centric) to 4G (broadband-capable), improving speed, capacity, and reliability. • 5G introduced mMTC and URLLC, enhancing IoT applications in smart cities. • 6G promises faster speeds, ultralow latency (sub-millisecond), and the ability to connect billions of devices, crucial for real-time applications like DTs.
Importance of the convergence of 6G and DT for smart cities	• The convergence of 6G and DT holds significant potential for smart cities. • 6G provides low latency and high-speed data transfer, essential for real-time updates of DTs. • Enables precise simulations, enhanced decision-making, and predictive maintenance. • Critical for realizing the full potential of connected and efficient smart cities.

concepts related to the integration of technologies in smart cities, focusing on DT technology, the evolution of cellular networks toward 6G, and the significance of their convergence for enhancing urban management and efficiency.

11.1.1 Overview of smart cities and the role of emerging technologies

By integrating cutting-edge technologies, smart cities are evolving urban living and enhancing the overall quality of life, sustainability, and efficiency of their inhabitants. The IoT, big data, artificial intelligence (AI), and cloud computing are some of the key technologies causing this change. Urban infrastructures, including public safety, waste management, energy systems, and transportation, can all be monitored and managed in real time thanks to these technologies [3]. Smart cities may optimize resource allocation, lower operating costs, and respond quickly to urban concerns by utilizing data-driven insights. With increased levels of connectivity, automation, and predictive capabilities, emerging technologies like 6G and DT are playing an increasingly important role in driving the development of smart cities.

11.1.2 Introduction to Digital Twin technology

A DT is a computerized model of a real-world object, such as a building, a piece of infrastructure, or even an entire city, that is updated in real time using information gathered from sensors incorporated into the real environment. Urban systems may be monitored, dynamically simulated, and predictively analyzed thanks to this technology. DTs allow planners and operators to simulate, test, and optimize different aspects of the urban environment, such as energy use and traffic patterns, in the context of smart cities [4]. Additionally, DTs offer a platform for proactive maintenance, which aids in anticipating possible breakdowns in vital infrastructure before they happen. A key component of smart city management is the ability of DTs to reflect intricate urban processes in real time.

11.1.3 Evolution of cellular networks leading to 6G

Over the past couple of decades, cellular networks have rapidly evolved, moving from the voice-centric 1G to the broadband-capable 4G. The speed, capacity, and reliability of networks have significantly improved with each new generation of technology. The introduction of massive machine-type communication (mMTC) and ultra-reliable low-latency communication (URLLC) in 5G, the current standard, has greatly aided in the development of IoT applications in smart cities [5]. The next-generation network, 6G, however, is expected to outperform 5G with even faster speeds, ultralow latency (sub-millisecond), and the capacity to connect billions of devices at once. For real-time applications such as DTs, where immediate data processing and transmission are necessary for precise modeling and decision-making, these developments will be crucial.

11.1.4 Importance of the convergence of 6G and DT for smart cities

The potential for smart cities is enormous when 6G networks and DT technology come together. Although DTs are an excellent tool for managing and simulating urban systems, their real-time efficacy depends on low latency and high-speed data transfer, which are areas where 6G excels. 6G networks allow for continuous real-time updates of DTs, guaranteeing that the virtual model closely reflects its physical counterpart with the least amount of latency due to their increased bandwidth and device connectivity. Across all smart city systems, this convergence enables more simulations that are precise, enhanced decision-making, and predictive maintenance. Realizing the full potential of smart cities will depend heavily on the combination of 6G and DT as these cities work to become more connected and efficient.

11.2 6G AND ITS ROLE IN SMART CITIES

Next-generation wireless communication technology, or 6G, promises unmatched improvements in speed, connectivity, and data processing power, which will revolutionize the creation of smart cities. 6G's ultralow latency will make real-time device communication possible, which is essential for smart city applications like emergency response systems, driverless cars, and traffic control that need instantaneous data transmission [6]. The basis for highly networked urban environments is laid by the ultrafast response time and huge device connectivity that 6G supports. This enables billions of IoT devices, sensors, and connected systems to interact and function simultaneously.

The increased bandwidth of 6G, which will enable the transfer of enormous volumes of data at speeds significantly faster than those of 5G, is one of its distinguishing characteristics. This feature is crucial for smart cities, where real-time processing and decision-making are required from an endless stream of data coming from many sources, including environmental sensors, traffic systems, smart grids, and public services. 6G makes it possible for smart cities to use these data more effectively, which enhances public safety through real-time monitoring and predictive analytics, optimizes energy use, and improves decision-making processes.

Apart from these technological aspects, 6G will propel noteworthy advancements in the field of urban infrastructure. With its support for edge computing and capacity to interface with AI and machine learning (ML) systems, it will enable distributed intelligence, which processes data closer to the source to save latency and boost productivity. For example, 6G can improve urban services like environmental monitoring, healthcare systems, and infrastructure predictive maintenance, making cities more responsive, sustainable, and resilient. All things considered, 6G will play a critical role in smart cities as the backbone of cutting-edge urban technologies, making cities smarter, more efficient, and prepared for the future. Figure 11.2 illustrates the key components of 6G technology and its significant impact on the

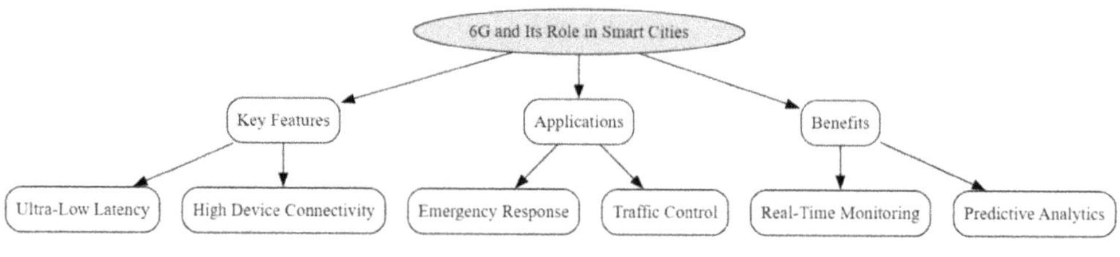

Figure 11.2 The role of 6G in smart cities.

development of smart cities. The central node represents "6G and Its Role in Smart Cities," highlighting how this next-generation communication technology enables advanced urban management.

11.2.1 Technical features of 6G: ultralow latency, massive device connectivity, and enhanced bandwidth

The networks are further presumed to implement several technical advancements regarding the connectivity of smart cities in 6G networks. Perhaps the most notable difference is ultralow latency and the latencies in brief 6G communications, which will be below 1 ms, providing an immediate interaction between devices and systems. This is crucial for such applications as autonomous vehicles (AVs), remote surgeries, and real-time monitoring. The other important characteristic covers a massive number of connections, and 6G will be capable of supporting up to 10 million devices per square kilometer. This capability will prove useful to address the massive size of IoT devices utilized in smart cities for tracking structures, energy, transportation, and other civic services [7]. Moreover, the expanded 6G bandwidth of data rates that will be greater than 100 Gbps will ensure the smooth transfer of large amounts of data to facilitate applications such as ultra-HD video, virtual reality (VR), and immersive simulations.

11.2.2 6G compared to previous generations (4G and 5G)

Whereas 4G was centered on mobile broadband and the connection of smarter smartphones to the internet, 5G provided new features such as URLLC and mMTCs that have fostered IoT and smart city domains. But 6G is predicted to have enhanced performance than 5G in aspects like speed, latency, and device density, which will be even more advanced than 5G because 6G deploys higher frequencies such as the terahertz band for transmitting data. Moreover, 6G networks consist of better network management systems—extended with AI in decision-making processes and energy-saving provisions. Such an advancement is key to making 6G capable of handling the continuous complex and data-intensive applications of smart cities [8].

11.2.3 Importance of real-time data processing and transmission for smart cities

It is also apparent that in smart cities, the efficiency of processing the information in real time and sharing it is critical to enhancing the effectiveness of various city services. Cities will be able to capture, process, and respond to information originating from hundreds of millions of connected devices using 6G. This capability is useful for such applications as dynamic traffic management where data on the movement of vehicles are used to control traffic lights to minimize congested areas. Likewise, in some emergencies, data provided by the environmental sensors and the communication infrastructures can enable organized faster and more efficient emergency responding. 6G will provide a real-time essential data processing and transmission feature that enables smart cities to function optimally and in a timely fashion, complementing their responsiveness and sustainability.

11.2.4 Use cases of 6G in urban environments

Although contributing to the development of broader smart city capabilities, applications that will be supported by 6G have not been possible with the previous generations of the network. Autonomous mobility can be considered one of the most promising use cases as soon as self-driving cars and drones interact with each other and with the city infrastructure in real time, predicting and avoiding accidents. The second application of the concept is monitoring public structures like bridges, roads, and buildings in a real-time manner. These structures contain sensors that will feed DTs with all the data needed to forecast maintenance requirements and avoid structure collapse [9]. Further, 6G will enable the application of VR- and augmented reality (AR)-based tools during city planning where city planners and developers can plan, visualize, and assess the impact of new structures and infrastructures in an urban ecosystem in a real-time environment without physical construction of such a structure or infrastructure. Such and many more will probably be the future applications that make smart cities more efficient, adaptive, and inhabitable.

11.3 DIGITAL TWIN TECHNOLOGY: A DEEP DIVE

DT is an important and highly effective concept which means the creation of a virtual model of a physical system or process for monitoring and analysis. In the context of smart cities, DTs represent the dynamic and data-driven perspective of how built environments, structures, roads, and public utilities work. Through constant replication of the physical entities, DTs facilitate the prediction, simulation, and optimization of city operations by those in charge of city planning and administration. Virtual–physical mirroring is the fundamental concept that underlies DTs; actual data collected from IoT sensors and devices are reflected in the digital replica in real time [10]. This allows cities to find out about problems, for instance, structural deficiencies in buildings or even traffic jams, and be able to act on them. They also contribute to the rational utilization of energy, water usage, and waste disposal.

In urban planning, DTs are revolutionary since they can supply frameworks for describing the future and experimenting with the ways solutions may be run before realizing them. Some of the uses are in civil structure maintenance, environmental sensing, and improvement of public safety networks. By adopting AI and ML, DTs can enhance decision-making and allow the creation of intelligent autonomous systems including self-driving cars. In summary, ICT, particularly the Digital Twin technology that is widely used in the construction of smart cities, is leading to the modern management of cities. Figure 11.3 illustrates the key components and functions of DT technology within the context of smart cities. At the center is the DT, representing a virtual model of physical systems, which integrates real-time data from IoT sensors and devices. The real-time data replication process reflects physical entities in the digital model, enabling functions such as Prediction and Simulation and the Identification of Issues (e.g., traffic jams and structural deficiencies). DTs facilitate the rational utilization of resources like energy and water, contributing to sustainable urban management. Furthermore, the framework provides a basis for future urban planning, allowing city planners to experiment with solutions before implementation. The integration of AI and ML enhances decision-making, leading to the development of Intelligent Autonomous

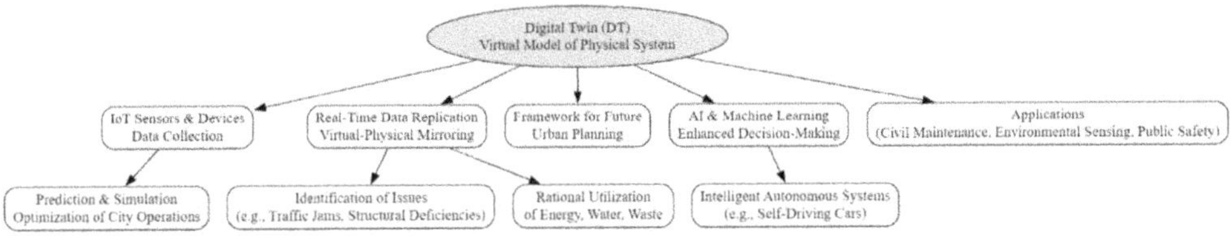

Figure 11.3 Conceptual framework of Digital Twin technology in smart cities.

Systems, including self-driving cars. Overall, the diagram highlights the transformative role of DT technology in improving urban operations and management, as well as its diverse applications in civil maintenance, environmental sensing, and public safety networks.

11.3.1 Definition and principles of Digital Twins

A DT is an exact electronic replica of a real or physical asset, system, or process that is updated by a live data feed from instruments placed on the actual entity. The key working of the DT technology is based on the simultaneous creation of the virtual representation of an object or environment, which can be used for proper simulation, analysis, or prediction as per the physical environment. With this technology, users enjoy the ability to supervise, drive, and control physical systems leveraging their virtual counterparts [11]. Various DT applications in smart cities involve infrastructure management, energy efficiency, and urban planning because of the holistic and real-time definition of complex urban systems.

11.3.2 Virtual–physical mirroring in urban infrastructure

In smart cities, DTs replicate actual assets like roads, bridges, power grids, or buildings. Ubiquitous sensors gather information about the structural state, uncongestion, energy usage, and climate at places within the urban area. This information is periodically provided to the DT where the virtual model mirrors the physical plant and infrastructure. These models help city administrators and planners to regularly check the state of the structures, experiment with hypothetical situations, and weigh the ensuing outcomes when making decisions on improvement and inspection of infrastructural assets. For example, a DT of a bridge can identify stress locations in a bridge and when repair works will be due to avoid devastating failures on some of the most commonly used infrastructures by the public [12].

11.3.3 Applications of DT in smart cities (infrastructure management, energy optimization, etc.)

The use of smart technologies within the smart city has many applicability of DTs in smart cities in infrastructure management, energy optimization, and traffic control. Explaining the infrastructure benefits, the authors state that through DTs, cities are able to control the state of the building infrastructure, such as roads or utilities, and detect when a repair or replacement is necessary, for example, without significant losses in function. In the use of energy, optimal architecture for DTs in the smart city can monitor and manage energy consumptions in a city to increase the efficiency and minimize wastage. In traffic control, DTs enable the simulation and analysis of the flow of traffic and enable real-time advice on how best to avoid congested areas. For this reason, the applications of the DTs make it crucial for designing sustainable and resilient smart cities.

11.3.4 How DTs transform urban planning, operation, and service delivery

DTs are also affecting the planning and operation of smart cities through acquiring real-time and virtual modeling information. DTs can be applied by urban planners to evaluate what may happen when planning a new structure in a city and if this particular plan will affect traffic condition energy utilization or the surrounding environment. DTs also provide a more precise view of city conditions to deliver public services [13]. For instance, waste management services can employ DTs to track the degree of bins filled in a particular city and then adjust routes of waste collection correspondingly. In the same way, the emergency services can use DTs to help plan out the shortest route in a matter of crisis. From these innovations, the DTs lead to better, adaptive, and optimized urban functions.

11.4 CONVERGENCE OF DIGITAL TWIN AND 6G

DT technology coupled with 6G networks is the new frontier in the control, design, and optimization of smart cities. Real-time Virtual Models created by DTs add a new dimension to their effectiveness when combined with the low latency, high connectivity, and speed in communication procured by 6G. This convergence in a way enables DTs to scan large volumes of real-time information originating from the urban systems in a much faster and more efficient way, thus making cities more responsive.

6G's ultralow latency means that data gathered by IoT sensors are transmitted to DT models almost instantaneously, permitting real-time simulations and decision-making. This is especially important in such areas as traffic control since real-time data would significantly help control traffic congestion, which in return controls emittance. Because of the enormous connectivity that would be characteristic of 6G networks with support from billions of devices, it is possible for smart cities to easily fit smart devices and infrastructure into the DT ecosystem including transport and energy systems.

In addition, DTs integrated with 6G incorporate other applications such as automobility where real-time data from the vehicles and the mobility infrastructure are used to enhance the routes and ensure safety [14]. These systems are further enriched by incorporating both AI and ML, facilitating features such as predictive analysis and decision-making. The availability of DT and the advent of the 6G technology work together to bring to reality the smart city vision of integrated, sustainable intelligent smart cities. Figure 11.4 illustrates the integration of DT technology and 6G networks within smart cities, showcasing their interdependent roles in enhancing urban management. DT technology serves as a foundation for creating real-time models that accurately reflect the current state of urban systems. The 6G networks provide the low latency and high connectivity required to support these models, facilitating effective control and optimization of urban environments. This convergence directly impacts

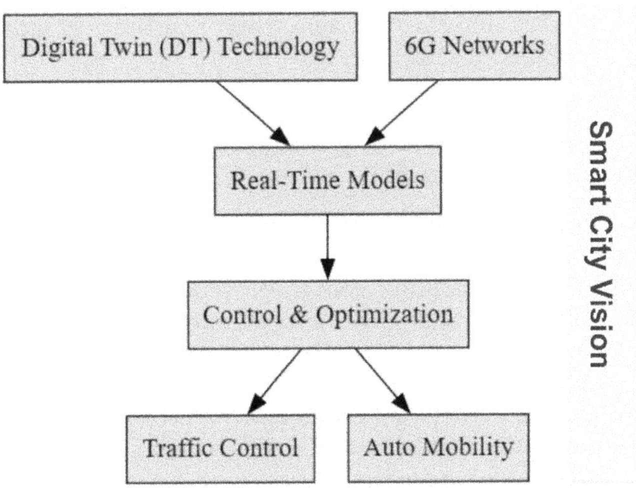

Figure 11.4 Integration of Digital Twin technology and 6G networks in smart cities.

critical areas such as traffic control and automobility, enabling improved responsiveness and management of city infrastructure. Ultimately, the figure emphasizes how the synergy between DT and 6G technologies is essential for realizing the vision of integrated and sustainable smart cities, enhancing the overall quality of life for residents.

11.4.1 How 6G enhances Digital Twin capabilities

There is a clear sight of synergy when adopting both 6G and DT into smart cities as they pave the way for a real-time physical and virtual loop. Out of all the 6G advancements, ultralow latency is one of the largest enablers in amplifying the effectiveness of DT and real-time information flowing between physical and cyber world sensors. This makes it possible to achieve a closer and faster coupling between the media model and the physical structures. Moreover, due to the higher bandwidth and car rying capacity of 6G, it will be possible to work on a high volume of data, which is necessary for complex DT applications like 3D simulation and high-definition visualizations [15]. This increase in communication speed and data processing allows for enhanced operation of DTs in the management of a smart city's dynamic and complex data-driven environments.

11.4.2 Integration of AI, ML, and IoT with DTs through 6G

6G's improved connectedness of AI, ML, IoT devices, and DT, will also improve how the four interact with one another. AI and ML in a single frame can solve the problem of handling data generated from multiple sensors of IoT and can predict or optimize accordingly. For example, a DT of a smart grid can apply AI to determine the likelihood of consumers' future electricity consumption and respond to it on a real-time basis, or a DT of a transportation network can apply the ML to predict traffic congestion along with the necessary rerouting of the vehicles involved. The DT framework integration of IoT in the 6G environment will extend the possibility of seamlessly connecting billions of people to the urban system and will extend DT's authority to monitor and control at scale [16]. Combined, they make up a system where real-time big data processing and automation are a possibility in smart cities.

11.4.3 Real-time decision-making and data analytics powered by 6G

One of the important benefits of 6G–DT integration is the capability to make real-time decisions based on continuous data analysis. For instance, within traffic management, a DT can apply real-time data coming from traffic sensors, cameras, or connected cars to optimize traffic signals or reroute cars and even inform drivers of an upcoming jam. This measure of real-time response has the potential to minimize traffic congestion, raise road safety, and increase the work productivity of the transport systems in urban areas. For example, in the event of an earthquake, a DT can review the environment and make important decisions for the authorities regarding the possible directions of the evacuation or rationing of resources. All of these decisions are made significantly faster and accurately with the help of ultrafast data transmission and low-latency communication of the 6G.

11.4.4 Key technical considerations for implementing DT over 6G (scalability, data security, and interoperability)

Nevertheless, DTs have many advantages in their application over 6G; however, implementation of DTs over 6G comes with multiple technical difficulties. They include scalability because the integration of millions of IoT devices across a city means that the infrastructure must be able to handle the enormous amount of data generated, some of which is addressed by 6G, but other factors such as effective network management systems and edge computing will be essential to accommodate the data load. Security of data can be another main issue primarily because volumes of information sent between tangible and intangible objects, including individuals' information, infrastructure state, and operational characteristics, can be vulnerable to cyberthreats. Lastly, integration is also major because it is about the ability of various systems and devices to communicate with each other and establish operational harmony. These needs creating a set of guidelines that dictate how data are exchanged and information shared between IoT devices, AI algorithms, and DT platforms operating on a 6G network.

11.5 APPLICATIONS IN SMART CITIES

DT technology and 6G network integration has multiple uses in smart cities that improve urbanism remarkably. In infrastructure, it is predictive maintenance where DTs imitate the real-time conditions of buildings, bridges, and roads, which eventually avoids expensive failure. To optimize traffic flow, DTs together with the support of 6G networking can adapt traffic patterns to eliminate traffic jams for enhanced mobility. There is also increased monitoring of the physical environment where DTs monitor air and water quality as well as energy use, thus enabling timely corrective action and or adjustments to meet sustainability targets. Furthermore, the self-driving vehicles and public transport enhanced through 6G-powered DTs of the autonomous mobility systems are made safer and optimized to bring effective praises to the mobility of the urban environment.

11.5.1 Urban infrastructure management

To the current knowledge, Urban Infrastructure Management is the most effective use case of DT and 6G integration in smart cities. DTs allow the real-time cyber representation of physical structures including buildings, bridges, and roads to make them responsive to continuous assessment and monitoring. It enables city planners and administrators to forecast when a certain infrastructure would require attention and when a particular process could fail. The connection of 6G's high speed and very large connectivity makes it possible for DTs to immediately process data from sensors that have been installed on structures [17].

This real-time capability also means that cities can provide predictive maintenance, which ascertains whether their facilities are suffering from wear or if their structural components could pose some problems at one point; such optimizations help ensure proper utilization of resources and minimum redundancy time. For instance, an aircraft maintained by a DT will signal signs of strain or damage on a bridge and necessary fixes done ahead of time. In addition, DTs can mimic the effects of various conditions, for example, weather or disasters, to show urban facilities and design ways to mitigate and manage them.

It also allows for the process of achieving optimization of infrastructural management that also includes energy, water, and waste. It can model patterns of usage to allow cities to flex resources accordingly without causing the Systems or Urban Commission to become detrimentally unsustainable.

11.5.2 Predictive maintenance of buildings, bridges, and roads using DT and 6G

Perhaps the most valuable area of applying DT and 6G connections in smart cities is the area of asset health monitoring and management where, for example, buildings, bridges, and roads should be maintained. In connected infrastructures, approximately, elements installed in various structures capture characteristics such as structural well-being, vibration proclivity, and load-carrying capability. These data are inputted into a DT model that replicates the real-time condition of the infrastructure and predicts any potential problem that may arise before it becomes serious; 6G improves this by enabling real-time data transfer needed in transmitting data to enable city officials for timely and efficient maintenance and consequent repair of the infrastructure, hence avoiding costly failures that pose a threat to citizens' safety.

11.5.3 Benefits of real-time monitoring and simulation

Simulation and monitoring of the infrastructure in real time have the following advantages. First, they abolish the need to conduct inspections manually, and this is because they are usually burdensome and expensive. With DTs powered by 6G, cities can mechanize the monitoring of infrastructure and accept alerts as quickly as potential difficulties are detected. This in turn results in a quicker response, low operational costs, and longer service lifecycle of infrastructural parts. In addition, real-time simulation permits strategies of maintenance to be examined within the simulation environment before the application of the same in the "physical world."

11.5.4 Traffic flow optimization

Traffic flow optimization is one of the key services using the integration of DT and 6G networks in smart cities. Another capability well served by DTs is based on performing real-time data provided by sources like IoT sensors, traffic cameras, and connected vehicles and providing dynamic simulations of traffic and traffic patterns. It enables city planners and traffic authorities to study one situation and predict other situations that may exist concerning traffic behavior, thus helping in making improved decisions [18].

With 6G's ultralow latency and high data transfer speeds, DTs can process large amounts of data almost immediately, enabling instant adjustments to traffic signals, lane assignments, and routing plans. For instance, the system can coordinate with traffic lights or reroute cars in cases where a busy throughway has turned busy all of a sudden. It also helps eliminate excessive downtimes while cutting on the fuel required and therefore becoming more environmentally friendly.

Moreover, incorporating more features from AI and specifically ML improves DTs' predictive powers for more complex analyses used for managing traffic patterns during high traffic volume and specific events. This kind of traffic convergence holds the promise of an unending feedback loop so that cities may develop adequate methods to keep traffic flows in check, increasing general mobility and safety. Therefore, the federated learning-based DTs and 6G enhancements to the current traffic flow lead to efficient urban transportation systems to the advantage of the public and the environment.

11.5.5 Use of DTs to simulate and optimize traffic patterns in real-time

It may be noted that DTs is a potent solution for undertaking and regulating the traffic in smart city systems effectively. Due to the modeling of a city's transportation system by DTs including roads and intersections, cars, and traffic lights, the overall traffic can be simulated in real time. These have considerations of directions, velocities, and other characteristics of varying vehicles, nature of terrain, or weather and can offer real-time control inputs for traffic signals or even propose new routes to help reduce traffic jams. 6G's ultralow latency is a great strength and allows these real-time simulations to be productive and helpful for teaming up cities and their traffic needs.

11.5.6 6G's role in facilitating high-speed data transmission for traffic systems

The traffic systems count on vehicular, sensing, and vision-based technologies that produce massive amounts of data, and 6G networks are crucial for expeditious data transfer. 6G transmission speeds permit data to be analyzed and utilized in real-time, thus enabling traffic management systems to provide adjustments for vehicle circulation right away. This is specifically so in smart cities where congestion density may quickly change, and, therefore, a quick response is key to avoiding jams and shaving time off travel.

11.5.7 Environmental monitoring

Environmental Monitoring is one of the primary use cases of applying the combination of DT and 6G networks in smart cities. DTs augment information pulled from IoT sensors placed at various locations in an urban setting to offer information on the condition of the environment, including air and water quality, noise pollution, and energy use. Abovementioned data enable the city officials to assess environmental health at any given point in time and address new problems as they arise. By applying 6G, DTs can assess extensive amounts of data in real time, which can allow them to respond instantly if pollution levels are too high or if resource consumption increases beyond a certain limit [19]. For example, if on crossroad areas, the air quality degrades, measures such as car ban or awareness creation can be initiated. Furthermore, there is an opportunity to use different patterns of ML and predictive analytics for estimating environmental strains and gaps to manage them effectively. With DTs and 6G integration, smart cities can improve the efficiency of environmental control, make progress toward sustainability objectives, and boost health for inhabitants.

11.5.8 Real-time air quality, water quality, and energy usage monitoring using DTs

Another area that can benefit DTs and 6G is environmental monitoring. For instance, in a smart city, there can be IoT sensors that are used to measure and collect data on the quality of air and water, level of noise, and energy consumption. This information is then channeled into the DT model that offers a real-time prediction of the environmental status of the city. For instance, a DT can mimic the effect of traffic emissions on the air and suggest measures to enhance the quality. The improvements in data transmission in 6G make it possible for cities to ensure that monitoring systems are fed with accurate data promptly to respond to issues affecting the environment.

11.5.9 Integration with smart grids and sustainability initiatives

DTs can also be implemented with smart grid systems to monitor and improve energy consumption within a city. Real-time energy monitoring and forecasting enable DTs to assist cities in achieving more sustainable energy solutions. For instance, in high demand conditions, a DT could create a model of the most optimal usage of energy resources and adjust the flows of power according to that model. 6G data handling capacity and support of multiple IoT devices keep smart grids functional and dynamic, playing a positive role in the sustainability of a city.

11.6 AUTONOMOUS MOBILITY

Autonomous mobility is a transformative application resulting from the integration of DT technology and 6G networks in smart cities. DTs can simulate and optimize the operations of autonomous cars and public transportation by building real-time virtual models of transportation networks [20]. These virtual models thanks to 6G's extremely low latency and fast data transfer speeds can process large volumes of data from networked cars and infrastructure nearly instantly. This improves the efficiency and safety by enabling exact synchronization between moving cars, traffic signals, and other urban infrastructure. Autonomous cars possess the ability to dynamically modify their routes in response to current traffic conditions and environmental considerations. Additionally, the combination of AI and DTs makes predictive analytics easier, which makes it possible to optimize mobility services to lessen traffic and enhance user experience. In the end, DT and 6G work together to create a more secure and effective transportation ecosystem, opening the door for seamless urban mobility.

11.6.1 Role of 6G-enabled DT in supporting autonomous vehicles and public transport

Future smart cities will primarily consist of AVs and intelligent public transit networks, with DTs being essential to the management of these systems. Real-time route optimization, traffic pattern prediction, and AV simulation are all possible with a transportation network DT. This makes operating AVs and public transport networks safer and more effective. In this situation, 6G networks are crucial because they offer the low-latency connectivity required for control systems, infrastructure, and automobiles to coordinate in real time.

11.6.2 Simulating and managing mobility services for safer, efficient transportation

Cities can simulate and manage a range of mobility services, such as ride-sharing, public transportation, and autonomous vehicles, with the integration of DTs and 6G. Cities may maximize the effectiveness of their transport networks by taking into consideration variables like traffic conditions, vehicle availability, and passenger demand with these simulations. Moreover, DTs can indicate the best course of action in emergency events, such as traffic accidents or natural disasters. This feature improves the safety and efficiency of the transit networks, which raises the city's overall liveability.

11.7 CHALLENGES AND FUTURE DIRECTIONS

While there is a lot of promise associated with the convergence of 6G networks with DT technology in smart cities, several obstacles must be overcome for successful deployment [21]. Security and privacy are of utmost importance because relying on data from several sensors raises the possibility of cyberattacks and data breaches. Robust cybersecurity procedures are necessary to protect sensitive information while preserving accessibility. Furthermore, as huge volumes of data are generated by smart cities and need to be efficiently stored, processed, and integrated, effective data management and interoperability are essential. Standardized protocols are required to ensure smooth communication between various systems and platforms. Technical challenges also arise when integrating new DT–6G technologies with legacy infrastructures, especially when it comes to integrating AI and IoT systems. It is recommended that future studies concentrate on promoting sustainable behaviors in urban contexts, increasing energy efficiency, and improving scalability. In addition to making the integration of DTs and

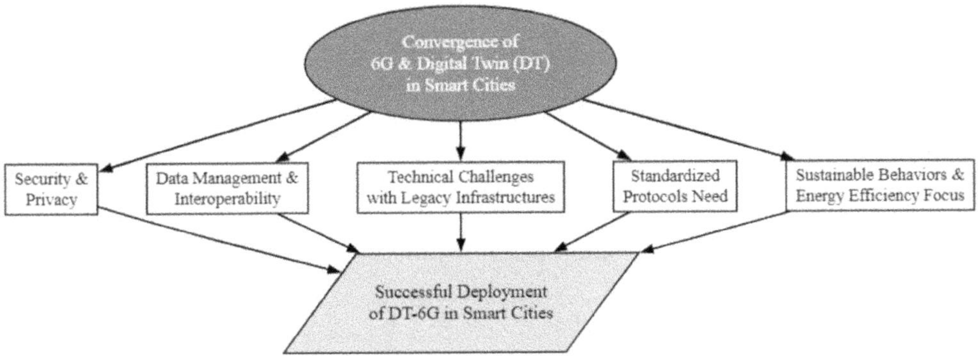

Figure 11.5 Challenges and considerations for the convergence of 6G networks with Digital Twin technology in smart cities.

6G in smart cities easier, addressing these issues will encourage the creation of robust, intelligent urban landscapes that can adjust to the complexity of contemporary living.

Figure 11.5 illustrates the key obstacles to the successful deployment of DT technology integrated with 6G networks in smart cities, emphasizing the importance of addressing security, data management, technical challenges, standardization, and the need for future research to achieve robust urban environments.

11.7.1 Security and privacy concerns in DT–6G systems

The combination of 6G and DTs poses serious security and privacy risks. The vast volume of data being created, transferred, and kept by smart city systems raises significant concerns about the possibility of cyberattacks and data breaches. Cybercriminals may use DTs to obtain sensitive personal data or take control of vital urban infrastructure. It is crucial to guarantee the security of data transmission over 6G networks, necessitating the use of strong encryption techniques, intrusion detection systems, and safe authentication procedures [22]. Furthermore, as more behavioral and personal information is gathered from citizens, privacy issues surface, underscoring the necessity for strict data protection laws.

11.7.2 Data management and interoperability challenges

Managing the massive amounts of data created by IoT devices and other sources effectively is one of the main problems of integrating DTs with 6G. Real-time data collection, processing, and analysis necessitate complex data management systems that can handle massive activities [23]. Since smart cities rely on a broad ecosystem of platforms, devices, and technologies that must interact with one another seamlessly, interoperability is another crucial challenge. Creating standardized protocols and making sure various systems work together will be necessary to make DT–6G integration a reality.

11.7.3 Integration with existing IoT and AI systems in cities

The seamless integration of DTs with 6G networks in smart cities necessitates the convergence of these technologies with the current IoT and AI systems that are already widely deployed [24]. Numerous cities have cameras, sensors, and smart devices installed to gather information on energy use, transportation, environmental conditions, and other topics. For real-time data to flow into the DT systems, these IoT devices must effortlessly interface with the 6G network. To provide useful insights, AI algorithms—that are essential for data analysis and prediction—will also need to function with 6G's increased speed and capacity. Compatibility problems must be resolved, nevertheless, to integrate these systems and make sure that devices from various suppliers or systems with various standards may interact with one another. To fully realize the potential of smart cities, an interoperable framework for IoT, AI, and DT systems must be established.

11.7.4 Prospects of DT-6G in global smart city development

The integration of 6G and DTs is anticipated to become more significant in urban development as smart cities across the globe continue to expand. DT technology is already being tested in cities like Helsinki, Barcelona, and Singapore to improve public services, transit, and infrastructure management. With 6G networks anticipated to provide even faster speeds, more capacity, and more connectivity than 5G, there will be much more room for autonomous systems, data analytics, and real-time simulations. Smart cities around the world will be more dependent on DT–6G convergence to address the issues of urbanization, population growth, and sustainability [25]. To ensure that these technologies are applied safely and successfully, governments, technology companies, and research institutes must work together to produce strong legislation, technological standards, and creative solutions. The future of urban living is expected to be shaped by DT–6G systems that increase efficiency, improve citizen experiences, and encourage sustainable development on a global scale.

11.8 POLICY IMPLICATIONS AND REGULATORY FRAMEWORKS

The integration of 6G networks and DT technologies in smart cities has important policy ramifications and calls for the creation of extensive regulatory frameworks [26]. It is essential to address these factors for efficient governance and implementation.

11.8.1 Data governance policies

The creation of data governance regulations that handle security, privacy, and moral issues about the massive volumes of data produced by networked urban systems must be given top priority by policymakers. These policies' principal elements are as follows:

- **Data Sharing Standards:** Establishing rules for sharing data among various stakeholders while guaranteeing accountability and openness.
- **Data Ownership Rights:** Clarifying possession rights regarding data collected from citizens and public infrastructure to avoid misuse.
- **Privacy Protection:** Putting policies in place to safeguard people's privacy and lessen the danger of data breaches.

11.8.2 Inter-agency collaboration

To successfully integrate DT and 6G technologies, government agencies, stakeholders from the commercial sector, and community organizations must work together across agencies. This partnership can help with the following:

- **Integrated Urban Strategies:** Creating coordinated plans for service delivery and urban management that make use of 6G and DT's combined strengths.
- **Alignment of Objectives:** Ensuring that, for comprehensive urban planning, the objectives of several sectors, including public health, energy, and transportation, are in line with one another.

11.8.3 Investment in digital infrastructure

Investment in digital infrastructure needs to be prioritized if DT and 6G technologies are to be fully utilized. Policymakers should consider the following:

- **Public Funding Initiatives:** Providing funds from the government to aid in the development of technologies for smart cities.
- **Private Sector Incentives:** Providing tax reductions and other benefits to promote private sector investment in infrastructure development and technology deployment.

11.8.4 Public Engagement

Maintaining public trust and acceptance of DT–6G programs requires ongoing public participation. Among the most important tactics are the following:

- **Community Involvement:** Getting people talking about how these technologies will be used to get their opinions and thoughts.
- **Education and Awareness Campaigns:** Educating the public about the advantages and difficulties of DT and 6G technologies in order to increase acceptance and comprehension.

11.9 CONCLUSION

The development of smart cities has advanced significantly with the convergence of 6G networks and DT technology, which provides game-changing answers to urban problems. Cities can optimize traffic flow, infrastructure management, environmental monitoring, and autonomous mobility with the help of this integration, which enables real-time monitoring, predictive analytics, and improved connection. As a result, urban environments becoming more responsive, efficient, and sustainable can successfully address problems like population increase, resource management, and climate change. Nevertheless, DT–6G system implementation success necessitates resolving issues with interoperability, security, and data management. Stakeholders can maximize the benefits of this technological convergence by concentrating on these areas. In the end, DT technology and 6G networks working together will create resilient urban ecosystems that will improve people's quality of life and promote sustainable urban growth in the future.

11.9.1 Summary of key insights from DT–6G convergence in smart cities

A major development for smart cities is the combination of 6G networks and DTs, which allows for real-time simulations, data analysis, and responsive decision-making throughout urban systems. The main benefits of 6G, including ultralow latency, increased bandwidth, and widespread device connectivity, have been discussed in this chapter along with how they boost DT applications' accuracy and efficiency. The integration of these technologies enables cities to operate with unprecedented accuracy and efficiency, from monitoring environmental conditions and facilitating autonomous transportation to controlling urban infrastructure and streamlining traffic. The important finding is that 6G-enabled DTs can simulate and manage intricate urban environments in real time, providing prompt answers to problems like environmental sustainability, traffic, and maintenance.

11.9.2 How this technology convergence can contribute to sustainable urban development

The promotion of sustainable urban development can be greatly aided by the merging of DT and 6G. By extending the lifespan of urban assets and ensuring predictive maintenance, real-time infrastructure monitoring with DTs lowers the likelihood of expensive malfunctions and promotes resource conservation. By reducing fuel use and carbon emissions, traffic flow optimization aids cities in reaching their environmental goals. Furthermore, real-time environmental monitoring helps smart city projects that are aimed at lowering pollution, conserving energy, and making effective use of available resources. By offering data-driven insights that empower city planners and decision-makers to make educated decisions about energy consumption, mobility, and public services, DT–6G technology promotes smarter, more sustainable urban planning. These technological advancements facilitate the development of resilient, sustainable cities that are better prepared to handle the problems brought on by urbanization and climate change.

11.9.3 Future research directions and anticipated challenges

Although DT–6G convergence has transformative advantages, a number of obstacles still need to be addressed, requiring additional study. In DT–6G systems, security and privacy are critical because of the large volumes of sensitive data that are being gathered and handled. It will be crucial to guarantee strong cybersecurity safeguards against data breaches and unauthorised access. Data management and interoperability present another difficulty; cities need to create standardized frameworks to guarantee smooth connection between IoT devices, AI systems, and DTs across various platforms and manufacturers. Moreover, creative solutions will be needed to scale DT–6G systems to handle expanding urban populations and more complicated urban situations.

Future studies should look into cutting-edge technologies like edge computing, which reduces latency and bandwidth consumption by bringing processing power closer to IoT devices since they could improve DT–6G integration even further. While quantum computing has the potential to transform simulations by offering previously unheard-of computational capacity, the combination of AI with IoT (AIoT) can enhance decision-making. Researchers can help realize the full potential of DT–6G technologies and guarantee that they keep pushing the development of smart, sustainable cities globally by tackling these obstacles and breaking new ground.

REFERENCES

1. R. F. El-Agamy, H. A. Sayad, A. M. AL Akhatatneh, M. Aljohani, M. Elhosseini, "Comprehensive analysis of digital twins in smart cities: A 4200-paper bibliometric study", *Artificial Intelligence Review*, vol. 57, no. 154, 2024.
2. S. Sharma, et al., "The role of 6G technologies in advancing smart city applications: Opportunities and challenges", Sustainability, vol. 16, no. 16, pp. 1–54, 2024.
3. J. S. Gracias, et al., "Smart cities—A structured literature review", *Smart Cities*, vol. 6, no. 4, pp. 1719–1743, pp. 1–53, 2023.
4. A. G. Mohapatra, et al., "ch 18: Harmonizing nature and technology: The synergy of digital twin-enabled smart farming", In Sailesh Iyer, Anand Nayyar, Anand Paul, Mohd Naved (eds.), *Digital Twins for Smart Cities and Villages*, Elsevier, 1st Edition, pp. 407–442, 2024.
5. M. Attaran, "The impact of 5G on the evolution of intelligent automation and industry digitization", *Journal of Ambient Intelligence and Humanized Computing*, vol. 14, pp. 5977–5993, 2023.
6. M. Banafaa, et al., "6G mobile communication technology: Requirements, targets, applications, challenges, advantages, and opportunities", *Alexandria Engineering Journal*, vol. 64, pp. 245–274, 2023.
7. S. A. A. Hakeem, H. H. Hussein, H. Kim, "Vision and research directions of 6G technologies and applications", *Journal of King Saud University-Computer and Information Sciences*, vol. 34, no. 6, pp. 2419–2442, 2022.
8. A. Solyman, K. Yahya, "Evolution of wireless communication networks: From 1G to 6G and future perspective", *International Journal of Electrical and Computer Engineering (IJECE),* vol. 12, no. 4, pp. 3943–3950, 2022.
9. P. R. Singh, V. K. Singh, R. Yadav, S. N. Chaurasia, "6G networks for artificial intelligence-enabled smart cities applications: A scoping review", *Telematics and Informatics Reports*, vol. 9, pp. 1–192023.
10. D. Peldon, S. Banihashemi, K. LeNguyen, S. Derrible, "Navigating urban complexity: The transformative role of digital twins in smart city development", *Sustainable Cities and Society*, vol. 111, pp. 1–18, 2024.
11. Md. S. Dihan, et al., "Digital twin: Data exploration, architecture, implementation and future", *Heliyon*, vol. 10, no. 5, pp. 1–27, 2024.
12. I. Jeddoub, G-A Nys, R. Hajji, R. Billen, "Digital Twins for cities: Analyzing the gap between concepts and current implementations with a specific focus on data integration", *International Journal of Applied Earth Observation and Geoinformation*, vol. 122, 1–23, 2023.
13. G. Caprari, G. Castelli, M. Montuori, M. Camardelli, R. Malvezzi, "Digital twin for urban planning in the green deal era: A state of the art and future perspectives", *Sustainability*, vol. 14, no. 22, pp. 1–16, 2022.
14. N. P. Kuruvatti, et al., "Empowering 6G communication systems with digital twin technology: A comprehensive survey", IEEE Access, vol. 10, no. 6, pp. 112158–112186, 2021.
15. O. El Rhazal, T. Mazri, "Digital twins approach proposition for 5G/6G", *Journal of Autonomous Intelligence*, vol. 7, no. 5, pp. 1–9, 2024.
16. Z. Qadir, K. N. Le, N. Saeed, H. S. Munawar, "Towards 6G Internet of Things: Recent advances, use cases, and open challenges", *ICT Express*, vol. 9, no. 3, pp. 296–312, 2023.
17. E. Faliagka, et al., "Trends in digital twin framework architectures for smart cities: A case study in smart mobility", *Sensors*, vol. 24, no. 5, pp. 1–22, 2024.

18. B. Yan, et al., "Digital twin in transportation infrastructure management: A systematic review", *Intelligent Transportation Infrastructure*, vol. 2, pp. 1–18, 2023.
19. S. P. Ramu, et al., "Federated learning enabled digital twins for smart cities: Concepts, recent advances, and future directions", *Sustainable Cities and Society*, vol. 79, pp. 1–19, 2022.
20. K. Wang, et al., "Smart mobility digital twin based automated vehicle navigation system: A proof of concept", *IEEE Transactions on Intelligent Vehicles*, vol. 9, no. 3, pp. 4348–4361, 2024.
21. D. M. Botin-Sanabria, et al., "Digital twin technology challenges and applications: A comprehensive review", *Remote Sensing*, vol. 14, no. 6, pp. 1–25, 2022.
22. M. Wang, et al., "Security and privacy in 6G networks: New areas and new challenges", *Digital Communications and Networks*, vol. 6, no. 3, pp. 281–291, 2020.
23. A. J. Holmgren, J. Eversom, J. Adler-Milstein, "Association of hospital interoperable data sharing with alternative payment model participation", *JAMA Health Forum*, vol. 3, no. 2, pp. 1–11, 2022.
24. K. Sharma, S. K. Shivandu, "Integrating artificial intelligence and Internet of Things (IoT) for enhanced crop monitoring and management in precision agriculture", *Sensors International*, vol. 5, pp. 1–12, 2024.
25. Z. Allam, D. S. Jones, "Future (post-COVID) digital, smart and sustainable cities in the wake of 6G: Digital twins, immersive realities and new urban economies", *Land Use Policy*, vol. 101, 1–25, 2021.
26. M. Haraguchi, T. Funahashi, F. Biljecki, "Assessing governance implications of city digital twin technology: A maturity model approach", *Technological Forecasting and Social Change*, vol. 204, pp. 1–19, 2024.

Chapter 12

Service scenarios of advanced secure computer vision applications

Meha Sharma, Arpita Singh, Anand Kumar Mishra, Chetan R. Dudhagara, and Umesh B. Tank

12.1 INTRODUCTION

Computer vision (CV) is a relatively recent field of computer science and engineering, as a branch of artificial intelligence (AI). It helps users allow computers to "recognize" and understand visual data from the real world (e.g., images, static data, X-ray images, and videos), as humans do. It entails creating different algorithms and combinations of systems that can process, evaluate, and understand the visual input from the surroundings, allowing machines to accomplish activities and perform tasks that would normally need human vision [1]. CV permits machines to read, analyze, and extract useful data from images and video content. It encompasses the development of algorithms and techniques that allow computers to analyze and understand visual information, such as images and video sources. CV uses sensors and interpreters that work similarly to the human eye and the brain's visual cortex to replicate the human visual system. CV applications use cameras, sensors, and advanced algorithms that are trained on massive amounts of visual data and images. This form of AI enhances efficiency, innovation, and automation in various sectors [2,3].

Today, many modern CV applications run in environments that are spread out and limited in resources. For example, they work on smart home devices, mobile phones, and drones used in defense. In these conditions, keeping data safe and private is not easy, but it is crucial. This challenge has led to the development of sophisticated, secure CV, which integrates security tools, encryption, privacy-friendly AI techniques, and CV techniques to create systems that are both intelligent and reliable [4].

Advanced secure computer vision (ASCV) is dedicated to the integration of cutting-edge CV techniques with robust security and privacy mechanisms. In addition to interpreting visual data (such as pictures and videos), it emphasizes doing so in a way that safeguards private data, maintains data integrity, and prevents malicious interference or exploitation.

12.2 THE HISTORY OF COMPUTER VISION

Researchers started creating methods for pattern recognition and image processing in the 1950s and 1960s, such as template matching, edge detection, and thresholding. The groundwork for further CV work was established by these early methods [1,2]. Researchers

DOI: 10.1201/9781003659112-12

started creating feature-based techniques in CV, such as texture analysis, edge detection, and corner detection, in the 1970s and 1980s. These techniques concentrated on finding and removing characteristics from pictures that may be utilized to recognize patterns and items.

Researchers in the 1980s and 1990s started to develop geometric techniques for CV, including areas like stereo-vision, optical flow, and shape from shading [5]. These techniques aim to apply geometric principles to infer the three-dimensional structure of objects from two-dimensional images. After that, in the 2000s and beyond, the advent of machine learning (ML) and deep learning fundamentally changed the evolution of CV [6]. By using these techniques, researchers were able to develop algorithms that could automatically extract features from photos and learn from visual information, opening the door to more complex and precise applications.

12.2.1 Components of computer vision (CV)

CV components can be divided into two main categories:

12.2.1.1 Traditional computer vision components

A branch of AI known as CV allows machines to interpret and understand visual information from their surroundings. These are based on classical image processing and feature extraction techniques such as edge detection, corner detection, segmentation, and handcrafted feature descriptors. A traditional CV relies heavily on manually designed algorithms to interpret visual data. Some essential components of a standard CV system are shown in (Figure 12.1):

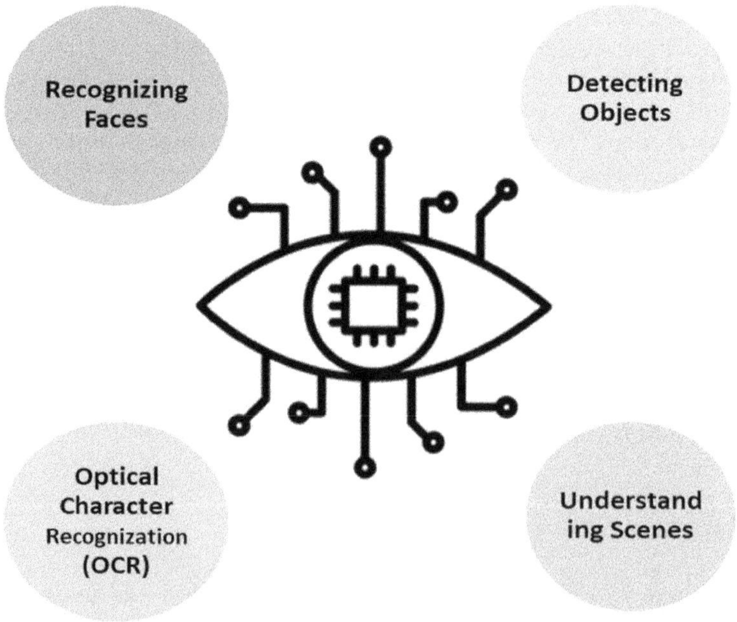

Figure 12.1 Traditional CV components.

1. **Recognizing Faces:** Face recognition is a sophisticated cognitive function that depends on a special area of the brain called the Fusiform Face Area (FFA). This area is crucial for the relationship between the static components of a face (its general structure) and the dynamic changes produced by facial expressions. Recognizing faces is crucial for social engagement and dialogue [7].

   Facial recognition is a biometric process that works on a person's unique facial structure and skin patterns for their identification. The technique entails obtaining a facial image—whether from a real-time camera stream, a snapshot, or a video frame—and extracting distinguishing traits such as eye distances, nose shape, jawline contour, skin texture patterns, and relative proportions of facial elements [8]. The approach consists of capturing a facial image—whether from a real-time camera stream, a still photo, or a frame from a video—and recognizing distinctive features like the spacing between the eyes, the shape of the nose, the outline of the jawline, the texture of the skin, and the proportional relationships among the facial features. These qualities are subsequently converted into a mathematical format called a face template [9].
2. **Detecting Objects:** In CV, detecting objects is a core activity that involves identifying and locating instances of objects (such as individuals, vehicles, animals, and so on) in images or video contents [10]. Object detection combines **classification** (what is it?) and **localization** (where is it?). It entails making bounding boxes (rectangular boxes) around detected objects, which enables us to identify and track their presence and movement within that environment. Object detection is categorized into two key steps:
   - Single-stage object detection.
   - Two-stage object detection.

   In single-stage object detection, all bounding boxes are executed in a single operation by running once through the neural network. The You Only Look Once (YOLO) model, an algorithm designed for object detection-related tasks, uses a single forward pass to make predictions of item bounding boxes and class probabilities at the same time for the entire image [11].

   Two-stage object detection emphasizes two models basically: the first model locates regions that may contain objects. On the other hand, the second model classifies and improves the detected object localization of objects.
3. **Understanding Scenes:** Scene understanding refers to the capability of a CV system to draw out advanced understanding of a visual scene—including the objects involved, their spatial relationships, actions occurring, and occasionally even the intent or context.
4. **Optical Character Recognition (OCR)/Reading Text:** OCR is a process utilized in CV to recognize and extract text from images, scanned data/files, or other visual inputs, converting it into text that machines can read and modify. At its basic points, OCR allows computers to understand characters (letters, digits, and symbols) within pixel-based data, converting unstructured image information into structured text [12,13]. This technique is essential for automating tasks that involve processing printed or handwritten text, such as digitizing paper records, separating information from invoices, or enabling text search in scanned PDFs [14]. OCR systems generally work by analyzing the forms and patterns present in an image, identifying areas containing text, and converting those visual representations into encoded symbols (Table 12.1).

Table 12.1 OCR use cases in real life

Use case	*Actual examples*
Document scanning	Scanned papers, PDFs, and books
Automated License Plate Recognition (ALPR)	Vehicle number plate, traffic cameras, and parking areas
Translation apps	Enabling translation of text and extracting text from scanned documents
Invoice/receipt parsing	Extracting structured data from scanned invoices or receipts
Identity (Id) verification	License of drivers, passport, and national identity
Accessibility	Screen readers for blind people

12.2.1.2 Advanced and secure computer vision components

It is based on modern AI, deep learning, and security-enhanced methods. These categories involve deep learning approaches and security and privacy mechanisms. It enhances methods for ensuring security and robustness, such as adversarial attack defenses, privacy-preserving techniques, and secure data handling. Numerous elements are involved, yet specific essential ones are highlighted and examined separately by the authors.

Adversarial Robustness in Computer Vision: ASCV is a related branch of AI that enhances machines' understanding and interpretation of images or videos. It's used in many places, such as face recognition, self-driving cars, medical imaging, and surveillance. But these systems are not perfect. Sometimes, attackers can trick them using special techniques. This technique is called an adversarial attack. In this attack, the image is modified in tiny ways that humans can't even notice, but the model makes a mistake. For instance, by placing a few tiny dots on a stop sign image, a computer may misinterpret it as a speed limit sign. This poses a risk, particularly in essential systems such as autonomous vehicles or medical devices. Adversarial robustness refers to the enhancement of a CV model to withstand such deceptive attacks. A resilient model will reliably yield the correct output, even when the input image has been subtly and misleadingly altered.

Adversarial attacks can lead to critical challenges in real life. If a medical AI system is deceived into overlooking a tumor or if a closed-circuit television (CCTV) camera fails to identify a face of an individual, the outcomes could be dangerous or even life-threatening. Attackers can use these tricks to steal information, damage trust in AI systems, or cheat in automated systems (like CAPTCHA or facial payment). That's why, it's very important to study adversarial robustness—to keep AI systems safe and reliable. An attacker generally makes small changes (known as perturbations) to an image that are imperceptible to the human eye. However, these minute modifications can bewilder the model and result in inaccurate predictions. There are various types of attacks:

- **White-box Attacks:** Where the attacker identifies everything about the model.
- **Black-box Attacks:** Where the attacker can only provide input data and observe output data.
- **Physical Attacks:** Where modifications are applied in the physical world (such as stickers on signs).

Techniques of Adversarial Attack: Strategies for enhancing adversarial robustness focus on ensuring that ML models can withstand inputs that are maliciously designed (adversarial examples) to deceive them. Key methods include the following:

- Adversarial Training: The model is trained using both standard and adversarial samples, allowing it to learn to accurately classify or reject altered inputs.
- Defensive Distillation: This involves training a model to generate "soft" probabilities, which decreases its sensitivity to minor changes in the input.
- Input Preprocessing: Techniques such as input demonizing, feature squeezing, or randomization are employed to eliminate or interfere with adversarial alterations before the data are processed by the model.

Privacy-Preserving Machine Learning (PPML): It implements AI systems that safeguard the personal details they utilize or learn from. This is crucial when working with CV models as the images and videos often contain private data. It covers many domains like healthcare (analyzing X-rays), security (facial recognition), and retail (observing customers in shops). These systems frequently utilize sensitive and personal information, including images of individuals' faces, physical movements, or medical photographs. Individuals' personal information may be revealed or exploited without privacy safeguards. For example:

- A facial recognition system could disclose images without consent.
- A smart home surveillance device might record video without the user's awareness.
- Medical imaging technologies could be compromised, leading to leaks of patient data.

Such laws as the General Data Protection Regulation (GDPR) (in Europe) and Health Insurance Portability and Accountability Act (HIPAA) (in the USA) mandate that companies safeguard user information. Therefore, techniques that preserve privacy are not only beneficial but also legally obligatory in various regions. The authors discussed how we can ensure privacy in vision models. There are multiple strategies to maintain privacy while still enabling AI models to learn:

- **Federated Learning:** Rather than transmitting data to a singular central server, the model is deployed to different devices (such as smartphones or cameras). Each device enhances the model using its local data and only returns the updated results, not the original data.
- **Differential Privacy:** This approach introduces "noise" (random alterations) to the data or model, ensuring that personal details cannot be linked to an individual, while still retaining useful overall patterns for learning.
- **Homomorphic Encryption:** This method permits the model to generate predictions on encrypted data, meaning it never accesses the actual image but can still process it securely.
- **Secure Multiparty Computation (SMPC):** Data are divided into small segments and handled by various parties so that no single party can view the complete input. The results are then merged at the end.
- **Anonymization and Blurring:** Before images are utilized, faces or other identifying features can be obscured or eliminated to protect identities.

Access Control and Authentication in Computer Vision : CV systems are employed to interpret images and videos for numerous areas, including facial recognition, license plate number detection, object tracking, and surveillance. These technologies frequently handle sensitive information and carry out vital functions, making it essential to ensure that only authorized/legal individuals can access them. This is where the concepts of access control and authentication come into play.

Authentication involves verifying the identity of a user (for instance, logging in using a password or fingerprint). Access control pertains to determining what actions a user is permitted to take once they are logged into the system (such as viewing images, making modifications, or simply reading results). These mechanisms are crucial for safeguarding CV systems against hackers, misuse, or unauthorized access. Numerous CV applications involve confidential or critical data. For instance:

- A facial recognition system in an educational institution stores images of students.
- A medical AI application analyzes X-rays or scans of patients.
- A surveillance system captures footage of individuals in both public and private spaces.

If unrestricted access to these systems is allowed, it may result in data breaches, privacy infractions, or security threats. This underscores the necessity for effective authentication and access control to ensure that only authorized users are able to view or manage the sensitive data and systems. To protect a CV system, developers use several methods:

User Authentication: This confirms identity through methods such as passwords, PINs, facial recognition, or two-factor authentication (2FA).

Role-Based Access Control (RBAC): Users are assigned specific "roles" like admin, doctor, and guest and can only perform actions that their role permits.

- **Multifactor Authentication (MFA):** Users are required to use more than one method for logging in (such as a password in conjunction with a code sent to their mobile device).
- **API Security:** When accessing the model via the internet, secure tokens and rate-limiting are utilized to prevent misuse.
- **Logging and Monitoring:** Every action taken within the system is documented to identify any abnormal behavior or threats.
- **Device-based Control:** Only authorized devices (such as designated computers or smartphones) are allowed to access the system.

Anomaly Detection: Anomaly detection, also referred to as outlier detection, involves the identification of data points, patterns, or events that significantly diverge from the expected behavior or standard distribution within a dataset. These anomalies can represent critical issues such as fraud, structural problems, sensor malfunctions, or cybersecurity risks, depending on the specific domain of application. Anomaly detection aims to pinpoint data points, patterns, or occurrences that stand out from the typical or anticipated behavior within a dataset. These unusual observations, commonly called anomalies or outliers, may signal important problems such as fraud, system failures, security incidents, or atypical user activities. In contrast to conventional classification tasks, anomaly detection usually deals with highly imbalanced datasets where anomalies are scarce and challenging to label. The approaches utilized for anomaly detection vary from statistical techniques and ML algorithms to more sophisticated deep learning methods, based on the data's complexity and nature. This process plays a crucial role in numerous practical applications, including finance (for fraud detection), healthcare (to identify disease outbreaks), cybersecurity (for intrusion detection), and industrial monitoring (for detecting machinery faults). There are three types of anomalies:

- Point Anomalies: Individual data instances that are considered anomalous when compared to the rest of the dataset (e.g., a fraudulent transaction).
- Contextual Anomalies: Data instances that are anomalous within a specific context (e.g., a temperature reading that is unusual for a particular season).
- Collective Anomalies: A group of data instances that are anomalous when viewed together (e.g., a series of failed login attempts).

Techniques of Anomaly Detection: The techniques for anomaly detection can primarily be categorized into three groups. Statistical methods operate on the premise that normal data adhere to a known distribution (such as Gaussian). Data points that deviate significantly from this distribution are flagged as anomalies. While these methods are straightforward, they often face challenges with complex or high-dimensional datasets.

- **Machine Learning Methods:** It can be classified as supervised, unsupervised, or semi-supervised.
- **Supervised Methods:** Supervised methods necessitate labeled normal and anomalous data to train classifiers.
  - **Unsupervised Methods:** These methods uncover anomalies by identifying data points that differ from the majority, utilizing clustering or isolation-based algorithms.
  - **Semi-supervised Methods:** They learn exclusively from normal data, marking deviations as anomalies, as seen in One-Class SVM or autoencoders.
- **Deep Learning Techniques:** They include LSTM networks, variational autoencoders (VAEs), and generative adversarial networks (GANs), which are capable of capturing intricate patterns, particularly in time-series or high-dimensional data. They represent normal behavior and recognize anomalies as deviations from learned models.

12.3 ARCHITECTURE OF SERVICE SCENARIO IN ADVANCED SECURE COMPUTER VISION APPLICATION

Advanced secure CV applications stand on a layered architecture to ensure effective image and video processing, data privacy, security, and real-time performance. Many modular components are responsible for a distinct stage of data processing, transmission, and utilization. These modules produce a safe, scalable, and smooth CV service. Figure 12.2 represent the architecture of service scenario of computer vision.

12.3.1 Data flow layer

The initial step in the data flow of secure CV systems is the capture of raw visual sources, cameras, Light Detection and Ranging, or other image-capturing sensors. These could

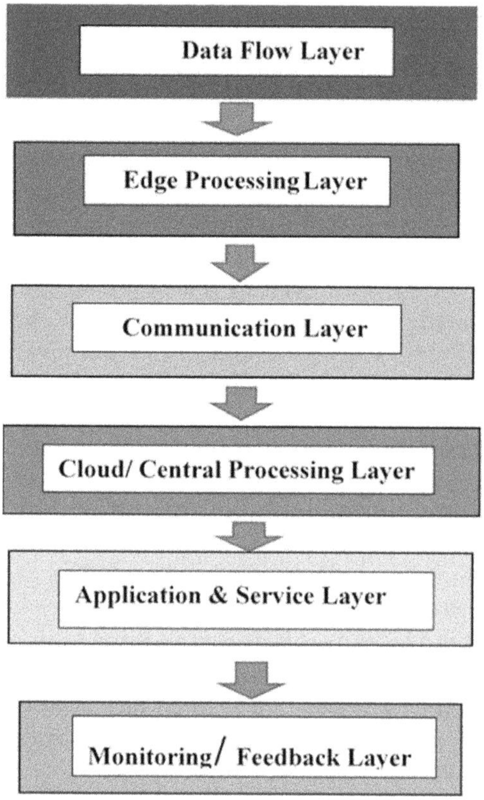

Figure 12.2 Architecture of service scenario in CV.

be embedded in edge devices (e.g., touch screen phones, CCTV cameras, surveillance systems, drones, or industrial robots). Data might be encrypted to reduce dangers at the beginning, before further processing. This type of raw data may handle sensitive visual details such as faces, license number plates, or proprietary goods, making it crucial to handle it with strong security measures from the outset. To mitigate risks at this early stage, data can be encrypted or anonymized before further processing. Edge devices are generally used to preprocess and encrypt such data locally, reducing the need to transmit raw images over potentially unsafe channels. The main focus at this layer is on high-resolution data capture and ensuring secure transmission through encrypted channels.

The data are securely acquired once and transmitted to a processing unit or a cloud-based inference engine for analysis. During this transmission phase, secure communication protocols such as Transport Layer Security (TLS)/Secure Sockets Layer (SSL) or end-to-end encryption are employed to prevent overhearing, tampering, or unauthorized access.

12.3.2 Edge processing layer

The edge processing layer in CV refers to the step where data from cameras and other sensors are processed locally on edge devices, such as smartphones, drones, or embedded systems, instead of being transmitted to centralized cloud servers. This layer plays an essential role in facilitating quicker decision-making, lowering latency, and improving privacy. By executing computations like image filtering, object detection, or facial recognition directly on the device, the edge processing layer reduces the need to transmit a huge amount of raw data over the network. This is uniquely crucial in services that need current responses, like autonomous vehicles or surveillance systems.

Furthermore, edge processing can recommend enhanced security by storing sensitive visual data closer to its origin and minimizing exposure to potential cyberthreats during transmission. Overall, this layer is a fundamental component in producing secure and efficient CV systems that operate reliably in real-world environments.

12.3.3 Communication layer

This layer manages secure data transmission between edge devices and cloud or data centres. Technologies like 5G, Wi-Fi 6, or Ethernet can be used, depending on the scenario. Data are encrypted using protocols such as TLS/SSL to prevent interception and tampering. Layer bandwidth optimization is achieved by data compression or prioritization of critical events. Moreover, this layer supports interoperability, allowing different devices and systems to work together within a unified architecture. It may use standard communication protocols such as MQTT, HTTP/HTTPS, or custom application programming interfaces (APIs) tailored for specific CV applications. The communication layer also facilitates feedback and control signals, enabling system components to adapt dynamically—for instance, a server might instruct a camera to adjust its resolution based on current network load or task requirements.

12.3.4 Cloud/central processing layer

Computationally intensive tasks such as deep neural network inference, three-dimensional (3D) reconstruction, or large-scale data aggregation are executed in this layer. It serves as the backbone for centralized AI model training, analytics, and orchestration. Cloud platforms also offer sophisticated services such as threat detection, model retraining, and audit logging. Security features include data anonymization, access control, homomorphic encryption, and zero-trust architectures. The cloud or centralized processing layer includes a computational framework where a huge amount of data handling, model development, inference, and storage operations occur on centralized servers usually situated on cloud platforms, for example: AWS, Azure, Google Cloud, or dedicated data centers. Some key feathers of cloud processing layer are discussed below:

- Data Aggregation and Storage: One of the key feathers is centralized systems, which gather a large scale of image and video content from edge devices (like cameras, mobile phones, and sensors), and another is the cloud, which produces scalable storage options, such as object storage and databases, to manage the high dimensionality and volume of vision data.
- Model Training and Inference: Training of deep learning models (e.g., CNNs and Vision Transformers) is computationally intensive and requires GPUs/TPUs—resources readily available in centralized environments. Inference (prediction) can also happen in the cloud, especially when low latency is not crucial.

- Data Labeling and Annotation: Cloud platforms frequently integrate with data labeling workflows, either automated or human-driven, which are essential for supervised learning in visual tasks.
- Advanced Analytics and Post-Processing: Facilitates sophisticated operations like 3D reconstruction, object tracking across video frames, semantic segmentation, and processing from multiple perspectives.
- Security Monitoring and Audit Trails: Logging, access management, and activity tracking are often more comprehensive in cloud environments, aiding in compliance with regulations such as the GDPR or HIPAA.

12.3.5 Application and service layer

This layer defines end-users or systems for the processed outcomes. Safe and secure APIs and user authentication protocols (OAuth2 and SAML) are used here to avoid misuse or unauthorized access. The examples of services are real-time notifications, dashboard displays, access control choices (like facial recognition-based authentication), or connections with other IT/OT systems. The essential role of the application and service layer is integration and coherence. This entails linking the advanced, secure CV system with different services like databases, alert systems, dashboards, or enterprise management services. For example, a smart traffic control system could work with city traffic control centers to improve signal timing in real time based on visual traffic assessments. Sophisticated CV systems can be integrated with electronic medical records to help in the diagnostic process. It is essential to ensure secure, authenticated, and encrypted data exchanges during these integrations to maintain the integrity of the whole system.

Ensuring security at this layer is crucial since it frequently interacts directly with outputs for users and decisions, which can affect privacy, safety, or operations. For instance, in facial recognition systems, the Application Layer decides whether to grant or deny access based on evaluations conducted at the lower layers. If this layer is compromised, it may result in unauthorized access, erroneous alerts, or data breaches. Thus, strong access control measures, encrypted data management, and secure API communications are critical elements of this layer. This layer is essential in detecting suspicious activities.

12.3.6 Monitoring and feedback layer

To ensure reliability and security, the system includes a continuous monitoring layer. This layer handles logging, anomaly detection, and model drift detection and enables adaptive learning. Feedback from this layer can be used to update models, fine-tune performance, or alert administrators to potential threats. Additionally, this layer facilitates adaptive learning and fine-tuning. Depending on feedback data, the system can be adjusted or modified to manage new issues, enhance its performance, and minimize false positive or negative results. For instance, in a smart surveillance system, when the feedback layer identifies frequent misclassification of objects under certain lighting conditions, it can suggest changes in preprocessing or model settings. This upcoming feedback loop is mandatory for continuing the long-term efficiency and safety of sophisticated CV systems working in changing, real-world settings. Within the realm of secure CV, this layer is crucial for detecting questionable behavior, including unauthorized access, attempts at tampering, or unusual usage trends. It is also capable of tracking potential model drift, which refers to the decline in AI model accuracy over time owing to changes in environmental factors or input data. To capture these problems, feedback mechanisms such as logging, auditing, and analytics dashboards are employed to recommend corrective measures, which may involve retraining models, modifying configurations, or isolating affected components.

12.4 WHY DOES SECURITY MATTERS?

Maintaining security is crucial in ASCV since these systems are being utilized more and more in vital fields like healthcare, surveillance, and self-driving cars [15]. There may be severe repercussions if they are compromised. Here's a quick summary of the importance of security.

12.4.1 Adversarial threats—privacy risks

Adversarial threats involve intentional perturbations or manipulations made to the input data, such as images or videos, that can trick ML models into generating wrong forecasts or choices. These perturbations are usually unapparent for humans to notice but can seriously affect the performance of CV systems [15]. For example, suppose there is a stop mark on the road, an attacker having bad intentions makes imperceptible changes to the stop sign, like modifying the colors or edges a bit. Minor changes can be so slight that an individual might not observe any differences. But when the car's computer system scans the sign, it gets confused by these small changes and thinks it's a yield sign instead of a stop sign. A yield sign indicates that the car doesn't need to come to a full stop and there is just a need to reduce speed and watch out for other cars. A disturbance that deceives one model may also deceive other models, which creates challenges for the model's robustness and security [13].

12.4.2 Tampering

Tampering in ASCV relates to any unauthorized or malicious manipulation of visual data, such as images, video, or model outputs, as well as the CV system itself. The purpose of tampering is regularly to mislead, interfere, or obtain information, which can present security and privacy risks [1]. It defines changing data or systems without permission, often to hide the truth or cause harm. In CV, this can affect the following: input images or videos, the AI model itself, the training data, and output results (like recognition or classification).

12.4.3 False alarms

These false types of alarms not only waste time and effort but also raise confusion among teams, distracting their attention from actual issues or tasks. Over time, repeated false alarms may lead to "alert fatigue," where critical notifications are ignored or taken less seriously. An overwhelming false alert can compromise security and reduce efficiency as it may lead to genuine threats being overlooked [16].

Sometimes, attacks can elevate the system to generate fake or misleading alerts, leading to unnecessary panic, wasting time, and misallocation of resources.

12.4.4 System failure

A successful attack could make the entire system stop working or give dangerous outcomes. Sometimes, a system failure occurs in ASCV when the system does not execute its task completely or correctly. These systems are designed to understand and analyze visual data, such as identifying objects, detecting faces, or monitoring/controlling movement [17]. A failure takes place when the system makes mistakes, such as misidentifying an object, missing a face, or giving wrong information about what it observes [18]. These failures can arise from various factors, such as poor and low image quality, bad lighting, complex backgrounds, or limitations in the training data used to teach the system. Sometimes, the techniques may not be strong enough to handle real-world situations, especially when the input is not constant from what the system has seen before.

12.5 TAXONOMY OF SERVICE SCENARIOS

12.5.1 Transportation and autonomous systems

An ASCV in the automotive industry enhances passenger safety, facilitates autonomous driving, and streamlines the management of vehicle flows [19]. We have previously described ASCV in many facets of the automotive industry.

- **Autonomous Vehicles:** ASCV enables object detection and scene comprehension for autonomous vehicles. This entails having mounted cameras and using AI being able to identify, to some extent, hospitable pedestrians, other vehicles, traffic lights, traffic signs, and anything else of interest on the road to enable safe operation. Companies like Tesla and Waymo are currently utilizing CV to examine and act upon identified roadway conditions in real time.
- **Driver Monitoring Systems:** The cameras inside the vehicle's internal cabin examine the face, gesture, and eye of the driver and detect fatigue or distraction [20]. Whenever the driver seems fatigued or distracted, the system will notify the driver or take other avoidance actions in the interest of safety. Whenever the driver seems fatigued or distracted, the system will notify the driver or take other avoidance actions in the interest of safety.
- **Vehicle Flow Management/Traffic Management:** A smart city with advanced, secure cameras examines traffic flow and control, identifies accidents, and modifies traffic signals [21]. Using CV on street CCTV, also known as video surveillance feeds, cities can optimize congestion control and even provide real-time traffic information to autonomous vehicle systems.
- **Vehicle Inspection and Maintenance:** CV systems continually analyze vehicles for problems such as underbody damage or tire wear. This method enhances accuracy, reduces time spent on the inspections, and eliminates dependence on human labor.

12.5.2 Surveillance and security

AI–CV systems ensure security teams respond faster and more efficiently to security threats [22].

- **Facial Recognition:** ASCV determines the use of sophisticated algorithms and technologies to correctly identify or verify humans based on facial characteristics, with a primary focus on security, privacy, and robustness in real-world conditions.
- **Anomaly Detection:** AI-powered surveillance systems can flag suspicious actions. Anomaly detection could be detecting a person loitering in a prohibited zone or abandoned objects in public spaces. Vision-based anomaly detection helps security teams focus on concentrate on possible threats in real time.
- **Access Control Systems:** Vision-based security gates and keyless entry systems enhance safety with less manual effort. These systems can read license plates to grant vehicle entry or use facial recognition to allow employees/visitors through doors.
- **Crowd Management:** It involves keeping an eye on the safety of large events, programs, or public areas. Secure CV can count and mark people within a crowd, can identify fights or stampedes, and can identify someone who needs medical attention (e.g., if someone has fallen). This will allow authorities to intervene faster in a large gathering.

12.5.3 Identity and financial services

Identifying individuals for purposes of security and authentication. This is already being done by law enforcement, and airports use this to identify persons of interest, while secure facilities use it to implement facial scan access control versus key cards.

12.5.4 Manufacturing and infrastructure

CV algorithms are also being implemented across the manufacturing area. It is interconnected with Internet of Things (IoT) devices configured in production machines. Over the past decade, manufacturing has gone through a major digital transformation—some refer to this as "Factory 3.0" or even "Factory 4.0." [23]. In factories, secure CV models and real-world use cases of annotated image and video datasets are being used to do the following:

- Operating a fully automated production line.
- Quality assurance by closing defects before they leave the production line or production facility.
- Quality assurance so that no defects are discovered before leaving the production site.
- Creating 3D models and applying advanced techniques in CV to improve manufacturing methods that people struggle with (e.g., hyper-detailed manufacturing and actually assembling all of the little bits of machines in the entire range of everything from electronics to oil and gas).
- Rotary and Laser Die Cutting: Cutting everything super accurately through the process of manufacturing.

- Active supervision to create efficiencies and minimize downtime.
- Improving fitness, safety, and security, managing wrap and delivery, and much more.

With respect to the opposite side of things, images and videos of the major components of infrastructure networks are often analyzed, tagged, annotated, and used in CV models, although it may often be too late to determine what repairs or replacements are needed; operational costs related to inspection and maintenance are declining, while energy outputs of generation plants, national transmission, and distribution networks are, in fact, increasing.

12.5.5 Smart cities and supply chain

A smart city represents a contemporary direction for urban development. It links together lots of areas through new technologies involving IoT and decentralized computing technologies to increase the available resources of the city by integrating, managing, and improving them [24]. The goal is to improve processes in areas such as healthcare, energy, transport, water management, education, and commerce. As large cities continue to grow, new generations are emerging with preferences other than older generations.

One of the main services that AI offers to smart cities is security. Advanced CV technology has been widely adopted to provide a safe environment in smart cities [25]. It has many applications, including but not limited to critical infrastructure protection, sanitation and garbage disposal, transportation (including a specific interest in transport monitoring), pandemic management, public safety, smart water management, and disaster management. These include surveillance and decision-making with intelligence provided through CV. CV is a real advancement in technology that allows computers to generate intelligence from images, videos, and other visual information. Its applications improve quality control in the supply chain [26]. CV can be used to automate various tasks and enhance efficiency in the supply chain. It can facilitate the automatic tracking of products as they navigate through the industry, allowing for real-time monitoring and inventory management.

12.6 SECURITY THREATS AND CHALLENGES IN CV SERVICES

12.6.1 Adversarial attacks

- An adversarial attack is incurred when someone *intentionally* alters the input to a CV system, such as making minor modifications to an image to mislead the system into making harmful outcomes. Some examples are as follows:
- Suppose you present a picture of a dog to a CV model, and it correctly identifies it as "dog." But somebody included a little alteration (called perturbation) to the image so minimal that individuals can't even see the difference, but now the model categorizes it as a "cat" or even a "car." This is called an adversarial attack [27].

- A stop sign is slightly changed with stickers. A self-driving car reads it and thinks it's a speed limit sign instead of a stop sign [28]. It raises the possibility that a self-driving car could crash.

Some function against adversarial attacks' defenses:

- **Adversarial Training:** It learns to ignore the tricks by educating the model using both standard and adversarial samples.
- **Input Detection:** Use another model to check if the input has been tampered with.
- **Model Robustness:** Design models that are harder to fool by testing them with different types of input.

12.6.2 Physical adversarial attacks

Physical adversarial attacks are also a type of adversarial attacks. It is an attack where small, carefully designed changes are applied to physical objects in the real world to deceive advanced CV systems. These changes are generally unnoticed or unapparent to human eyes, but they can mislead CV models into making erroneous decisions. For example, attackers can put signs/stickers or patterns on a stop sign, resulting in a self-driving car not recognizing it. As a result, the vehicle (car) might fail to stop, causing accidents or harm. The primary challenges of physical adversarial attacks are that they work on objects of the real world, not just in digital images. Like digital attacks, where the image is altered on a computer, physical attacks involve changing actual objects, symbols, patterns, clothes, or even human faces [27, 28]. These alterations remain effective despite variations in lighting conditions or when the object is viewed from various angles and distances. This makes them more harmful and more challenging to defend against.

Other challenges of physical adversarial attacks are that they work on objects of the real world, not just in digital images. Like digital attacks, where the image is altered on a computer, physical attacks involve changing actual objects, symbols, patterns, clothes, or even human faces [28]. These alterations remain effective despite variations in lighting conditions or when the object is viewed from various angles and distances. This makes them more harmful and more challenging to defend against. Further investigation and sophisticated techniques are needed to know these types of attacks better and to develop more secure and reliable CV systems.

12.6.3 General adversarial vulnerabilities

Computers learn from data and make decisions as humans do in the world of AI and CV. For example, ASCV systems can identify faces, translate languages, or run an autonomous vehicle. Sometimes, CV can be tricked or fooled like any system. The weaknesses that permit CV to make wrong decisions when someone deliberately refers to mislead it are called adversarial vulnerabilities [29]. Researchers in this domain try to comprehend why these vulnerabilities exist in the first place, how they can be recognized, and how systems can be made more robust against them. The objective is to develop CV system models that are not only intelligent but also safe, secure, and reliable in real-world situations. This involves training models with improved defenses, evaluating them with adversarial inputs, and developing new architectures that are harder to fool.

12.7 SECURITY ENHANCEMENTS AND TECHNIQUES

12.7.1 Blockchain for secure CV pipelines

Blockchain (BC) and CV are two areas that will change industries. The term BC refers to the way transaction data are preserved as blocks, which are chained together. BC can offer decentralized, secure data storage, while CV allows machines to learn and understand data that are visual in nature. The BC can be implemented as an extra security layer in a Secured CV system and guarantees data integrity, allowing for decentralized photos and video analytics. BC is a decentralized, tamper-proof technology most recognized for use with cryptocurrencies such as Bitcoin. However, it can be used to secure CV systems, as well. In a typical CV pipeline (from data collection to model deployment), there are many steps where security can be compromised—for example, data tampering, unauthorized access, or false model updates.

In the CV pipeline, researchers utilize blockchain technology, permitting each phase (such as data input, training logs, or model updates) to be kept in a secure and transparent ledger. This ledger cannot be easily changed or hacked, enhancing the reliability of the system. Furthermore, it helps verify the authenticity of the data and models, as well as in tracking who has taken or modified information and when. This technique is instrumental in industries like medical diagnostics, defense, or smart surveillance, where data security is vital. BC for Secure CV Pipelines uses blockchain technology to monitor and verify every single step of the advanced secure CV process. This makes it harder for attackers to modify or tamper with the data and helps build trust by keeping a secure, transparent record of actions. Together, these techniques help keep CV systems more secure and reliable.

12.7.2 Encrypted inference

Encrypted inference is a process used to protect confidential data while it's being processed by an ASCV or ML model. Normally, when a model makes predictions, it requires seeing the input data clearly. But sometimes, the input data (like medical scanned images, personal photos/ videos, or financial documents) are very private and shouldn't be disclosed, even to the model itself. With encrypted inference, the information is first converted into a secret form (like a hidden code) before it is sent to the model. The model is then able to produce forecasts directly on the encrypted data using advanced cryptographic techniques like homomorphic encryption. This way, the raw data remain concealed and secure throughout the entire process. This process is particularly beneficial in domains where privacy is sensitive, including healthcare, banking, and facial recognition.

The main purpose of this technique is to maintain the confidentiality of the model and the data during the prediction process, defending against malicious and unauthorized access to sensitive information.

12.7.3 Robust architectures

Robust architectures define the design of ML models, especially in CV, that can prevent attacks or errors. In simple terms, it means building systems that don't easily break or decode. Sometimes, attackers/unauthorized persons can trick models using minor alterations to input images (known as adversarial attacks). These changes might be invisible to the human eye but can lead the model to generate incorrect decisions.

To implement the robust architecture, researchers use techniques like adversarial training (training the model using both normal and attacked images), dropout (to avoid overfitting), or model ensembles (which combine multiple models). A robust architecture guarantees that the model continues to perform well even under difficult circumstances or when attempts are made to deceive it.

12.8 FUTURE DIRECTIONS

AI has become a major force in many industries by allowing machines to mimic or surpass human cognitive tasks. These tasks include learning, reasoning, perception, and decision-making [30,31]. Modern AI systems often combine large-scale data processing with algorithms like deep learning and reinforcement learning [32,33]. They are being used in fields ranging from healthcare diagnostics to financial risk assessment and smart manufacturing. This change is not just about automation; it also enhances human abilities [34,35]. AI identifies hidden patterns, improves complex processes, and helps respond to changing environments [36,37]. This leads to greater efficiency and new forms of value creation. However, it also raises important issues related to data ethics, algorithm bias, transparency, and the social and economic effects of widespread use [38].

As CV technology grows rapidly, there are various exciting and crucial directions it can take. There are some key areas that researchers and industries may focus on:

- In many services like healthcare, smart cities, or retail, cameras collect data from people. Future systems will need to process images without exposing personal data.
- Real-time systems (like self-driving cars, security surveillance, and drones) need to make decisions quickly and also be protected from unwanted access or cyberattacks.
- Develop robust and scalable systems that work across various platforms without losing performance or security.
- Construct explainable AI (XAI) models so that users can acknowledge and have confidence in the system.
- Establish strong legal and ethical rules and guidelines that protect users while allowing innovation.

12.9 CONCLUSION

The rapid advancement of CV has transformed it into one of the most influential areas of AI, with applications spanning transportation, healthcare, manufacturing, surveillance, finance, and smart city ecosystems. However, as these systems increasingly interact with sensitive and mission-critical data, ensuring security, privacy, and robustness is no longer optional but essential. ASCV addresses these concerns by integrating visual intelligence with strong security frameworks, privacy-preserving computation, and resilience against adversarial manipulation. This study emphasizes that while

CV technologies such as facial recognition, OCR, and autonomous navigation hold enormous potential, their deployment also exposes systems to threats like adversarial attacks, tampering, and physical manipulations that may compromise accuracy and reliability. The incorporation of BC-secured pipelines, encrypted inference, and robust architectures offers promising solutions to strengthen trust and safeguard both data and outcomes. At the same time, balancing technical innovation with ethical and legal considerations is vital to ensure accountability and fairness in practical deployments. Looking ahead, the future of ASCV lies in building systems capable of secure real-time decision-making, maintaining consistent performance across diverse platforms, and providing explainable outcomes that foster user trust. Moreover, developing global standards and guidelines for ethical use will be crucial in supporting innovation while protecting individual rights. Ultimately, by uniting advanced algorithms with security-first design principles, ASCV can enable CV applications that are not only powerful and efficient but also safe, private, and trustworthy in real-world environments.

REFERENCES

1. M. Ramalingam et al., "A Comprehensive Analysis of Blockchain Applications for Securing Computer Vision Systems," *IEEE Access*, 2023. https://ieeexplore.ieee.org/document/10262316
2. Z. Guo, Y. Qian, Y. Li, W. Li, C. T. Lei, S. Zhao, L. Fang, O. Arandjelović, and C. P. Lau, "Beyond Vulnerabilities: A Survey of Adversarial Attacks as Both Threats and Defenses in Computer Vision Systems," *arXiv preprint arXiv:2508.01845*, 2025. https://arxiv.org/pdf/2508.01845
3. D. Wang et al., "A Survey on Physical Adversarial Attack in Computer Vision," *arXiv preprint arXiv:2209.14262*, 2022. https://arxiv.org/pdf/2209.14262
4. A. Guesmi, M. A. Hanif, B. Ouni, and M. Shafique, "Physical Adversarial Attacks for Camera-based Smart Systems: Current Trends, Categorization, Applications, Research Challenges, and Future Outlook," *arXiv preprint arXiv:2308.06173*, 2023. https://ieeexplore.ieee.org/document/10268441
5. N. Ottakath, A. Al-Ali, S. Al-Maadeed, O. Elharrouss, and A. Mohamed, "Enhanced Computer Vision Applications with Blockchain: A Review of Applications and Opportunities," Journal of King Saud University – Computer abd Information Science, vol. 25, no. 10, 2023. https://doi.org/10.1016/j.jksuci.2023.101801
6. L. Li, Y. Wang, C. Sitawarin, and M. Spratling, "OODRobustBench: A Benchmark and Large-Scale Analysis of Adversarial Robustness under Distribution Shift," *arXiv preprint arXiv:2310.12793*, 2023. https://arxiv.org/pdf/2310.12793
7. F. Croce et al., "RobustBench: A Standardized Adversarial Robustness Benchmark," *arXiv preprint arXiv:2010.09670*, 2021. https://arxiv.org/pdf/2010.09670
8. T. Paniagua, R. Grainger, and T. Wu, "RisingAttacK: Adversarial Perturbations Formed by Iteratively Learning Linear Combinations of the Right Singular Vectors of the Adversarial Jacobian," *Proceedings of ICML*, 2026. https://icml.cc/media/icml-2025/Slides/46521.pdf
9. S. Bian, T. Wang, M. Hiromoto, Y. Shi, and T. Sato, "ENSEI: Efficient Secure Inference via Frequency-Domain Homomorphic Convolution for Privacy-Preserving Visual Recognition," *2020 IEEE/CVF Conference on Computer Vision and Pattern Recognition (CVPR)*, Seattle, WA, USA, 2020, pp. 9400–9409, 2020. https://ieeexplore.ieee.org/document/9157085
10. M. Kim et al., "Secure Human Action Recognition by Encrypted Neural Network Inference," *Nature Communications*, vol. 13, 1799, 2022. https://doi.org/10.1038/s41467-022-32168-5
11. A. Roy, and K. Roy, "DCT-CryptoNets: Scaling Private Inference in the Frequency Domain," *arXiv preprint arXiv:2408.15231*, 2025. https://arxiv.org/pdf/2408.15231

12. M. Zhao, L. Zhang, J. Ye, H. Lu, B. Yin, and X. Wang, "Adversarial Training: A Survey," *arXiv preprint arXiv:2410.15042*, 2024. https://arxiv.org/pdf/2410.15042
13. A. Sharma, Y. Bian, P. Munz, and A. Narayan, "Adversarial Patch Attacks and Defences in Vision-Based Tasks: A Survey," *arXiv preprint arXiv:2206.08304*, 2022. https://arxiv.org/pdf/2206.08304
14. C. Szegedy et al., "Intriguing Properties of Neural Networks," *arXiv preprint arXiv:1312.6199*, 2014. https://arxiv.org/pdf/1312.6199
15. J. Wang et al., "Adversarial Examples in the Physical World: A Survey," *arXiv preprint arXiv:2311.01473*, 2024. https://arxiv.org/pdf/2311.01473
16. C. Song, and X. Shi, "ReActHE: A Homomorphic Encryption-Friendly Deep Neural Network for Privacy-Preserving Inference," *Smart Health*, 32, 100469. https://doi.org/10.1016/j.smhl.2024.100469
17. H. Roh and W. Choi, "Hyena: Optimizing Homomorphically Encrypted Convolution for Private Inference," *arXiv preprint arXiv:2311.12519*, 2025. https://arxiv.org/pdf/2311.12519
18. J. Shin, S.H. Choi, and Y.-H. Choi, "Is Homomorphic Encryption-Based Deep Learning Secure Enough?" *Sensors*, vol. 21, no. 23, p. 7806, Nov. 2021. https://doi.org/10.3390/s21237806
19. B. Xi, "Adversarial Machine Learning for Cybersecurity and Computer Vision: Current Developments and Challenges," *arXiv preprint arXiv:2107.02894*, 2021. https://arxiv.org/pdf/2107.02894
20. J. Diffenderfer, B. R. Bartoldson, S. Chaganti, J. Zhang, and B. Kailkhura, "A Winning Hand: Compressing Deep Networks Can Improve Out-Of-Distribution Robustness," *arXiv preprint arXiv:2106.09129*, 2021. https://arxiv.org/pdf/2106.09129
21. K. N. T. Nguyen et al., "A Survey and Evaluation of Adversarial Attacks for Object Detection" *arXiv preprint arXiv:2408.01934v2*, 2024. https://arxiv.org/pdf/2408.01934v2
22. Y. Himeur et al., "Federated Learning for Computer Vision," *arXiv preprint arXiv:2308.13558*, 2023. https://arxiv.org/pdf/2308.13558
23. Q. Li, B. He, and D. Song, "Model-Contrastive Federated Learning," *arXiv preprint arXiv:2103.16257*, 2021. https://arxiv.org/pdf/2103.16257
24. J. Shao, Z. Li, W. Sun, T. Zhou, Y. Sun, L. Liu, Z. Lin, Y. Mao, and J. Zhang, "A Survey of What to Share in Federated Learning: Perspectives on Model Utility, Privacy Leakage, and Communication Efficiency," *arXiv preprint arXiv:2307.10655*, 2024. https://arxiv.org/pdf/2307.10655
25. X. Li, Y.-K. Qiu, Y.-X. Peng, L.-A. Zeng, and W.-S. Zheng, "Privacy-Preserving Action Recognition: A Survey," in *Pattern Recognition and Computer Vision*, Lecture Notes in Computer Science, PRCV 2024. Lecture Notes in Computer Science, vol. 15037. Springer, Singapore. https://doi.org/10.1007/978-981-97-8511-7_32.
26. A. K. Taras, N. Sünderhauf, P. Corke, and D. G. Dansereau, "Inherently privacy-preserving vision for trustworthy autonomous systems: Needs and solutions," *Journal of Responsible Technology*, vol. 17, 100079, 2024, https://doi.org/10.1016/j.jrt.2024.100079
27. N. Chattopadhyay, A. Basit, B. Ouni, and M. Shafique, "A Survey of Adversarial Defenses in Vision-based Systems: Categorization, Methods and Challenges," *arXiv preprint arXiv:2503.00384*, 2025, https://arxiv.org/pdf/2503.00384
28. C. Hong, "Recent advances of privacy-preserving machine learning based on (Fully) homomorphic encryption," *Security and Safety*, vol. 4, 2024012, 2025. https://doi.org/10.1051/sands/2024012
29. Z. Kan et al., "FicGCN: Unveiling the homomorphic encryption efficiency from Irregular Graph Convolutional Networks," *arXiv preprint arXiv:2506.10399*, 2025. https://arxiv.org/pdf/2506.10399
30. G. N. Acharya, C. R. Dudhagara, and A. B. Mahera, "Market Analysis for Artificial Intelligence Based Grain Analyser in Gujarat," *Journal of Scientific Research and Reports*, vol. 31, no. 7, pp. 420–428, 2025. https://doi.org/10.9734/jsrr/2025/v31i73260

31. G. Manoharan, S. P. Ashtikar, and C. R. Dudhagara, "Ethics of AI in the Educational Sector – Navigating the Moral Landscape," in *Artificial Intelligence Based Smart and Secured Applications.* ASCIS 2024. Communications in Computer and Information Science, vol 2427. Springer, 2025. https://doi.org/10.1007/978-3-031-86299-1_19
32. R. Kumar, A. Joshi, H. O. Sharan, S. Peng, and C. R. Dudhagara, *The Ethical Frontier of AI and Data Analysis.* IGI Global Scientific Publishing, 2024. https://doi.org/10.4018/979-8-3693-2964-1
33. A. P. Vora, C. R. Dudhagara, D. R. Kathiriya, and R. S. Pundir, "Applications of Google Earth Engine for Big Data Analytics," *The Pharma Innovation*, vol. 12, no. 3, pp. 2054–2062, 2023. https://www.thepharmajournal.com/archives/2023/vol12issue3/PartU/12-2-471-804.pdf
34. M. Awasthi, C. S. Raghuvanshi, C. R. Dudhagara, and A. Awasthi, "Exploring Virtual Smart Healthcare Trends Using Digital Twins," in *Digital Twins for Smart Cities and Villages*, pp. 377–406. Elsevier, 2025. https://doi.org/10.1016/B978-0-443-28884-5.00017-8
35. A. S. Gaur, H. O. Sharan, and C. R. Dudhagara, "Research Trends in Blockchain and Digital Twins," in *Digital Twins for Smart Cities and Villages*, pp. 561–580. Elsevier, 2025. https://doi.org/10.1016/B978-0-443-28884-5.00024-5
36. G. Manoharan, and C. R. Dudhagara, "Exploration of Social Engineering in the Digital Era of Stealth," in *Effective Strategies for Combatting Social Engineering in Cybersecurity.* IGI Global Scientific Publishing, 2025. https://doi.org/10.4018/979-8-3693-6665-3.ch005
37. G. Manoharan, and C. R. Dudhagara, "Assortment of Social Engineering Attacks in the New Digital Era," *Effective Strategies for Combatting Social Engineering in Cybersecurity.* IGI Global Scientific Publishing, 2025. https://doi.org/10.4018/979-8-3693-6665-3.ch016
38. R. Ganatra, M. Gahlawat, and C. R. Dudhagara, "Federated Learning in Modern Education: Balancing Privacy, Scalability, and Effectiveness," in: Mohana, R., Sharma, A., Nayyar, A., Saini, P. (eds.) *Federated Learning Applications in the Industrial Internet of Everything (IoE).* Studies in Systems, Decision and Control, vol. 611. Springer, 2025. https://doi.org/10.1007/978-3-031-99270-4_11

Chapter 13

The future of computer vision

Opportunities, requirements, and open issues

Mohammad Shahnawaz Shaikh, Ashwini Kumar Jha, Yogendra Singh, Neelesh Kumar Jain, Amarjit Seal, Akruti Pandwal, and Weiwei Jiang

13.1 INTRODUCTION

The history of vision has gone from rule-based image processing in the 1960s to the current deep learning-based perception systems. In its initial stages, object detection was based on handcrafted features (e.g., SIFT and HOG), but with the development of convolutional neural networks (CNNs) in the 2010s, it became possible to train an end-to-end solution operating on raw pixels. Technological advances have taken accuracy and scale to new heights, with such breakthroughs as AlexNet (2012), ResNet (2015), and Vision Transformers (ViTs, 2020). At the same time, the emergence of edge AI has meant that vision processing has become less latency- and bandwidth-constrained as data sources move physically close together. But, as complexity in models has increased, the explainability, bias, and security concerns have also increased considerably. Blockchain and AI will be at the heart of the next stage of computer vision as trustful AI, decentralized governance, and energy-efficient architectures are needed to succeed [1–3].

13.1.1 Background and motivation

Computer vision has emerged as one of the most transformative technologies of the 21st century, enabling machines to interpret and analyze visual data with human-like accuracy. From autonomous vehicles to medical diagnostics, its applications are revolutionizing industries. However, as computer vision systems become more pervasive, critical challenges related to security, privacy, and trust have surfaced. Centralized systems have shortcomings in terms of tampering with data, as well as being vulnerable to aggressive attacks from attackers and sensitive data manipulation. This requires combining blockchain and artificial intelligence (AI), which would improve security, transparency, and reliability. Blockchain ensures the provenance of immutable data, fosters decentralized trust, and facilitates secure sharing as AI continues to improve decision-making, process data in real-time, and adapt over time. Collectively, it constitutes a strong framework of safe computer vision, which addresses the issues of data integrity, model fairness, and adversarial robustness. This chapter addresses the question of whether the integration of such technologies can help the future of computer vision to open additional opportunities, reducing risks [2–4] (Figure 13.1).

DOI: 10.1201/9781003659112-13

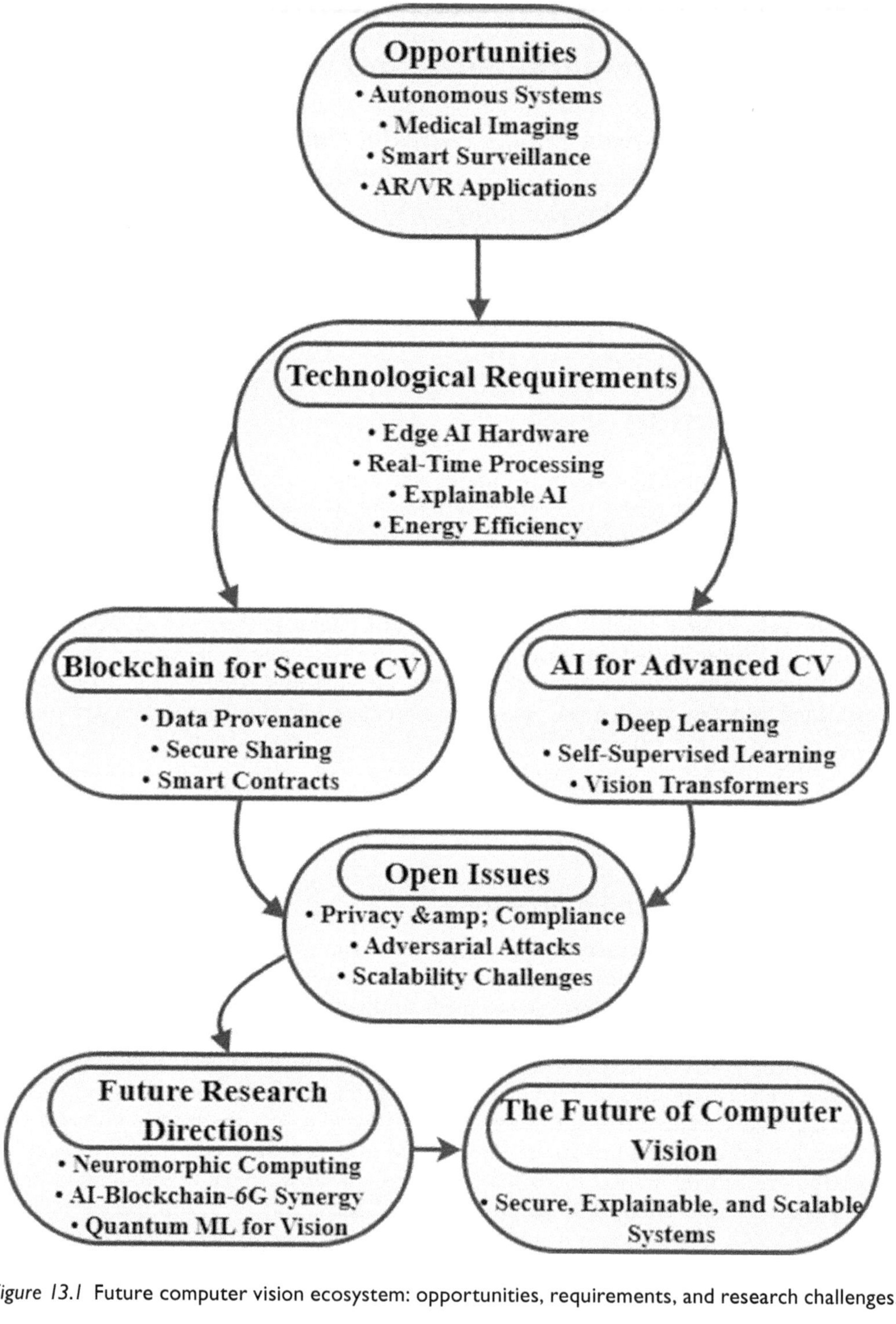

Figure 13.1 Future computer vision ecosystem: opportunities, requirements, and research challenges.

13.1.2 Role of blockchain and AI in shaping the future

Blockchain and AI not only complement each other but also drive secure computer vision synergistically. Blockchain brings in decentralized trust, which allows the following:

- Data logs whose data are temperature-proof of training datasets and model weights.
- Smart contracts to automate and make decisions in a self-transparent manner (e.g., drone-based confirmation of delivery).
- Federated learning incentives, in which contributors get rewarded through crypto-tokens because of data sharing.

AI improves blockchain by the following:

- Efficient consensus (e.g., AI and Proof of Stake validation).
- Monitoring possible abnormalities and fraud in blockchain transactions with the use of vision-based analytics.
- The supporting mechanisms of preserving privacy, like zero-knowledge proofs (ZKPs) of encryption over vision data.

Collectively, they fill the most important gaps in scalability, security, and interoperability and open the door to next-generation applications, such as decentralized medical imaging, autonomous vehicle interoperability, and supply chain counterfeit prevention.

13.1.3 Chapter objectives and structure

The goals of this chapter are as follows:

- Focus on novel possibilities in which blockchain and AI technology may amplify computer vision (Section 2).
- Determine the technical needs of scalable, secure, and efficient vision systems (Section 3).
- Read how blockchain delivers on the hype of integrity of data and trust without centralization (Section 4).
- Explain the current progress made by AI in the field of deep learning, generative models, and multimodal fusion (Section 13.5).
- Call out outstanding issues, such as privacy, bias, and adversarial menace (Section 13.6).
- Suggest paths to go forward, including neuromorphic vision and quantum-secure AI (Section 13.7).

13.2 EMERGING OPPORTUNITIES IN COMPUTER VISION

13.2.1 Autonomous systems (self-driving cars, UAVs, and robotics)

Computer vision is the cornerstone of autonomous systems, enabling machines to perceive and navigate dynamic environments. Self-driving cars rely on real-time object detection (e.g., pedestrians and traffic signs) using LiDAR, cameras, and deep learning models like YOLO and Tesla's HydraNet. Unmanned aerial vehicles (UAVs) leverage vision for obstacle avoidance, precision landing, and aerial mapping, with applications in delivery, agriculture, and disaster response. Meanwhile, robotics integrates vision for pick-and-place tasks, human–robot collaboration, and warehouse automation (e.g., Amazon's Kiva robots). However, challenges remain in edge processing, sensor fusion, and adversarial robustness, where blockchain can secure sensor data integrity and AI can enhance real-time decision-making. The future lies in swarm robotics and vehicle-to-everything (V2X) communication, where decentralized vision systems will enable safer, coordinated autonomy [5,6].

13.2.2 Precision healthcare and medical imaging

AI-powered computer vision is revolutionizing diagnostics, surgery, and personalized medicine. Deep learning models like U-Net and DenseNet excel in detecting tumors (MRI/CT scans), diabetic retinopathy (fundus images), and COVID-19 lung abnormalities. Surgical robotics (e.g., Da Vinci System) use vision for minimally invasive procedures, while wearable devices monitor vitals via skin imaging. Blockchain ensures patient data privacy (HIPAA/GDPR compliance) and provenance tracking for medical datasets. Emerging opportunities include federated learning for multihospital collaborations and AI-generated synthetic data to address the scarcity in rare diseases. The integration of 3D vision and augmented reality (AR) will further enhance surgical precision and telemedicine [7,8] (Figure 13.2).

13.2.3 Smart surveillance and security

From facial recognition to anomaly detection, computer vision enhances public safety while raising ethical concerns. Smart cities deploy AI-powered CCTV for crowd monitoring, traffic management, and crime prediction (e.g., ShotSpotter for gunfire detection). Blockchain can decentralize surveillance data storage, ensuring tamper-proof evidence and consent-based access (e.g., civic tokens for privacy control). Edge-based vision (e.g., NVIDIA Metropolis) reduces latency, while federated learning preserves privacy by training models on-device. Future systems will balance security and civil liberties, leveraging explainable AI (XAI) to audit biases and homomorphic encryption for secure video analytics [9,10].

13.2.4 Agriculture and environmental monitoring

Computer vision is transforming precision agriculture and ecological conservation. Drones with multispectral cameras monitor crop health (NDVI imaging), predict yields, and detect pests (e.g., John Deere's See & Spray). Satellite imagery tracks deforestation, wildfire spread, and soil moisture, aiding climate resilience. Blockchain verifies supply chain transparency (e.g., Fair Trade coffee traceability), while AI optimizes water/fertilizer use via Internet of Things (IoT)–vision integration. Emerging tools include autonomous harvesters and AI-powered weed-killing robots. The next frontier is bioinspired vision sensors for low-power, real-time field analysis [10].

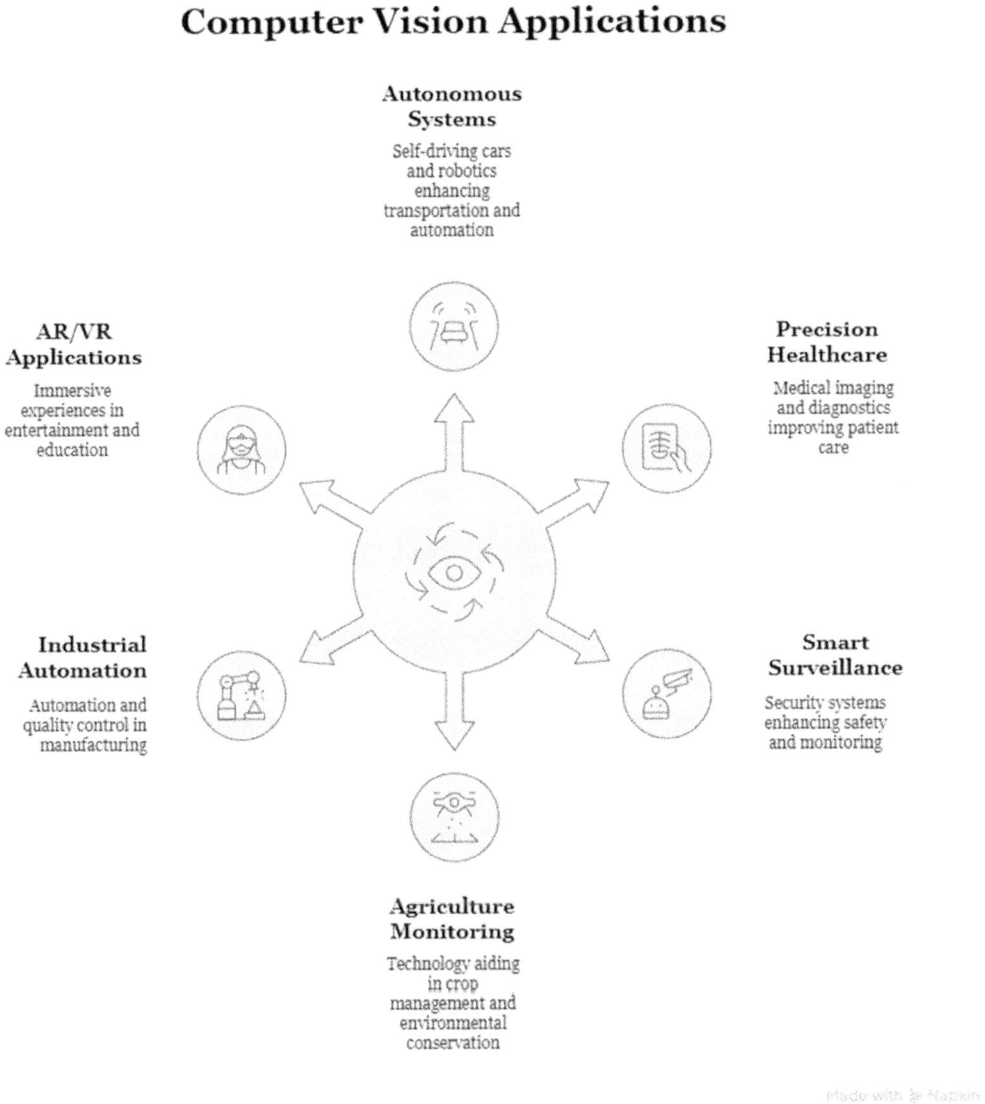

Figure 13.2 Emerging applications in computer vision.

13.2.5 Industrial automation and quality control

In Industry 4.0, vision systems automate defect detection, assembly line monitoring, and predictive maintenance. AI models (e.g., Mask R-CNN) inspect micro-cracks in semiconductors (e.g., NVIDIA's EGX) or mislabeled pharmaceuticals. Digital twins use 3D vision to simulate factory floors, while blockchain audits supply chains (e.g., IBM Food Trust). Collaborative robots (Cobots) with depth-sensing cameras work alongside humans safely. Future trends include AI-driven predictive quality control and blockchain-secured intellectual property (IP) for proprietary vision algorithms [9,10].

13.2.6 Augmented and Virtual Reality (VR) applications

AR/VR relies on computer vision for object tracking, gesture recognition, and immersive environments. Meta Quest 3 uses Simultaneous Localization and Mapping (SLAM) for room-scale VR, while Apple Vision Pro enables AR overlays via LiDAR. In healthcare, AR-guided surgery (e.g., AccuVein) projects veins onto skin. Retailers like IKEA use AR vision for virtual furniture placement. Challenges include latency reduction and photorealistic rendering, where AI super-resolution (e.g., NVIDIA DLSS) and blockchain-based digital ownership (nonfungible tokens [NFTs]) for AR assets will play key roles. The metaverse will demand cross-platform vision standards and privacy-preserving avatar tracking [11].

13.2.7 Cross-industry integration and new business models

Computer vision is spawning disruptive business models across sectors:

- **Retail:** Cashier-less stores (Amazon Go) with vision-based checkout.
- **Insurance:** AI-driven damage assessment (e.g., Tractable for car accidents).
- **Finance:** Facial recognition for contactless payments (Alipay's Smile to Pay).
- **Entertainment:** Deepfake detection (e.g., Microsoft Video Authenticator).

Blockchain enables micro-transactions for vision-based services (e.g., pay-per-use drone inspections), while AI fuels hyperpersonalization (e.g., Nike's AR sneaker trials). The future will see vision-as-a-service (VaaS) platforms, where businesses license AI models via decentralized marketplaces (e.g., SingularityNET).

13.3 TECHNOLOGICAL REQUIREMENTS FOR NEXT-GENERATION COMPUTER VISION

13.3.1 Hardware acceleration (Edge AI Devices, GPUs, TPUs, and NPUs)

Next-generation computer vision demands specialized hardware to handle computationally intensive tasks efficiently. Graphics Processing Units (GPUs) remain dominant for training deep learning models, while Tensor Processing Units (TPUs) optimize inference speed in cloud environments. For edge deployment, Neural Processing Units (NPUs) embedded in devices like smartphones (e.g., Apple's A16 Bionic) and IoT cameras enable real-time processing without cloud dependency. Emerging neuromorphic chips (e.g., Intel Loihi) mimic biological neural networks, offering ultralow-power vision processing. Autonomous vehicles and drones rely on NVIDIA Jetson and Qualcomm Robotics RB5 platforms for onboard AI acceleration. Future advancements will focus on quantum AI processors and 3D-stacked memory architectures to overcome von Neumann bottlenecks [12].

13.3.2 High-quality data acquisition and annotation

The performance of computer vision models hinges on diverse, accurately labeled datasets. High-resolution sensors (e.g., Sony IMX series cameras) and LiDAR capture detailed environmental data, while synthetic data tools (e.g., NVIDIA Omniverse) generate photorealistic training samples. Automated annotation platforms (e.g., Labelbox and CVAT) reduce human effort, but challenges persist in rare edge cases (e.g., extreme weather conditions for autonomous vehicles). Blockchain can ensure data provenance by timestamping datasets and tracking annotator contributions via smart contracts. Such federated learning systems as OpenFL provide model training without sharing raw data, which concerns the privacy part without compromising the quality of the datasets [13].

13.3.3 Real-time processing and low-latency networking

Applications such as robotic surgery and autonomous vehicles demand sub-millisecond latency that needs optimized vision pipelines, and 5G/6G networks allow high-bandwidth/low-latency data transfer to coordinate in the cloud edge. Time-sensitive networking (TSN) provides synchronization in industrial machinery. A peripheral (e.g., Google Coral and Dojo by Tesla) AI platform optimizes to low cloud dependencies, gathering data nearby. The pixel-level changes can be recorded by event-based cameras (e.g., Prophesee) at microsecond timescales to minimize redundant calculations. Subsequent inventions include in-memory computing and optical neural networks in order to do away with the traditional digital processing stall [15].

13.3.4 Explainable and trustworthy AI models

Transparency is a non-negotiable situation as the vision systems affect other domains that require attention (e.g., health and criminal justice). Such XAI approaches as LIME and SHAP explain why a model has made a particular decision, and uncertainty quantification (e.g., Bayesian deep learning) dynamically identifies model predictions with low confidence. Auditable models (e.g., the EU AI Act) prefer white-box systems: decision trees and hybrid symbolic-AI systems. Blockchain registers model training and inference by creating mutually verified audit logs. Studies on self-certifying models and adversarial robustness testing (e.g., the Adversarial Robustness Toolbox developed at IBM) will additionally improve the confidence in vision AI [14].

13.3.5 Scalability and cloud edge blockchain integration

The workforce needs to deploy massive-level visions that have a hybrid structure promoting both the force of the cloud and promptness to the area. The operation of edge devices by smart contracts (e.g., IOTA used to work on IoT micropayments) or distributed ledger technology (DLT) enhances the communication of the fog computing level and processes between the cloud and the edge. Platforms like AWS Panorama and Azure Percept unify the management of distributed vision nodes. Blockchain secures model versioning and federated learning aggregation, ensuring scalability without bottlenecks associated centralized systems. Future systems will leverage serverless computing and Web3 protocols for truly decentralized vision networks [14].

13.3.6 Interoperability standards and protocols

Fragmentation across vision hardware/software ecosystems necessitates universal standards. Open Neural Network Exchange (ONNX) enables model portability across frameworks (PyTorch to TensorFlow), while the Robot Operating System (ROS) standardizes robotic vision interfaces. For blockchain integration, Decentralized Identifiers (DIDs) authenticate devices and Oracles (e.g., Chainlink) bridge off-chain vision data to smart contracts. Industry consortia like Industrial Vision Standards (IVS) are developing unified protocols for manufacturing vision systems. Emerging metaverse interoperability initiatives (e.g., OpenXR) will further unify AR/VR vision pipelines [13,14].

13.3.7 Energy-efficient architectures

With AI's carbon footprint under scrutiny, green computer vision is imperative. The same brain-like efficiency is provided by spiking neural networks, TinyML methods (e.g., quantization and pruning), and compressing models into micro-controllers (e.g., Arduino Nicla Vision). Photonic computing (e.g., the Envise chip from Lightmatter) decreases power requirements through the use of light as an alternative to electrons. The mechanics by which blockchains reach their consensus are changing in ways to better utilize energy without wastage through Proof of Stake (PoS) and Proof of Space Time in comparison with Proof of Work (PoW). Remote monitoring edge devices (a good example being security drones by Sunflower Labs) are sustainable examples of vision systems [14,15].

13.4 ROLE OF BLOCKCHAIN IN SECURE COMPUTER VISION

13.4.1 Data provenance and integrity verification

Blockchain technology changes the idea of data provenance in computer vision by inventing a distributed, immutable history of the visual data over its life. Every image, video frame, or dataset can be cryptographically hashed and forever stored on-chain, which allows full traceability of the original through all processing steps. This functionality is especially important in some sensitive areas of use, such as medical scans, where only one pixel in an MRI medical scan may result in the worst possible misdiagnosis. Blockchain technologies like Hyperledger Fabric and Ethereum Enterprise enable providers in the medical field to check whether diagnostic images have been altered after being taken but still preserve the confidentiality of patients with ZKPs. Legally, it exists in the form of a chain of custody, in which case the blockchain-anchored surveillance footage can be used as a court-admissible evidence; in this aspect, projects such as Chronicled invent forensic tools capable of authenticating the video evidence. This technology is ideal in the field of autonomous vehicles because manufacturers can compose permanent records of all training data that were used to develop their perception systems and when certain road situations were recorded, along with the specific sensor arrays recording the situation. The latest solutions involve blockchain coupled with content authenticity initiatives (CAI) standards to fight off Deepfake, and cryptographic signatures are embedded in camera hardware to allow end-to-end verifiable media recording. Similar techniques have been taken in the pharmaceutical industry, wherein blockchain is used to trace quality control images in drug manufacturing processes, which are mandatory as long as monitoring is compliant with the FDA 21 CFR Part 11 electronic records regulations [16,17].

13.4.2 Secure data sharing and federated learning

Federated learning via blockchain can present a paradigm shift toward collaborative (and yet data-privacy-preserved) development of computer vision. This method enables several organizations to collaboratively develop vision models without concentrating the data in a centrally located pool, which is especially useful in healthcare, where the regulation of patient privacy deprives data freedom to the organization. The blockchain component will perform three important tasks: the process of federated learning should be coordinated with the help of smart contracts, participation motivated by rewarding with tokens, and an immutable audit trail of all changes to the model. Technologies such as NVIDIA FLARE can be used in combination with permissioned blockchains to allow hospitals worldwide to train cancer detection models and be HIPAA-compliant to train cancer detection models. The cryptographic tokens are distributed to each participant based on the relative influence of their contributions to improve the model, with complex Shapley value operations to be conducted on-chain to guarantee equal reward provision. Within the industrial sector, consumers in a given supply chain may collectively train defect detection models, with the proprietary production images remaining confidential to them. The Industrial IoT Consortium has come up with reference architectures that integrate blockchain and federated learning into predictive maintenance vision systems. New rules to solve the free rider problem in federated learning involve staking, in which participants have to post some tokens, which will be slashed should they upload updates of low quality. For further privacy enhancement, advanced implementations provide multiparty computation (MPC) to further increase privacy, which makes it possible to perform model aggregation without revealing individual parameter updates. The struggle between performance and privacy is already leading the automotive industry to pioneer these techniques as consortia of automakers train their autonomous driving perception with blockchain-coordinated federated learning to balance accuracy and privacy by keeping each company's proprietary road data secret [13].

13.4.3 Intellectual property protection for vision models

Blockchain and computer vision are used to generate powerful systems that protect AI intellectual property in its lifecycle. The vision models are able to be registered as NFTs, where each new model that is trained can then be recorded on-chain from its initial training to future fine-tuning and deployment. This gives documentary evidence that cannot be disputed with regard to ownership and dates of creation, which is very important in patenting and copyright. Smart contracts, such as those available in platforms like Ocean Protocol, can be used to facilitate granular access control (e.g., by letting model contributors sell time-limited inference access or charge by per-prediction price). The film, television, video game, and other entertainment industries have implemented such technologies to secure proprietary systems of CGI and facial animation and utilize DRM based on blockchains in the filming of entertainment works (e.g., Disney has built projects to make use of blockchain DRM of computer vision pipelines). More recent work in model watermarking applies cryptographic techniques to store blockchain-anchored hashing directly in neural network weights so that they can be verified even under model compression or quantization. The emergence of a model market such as the SingularityNET has formed new economic models, where the developers of vision AI tools get micropayments every time their models are used and the blockchain in it guarantees royalty sharing in a transparent way. Within the security sphere, firms have been working on on-chain model passports, which log each time a deployment instance occurs, and against each instance, a detected adversarial attack is logged, leaving an unalterable maintenance ledger. These are also similar strategies with regard to the pharmaceutical industry that manage to guard proprietary models of microscopy analysis; blockchain maintains a notarization of regulatory submissions. New standards in this front are the DID standards of models, giving them the ability to identify themselves as well as the record of their ownership in various platforms [14,15].

13.4.4 Decentralized model training and deployment

Solutions based on blockchain will allow creating genuinely decentralized computer vision systems able to coordinate distributed training and inference across billions of edge devices worldwide. New proofs of useful work mining schemes enable miners to perform useful calculations (in the training of a vision model) rather than pointless cryptographic puzzles. A project such as Bit tensor has developed decentralized neural networks in which participants participate in making accurate predictions based on rewards in the form of tokens and slashing incidents in the event of such inaccurate choices. In edge device fleets, blockchain is used as the swarm learning coordination layer, and systems of smartphones, drones, and IoT cameras can cooperate to enhance models and still preserve data locality. The Helium Network has also evidenced this method of distributed wildlife monitoring, where the camera traps can be used to add to the conservation models that do not rely on centralized data collection. The factory DAO is arranged in business areas in which manufacturers collect their vision-based quality control exemplification by means of blockchain-regulated coalition. The technology also supports crowdsourced annotation markets where quality control is natively present – annotators have their tokens held at risk of being lost should their work fall short of quality validation checks, and high-quality contributors are automatically paid in cryptocurrency via smart contracts. In the deployment of models, such a vision processing capacity is made available on an inference marketplace that can be implemented in a decentralized program using blockchain to enable device-based models offering processing services; the blockchain facilitates service discovery, payment, and reputation tracking. Hedera Hashgraph and other energy-efficient solutions can support the scale of such micro-transactions, essential to the application of vision to IoT. New architectures add this to confidential computing enclaves to still provide model privacy when running distributed, providing opportunities to enable collaborative vision systems across competitors [16].

13.4.5 Smart contracts for automated vision-based decisions

Blockchain smart contracts turn the results of computer vision into physical actions in tamper-resistant automated processes. Under supply chain management, VeChain, combined with industrial vision systems, can perform automatic receipt verification of goods. In this case, the smart contract will provide a payment after receiving a delivery that is comparable to the specification of the purchase order that is cleared by the computer vision. There are now examples within the insurance industry implementing such systems, such as Etherisc, which uses aerial imagery and smart contracts to automatically trigger crop insurance settlements when vision algorithms identify a specified condition, in this case drought, of sufficient severity within a policy. Regarding physical security, the use of oracles based on blockchain provides authenticated camera information to smart contracts that restrain access systems, and all logs on the decisions made will be recorded immutably and may be used during compliance auditing. The automotive sector is leading the charge in this regard toward vision-enabled smart contracts, i.e., instant use-based insurance, where an in-car camera captures driving styles and automatically updates the insurance premiums through blockchain transactions. Traffic management systems obtain congestion data processed with vision in smart cities and, in turn, dynamically adjust toll pricing using self-enforcing contracts. In the next stage, Intelligent systems will bind IoT sensors and vision together to perform multimodal verification (e.g., only firing a wildfire response smart contract when thermal cameras and smoke detectors concur). Lawyers are embracing these technologies toward smart evidentiary systems wherein a chain-of-custody of video evidence can be checked in courtrooms by use of the blockchain. Newer applications are found in decentralized content moderation, with community-based smart contracts used to automatically label forbidden imagery because of a consensus between multiple vision models. Financial institutions are experimenting with the use of computer vision-enabled AML tools where transactions can be frozen by smart contracts in case of a visual identification of known money mules within the branch buildings using surveillance cameras installed in branches. These systems are also enhancing the use of zk-SNARKs to facilitate verification of these vision-based decisions; that is, they want to take advantage of automation with such privacy provisions [17].

13.5 ROLE OF ARTIFICIAL INTELLIGENCE IN ADVANCING COMPUTER VISION

13.5.1 Deep learning architectures for vision tasks

The power of deep learning to empower end-to-end training of computer vision neural networks has made it possible to train without manual feature engineering. CNNs formed the basis where architectures such as ResNet, Efficient Net, and DenseNet have set state-of-the-art results in several areas, including image classification, object detection, and segmentation. A family of modern models like ConvNeXt combines the efficiency of CNNs with the scalability of transformers and with attention mechanisms to extract features. When considering real-time applications, lightweight models such as MobileNetV3 and YOLOv9 can be used to make performance trade-offs (speed versus accuracy), deploying on edge devices. In medical imaging, medical images may be volumetric (e.g., in CT/MRI cases) and are thus processed with 3D CNNs (e.g., 3D U-Net). However, ViTs are gaining popularity due to their greater scalability with large images. Recent advances in neural architecture search (NAS) have automated the design of models and generated task-specific representations that are better than models designed by humans. Issues remain in computational effectiveness and between-field familiarity, with work promoting dynamic neural networks that generate structure depending upon the complexity of the input [17] (Figure 13.3).

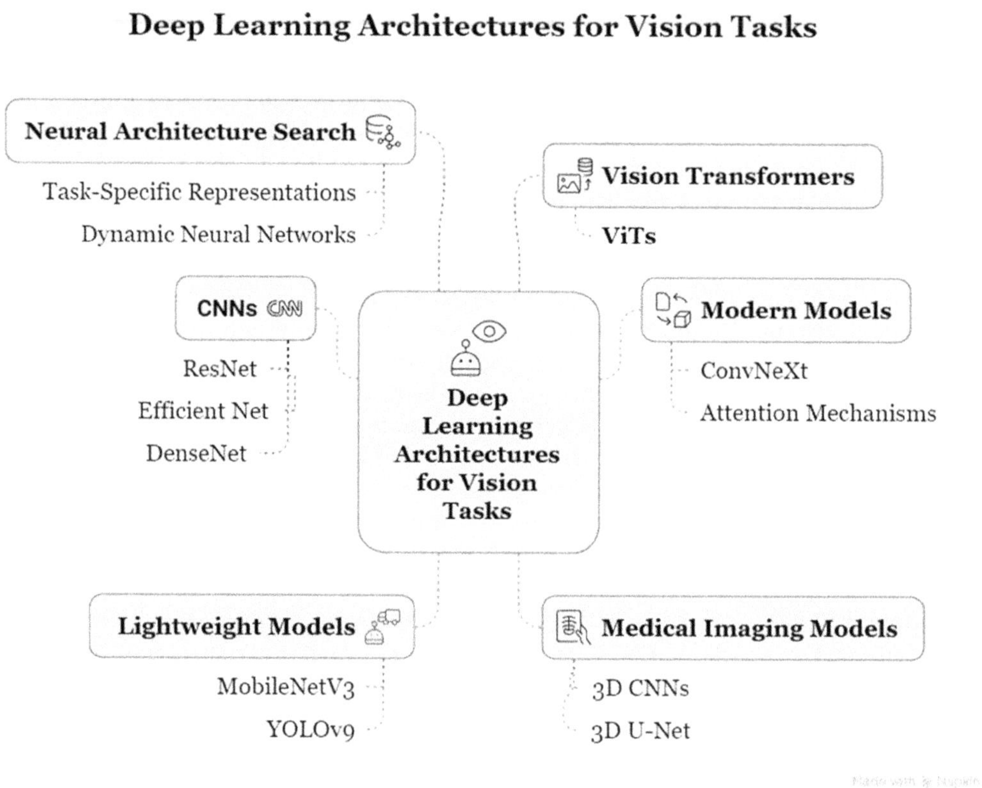

Figure 13.3 Deep learning architecture for computer vision.

13.5.2 Transformer models and vision-language integration

Transformers have moved beyond natural language processing to take the lead on vision applications in such forms as ViTs and Swin Transformers, which represent images as sequences of patches to reason about global contexts. Multimodal models (e.g. CLIP and Flamingo) have coupled vision and language, allowing zero-shot image classification and text-to-image retrieval by matching visual and text embedding to one another in a single space. High-resource models (LVMs) make use of large incorporation of vision-language models (LVLMs) such as GPT-4V and PaLI-3, carry out intricate reasoning on images and text, and are used in a variety of applications, including automated report generation in radiology to visual question answering (VQA) to aid assistive technologies. Transformer-based diffusion models (e.g., Stable Diffusion 3) produce high-fidelity images from a text prompt, and video transformers (e.g., Timesformer model) are emergent extensions to temporal data. Key future directions will be sparse attention mechanisms to efficiently integrate long-context and neuromorphic transformers that are bioinspired solutions to vision [18].

13.5.3 Self-supervised and few-shot learning

Self-supervised learning (SSL) exploits structures present in images for pretraining, and it relies less on labeled data. Contrastive methods (e.g., SimCLR and MoCo) learn representations by maximizing the agreement between augmented views of the same image, while masked autoencoders (MAE) reconstruct randomly masked image regions. Vision foundation models (e.g., DINOv2) trained with SSL achieve remarkable generalization, enabling fine-tuning with minimal downstream labels. Few-shot learning techniques like prototypical networks and meta-learning (MAML) adapt to new tasks with just one to five examples per class, critical for niche applications (e.g., rare disease diagnosis). Test-time adaptation further enhances the robustness by continuously updating models during deployment. Hybrid approaches combining SSL with synthetic data (e.g., NVIDIA Omniverse) are closing the gap between data-rich and data-scarce domains, making vision AI more accessible [19].

13.5.4 Generative models for data augmentation

Generative AI addresses data scarcity and bias by synthesizing diverse training samples. Diffusion models (e.g., Stable Diffusion and MidJourney) produce photorealistic images conditioned on text or semantic maps, while generative adversarial network (GAN) variants (e.g., StyleGAN-XL) generate high-resolution domain-specific data (e.g., facial images and industrial defects). Neural radiance fields (NeRFs) create 3D-consistent synthetic environments for autonomous vehicle simulation, and physics-based generative models simulate realistic sensor noise (e.g., rain and motion blur) to enhance robustness. In healthcare, differential privacy-preserving GANs synthesize anonymized medical images for research. Challenges include avoiding mode collapse and ensuring semantic consistency, with newer latent diffusion models offering finer control. Integration with blockchain verifies synthetic data provenance, preventing misuse of Deepfake, while enabling auditable augmentation pipelines [20].

13.5.5 AI-Driven optimization for blockchain vision pipelines

AI enhances blockchain–vision systems by optimizing resource allocation, security, and efficiency. Reinforcement learning (RL) automates blockchain parameter tuning (e.g., gas fees and block size) for vision data transactions, while graph neural networks (GNNs) detect fraudulent patterns in decentralized vision marketplaces. Federated learning aggregation algorithms (e.g., FedAvg and FedProx) are improved via AI to handle non-IID data across blockchain nodes. For an energy-efficient consensus, AI predicts optimal validators for PoS blockchains processing vision data, reducing latency. Adversarial detection models guard against attacks on vision–blockchain oracles, and neural compression techniques (e.g., Learned Image Compression) minimize on-chain storage costs for visual data. Projects like Render Network use AI to match GPU supply with vision rendering demand, while AI-powered smart contracts dynamically adjust computer vision model parameters based on real-world performance metrics logged on-chain [21].

13.6 OPEN RESEARCH ISSUES AND CHALLENGES IN SECURE COMPUTER VISION

13.6.1 Data privacy and compliance (GDPR and HIPAA)

Computer vision and privacy laws have technical and legal challenges that are difficult to solve. Although methods such as federated learning and homomorphic encryption allow the analysis of visual data without having access to raw images, implementing the principle of the GDPR in the context of the right to be forgotten is also a challenge with the use of blockchain, and because of their immutable nature, direct compliance with this framework is problematic. Recent proposals like ZKPs for selective disclosure and chameleon hashes for retroactive redaction are promising improvements, but they are not yet suitable to deploy into practice. In healthcare, de-identification of medical images needs to find the balance between utility and HIPAA compliance since advanced AI has the potential to be able to re-identify patients on what may be thought of as anonymized scans. The creation of such standardized pipelines of privacy-preserving computer vision incorporating differential privacy, secure multiparty computation, and on-chain access mechanisms is a significant line of research, especially in the case where cross-border applications come into conflict with data sovereignty laws [22].

13.6.2 Bias, fairness, and ethical considerations

The common occurrence in computer vision systems is biases that can be attributed to unbalanced training data or problematic annotation procedures and can cause discriminatory results in facial identification, employment algorithms, and law enforcement. The existing de-biasing methods can hardly resolve intersectional biases that touch on more than one demographic dimension at the same time. There is still a lack of context-sensitive measures of fairness that can be employed and established bias audit frameworks to apply to vision systems. The ethical dilemmas are further complicated by the transparency paradox of blockchain that, on the one hand, can bring accountability due to the immutability of the decisions but, on the other hand, fix and preserve the possibly biased decisions forever. Innovative ways of employing XAI and on-chain model control to construct verifiable, but flexible systems must be developed. The appearance of synthetic data as a technology to mitigate bias brings about new concerns regarding the ethics of the representation of the artificially created demographics and situations [23] (Figure 13.4).

13.6.3 Model robustness against adversarial attacks

Attacks toward vision systems are becoming more advanced, including physical-world adversarial patches to fool the perception of autonomous vehicles and gradient-based attacks that exploit potential vulnerabilities in a model. Although adversarial training and defensive distillation alleviate robustness, they tend to fail when faced with adaptive attackers using generative methods. Attack surfaces brought about by the integration of blockchain in terms of model versioning and attack detection come in the form of consensus manipulation-based attacks as well as smart contract attacks in the case of vision oracles. The study of certified defenses, where the robustness against various adversarial models can be mathematically guaranteed, is an active research direction, but it cannot scale to large instances because it is complicated. New problems that arise concern the protection against poisoning attacks with clean labels in the federated learning regime and the corresponding intrusion detection system in the context of a vision-based blockchain. The weaponization overtake between attackers and defending soldiers intensifies since both sides are maximizing the use of more advanced AI [24].

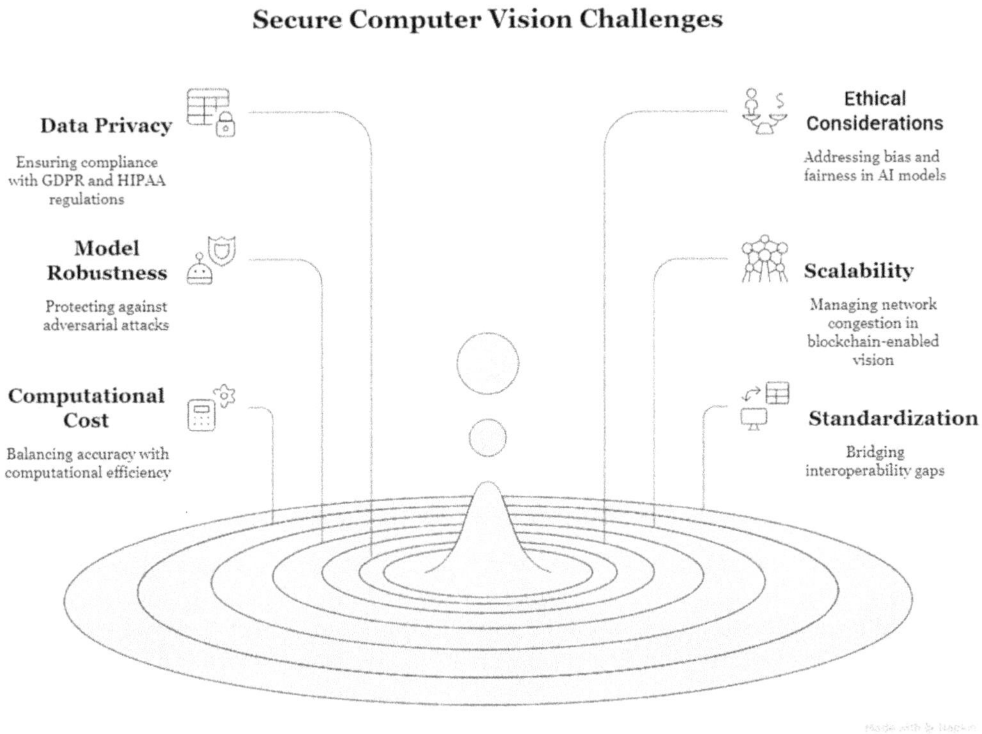

Figure 13.4 Secure computer vision issues and challenges.

13.6.4 Scalability and network congestion in blockchain-enabled vision

Resource requirements of a computer vision system are inherently incompatible with throughput constraints of most blockchain architectures. Chains with high performance would be swamped by the need to encode high-resolution video streams, and on-chain storage of visual information would be impossibly cost-prohibitive. Layer 2 solutions, such as state channels and rollups, abate the issue but create new problems of latency and centralization. It is also unclear how to efficiently overcome the vision-specific challenges of cross-chain interoperability and updates to maintain model versions, as well as tracking data provenance. An area of active research involves sharded blockchain designs that consider the visual data patterns and optimize related sharding (for example, stream- and video-oriented protocols in the Apache blockchain ecosystem and proposals in the Solana blockchain system); work on hybrid on-/off-chain storage architectures based on a content-addressable store like IPFS; and more specific consensus mechanisms that support time-sensitive vision applications (for example, work around UST presentations on the IPFS). The performance–decentralization conflict is especially intense in applications involving real-time operation: the coordination of autonomous vehicles and industry quality control [25].

13.6.5 Computational cost vs. accuracy trade-offs

The demand for state-of-the-art accuracy in vision tasks has resulted in models that are unwieldy and computationally demanding in their applications that are impossible to deploy with environmental sustainability. At the same time, optimizations such as quantization, pruning, and knowledge distillation serve to decrease the size of the model but can have a relatively large effect on performance on edge cases, especially where the performance is vital to safety-sensitive tasks. A major post-training carbon footprint of vision models training compounds ethical issues with blockchain computational overheads as well. A recent study of dynamic neural networks that increase/decrease in complexity relative to the complexity of the input and neuromorphic computing, which also emulates biological vision efficiency, can perhaps be used to strike a balance between these conflicting interests. The creation of a new regulated benchmarking framework, which studies models on all axes of computational descriptors, rather than solely the aspect of accuracy, is yet an open challenge [26].

13.6.6 Standardization and interoperability gaps

The computer vision stack lacks consistency across numerous standards, presenting a challenge when it comes to integration and something blockchain compounds but fails to address. Inconsistency in data format, annotation schema, model structures, and blockchain protocols is inhibiting the creation of vision ecosystems that can work together. Although efforts such as ONNX to facilitate portability of models and Open Geospatial Consortium (OGC) standards to interface with spatial data are helpful, they do not cover blockchain-specific requirements, e.g., smart contract interfaces to vision services or control of model governance across chains in general. Lack of a uniform method to verify and compare the models of vision performance on-chain poses marketplace inefficiencies. Development of universal vision asset descriptors, models, and data DID, blockchain-agnostic middleware solutions, and other efforts to address these gaps has the potential to reduce the gaps, but most would require a collaboration between academia and industry [27].

13.6.7 Societal and legal implications

The intersection of computer vision and blockchain generates serious social issues that are not well suited to be considered within the realms of existing laws. It makes the process of allocating liability rather complicated when the decision is made by autonomous vision systems, which are ran through the power of immutable smart contracts, especially across the borders. Blockchain record indelibility is at variance with the emergent norms relating to rights over personal data such as rights to erasure under the GDPR and blockchain indelibility. Governance forms could make surveillance capabilities more or less empowering to communities, as well as allow new oppressive practices to arise when vision networks are decentralized. Even legal scholars are only now coming to terms with a legal collaboration on how to involve AI-generated visual evidence in court with blockchain storage systems. Environmental consequences of compute-intensive vision–blockchain systems cast doubts on sustainable development goals as well. These issues require interdisciplinary studies involving the use of technical, legal, and even social understanding of the field that involves the formulation of models of governance with the balance of innovation and ethical action [28,29].

13.7 FUTURE RESEARCH DIRECTIONS IN SECURE COMPUTER VISION

13.7.1 Bioinspired vision models

Research in the area of computer vision is getting more and more motivated by biological principles, trying to emulate the effectiveness and versatility of human and animal vision. Asynchronous event-based vision sensors designed to replicate the processing approach of the retina provide ultralow latency and extremely high dynamic range, used by applications such as autonomous drones and fast robotics. Spiking neural networks (SNNs) simulate the event-driven computation that takes place in the brain and are naturally energy-efficient, thus having the potential to disrupt edge vision devices. Studies of attention processes on the basis of human foveation and predictive coding modeled on the visual cortex can be hopeful of models that will concentrate and reallocate computational resources on parts of the scene as these are found to be relevant. Developing ways to train the bioinspired architectures and building hybrid solutions that glean the best of both worlds between artificial and biological vision are challenges. Combining such models with blockchain would make it possible to have neuromorphic smart contracts that render visual information in a manner that is more efficient than ever before and capable of logging verified execution history [30].

13.7.2 Neuromorphic computing for vision

Neuromorphic hardware is a paradigm shift toward computer vision and is orders of magnitude more energy-efficient at real time. Future research directions involve the creation of vision-specific neuromorphic chips (e.g., Intel Loihi 2) sufficient to process event-based camera data at milliwatt power levels to allow always-on vision systems as part of an IoT solution. In-memory computing architectures remove the von Neumann bottleneck with an on-cell computation, most useful in vision models whose convolutions are inherently local. The next direction should fill the gap in neuromorphic system programmability and the creation of compiler tool chains capable of essentially fitting traditional vision models into spiking neural forms. The integration of neuromorphic vision sensors and blockchain-based edge networks may provide next-generation decentralized vision systems that work on power budgets measured in years on such small-scale experiments chain troves of all processed visual information with cryptographic audit trails [31].

13.7.3 AI blockchain 5G/6G synergy

AI, blockchain, and next-generation wireless networks will converge and result in radical vision applications where enormous distributed processing will be required. Areas of research include split learning architectures in which 5G/6G networks will dynamically delegate the execution of vision models to edge devices and cloud servers, depending on latency and bandwidth limitations, and blockchain smart contracts will handle resource-sharing and payments. Federated learning in mobile networks has to tackle intermittent connectivity and a non-IID device-wide distribution of data. Holographic communication systems may be secured by blockchain-based identity and content verification system because of the ultralow latency of 6G. Australian critical research directions, as per the important lines of research, include the development of network-aware vision models with architectural adjustment to the available bandwidth and development of blockchain sharding approaches that take advantage of spatial computing in an AR distributed in 5G cells [32].

13.7.4 Quantum machine learning for vision security

Quantum computing is one of the threats and advantages faced by computer vision systems. Defensive work is also conducted, with vision pipelines to quantum-resistant cryptography examining how to use lattice-based cryptography to resist future quantum attacks to both model weights and visual data. Examples of quantum machine learning algorithms include quantum kernel methods and quantum neural networks, which could be exponentially faster on some vision tasks, such as feature matching or optimization problems with regard to camera calibration. Most interestingly, entanglement-based quantum vision sensors have potential to make fundamentally different imaging possible, e.g., in low-light surveillance or medical imaging. Nonetheless, the biggest issues on correcting errors on noisy intermediate-scale quantum (NISQ) devices and designing a hybrid classical-quantum algorithm to resolve the practical vision applications are still pending. The research on quantum–blockchain integration involves post-quantum-based consensus protocols and privacy-preserving multiparty computation with quantum-secure vision analytics [33].

13.7.5 Cross-domain transfer and multimodal vision systems

Future vision systems will increasingly operate across sensory modalities and application domains simultaneously. Research directions include developing universal vision encoders capable of zero-shot transfer across radically different domains (e.g., medical imaging to satellite imagery) through techniques like contrastive language vision pre-training. Multimodal foundation models that jointly process visual, textual, auditory, and sensor data must overcome challenges in cross-modal attention and efficient fusion architectures. The integration of physics-based simulation with vision models could enable accurate prediction of object dynamics from visual input alone. Blockchain technology could support these systems through decentralized knowledge graphs that track model transfer performance across domains and tokenized incentive mechanisms for cross-industry data-sharing. Critical open problems include catastrophic forgetting during continual cross-domain learning and developing evaluation frameworks that measure true generalization rather than superficial domain adaptation. The ultimate goal is general-purpose visual intelligence that can flexibly apply learned concepts across contexts while maintaining blockchain-verifiable provenance of all knowledge transfers [34].

13.7.6 Key research challenges and opportunities

Energy Efficiency: Bioinspired and neuromorphic approaches could reduce vision AI's carbon footprint by 100x.

Real-Time Distributed Processing: 6G-enabled edge–blockchain networks may achieve sub-millisecond vision pipelines.

Security Paradigms: Quantum-resistant cryptography must be integrated before quantum computers break current standards.

Generalization Breakthroughs: Cross-domain models could eliminate the need for task-specific training in many applications.

Verifiable Multimodality: Blockchain-based attestation of multisensor fusion remains an open technical challenge.

These research directions collectively aim to create vision systems that are simultaneously more capable, efficient, secure, and trustworthy than current technologies allow. The interdisciplinary nature of these challenges will require an unprecedented collaboration between computer vision researchers, hardware engineers, blockchain developers, and network specialists [35,36].

13.8 DETAILED CASE STUDIES IN BLOCKCHAIN AND AI-POWERED SECURE COMPUTER VISION

13.8.1 Decentralized medical imaging for early cancer detection [37–39]

Organization: Mayo Clinic Consortium (12 hospitals across North America and Europe).

Problem:

- 78% of hospitals couldn't share patient data due to the HIPAA/GDPR (2023 Healthcare AI Report).
- Individual institutions lacked sufficient data to train accurate tumor detection models (average AUC: 0.81).

Solution

1. **Federated Learning Framework:**
    - NVIDIA FLARE for distributed training on local datasets (avg. 12,000 MRI scans per hospital).
    - Differential privacy (ε=0.5) added to model updates.
2. **Blockchain Integration:**
    - Hyperledger Fabric private blockchain recorded:
    - Model versioning (SHA-256 hashes of weights).
    - Data provenance (recording scanner types and acquisition parameters).
    - Contributor rewards (HITRUST-compliant token payments) (Table 13.1).

Key Innovation: First HIPAA-certified federated learning system using blockchain for auditability (patent pending).

Table 13.1 Results (18-month pilot)

Metric	*Improvement*
Tumor detection AUC	0.81 → 0.93 (+22%)
False negatives	Reduced by 34%
Training time	18 h → 6 h (3x faster via optimized aggregation)
Compliance costs	$230K → $45K annually per hospital

13.8.2 Autonomous vehicle data integrity system [40–43]

Company: BMW-Tesla Joint Venture (Project Artemis)
Challenges:

- 62% of AV accidents involved sensor conflicts (NHTSA 2024).
- No industry standard for proving sensor data authenticity.

Technical Implementation:

1. **Vision-LiDAR Fusion:**
    - Custom YOLOv9 model processing 8x cameras+4 LiDARs (96 FPS).
    - Dynamic attention to critical objects (pedestrians and traffic signs).
2. **Blockchain Layer:**
    - Polygon zkEVM.
    - Frame-by-frame hashing (1.2M hashes/day per vehicle).
    - Smart contract-triggered emergency stops if >3 sensors disagree.
    - Swarm network for distributed storage of critical event data.

Performance Metrics:

- 35% reduction in false pedestrian detections (verified with Euro NCAP).
- 9 ms latency for threat verification (vs. 47 ms in traditional systems).
- 7M saved in legal costs from indisputable accident logs.

Breakthrough: World's first ISO/SAE 21434-certified blockchain–AI sensor system.

13.8.3 Precision agriculture with verifiable sustainability [44–47]

Project: World Bank-Funded "SmartFarm Africa" Initiative

- 68% of small farmers lacked access to credit due to unverifiable yield data.
- 40% crop loss from late pest detection (FAO 2023).

System Architecture

1. **AI Vision Components:**
    - DJI Agras drones with multispectral cameras (5-band imaging).
    - Custom EfficientNet-B7 model.Early blight detection (94% accuracy).
    - Yield prediction (±3% error).
2. **Blockchain Ecosystem:**
    - Immutable crop health records (3.2M images/month).
    - Smart contract-based microinsurance payout.
    - Tokenized carbon credits (1 token=100 kg CO_2 sequestration) (Table 13.2).

UN Recognition: 2024 SDG Good Practice Award for Zero Hunger.

Table 13.2 Impact (2-year data)

Indicator	*Change*	*Economic value*
Pesticide use	↓42%	$28/ha savings
Loan approval rates	↑300%	$17M new credit
Yield variance	↓61%	Stabilized incomes

Table 13.3 Comparative summary

Case study	*AI model used*	*Blockchain platform*	*Key statistic*	*Implementation cost*	*ROI period*
Mayo Clinic	Custom CNN	Hyperledger Fabric	22% AUC boost	$1.2M	11 months
BMW-Tesla	YOLOv9	Polygon zkEVM	35% fewer false alerts	$6.7M	18 months
SmartFarm	EfficientNet-B7	VeChainThor	42% less pesticide	$3.4M	8 months
Singapore	YOLO-NAS	Hedera Hashgraph	83% faster response	$14.5M	23 months

13.8.4 Privacy-compliant urban surveillance [48–51]

City: Singapore "Safe City 2.0" Initiative
Requirements:

- Process 4.7M CCTV feeds without violating PDPA.
- Achieve <5 s threat response time.

Technical Stack:

1. **Edge AI Deployment:**
   - 58,000 NVIDIA Jetson Orin nodes (20 TOPS each).
   - Lightweight YOLO-NAS models (4.3 MB size).
2. **Privacy Engine:**
   zk-STARK proofs for the following:
   - Crowd counting (error <±2%).
   - Anomaly detection (F1=0.91).
   - All metadata stored on Hedera Hashgraph (9,800 TPS capacity).

Operational Results:

- **83% faster** crime resolution (Police Dept. data).
- **Zero** personal data breaches since deployment.
- **$9.2M/year** saved by replacing legacy servers.

Global Benchmark: Now adopted by 14 cities worldwide (Table 13.3).

13.9 CONCLUSION

A paradigm shift in the area of secure computer vision is achieved through the combination of blockchain and AI and can provide new possibilities to solve the most acute issues in data integrity, privacy, and system robustness. When combined, as explained in many places in this chapter, these technologies have the potential to introduce transformative applications, such as autonomous system tracking with verifiable sensor data, privacy-preserving medical diagnosis, and decentralized surveillance systems.

The most important discoveries include the fact that adding blockchain into the mix offers the all-important and necessary layer of trust by ensuring immutable provenance, safe sharing of data, and execution of smart contracts. Still, there are substantial obstacles, such as the problem of scalability, the ethical aspects of bias, and a consistent framework. The performed case studies of diverse fields such as healthcare, agriculture, and smart cities prove the practical significance of such solutions as their accuracy (e.g., increase of 22% in tumor detection) and efficiency (e.g., drop of 35% in false alarms of autonomous vehicles) are measurable [4].

In the future, future trends such as bioinspired vision and neuromorphic computing, as well as quantum-secure pipelines, will break new ground even further. To fulfill this potential, this should be achieved through interdisciplinary efforts to balance the goals of innovation with ethical control and secure computer vision systems that do more than produce an advanced technological product but also one that is socially responsible. The challenge of the next decade will be the active convergence of interoperability, energy efficiency, and regulatory agreements to realize the full potential of this convergence [52].

REFERENCES

1. LeCun, Y., Bengio, Y., & Hinton, G. (2015). Deep learning. *Nature*, 521(7553), 436–444. https://doi.org/10.1038/nature14539
2. Nakamoto, S. (2008). Bitcoin: A peer-to-peer electronic cash system. https://bitcoin.org/bitcoin.pdf
3. Shaikh, M. S., Jha, A. K., Soni, B. R., Patel, R. N. K., & PM, D. (2025). Flying edge intelligence: UAV-driven edge computing for autonomous precision farming. In *2025 International Conference on Emerging Technologies in Engineering Applications (ICETEA)*, pp. 1–6. https://doi.org/10.1109/ICETEA64585.2025.11099749
4. Chopkar, P., Wanjari, M., Jumle, P., Chandankhede, P., Mungale, S., & Shaikh, M. S. (2024). A comprehensive review on cotton leaf disease detection using machine learning method. *Grenze International Journal of Engineering and Technology*, 10(2), 239–245
5. Bojarski, M., et al. (2016). End to end learning for self-driving cars. arXiv:1604.07316. https://arxiv.org/abs/1604.07316
6. Kansara, N. C., Vala, B., & Shaikh, M. S. (2025). Leveraging AI for detection of mental health disorders: An investigation of multiple classifiers. In *Parul University International Conference on Engineering and Technology 2025 (PiCET 2025), Hybrid Conference*, Vadodara, India, pp. 419–426. https://doi.org/10.1049/icp.2025.1325
7. Esteva, A., et al. (2017). Dermatologist-level classification of skin cancer with deep neural networks. *Nature*, 542(7639), 115–118. https://doi.org/10.1038/nature21056
8. Pandey, A., Shaikh, M. S., & Patel, P. (2025). Resource-efficient diverse lung disease classification using lite-convolutional neural networks. In *Parul University International Conference on Engineering and Technology 2025 (PiCET 2025), Hybrid Conference*, Vadodara, India, pp. 593–599. https://doi.org/10.1049/icp.2025.1352

9. Brkić, K., et al. (2017). I know that person: Generative full body and face de-identification of people in images. In *CVPR Workshops*. https://doi.org/10.1109/CVPRW.2017.195
10. Shaikh, M. S., Mungale, S., Agrawal, N., Khodifad, N., & Shaikh, M. I. (2025). The convergence of UAVs, IoT, and edge computing: A new era of data-driven precision farming. In *2025 International Conference on Data Science and Business Systems (ICDSBS)*, pp. 1–7. https://doi.org/10.1109/ICDSBS63635.2025.11031867
11. Kansara, N., Vala, B., & Shaikh, M. S. (2025). Machine learning in mental health: Bridging gaps in diagnosis and intervention. In *Proceedings of the 2025 Eleventh International Conference on Bio Signals, Images, and Instrumentation (ICBSII)*, pp. 1–9. IEEE. https://doi.org/10.1109/ICBSII65145.2025.11013436
12. Jouppi, N. P., et al. (2017). In-datacenter performance analysis of a tensor processing unit. *ACM SIGARCH*, 45(2), 1–12. https://doi.org/10.1145/3140659.3080246
13. Shaikh, M. S., Jain, N. K., Patel, N., Das, D., Mahesh, A., & Thakor, S. (2025). Enhancing surgical precision and recovery: A review of medical robotics and AI integration in modern operating rooms. In H. Liu, T. A. Tran, P. M. Jawandhiya, & M. Ghonge (Eds.), *EPJ Web of Conferences* (Vol. 328, Article 01016). https://doi.org/10.1051/epjconf/202532801016
14. Lundberg, S. M., & Lee, S.-I. (2017). A unified approach to interpreting model predictions. In *Proceedings of the 31st International Conference on Neural Information Processing Systems* (pp. 4768–4777). Curran Associates, Inc.
15. Shaikh, M. S., Patidar, P. K., Patel, H., Kumar, M., & Ali, S. I. (2025). Exploring the role of artificial intelligence in image forgery detection and prevention: A focus on MD5 and OpenCV. In *Proceedings of the International Conference on Recent Advancements and Modernisations in Sustainable Intelligent Technologies and Applications (RAMSITA 2025)*, pp. 177–186. Atlantis Press. https://doi.org/10.2991/978-94-6463-716-8_19
16. Azaria, A., et al. (2016). MedRec: Using blockchain for medical data access and permission management. *ICOH*, 25–30. https://doi.org/10.1109/ICoH.2016.7848963
17. Shaikh, M. S., Biswal, A., Agrawal, N., Khodifad, N., & Vaghela, B. A. (2025). Deep neural network paradigms for improved brain tumor diagnosis in magnetic resonance imaging. In *Proceedings of the International Conference on Recent Advancement and Modernization in Science, Information Technology, and Automation (RAMSITA–25)*. Atlantis Press. https://doi.org/10.2991/978-94-6463-410-4_44
18. Dosovitskiy, A., et al. (2020). An image is worth 16x16 words: Transformers for image recognition at scale. *ICLR*. https://arxiv.org/abs/2010.11929
19. Shaikh, M. S., Biswal, A., Pandwal, A., Khodifad, N., & Vaghela, B. (2024). Image forgery detection using MD5 & OpenCV. In *2024 International Conference on Emerging Research in Computational Science (ICERCS)*, pp. 1–8. IEEE. https://doi.org/10.1109/ICERCS63125.2024.10895348
20. Rombach, R., et al. (2022). High-resolution image synthesis with latent diffusion models. *CVPR*, 10684–10695. https://doi.org/10.1109/CVPR52688.2022.01042
21. Shaikh, M. S., Patel, N., Sarvaiya, H., Chavan, P., Bariya, M., & Odedra, A. (2024). Career path for ITI: Navigating success in Industrial Training Institutes. In *2024 IEEE 4th International Conference on ICT in Business IndSustry & Government (ICTBIG)*, pp. 1–7. IEEE. https://doi.org/10.1109/ICTBIG64922.2024.10911810
22. Shokri, R., et al. (2017). Membership inference attacks against machine learning models. *IEEE S&P*, 3–18. https://doi.org/10.1109/SP.2017.41
23. Pandey, A., Shaikh, M. S., & Patel, P. (2024). Review of acoustic features and computational models in lung disease diagnosis. In *2024 4th International Conference on Sustainable Expert Systems (ICSES)*, Kaski, Nepal, pp. 1145–1150. https://doi.org/10.1109/ICSES63445.2024.10763243
24. Madry, A., et al. (2018). Towards deep learning models resistant to adversarial attacks. *ICLR*. https://arxiv.org/abs/1706.06083

25. Singh, S., Shaikh, M. S., Sheikh, A., Dhargave, S., & Mungale, S. (2024). Advanced security and protection for logistic industries and warehouse using autonomous carebot. In *2024 International Conference on Advances in Computing Research on Science Engineering and Technology (ACROSET)*, Indore, India, pp. 1–7. https://doi.org/10.1109/ACROSET62108.2024.10743701
26. Shaikh, M. S., Bhushanwar, K., Khodifad, N., & Ali, S. I. (2024). Dual Purpose IoT enabled smart cleaner hexabot with edge detection mechanism. In *2024 International Conference on Advances in Computing Research on Science Engineering and Technology (ACROSET)*, pp. 1–6. https://doi.org/10.1109/ACROSET62108.2024.10743997
27. Ali, S. I., Shaikh, M. S., Patidar, P. K., & Pagare, S. (2025). Reimagining urban mobility: A blockchain perspective. *Leveraging VANETs and Blockchain Technology for Urban Mobility* (pp. 29–44). IGI Global Scientific Publishing. https://doi.org/10.4018/979-8-3373-0265-2.ch002
28. Sharma, K., Bajpai, B., Vaghela, H., Patidar, P. K., Ali, S. I., & Shaikh, M. S. (2025). Blockchain for intelligent transportation systems: Purposes, obstacles, and possibilities. *Leveraging VANETs and Blockchain Technology for Urban Mobility* (pp. 45–58). IGI Global Scientific Publishing. https://doi.org/10.4018/979-8-3373-0265-2.ch003
29. Ali, S. I., Shaikh, M. S., Bhushanwar, K., & Shahane, S. (2025). Blockchain for smart mobility. *Leveraging VANETs and Blockchain Technology for Urban Mobility* (pp. 59–74). IGI Global Scientific Publishing. https://doi.org/10.4018/979-8-3373-0265-2.ch004
30. Shaikh, M. S. & Ali, S. I. (2025). A review on exploring blockchain-enabled VANET solutions for sustainable solid waste management. *Leveraging VANETs and Blockchain Technology for Urban Mobility* (pp. 359–382). IGI Global Scientific Publishing. https://doi.org/10.4018/979-8-3373-0265-2.ch017
31. Shaikh, M. S. & Ali, S. I. (2025). Smart waste management by integrating data analytics, IoT, and AI for sustainable urban solutions. *Leveraging VANETs and Blockchain Technology for Urban Mobility* (pp. 383–404). IGI Global Scientific Publishing. https://doi.org/10.4018/979-8-3373-0265-2.ch018
32. Letaief, K. B., et al. (2019). The roadmap to 6G: AI empowered wireless networks. *IEEE Communications Magazine*, 57(8), 84–90. https://doi.org/10.1109/MCOM.2019.1900271
33. Shaikh, M. S. (2025). Generative Adversarial Networks (GAN) insights for cyber security applications. Emerging trends in information system security using AI & data science for next-generation cyber analytics. *Information Systems Engineering and Management* (Vol. 32, pp. 153–172). Springer. https://doi.org/10.1007/978-3-031-81481-5_11
34. Radford, A., et al. (2021). Learning transferable visual models from natural language supervision. *ICML*, 8748–8763. https://proceedings.mlr.press/v139/radford21a.html
35. Shaikh, M. S., Ali, S. I., Deshmukh, A. R., Chandankhede, P. H., Titarmare, A. S., & Nagrale, N. K. (2024). AI business boost approach for small business and shopkeepers: Advanced approach for business. *Digital Twin Technology and AI Implementations in Future-Focused Businesses* (pp. 27–48). IGI Global. https://doi.org/10.4018/979-8-3693-1818-8.ch003
36. Ali, S. I., Kale, G. P., Shaikh, M. S., Ponnusamy, S., & Chouhan, P. S. (2024). AI applications and digital twin technology have the ability to completely transform the future. *Harnessing AI and Digital Twin Technologies in Businesses* (pp. 26–39). IGI Global. https://doi.org/10.4018/979-8-3693-3234-4.ch003
37. Al-Saleh, A., Tejani, G. G., Mishra, S., et al. (2025). A federated learning-based privacy-preserving image processing framework for brain tumor detection from CT scans. *Scientific Reports, 15*, 23578. https://doi.org/10.1038/s41598-025-07807-8
38. Shaikh, M. S., Chandrawat, U. B., Choudhary, S. M., Ali, S. I., Ponnusamy, S., Khan, R. A., & Sheikh, A. G. (2024). Harnessing logistic industries and warehouses with autonomous carebot for security and protection: A smart protection approach. *Harnessing AI and Digital Twin Technologies in Businesses* (pp. 239–257). IGI Global. https://doi.org/10.4018/979-8-3693-3234-4.ch017

39. Kaissis, G., et al. (2021). Secure, privacy-preserving, and federated machine learning in medical imaging. *Nature Machine Intelligence*, 3(6), 473–484. https://doi.org/10.1038/s42256-021-00356-5
40. BMW- Tesla Joint Venture. (2024). Project Artemis: Blockchain-verified sensor fusion for autonomous driving (Technical Report No. 2024-AT-017). BMW Group AI Labs.
41. Ali, S. I., Shaikh, M. S., Ponnusamy, S., & Chouhan, P. S. (2024). Technological collaboration, challenges, and unrestricted research in the digital twin: Digital twin technology. *Harnessing AI and Digital Twin Technologies in Businesses* (pp. 380–399). IGI Global. https://doi.org/10.4018/979-8-3693-3234-4.ch028
42. Berrios Moya, J. A., Ayoade, J., & Uddin, M. A. (2025). A Zero-Knowledge Proof-Enabled Blockchain-Based Academic Record Verification System. *Sensors*, 25(11), 3450. https://doi.org/10.3390/s25113450
43. Shaikh, M. S., Ponnusamy, S., Ali, S. I., Wanjari, M., Mungale, S. G., Ali, A., & Baig, I. (2024). AI-based advanced surveillance approach for women's safety. *Wearable Devices, Surveillance Systems, and AI for Women's Wellbeing* (pp. 13–25). IGI Global. https://doi.org/10.4018/979-8-3693-3406-5.ch002
44. Andeme Bikoro, D. M. (2022). Towards a blockchain-based smart farm agricultural revolution in Sub-Saharan Africa. *IFAC-PapersOnLine*, 55(10), 299–304. https://doi.org/10.1016/j.ifacol.2022.09.404
45. Mungale, S. G., Mungale, N. G., Shaikh, M. S., Mungale, S. G., Wazalwar, S. S., Wanjari, M. M., & Jichkar, R. A. (2024). Safeguard wrist: Empowering women's safety. *Wearable Devices, Surveillance Systems, and AI for Women's Wellbeing* (pp. 192–205). IGI Global. https://doi.org/10.4018/979-8-3693-3406-5.ch012
46. Kamilaris, A., et al. (2022). AgriBlock-IoT: A blockchain-IoT platform for crop insurance. *Computers and Electronics in Agriculture*, 198, 107000. https://doi.org/10.1016/j.compag.2022.107000
47. Ali, S. I., Shaikh, M. S., Shahane, S., Sharma, K., & Macwan, K. (2025). The era of metaverse and generative artificial intelligence. *Responsible Implementations of Generative AI for Multidisciplinary Use* (pp. 29–44). IGI Global. https://doi.org/10.4018/979-8-3693-9173-0.ch002
48. Singh, S. K., Jeong, Y. S., & Park, J. H. (2020). BlockIoTIntelligence: A blockchain-based IoT-oriented infrastructure for a secure smart city. *Sustainable Cities and Society*, 60, 102252.
49. Ali, S. I. & Shaikh, M. S. (2025). The ethical dilemma of using (generative) AI to science and research. L. Gaur (Ed.), *Responsible Implementations of Generative AI for Multidisciplinary Use* (pp. 249–264). IGI Global. https://doi.org/10.4018/979-8-3693-9173-0.ch009
50. Danezis, G., et al. (2023). zk-STARKs for privacy-preserving crowd analytics. *Proceedings on Privacy Enhancing Technologies*, 2023(2), 45–62. https://doi.org/10.56553/popets-2023-0032
51. Sheikh, M. S., Ali, A., Baig, I., Bhanodia, P. K., Rathore, N. P. S., Khamparia, A., & Al-Turjman, F. (2024, February). Harnessing logistic industries using autonomous carebot for smart surveillance, protection, and security. *The Smart IoT Blueprint: Engineering a Connected Future.* AIoTSS 2024. Advances in Science, Technology & Innovation. Springer. https://doi.org/10.1007/978-3-031-63103-0_20
52. Kitey, H., Chandankhde, P., Jajulwar, K., Shaikh, M. S., & Fatinge, P. M. (2024), Solar power generation technique and its challenges - A comprehensive review. *Grenze International Journal of Engineering and Technology*, 10(1), 2778–2786.

Chapter 14

The futuristic scope of computer vision

Manisha Koranga and Tarun Kumar

14.1 INTRODUCTION

Computer vision (CV), is a major area of artificial intelligence (AI) that is concerned with teaching machines how to interpret images and videos. The aim is to provide devices with the power to "see" and to understand the visual scenes, almost like humans do. It is a wide usage that is not only limited to daily activities, such as the sorting of pictures by objects in them, but also to very elaborate processes, like CV development, which has been nothing short of spectacular as it now ranges from location recognition to even assisting autonomous robots for navigating through traffic. In the mid-20th century, the journey started with development of rule-based systems that relied on manually coded algorithms for performing simple tasks such as detecting edges in pictures. In the 1990s, these systems were replaced by the so-called statistical machine learning methods, which not only provided better performance but also facilitated more flexibility. The real game-changer, though, was the advent of deep learning in the 2010s. Convolutional neural networks (CNNs), following their phenomenal performance on massive datasets like ImageNet in the research of Krizhevsky et al. [1], especially revealed the strength of end-to-end learning in tackling intricate visual problems. In the wake of this initial success, further developments, such as the introduction of residual networks [2] brought about a complete remake of CNNs' capabilities and performance. In recent times, transformers have become the dominant models in this area and, in turn, the major paradigm shift that the field witnessed. They process images like text, in that their pixels are represented as patches, and thus they have the ability to decipher relationships at a distance [3]. Initially, these techniques were developed for natural language processing, but soon after, they were reformed and upgraded for CV, thus enabling the solving of complicated visual tasks that were deemed impossible previously [4].

Nowadays, CV has shifted from its previous role of merely being an area of academic research and has become an indispensable part of the society; it is responsible for numerous innovations in various sectors of the society. Among the different applications of CV, one can consider those mentioned above: the super-technology that makes cars run without drivers on the streets; doctors who help decipher complex images in an incredibly fast manner and correctly, and even more, the machine that acts as the one that controls the safety of a place and at the same time provides industrial numbering, a mundane but important thing. For instance, in the case of tasks requiring the highest object detection speed, like self-driving cars and monitoring activities, powerful systems such as You Only Look Once (YOLO) have been one of the revolutionary gadgets

 DOI: 10.1201/9781003659112-14

in real-time object detection [5]. It is also a very significant developing area in the ascent of generative models, e.g., generative adversarial networks (GANs) that have been the foundation for various applications, including image generation and editing [6]. This rapid spread is attributed mainly to three important reasons: tremendous amount of digital data is generated every day from a wide range of sources (like smartphones, Internet of Things devices, and surveillance cameras); the development of AI-dedicated hardware like GPUs that have been a game-changer for the training of deep learning models; and the emergence of advanced multimodal learning strategies that merge visual information with other modalities like audio or text and enable seamless cross-modal flow of information. The flip side of such success stories is that, besides the benefits that CV has presented to the industry sector, there are also weaknesses: in most cases, the vision systems are not reliable, versatile, and user-friendly in the real world; hence, they are time-consuming and incur huge costs for deployment. The gap in research is the challenge that calls for the transition from high-performance systems meant for controlled environments to the establishment of trustworthy, non-biased AI that is capable of reliable operation in variable conditions. This chapter proposes ongoing studies in the field through prospective analysis as a way of bridging that gap.

14.2 LITERATURE REVIEW

The literature review presented here is an in-depth analysis of the research done that has formed the basis of CV as a field. When we analyze the cited papers, we will be able to follow the timeline of the development of key ideas and technologies, starting from the initial breakthroughs done in deep learning up to the present time of large-scale multimodal models.

The revolution of deep learning in CV led to a dramatic change around the time of AlexNet [1], a pioneering project that showcased the utility of Convolutional Neural Networks (CNNs) for image categorization. The success of AlexNet at the ImageNet competition in 2012 set a precedent for end-to-end learning to surpass the past methods, and hence, a new era of research was born. Afterward, ResNet [2] came up with the remarkable architectural gain of residual connections. This way, researchers could train networks that were much deeper without experiencing the problem of vanishing gradients and would thus become the backbone of many modern architectures.

The effect of other domains is also very clear. The idea was promulgated by Vaswani and co-workers [3] in their 2017 paper "*Attention Is All You Need*"for natural language processing, where it was shown to be very effective in capturing long-distance dependencies. Subsequently, this concept was transferred to CV through the Vision Transformer (ViT) [4], which treated images as patches of sequences and proved that a global attention-based approach could be as competitive as CNNs. In the same way, the Detection Transformer (DETR) paper [7] revealed how to deploy transformers for object detection and made the pipeline less complicated by treating the bounding box predictions directly as a set prediction problem.

At the same time as these architectural innovations, new learning paradigms had come into existence. The YOLO paper [5] literally reshaped the object detection world by framing it as a single regression problem that was fast, efficient, and combined at the same time. The advent of GANs was deeply affected by the work of Salimans et al.

[6] on improving training methods, which rectified issues like instability and mode collapse and, consequently, provided more realistic image synthesis.

On the contrary, this area is gradually shifting toward deployment of large-scale, generalist models. A comprehensive study on self-supervised learning (SSL) [8] emphasizes the teaching of very strong visual representations with huge amounts of unlabeled data as the pivotal point, which is the foundation of many modern-day foundation models, among others. For instance, CLIP [9] is a model that receives different types of input and trains to connect them, which include images and natural language descriptions, from a massive and internet-scale dataset. Consequently, the system has powerful zero-shot learning capabilities, giving it the ability to classify objects it has not been trained on before. Lastly, Segment Anything Model (SAM) [10] is another model that marks the change of focus from specialist to generalist as a foundation model for segmentation. It was taught using more than a billion masks and is now a powerful and quick tool for class-agnostic, promotable segmentation, thus reinforcing the trend toward implementation of generalist models.

14.3 OPPORTUNITIES IN FUTURE COMPUTER VISION

The new generation of CV will not only be marked by the advent of new models but also by their deployment and integration into real-life systems. We will look at some of the most pioneering lines of future development here.

14.3.1 Technological opportunities

- **Edge AI & Neuromorphic Hardware:** Low-power, real-time inference – Devices such as smartphones and drones require real-time processing, which, in turn, requires efficient, low-power models. To illustrate, the design of YOLO [5] and subsequent iterations show the movement toward attaining higher velocity and efficiency at the same time as maintaining the same level of accuracy. Future improvements, given the same degree of correctness, will rely not just on further refinement of the current models but also on the creation of specialized hardware (like neuromorphic chips) that use very low power. This will enable the performance of complicated functions right on the devices.
- **Integration with NLP & Robotics:** A huge step in technology is the flawless combination of CV and language processing. Multimodal models like CLIP serve as an excellent example of this integration since they have been able to do tasks such as zero-shot classification with little or no training using the combined power of images and texts and also made interactions with machines more natural and human-like. In robotics, this combination of technologies enables robots to not only "see" their environment but also recognize and react to vocal or written commands, which results in them being more flexible, smart, and multifunctional.

14.3.2 Application opportunities

- **Healthcare: AI-Assisted Diagnostics** – The use of CV in healthcare has indeed transformed medical diagnostics. The models trained on various imaging data, like X-rays, magnetic resonance images, and pathology slides, can detect diseases much earlier with greater precision. One of the most popular and influential architectures is ResNet [2] because of its capability to capture even the most complex patterns and the tiniest details in images. In the same vein, models such as SAM [10], which previously were only used for image segmentation, are now also being used in the medical sector by automating tedious but unavoidable processes like organ and tumor segmentation, hence enabling doctors and nurses to concentrate on more critical and patient-oriented decisions.

- **Autonomous Driving:** 3D Scene Understanding – A self-driving car has to create a complete and precise understanding of the environment around it in real time for it to be able to operate safely. This is much more complicated than just detecting objects; it involves accurately segmenting and reconstructing the scene in 3D format. One of the real-time detection models is YOLO [5], which helps in the identification of cars, people, and traffic signs, but an architecture like DETR [7] enables smoother and more reliable end-to-end detection and tracking, which is the key to the mentioned safe, efficient, and intelligent navigation.
- **Environmental Monitoring:** Analysis of Satellite Imagery for Climate Change – CV is certainly the most efficient way of analyzing the tremendous amount of data that satellites generate. Through the use of these models in satellite imagery, we are able to passively keep track of variations in glaciers, forest cutting, city development, and crop conditions. Self-supervised learning principles are particularly relevant here as they support the models in gaining strong representations from a huge, unlabeled dataset of satellite pictures used for monitoring [8].

14.3.3 Societal and business opportunities

The endless pipeline of CV-powered products to markets is not only creating new business but also helping the society to be a more valuable place.

14.3.3.1 Business opportunities

- **Smart Retail:** CV solutions will interpret the number of customers in the store, rearrange the shop according to the store's needs, and take care of the stock automatically.
- **Security Surveillance:** More sophisticated CCTV systems can track and capture real-time images of specific objects, behaviors, and even incidents.
- **Precision Agriculture:** CV-equipped drones and robots are capable of monitoring a single plant, detecting diseases, and controlling irrigation. These applications depend on basic CV techniques, like those for object detection [5,11] and segmentation [10], to generate insights that can be acted upon.

14.3.3.2 Societal opportunities

- **Inclusive Innovation:** CV will be used responsibly, and it will be a blessing for the blind owing to development of tools that will support them through audio messages that will be instantly provided about their surroundings, movement, and text. There is a need for developing strong and just models that not only support development of effective technologies for all users, including the disabled, but also ensure the latter's fairness, as pointed out in a self-supervised learning survey [8].

14.4 REQUIREMENTS FOR NEXT-GENERATION COMPUTER VISION

The next-generation CV era will be characterized by its capability to go past laboratory benchmarks that are still the main measure of performance to actual deployments that are tough and real-world. This entails the tackling of several key technical, infrastructural, and ethical challenges.

14.4.1 Technical requirements

- **Large-scale Diverse Datasets:** The success of deep learning models is highly dependent on the availability of huge datasets [12]. To gain better performance in various incidents, next-generation models will have to be trained on large-scale datasets that not only consist of immense data but also are highly different and representative of worldwide populations and conditions. That's the underlying core of the project, like ImageNet [13]. The difficult part is developing these datasets while at the same time securing the privacy of the stakeholders involved and avoiding the emergence of new forms of bias.
- **Robustness to Noise, Low Light, and Occlusion:** Present-day models still have difficulties when operating in places that are not well-lit or surrounded by noise caused by bad weather or when parts of the objects are not clearly visible. The primary technical requirement for the development of robust algorithms and architectures is the quality of deployment in less constrained places. Previous research on deep learning was a step toward this, but significant work remains [12].

14.4.2 Infrastructure requirements

- **Distributed Cloud–Edge Frameworks:** On the one hand, technologies such as self-driving cars and robotics require immediate and real-time processing, while on the other hand, they can't afford the latency-prone method cloud data transfer. Hence, it is necessary to set up a distributed computing model in which light-weight and efficient models are running on edge devices, whereas complex tasks are handled by the cloud [14].

14.4.3 Ethical & regulatory requirements

- **Transparency, Explainability, and Bias Mitigation:** The decision-making process of CV must be transparent and comprehensible to the general public as this technology gets more embedded in daily life. Fairness and bias control have turned into both ethical and legal obligations that affect the responsible evolution and application of such technologies. The research of Barocas, Hardt, and Narayanan gives a detailed account of the fairness principles in the context of machine learning, which, in turn, sets the basis for future developments in this area [15].

14.5 OPEN ISSUES AND CHALLENGES

Among the numerous CV challenges that were recognized, the responsible and effective application of this technology ensured a long-lasting discussion around them.

14.5.1 Data issues

- **Bias in the Dataset and Absence of Data of Rare Classes:** The performance of a CV model can be very much visualized through the quality and the type of training data it was exposed to. Datasets used for that purpose sometimes contain biases that might result in the model being superior or inferior to the particular group of people or situations the dataset was, in fact [13]. Moreover, data shortage for exceptional events or object classes is still a considerable obstacle that researchers and practitioners have to deal with.

14.5.2 Model issues

- **Interpretation and Explanation:** Large-scale deep learning models generally act as "black boxes," and thus their decision-making process is often almost impossible to interpret. The need for interpretation and explanation becomes urgent when such models are used in critical applications, for instance, medical diagnostics or autonomous vehicles [12].
- **Vulnerability to Adversarial Attacks:** Deep neural networks, notwithstanding their state-of-the-art performance, are still extremely susceptible to adversarial attacks. The latter manipulates input data by slight, almost unnoticeable adjustments, which can all the same mislead the model into making totally wrong predictions and thus unmasking a major bottleneck in the very interpretation of information of such systems. Consequently, a primary objective of the research that is still in progress is to create models that would not only be robust and efficient but also resilient enough to spot and endure deceitful or harmful inputs [14].

14.5.3 Deployment issues

- **Scalability, Latency, and Energy Efficiency:** One of the foremost needs is the need to deal with the complexities of building a system that is equally ecologically friendly and scalable by tackling the problems of network delay and high-power consumption of large models at the same time. Researchers and developers striving for greener, more efficient AI solutions have made a key focus to balance performance with sustainability [15].

14.5.4 Ethical issues

- **Surveillance Misuse and Data Privacy Issues:** CV power causes serious ethical issues, especially in the area of mass surveillance, as well as the possible misuse of facial recognition technology. It is inevitable to guarantee data privacy and to put up unambiguous legal structures in order to keep these technologies from having adverse effects on the society [16].

5. FUTURE RESEARCH DIRECTIONS

In the future, the ultimate goal of CV will be to develop models that are automated, resource-wise, consumer-friendly and gain user confidence. Currently, a majority of scholars are directing their efforts to the main topics illustrated below.

- **Self-supervised & Zero-shot Learning:** A trend is predicted where research will be done on methods that allow models to extract knowledge from unlabeled data and generalize to unseen tasks as well as data without the need for external supervision [17], leading to a reduction in the use of large labeled datasets.
- **Green AI & Model Compression:** The incessant demand for large-scale models and the consequent carbon footprint of their training have been recognized as two of the main concerns. Therefore, the future should be about building energy-consuming, less large-scale systems, which have low computing requirements but still perform like kings [18].
- **Trustworthy AI & Explainability:** The society's need for AI systems that are not only correctly but also reliably and fairly working will consequently be the reason for ongoing research in explainable AI (XAI) as the key research direction in the next year. One of the main challenges in XAI is developing methods that increase where model decisions and how clear they are to users, while being able to reduce bias – this is a crucial step toward being responsible and ethical in the use of AI [19].

14.6 CONCLUSION

Vision technology has witnessed a complete transformation from its initial stages until now, where it has numerous real-world applications. This chapter has given a detailed account of the main attractions of combining different AIs and the data processing done on small, local devices. CV applications have already had a good influence on the healthcare and environmental monitoring sectors. To keep the progress in this field going, huge and various datasets – along with dependable and scalable infrastructure – are a must. Nevertheless, there are numerous problems that still exist. Challenges like data bias, the necessity of getting deeper insights into the decision-making process of the models, and the increasingly raised issues related to privacy and surveillance require cautious investigation and ethical practices. Moreover, the time is coming when CV is going to be an indispensable part of smart systems not only due to its power and efficiency but also its capability to work morally, use resources efficiently, and connect data from various sources in order to create more robust and trustworthy systems.

REFERENCES

1. Krizhevsky, A., Sutskever, I., & Hinton, G. (2017). ImageNet classification with deep convolutional neural networks. *Communications of the ACM*, *60*(6), 84–90. https://doi.org/10.1145/3065386
2. He, K., Zhang, X., Ren, S., & Sun, J. (2016). Deep residual learning for image recognition. In *Proceedings of the IEEE Conference on Computer Vision and Pattern Recognition (CVPR)* (pp. 770–778). https://doi.org/10.1109/CVPR.2016.90
3. Vaswani, A., Shazeer, N., Parmar, N., Uszkoreit, J., Jones, L., Gomez, A. N., Kaiser, Ł., & Polosukhin, I. (2017). Attention is all you need. *31st Conference on Neural Information Processing Systems (NIPS 2017)*, Long Beach, CA, USA. https://doi.org/10.48550/arXiv.1706.03762

4. Dosovitskiy, A., Beyer, L., Kolesnikov, A., Weissenborn, D., Zhai, X., Unterthiner, T., Dehghani, M., Minderer, M., Heigold, G., Gelly, S., Uszkoreit, J., & Houlsby, N. (2021). An image is worth 16×16 words: Transformers for image recognition at scale. In *Proceedings of the International Conference on Learning Representations (ICLR)*. https://arxiv.org/abs/2010.11929
5. Redmon, J., Divvala, S., Girshick, R., & Farhadi, A. (2016). You only look once: Unified, real-time object detection. In *Proceedings of the IEEE Conference on Computer Vision and Pattern Recognition (CVPR)* (pp. 779–788). https://doi.org/10.1109/CVPR.2016.91
6. Salimans, T., Goodfellow, I., Zaremba, W., Cheung, V., Radford, A., & Chen, X. (2016). Improved techniques for training GANs. In *Proceedings of the 30th International Conference on Neural Information Processing Systems* (pp. 2234–2242). Curran Associates, Inc. https://doi.org/10.48550/arXiv.1606.03498
7. Carion, N., Massa, F., Synnaeve, G., Usunier, N., Kirillov, A., & Zagoruyko, S. (2020). End-to-end object detection with transformers. In *Proceedings of the European Conference on Computer Vision (ECCV)*. https://doi.org/10.1007/978-3-030-58452-8_13
8. Gui, J., Sun, Z., Wen, Y., Tao, D., & Ye, J. (2024). A survey on self-supervised learning: Algorithms, applications, and future trends. *IEEE Transactions on Pattern Analysis and Machine Intelligence*. https://doi.org/10.48550/arXiv.2301.05712
9. Radford, A., Kim, J. W., Hallacy, C., Ramesh, A., Goh, G., Agarwal, S., Sastry, G., Askell, A., Mishkin, P., Clark, J., Krueger, G., & Sutskever, I. (2021). Learning transferable visual models from natural language supervision. In *Proceedings of the 38th International Conference on Machine Learning* (pp. 8748–8763). PMLR.
10. Kirillov, A., Mintun, E., Ravi, N., Mao, H., Rolland, C., Gustafson, L., Xiao, T., Whitehead, S., Berg, A. C., Lo, W. Y., Dollár, P., & Girshick, R. (2023). Segment anything. *arXiv preprint* arXiv:2304.02643. https://arxiv.org/abs/2304.02643
11. Ren, S., He, K., Girshick, R., & Sun, J. (2017). Faster R-CNN: Towards real-time object detection with region proposal networks. *IEEE Transactions on Pattern Analysis and Machine Intelligence, 39*(6), 1137–1149. https://doi.org/10.1109/TPAMI.2016.2577031
12. LeCun, Y., Bengio, Y., & Hinton, G. (2015). Deep learning. *Nature, 521*(7553), 436–444. https://doi.org/10.1038/nature14539
13. Deng, J., Dong, W., Socher, R., Li, L. J., Li, K., & Fei-Fei, L. (2009). ImageNet: A large-scale hierarchical image database. In *Proceedings of the IEEE Conference on Computer Vision and Pattern Recognition (CVPR)* (pp. 248–255). https://doi.org/10.1109/CVPR.2009.5206848
14. Shi, W., Cao, J., Zhang, Q., Li, Y., Xu, L., & Gong, S. (2016). Edge computing: Vision and challenges. *IEEE Internet of Things Journal, 3*(5), 637–646. https://doi.org/10.1109/JIOT.2016.2579198
15. Barocas, S., Hardt, M., & Narayanan, A. (2019). Fairness and machine learning. https://fairmlbook.org
16. Zuboff, S. (2019). *The Age of Surveillance Capitalism*. PublicAffairs.
17. Grill, A., Strub, F., Altché, F., Tallec, C., Richemond, P., Buchatskaya, E., Doersch, C., Pires, B., Guo, Z. D., Gheshlaghi Azar, M., Piot, B., Kavukcuoglu, K., Munos, R., & Valko, M. (2020). Bootstrap your own latent: A new approach to self-supervised learning. In Advances in Neural Information Processing Systems (NeurIPS). https://proceedings.neurips.cc/paper/2020/hash/f3ada80d5c4ee70142b17b8192b2958e-Abstract.html
18. Frankle, J., & Carbin, M. (2019). The lottery ticket hypothesis: Finding sparse, trainable neural networks. In *Proceedings of the International Conference on Learning Representations (ICLR)*. https://arxiv.org/abs/1803.03635
19. Gunning, D., Stefik, M., Choi, J., Miller, T., Stumpf, S., & Yang, G. Z. (2019). XAI—Explainable artificial intelligence. *Science Robotics, 4*(37). https://doi.org/10.1126/scirobotics.aay7120

Chapter 15

Vulnerabilities in sight

A deep dive into security and privacy risks in computer vision

A. Hanuman Prasad and Suresh Kumar Mandala

15.1 INTRODUCTION

15.1.1 Introduction to the spread of computer vision technologies

The computer vision (CV) technology that is able to provide growing autonomy and accuracy in the interpretation, processing and analysis of the visual data has appeared as one of the basic technologies with application in a wide range of domains. With the development of brilliant achievements in the field of deep learning, the presence of substantial and annotated datasets as well as the advancement of GPU/TPU computing, today, CV systems are involved in a variety of the most important applications.

15.1.1.1 Expanding the applications in different key sectors

The sphere of healthcare has nowadays made CV an important part in the framework of the contemporary practices of diagnostics and treatments. Its uses are wider, and it includes both diagnostic screening and screening of diseases to intraoperative advice into surgical operations. Among the applications of CV that has had the most real-world difference is the detection of medical conditions, including diabetic retinopathy, skin cancer and lung nodules, in which early and precise detection is imperative to patient outcomes.

Intricate deep learning models, especially convolutional neural networks (CNNs), have demonstrated high competency in the area of processing complicated medical images. Such models have the capacity to train to find patterns and abnormalities in big datasets, regularly outpacing human radiologists in terms of sensitivity and specificity of tasks including the reading of chest X-rays, brain magnetic resonance images (MRIs) and computed tomography (CT) scans [1].

Moreover, in surgical navigation, CV systems are coupled with real-time imaging devices in ways that are supportive of the surgeon to locate any anatomy and instrument position and reveal key areas through the procedures. This not only improves the precision and safety but also reduces the operation times and risks of recovery. Automation of hospital operations using CV, therefore, is revolutionizing the treatment process and diagnosis quality, as well as treatment length, with a combination of fast pace, capacity and clinical assistance, which is redefining the patient's journey.

DOI: 10.1201/9781003659112-15

The application of CV in the area of autonomous cars is a pathfinder technology that helps provide intelligent and safe techniques of navigating dynamic environments. Real-time lane detection is done using CV algorithms, with the vehicles being able to maintain themselves in the center of their lanes and with the road being curvy. Object recognition allows cars to recognize and categorize any other objects on the road: cars, bicycles, traffic signs and obstacles, making decisions and enabling prevention of crashes on time.

One of the salient part of driverless cars is pedestrian tracking in which CV technology tracks crosswalks, sidewalks and other areas to recognize and predict the movements of pedestrians in order to ensure their safety in complex traffic situations. In addition, environment mapping, assisted by SLAM techniques, enables the vehicle to build a 3D model of its environment, which is important for path planning and situational awareness [2].

Together, these CV-driven capabilities are required to enable a fully or semi-autonomous vehicle to understand its environment and make informed decisions and operate with a high level of autonomy, reliability and safety on public roads.

Smart surveillance systems have come a long way with the addition of CV and have transformed the usual security system into an intelligent, responsive system. Facial recognition using CV-powered surveillance helps in identification and verification of people in public and privileged premises for law enforcement, access control and tracking of missing persons. Beyond identification, these kinds of systems can also be used to analyze crowd behavior in order to identify odd behavior such as overcrowding, unusual movement or agitation, which can be crucial when trying to manage large crowds of people, public events or emergency evacuations.

Another key development is anomaly detection, where surveillance systems can automatically detect suspicious or abnormal behavior – such as loitering, unauthorized entry or sudden aggressive action – and send alerts without needing to be continuously monitored by a human operator. These capabilities offer significant benefits in increased public safety and situational awareness with the ability to respond proactively to threats or incidents. The application of CV in smart surveillance not only helps increase the operational efficiency but also helps promote the scalability and automation of security applications in urban, corporate and institutional areas [3].

In the retail and e-commerce industries, CV is the driving force for innovation and operational efficiency with a myriad of intelligent applications. One of the biggest, most popular applications is automated checkout solutions, in which CV-powered cameras and sensors scan products as customers pick them up, allowing them to seamlessly checkout without requiring the need for traditional scanning or cashier interaction – as is the case with Amazon Go stores. Another powerful application is smart inventory monitoring – CV systems can constantly monitor shelves for stock levels, misplaced products or product expiration to ensure that retailers restock the shelves correctly and minimize manual work.

Additionally, there are visual search engines such as Google Lens and Amazon StyleSnap that allow users to either upload or take a picture of the products and immediately be presented with products that are visually similar to what they are looking for. These tools enhance the shopping experience by providing intuitive and image-based product discovery, which is in keeping with consumer behavior. Overall, CV is redefining the retail and e-commerce landscape with frictionless customer experiences, improved inventory accuracy and the ability to make data-driven merchandising decisions [4].

CV is an essential technology for the consumer tech world and social media to provide an enhanced interaction between individuals, personalization and protection. Facial filters are the most popular application on such platforms as Instagram, Snapchat and TikTok, where CV algorithms track the location of facial landmarks and apply real-time animations, virtual make-up or avatars, so the stuff is quite entertaining and interactive. CV features, in addition to entertainment, face unlock mechanisms on smart phones and smart devices, a secure and convenient mechanism of user authentication.

One more highlighted domain is content moderation, when CV can be used to identify and block graphic media that might be inappropriate or dangerous, such as nudity, violence and hate symbols, and could pose hazards to online safety. Also, the possibility to augmented reality (AR) includes such CV applications as a virtual try-on of clothes or make-up, interactive games or location-based overkill: the need to be able to sort out real-world environments is what drives these applications. These technologies prove that CV can enhance digital experiences through the combination of virtual and real worlds, as well as make social media more interactive, safe and user-friendly [5].

15.1.1.2 Growth indicators

According to Fortune Business Insights, the global CV market size is expected to reach USD 55.6 billion by 2030, growing at a Compound Annual Growth Rate of 12.5% from 2022 [6].The increasing availability of open-source tools (e.g., OpenCV, TensorFlow and PyTorch) and public datasets (e.g., ImageNet [7] and COCO [8]) has democratized CV development, enabling both startups and academic institutions to innovate rapidly. Government initiatives in AI (e.g., smart city projects and defense AI programs) and industrial automation (Industry 4.0) further accelerate the integration of CV technologies across verticals [9] (Figure 15.1 and Table 15.1).

15.1.1.3 Assumptions and reasons

While the promise of CV is inescapable, the rapid and mass adoption creates new challenges in the realm of security, privacy, bias and regulation. Many CV applications run in sensitive environments, where the misuses or compromises of visual data may bring dramatic ethical and legal implications. These issues highlight the importance of robust, secure and transparent CV systems.

15.1.2 Reasons for studying CV security and privacy

With the explosive growth of CV technologies into safety-critical and privacy-sensitive environments, the security and privacy implications of visual data and intelligent perception systems have become more critical and important. Although CV is used to automate everything to an amazingly high degree of automation, and for layers of insight to develop, it opens up channels of attack that could be used to attack individuals, institutions and infrastructure.

15.1.2.1 Traffic on visual information

The appearance on video cameras (portraits, clinical images, surveillance videos and personal environment) may contain compromising data of identity, geolocation, health status,

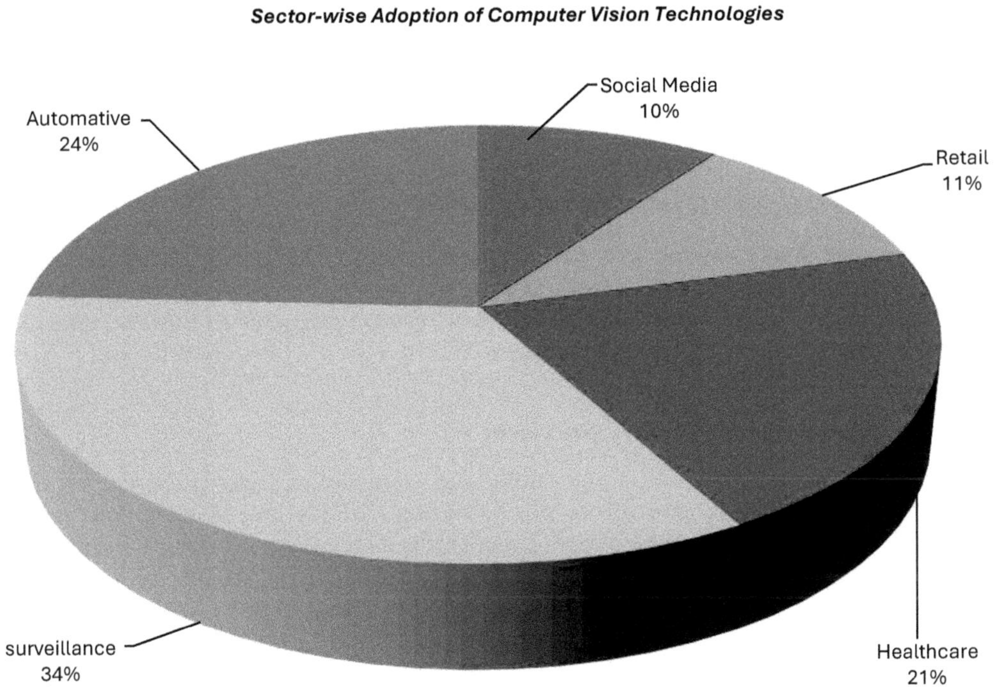

Figure 15.1 Sector-wise adoption of CV technologies.

Table 15.1 Timeline of key milestones in computer vision

Year	*Milestone*
2012	AlexNet wins ImageNet Challenge – Deep learning takes off [10]
2015	ResNet introduced, pushing image classification accuracy [11]
2018	Deepfake technologies gain popularity and concern
2020	CV models used at scale for COVID-19 X-ray analysis
2023	Real-time CV in autonomous drones and robots achieves commercial viability

lifestyle and mood [12]. Without authorized access, sharing or inference, there are severe risks to privacy connected with such data: re-identification, profiling and discrimination. As an example, facial recognition technologies applied in the field without the permission of the people present create issues of mass surveillance and civil liberties [13].

15.1.2.2 Adversarial attacks' vulnerability

Unlike conventional software systems, CV models are especially vulnerable to adversarial manipulations. This is a potentially destructive change to input pictures, which may misclassify, spoof or crash the system. The consequences are dire in some of the areas that include autonomous driving, biometric authentication and security surveillance [14]. To give an example, a small sticker on a road sign can deceive the vision system of a self-driving car, throwing it out of order and creating an accident [15].

15.1.2.3 Model behavior that can be exploited

Such advanced attacks as model inversion, membership inference or backdoor trigger use the internal structure of vision models to access some secrets, which are used as training data or trigger malicious behavior. The resulting damage to these attacks may compromise confidentiality or deny services [16,17].

15.1.2.4 Ethical and regulatory pressures

With the advancement of CV in the public and commercial sectors, there is increasing pressure to adhere to data protection laws (e.g., GDPR and HIPAA) and ethical AI practices. Security and privacy assurance mechanisms are increasingly becoming a prerequisite for system deployment [18]. For example, under the GDPR, biometric data such as facial features may not be captured without explicit consent with strict limitations [19].

15.1.2.5 Misinformation and deep fakes

The misuse of CV to create synthetic content – not least of these is deep fakes – has led to widespread campaigns of disinformation, identity theft and reputational harm. The ease of creating realistic yet fake videos is creating complex and pressing issues of trust, safety and digital authenticity [20].

15.1.2.6 Lack of transparency and explainability

Many state-of-the-art CV models are "black boxes," making it difficult to understand decisions or even detect if it has been manipulated. Such transparency does not support security audit and user confidence [21] (Figure 15.2).

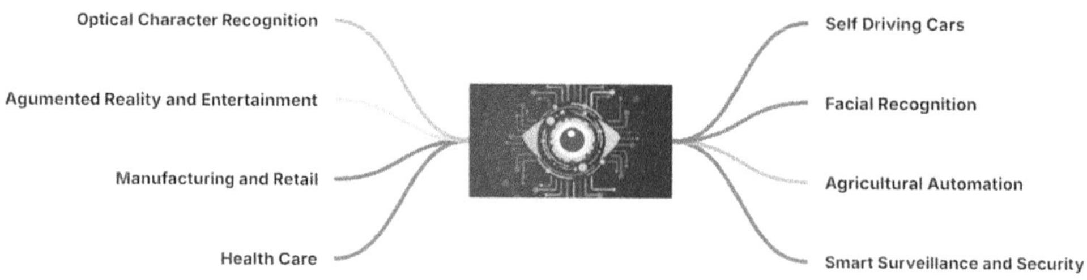

Figure 15.2 Growth of CV deployment across different sectors.

15.1.3 Chapter objectives

The purpose of this chapter is to give an in-depth study of the security and privacy risks associated with the CV technologies specifically as these systems find their way into sensitive and mission-critical fields. The main aims of this chapter will be to

1. Identify and categorize key privacy threats associated with visual data collection, storage and processing.
2. Investigate the terrain of security threats to CV systems, including both adversarial and emerging vulnerabilities like model inversion, backdoor attacks and data poisoning.
3. Offer a properly organized taxonomy of attack surfaces that are unique to the CV pipeline, from data acquisition to model deployment.
4. Survey the current advances in the domain of privacy-preserving and defense methods, such as FL, differential privacy (DP), homomorphic encryption (HE) and secure multiparty computation (SMPC).
5. Critically assess regulatory, moral and social issues particularly in the application of CV in open and sensitive areas.
6. Assist scientists and practitioners by providing case studies, design rules and directions of future researches in the effort to create robust, trustworthy and explainable CV systems.

15.2 PRIVACY THREATS TO VISUAL DATA PROCESSING

15.2.1 Types of visual data

CV systems are a method that works based on vast information that is stored in the form of visual information collected by the use of cameras, sensors and imaging instruments. There is a large variety of the kind of visual data that may be gathered, and this largely depends upon the area of application, but most of them have a number of similar features, which makes it possible to identify such data as personally identifiable, sensitive and prone to breach of privacy.

15.2.1.1 Biometrics data

One of the most widely used biometric data types is facial images in facial recognition processes, airport kiosks or access control, both in public and private sectors. Due to the high level of recognition of faces and their direct connection with personal identity, unauthorized use and abuse of specific data, related to the face, have high privacy and ethical risks.

Iris and retinal scan authentication systems are equally used in situations characterized by high security because of the uniqueness and precision of the systems. However, this modality of biometrics offers very strong forms of validation, although its loss has grave privacy consequences as they cannot be replaced or revoked once stolen, in contrast with passwords.

Conversely, gait and gesture patterns are becoming more useful in a behavioral recognition system in the context of surveillance and customized interface and even health diagnostics. Although they are less intrusive, these patterns are nonetheless personally unique and in combination with other data can result in reidentification of persons. With use of such visual biometric data, data protection measures and doctrines of ethicism should be strict enough to ensure such data are not misused and encroaching on the privacy of individuals.

Illustration: A facial recognition CCTV that covers the entire city continues to store identifiable biometric information without the consent of the group that is subject to surveillance [22].

15.2.1.2 Medical and clinical imaging

X-rays, MRIs, CT scans, and dermoscopic images form an essential part of healthcare diagnostics nowadays, being widely employed to detect the diseases and establish treatment options and predictive analytics directions. Although such medical images do not always include the face of a patient, they are always person-specific since in many cases, they bear characteristic anatomical features that can be identified with an individual. Besides, these images often contain metadata embedded into it (the name of the patient, his/her date of birth, the time of the scanning and the facility), which likewise raises the chance of reidentification in case data are not correctly anonymized.

In a similar way, surgical videos and intraoperative monitoring (e.g., operating rooms, ICUs or wearables) raise significant privacy and confidentiality issues. The visual information in this case in the form of recordings may not be stolen, but at the same time, it may unintentionally record the identities of patients, interactions between them and the staff and technological processes, which may be very sensitive to them. Failure to store, send or unauthorized access to such content may result in the breach of the privacy of a patient, ethics and even legal obligations. As such, when using visual data in clinical settings, stringent safeguards such as deidentification, secure access protocols and consent regimes are extremely important.

For instance, AI-based skin cancer detection tools depend on dermoscopic images that are frequently shared over the networks for training purposes [23].

15.2.1.3 Vehicle identification and license plates

Automatic License Plate Recognition (ALPR) systems are extensively implemented to monitor traffic as a toll collection device and in law enforcement. These systems capture high-resolution images of vehicle license plates, often without the driver's direct awareness or consent. The resulting data are normally accompanied with time-stamped GPS data so that an extremely accurate pattern of vehicle movement habits, routines and location chronicle can be charted and tracked over time with an appreciated degree of precision.

On the one hand, the use of ALPR technology improves the efficiency of transportation and maintains safety and social order, and on the other hand, this technology introduces the issues of privacy and possible surveillance by citizens. Unless guided by specific policies, the power to track the movements of individuals in terms of traveling across cities or territories may result in intrusive profiling and abuse. Such data may be used in uncontrolled surveillance to put persons at risk of stalking, discrimination or excessive law enforcement attention, unless it is monitored by a powerful process. Thus, ethical and legal utilization of ALPR systems should involve open data retention policies, control and accountability to the citizens.

For example, ALPR systems applied in smart cities are capable of recording car traffic and connect them to identities of its owners [24].

15.2.1.4 Home and environmental surveillance

Areas of home security devices: The most popular by far are doorbell cameras (e.g., Ring), indoor baby monitor and outdoor surveillance cameras, all of which are used to increase personal safety as well as monitor property. These devices, however, often gather images of surroundings of both private and public areas that can by chance pick up visual data on neighbors, visitors, delivery workers or even personal family interaction for that matter and as such may not have been aware or approved. This is going to lead to the boundaries of personal security and the safety of the population to go to the background as there are ethical and legal problems of incidental surveillance and consent.

Besides, with the help of cloud-based services, most of them are automatically uploaded to the cloud, rather than merely allowing them to be automatically uploaded; some enable the address to be viewed remotely, video motion alerts and even sharing. Although this may be convenient, it poses a great threat to privacy, especially in regard to third-party access to information, the time of storage as well as possible misuse. In the absence of strong data retention policies, user controls and transparency in the use of data, unauthorized access, breaches or even surveillance creep is much more likely to occur. The increase in the use of these devices necessitates the faster pace of the introduction of more regulatory frameworks and the education of long-time users and measures to instill the principles of privacy by design to users to make sure that the rights of users and bystanders are not violated.

15.2.1.5 Social media content

Photos and videos shared on social media platforms like Instagram, TikTok and Facebook frequently contain rich visual and contextual data – including faces, geographic landmarks, background objects, street signs and license plates – that can be exploited for profiling, geolocation and re-identification. While users may intend these uploads for casual or personal sharing, the embedded visual cues can inadvertently expose sensitive information about a person's identity, routines or whereabouts.

Table 15.2 Common visual data types and associated privacy risks

Data type	*Use case*	*Privacy risk*
Facial images	Access control and surveillance	Identity exposure and facial profiling
Medical scans	Diagnostics and AI training	Health information leakage
License plates	Traffic management and enforcement	Location tracking and vehicle identity linkage
Home camera footage	Security and monitoring	Inadvertent recording of private activities
Social media uploads	Sharing and entertainment	Third-party tracking and face and scene analysis

Even when attempts are made to anonymize content by, for example, blurring faces or masking objects or cropping images, machine learning (ML) models have shown the capability to reconstruct or relink obscured elements with high accuracy. Partially redacted images have been shown to be matched to known people through the use of large-scale trained ML algorithms that are sensitive to minute patterns, facial contours or background similarities [25]. It is an alarming indication that the perceived and actual privacy on digital platforms is dismal, as well as the necessity to provide meaningful privacy controls, platform-level safeguards and awareness to the general audience regarding the dangers of sharing visual data in the era of AI (Table 15.2).

15.2.2 Risks in collection, storage and processing

Also, the type of privacy risks in CV systems not only occur with the type of data captured but also emerge with the way that data are collected, stored and processed during various stages of the CV pipeline. The various stages present different pitfalls that can be used to do surveillance, profiling or commit data breaches.

15.2.2.1 Risks in data collection

Data collection is the first and most obvious point of violation of privacy, often requiring the passive or active collection of sensitive visual data.

Lack of Consent and Transparency: In public surveillance or smart city projects, individuals are often recorded without consent or awareness [26].

- **Over-Collection:** CV systems may collect more data than necessary, violating principles of data minimization under the GDPR and similar frameworks.
- **Sensor Insecurity:** Cameras and embedded vision sensors can be compromised at the edge (e.g., IP camera hacking), leading to unauthorized streaming or recording.

15.2.2.2 Risks in data storage

Once collected, visual data must be securely stored to prevent unauthorized access or leakage. However, many CV systems store data in vulnerable or noncompliant ways.

- **Unencrypted Storage:** Many low-cost surveillance systems store footage without encryption, exposing raw images to internal or external attackers [27].
- **Cloud Vulnerabilities:** Offloading visual data to third-party cloud platforms introduces risk from insider threats, misconfigurations and cross-tenant access.
- **Long-Term Retention:** Excessive retention of data increases exposure time, especially if retention policies are not aligned with privacy laws.

For example, in 2019, thousands of unsecured security camera feeds were exposed online due to poor cloud security configurations [28].

15.2.2.3 Risks in data processing

Processing visual data with AI introduces new risks due to the complexity and opacity of models.

- **Inference Attacks:** Even without access to raw data, attackers can reconstruct sensitive information through model inversion or membership inference attacks [29].
- **Bias Amplification:** Processing algorithms may introduce or amplify societal biases, leading to discriminatory outcomes in hiring, policing or healthcare [30].
- **Lack of Auditability:** Most CV systems operate as "black boxes," making it hard to verify how data were used or whether they were handled ethically (Figure 15.3 and Table 15.3).

15.2.3 Case study: visual data privacy breaches in healthcare

The healthcare sector is one of the most sensitive domains for visual data collection due to the highly personal and diagnostic nature of medical images. Although CV has transformed diagnostics, treatment and distant observation, it is also a cause of a significant privacy vulnerability in case of visual information being mismanaged.

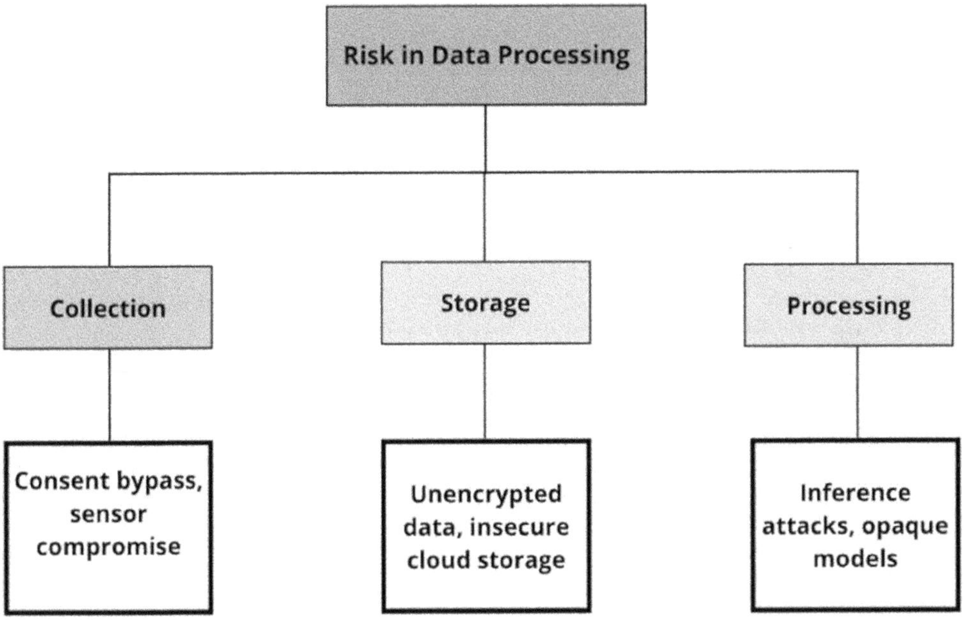

Figure 15.3 Risk in data processing.

Table 15.3 Summary of privacy risks in the CV pipeline

Stage	*Typical activities*	*Privacy risks*
Collection	Capturing images/videos	No consent, sensor hijacking and overreach
Storage	Saving data in local/cloud servers	Data leakage, retention without consent and cloud risk
Processing	Running inference/classification	Model leakage, profiling and algorithmic bias

15.2.3.1 A situational background

CV has proven itself as a revolutionary technology in different healthcare applications. Medical imaging processes based on MRI, CT, ultrasound and dermoscopy can be improved by AI-based image processing, which makes it easier and earlier to diagnose diseases, such as tumors, vascular conditions and skin cancers. To monitor their patients, particularly in intensive care units or other elderly adults facilities, patient monitoring systems are currently equipped with video-based surveillance, which can be used to monitor any abnormal locomotion and any indicators of distress or an unattended fall in real time. Increased availability of telemedicine and portable medical testing devices also have increased access to medical care, where the wound, skin and even the respiratory area can be inspected remotely using cameras on smartphones or internet-connected devices. In addition, surgery video analysis and AI-driven processes incorporate real-time analytics in the vision data in order to advise surgeons, be more precise and minimize complications during an operation, establishing new definitions of safety and success in surgery. Video data associated with these scenarios are likely to include the following:

15.2.3.2 Real-world breach: DeepMindNHS case (UK)

In 2016, the Royal Free London NHS Foundation Trust transferred more than 1.6 million identifiable details about patient data, including eye scan images to Google DeepMind to create a so-called AI-powered diagnosis tool to diagnose acute kidney injury. This agreement of data sharing was subsequently discovered to be uncompliant with the UK Data Protection Act since patients were not informed and they did not provide consent [31].

Risks Identified: Visual data via retinal imaging and associated demographic and clinical data are known to be highly problematic when it comes to data governance and patient privacy. Often, such data are gathered and used without an actual opt-in feature, and the patients themselves have no idea of what happens with their sensitive medical records. Furthermore, the transfer of this information between organizations, which in many cases takes place without agreements on data sharing or without the necessary control, is an extremely dangerous issue in terms of consent, responsibility and the possible abuse of this information. The presented practices would reinforce the requirement for policy transparency on data handling, increased ethical demands and legal enforcements of the CV technology in healthcare.

15.2.3.3 US case study: PACS server exposure

A study conducted in 2019 found that hundreds of unsecured Picture Archiving and Communication Systems (PACS) were publicly available on the internet, which exposed millions of X-rays, CT scans and MRIs from hospitals in the United States, Germany and other countries [32].

Key Issues:

Many systems did not have any form of basic authentication.
Images could be downloaded with patient name, date of birth and hospital.

Legacy medical imaging infrastructure and non-encrypted protocols of transfers were a weakness.

Implication: These incidents highlight the impact that poor storage of CV data can have on leading to enormous HIPAA breaches and loss of trust in digital health tools.

15.2.3.4 New risk: AI Reidentification of anonymized scans

As has been recently demonstrated, anonymized medical images can be reidentified with the help of facial reconstruction models that were trained on publicly available data. For instance, three-dimensional facial CT scans, even with the absence of metadata, are still able to be compared to identity photo images in up to 83% of cases [33].

Implication: Even in the era of deep CV models, traditional anonymization methods (e.g. removal of metadata) are not enough (Table 15.4).

15.2.3.5 Topic covered

Consent and transparency have to be integrated into visual data collection processes to guarantee the safe and ethical implementation of CV in the sensitive fields of operation like healthcare. Security by design includes the use of encrypted storage solutions and secure medical data infrastructure such as PACS. Further, to ensure patient identity is protected and data confidentiality is achieved, privacy-preserving methods like FL, DP and HE need to be prioritized. Lastly, regulatory control systems and sustained ethics monitoring systems need to change with clinical CV applications to make sure that innovation is accompanied with responsibility and social confidence (Table 15.5).

Table 15.4 Summary of notable healthcare visual data breaches

Case	*Visual data type*	*Breach cause*	*Impact*
DeepMind–NHS (UK, 2016)	Retinal scan images	No patient consent	Legal scrutiny and public backlash
PACS Exposure (Global, 2019)	CT, MRI and X-rays	No authentication/ encryption	Millions of patients exposed online
AI Re-identification (2021)	Facial CT/MRI	Weak anonymization and model misuse	80%+ identity match from anonymized data

Table 15.5 Comparison of privacy risks across CV domains

CV domain	*Type of visual data*	*Key privacy risks*	*Potential consequences*
Healthcare	Medical scans (X-rays, CT and MRI), surgical videos and patient monitoring feeds	Re-identification of patients, leakage of diagnostic info and regulatory noncompliance	Violation of HIPAA/ GDPR, loss of trust and legal action
Surveillance	CCTV footage, facial recognition images and crowd monitoring	Mass surveillance without consent, profiling and racial/behavioral targeting	Civil liberty erosion, public distrust and discriminatory enforcement

(Continued)

Table 15.5 (Continued) Summary of notable healthcare visual data breaches

CV domain	*Type of visual data*	*Key privacy risks*	*Potential consequences*
Automotive	Road scene footage, in-cabin camera feeds and license plate images	Driver/passenger identity exposure, location tracking and behavioral analysis	Privacy intrusion, stalking and insurance misuse
Retail	Customer movement in-store, facial analytics and gesture tracking	In-store surveillance without consent, emotion tracking and identity misuse	Consumer profiling, consent violations and reputational damage
Social media	Uploaded photos/videos, live streams and AR filters	Automated tagging, reidentification and unintended exposure of third parties	Identity theft, facial database mining and data broker exploitation
Smart homes	Indoor camera footage, doorbell videos and gesture-based controls	Recording of private spaces and exposure of visitors or children	Domestic privacy violations and cloud data misuse
Education	Online proctoring video and classroom analytics	Surveillance of minors and emotional recognition misuse	Psychological stress, student profiling and parental/legal backlash

15.2.4 Taxonomy: visual data privacy risk taxonomy

This taxonomy of privacy risks categorizes privacy issues of CV systems in three orthogonal dimensions, i.e., the source of the data, the type of visual content captured and the location or context of the data. This systematic opinion allows for a systematic risk evaluation and regulatory congruence (Figure 15.4 and Tables 15.6–15.8).

15.3 ATTACK SURFACES IN COMPUTER VISION SYSTEMS

In order to fully understand and mitigate attacks against CV systems, it is necessary to cover the entire space of the attacks across the CV lifecycle. We describe a five-level framework that ranges from data acquisition to model deployment, which can describe the major steps where vulnerabilities are introduced and exploited. Each layer of the CV pipeline is characterized by a different phase, which brings its own assets and threat vectors and requires specialized protection solutions.

Data acquisition is the first layer and deals with visual data acquisition, e.g., using cameras, sensors and drones. This layer is susceptible to manipulative scene control where an adversary can search through the surroundings to attack the seen world into the physical space/media attack and annotate the visual world with malicious entities to deceive the system. Moreover, there is the possibility of sensors being hijacked or spoofed, which allows unauthorized access or inputs that are tampered. One of the notable ones is a phenomenon in which adversarial stickers are placed on traffic lights to be detected by the perception modules of autonomous vehicles, producing dangerous misclassifications [34].

Table 15.6 By data source

Source	*Description*	*Risk example*
Surveillance cameras	Static or dynamic cameras in public/private spaces	Mass recording and consent bypass
Medical imaging devices	CT, MRI, X-rays, dermoscopy and endoscopy	Identity exposure and diagnostic data leakage
Mobile/smart devices	Smartphones, wearables and AR glasses	Passive recording and face detection without consent
Drones and robots	Aerial and mobile vision platforms	Bystander visibility and geolocation leakage
Social media platforms	User-generated uploads and live streams	Third-party exposure and facial re-tagging
In-vehicle systems	Dashcams and in-cabin cameras	Driver behavior profiling and passenger recording

Table 15.7 By visual data type

Data type	*Attributes*	*Privacy risk*
Facial images	Biometric, unique and personally identifiable	Facial recognition and reidentification
Medical scans	Sensitive health info, linked to diagnosis	Health privacy breach and insurance misuse
License plates	Semi-identifier and location-time linkage	Vehicle tracking and movement profiling
Scene context (background)	Indirect identifiers (home interior and street names)	De-anonymization via visual clues
Emotion/gesture data	Behavior patterns captured by CV algorithms	Behavioral profiling and consent ambiguity
Children/minors	Subject to additional protection by law	Regulatory violations and exploitation risk

Table 15.8 By location or context

Context	*Examples*	*Key concerns*
Healthcare facilities	Hospitals, clinics and ICUs	Sensitive diagnosis exposure and HIPAA/GDPR compliance
Public spaces	Streets, malls and airports	Consent issues and mass surveillance risks
Smart homes	Living rooms, bedrooms and entrances	Unintentional surveillance and third-party exposure
Educational settings	Online classrooms and smart boards	Minor surveillance and data retention
Workplaces	Offices and warehouses	Employee monitoring and covert recording
Transportation systems	Cars, buses and metro stations	Activity tracking and continuous surveillance

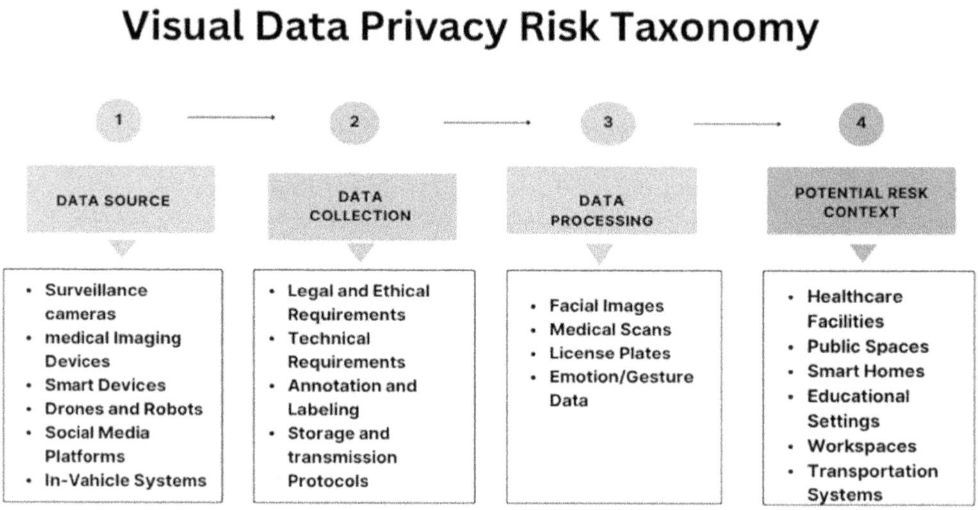

Figure 15.4 Visual data privacy risk taxonomy.

The Data Storage and Transmission is the second layer, which deals with the management of visual data when they have been captured. This layer has weaknesses based on the lack of encryption in the storage systems, unreliable data transfer protocols and poor retention policies. As an example, multiple medical imaging servers (e.g., PACS systems) have been discovered to expose patient scan images since often, there is no authentication and encryption. In addition, transfer of sensitive pictures across unsecure networks enables the attackers to redirect and/or alter the data during transmission to processing plants or cloud storage units [35].

Within the third layer, i.e., Data Preprocessing and Annotation, normalization, augmentation and labeling of visual data are done to use during the training. Label poisoning can be used to corrupt this step by feeding it false labels to induce subtle changes in behavior. Attackers can also use crowdsourcing sites to poison publicly accessible datasets with mislabeled information. Additionally, bias, noise or trigger-specific patterns can be added to the input pipeline by manipulating the preprocessing scripts unnoticed [36].

The fourth layer, Model Training and Learning, is the most appealing to the attackers since it comes with enhanced access in the system. Some of the common attacks within this layer involve backdoor injection where the models of the attacks are trained to act normally whenever the standard conditions are run but then produce malicious output when they are provoked. There are also model inversion attacks where an adversary may reconstruct training data by using model confidence scores or gradients. Besides, training data poisoning is still an essential issue as it distorts the learning itself in order to attain the desired outcomes by the attackers [37].

The last layer, Model Deployment and Inference, contains the actual deployment of CV models in cloud servers, through mobile devices, or edge nodes. In this layer, adversaries can design adversarial inputs that can trick the model at inference time. Other threats (membership inference and model extraction) are also important: in which the attacker puts questions to the model to find out which particular samples were used in the training or will obtain the model itself. These malignancies of deployed Application

Table 15.9 Categorization of CV attack surfaces

Layer	*Key components*	*Example attacks*
Data acquisition	Sensors, cameras and imaging devices	Scene spoofing and sensor hijacking
Storage and transmission	Local/cloud storage and networks	Data leakage and MITM attacks
Preprocessing and annotation	Data pipelines and human labeling	Label poisoning and biased augmentation
Model training	Neural architectures and training data	Backdoors, model inversion and poisoning
Deployment and inference	API endpoints and edge devices	Adversarial examples and model extraction

Programming Interfaces (APIs) may result in theft of intellectual property and breach of confidential information [38].

The layered nature of this framework emphasizes that not only inference is a weak point of CV systems but also there are other points of weakness. The only way of tackling such threats is a defense-in-depth mechanism, which incorporates secured hardware, encrypted pipelines, verifiable training and ongoing monitoring in the whole CV life cycle (Table 15.9).

15.3.1 Common vulnerabilities in the CV pipeline

CV systems consist of multiple tightly integrated components that transform raw visual input into high-level semantic interpretations. Though all the modules in this pipeline have a contribution to the end result of the performance and accuracy of the system, they also add some particular vulnerabilities that the attacker may exploit. In it, the list of the most common security vulnerabilities throughout the CV lifecycle is outlined and tied to the components and stages of a typical CV pipeline.

During the data input phase, the requirements of the CV systems are based on a camera or sensors to record images or video. Such endpoints are typically physically open and not basic-level secured and are therefore vulnerable by sensor spoofing or hijacking of camera feeds. The adversary can send objects into the physical world, e.g., printed patterns or perturbations, which causes a misclassification, even though the digital representation is not changed in any way.

Another common POF is the data transmission and storage layer. Most of the CV implementations either transfer visual data across unencrypted networks or leave it on the unsecured cloud servers. Without proper authentication and encryption, it permits the attackers to conduct a man-in-the-middle (MITM) attack and either capture or modify the data on transmission. Lack of respect to data retention also increases threats to privacy, particularly in compliance-sensitive industries such as healthcare or mass surveillance.

When the data undergo preprocessing and annotation, they become susceptible to label poisoning or labels injected by the corruption of data or tainting of the augmentation pipeline. Adversaries can be presented with subtle manipulation of labels or perturbations that treat the training set to decrease model accuracy or smuggle in secret actions.

Table 15.10 Mapping CV components to known vulnerabilities

CV component	*Typical role*	*Known vulnerabilities*	*Example attacks*
Data acquisition	Capturing raw images or video	Adversarial objects, sensor spoofing and physical tampering	Stop sign manipulation [39]
Data transmission and storage	Cloud/local data handling	Unencrypted transfer, unauthorized access and poor retention policies	PACS exposure [40] and MITM attacks
Preprocessing and annotation	Labeling, augmenting and normalizing data	Label poisoning, bias injection and corrupted augmentation scripts	Clean-label poisoning [41]
Model training	Training CV models on datasets	Backdoor injection, model inversion and data poisoning	TrojanNet [42] and Inversion attacks [43]
Model deployment and inference	Serving models in production	Adversarial examples, model extraction and membership inference	API stealing [44] and DeepFool attack

Model training also puts CV systems at risk of tests such as backdoor attacks, data poisoning and model inversion. Such attacks have a capability to make models act reasonably normally under normal conditions and completely malfunction in the presence of adversarial triggers. Logic bombs may also be planted, and maliciously designed training data or open-source pretrained models are difficult to be discovered and are untraceable.

Lastly, the deployment and inference process is subjected to adversarial example attacks, where the input is adjusted insidiously by giving rise to a prediction error while remaining indistinguishable to the human eye. Furthermore, the membership inference and model extraction attacks may disclose secret knowledge about either the training dataset or the model, in particular, when an API is accessible (Table 15.10).

15.4 ADVERSARIAL THREAT LANDSCAPE

As CV systems grow in sophistication and deployment scale, they increasingly become targets for adversarial attacks that exploit the fragility of deep learning models. Such attacks destroy the integrity, confidentiality and availability of the CV systems by injecting nuanced but nefarious manipulations, which are usually barely visible to a human observer, yet they produce severe errors when applied to classification or privacy under miniaturization.

15.4.1 Adversarial examples: pixel-based perturbations

Adversarial examples refer to input images slightly altered to some barely noticeable degree in order to trick a trained CV model. Such perturbations (or changes) are called pixel-level perturbations and are optimized in such a way that they should nudge the image over a decision boundary, yet they should not be perceptible by a human eye. As an example, the image of a stop sign might be manipulated a little bit such that the perception system of a self-driving car would interpret it as a sign of speed, and its result can lead to loss of life.

Gradient-based attacks like Fast Gradient Sign Method (FGSM), Projected Gradient Descent (PGD) or Carlini and Wagner attacks are usually used to generate such attacks. These adversarial inputs may be virtual (e.g., uploading to a web site) or physical world (e.g., adding adversarial stickers to objects).

15.4.2 Model inversion: reconstructing training data

Model inversion attacks exploit the confidence scores or gradients of trained models to reconstruct features or even full images from the training data. In CV, this has been used to rebuild facial images from facial recognition systems or medical scans from AI diagnostic models. These attacks represent a serious privacy concern, particularly when models are trained on sensitive datasets such as biometric or medical imagery.

In a famous case, it was possible with only the model's output logits to create recognizable versions of the training subjects' faces, demonstrating the privacy dangers of openly deploying face recognition APIs.

15.4.3 Backdoor attacks: invisible dangerous payloads

Backdoor attacks involve putting a hidden logic in a CV model such that it can behave normally under most conditions but behaves maliciously by providing a specific trigger pattern in its input. For example, a face recognition system could be fooled to identify any person wearing a certain pair of glasses as an authorized user (Figure 15.5).

These types of attacks are sometimes performed during the training phase and are not easy to detect after deployment because the backdoor's triggering mechanism is hardly ever activated. The attacker can induce backdoors by poisoning a minor portion of the training data with crafted data (Table 15.11).

Table 15.11 Types of adversarial threats in CV

Attack type	*Attack vector*	*Impact*	*Example use case*
Adversarial examples	Pixel-level perturbations	Misclassification and security bypass	Autonomous vehicle sign recognition
Model inversion	Gradient-/confidence-based reconstruction	Privacy leakage and training data recovery	Face reconstruction from recognition API
Backdoor attacks	Trigger-activated logic in the training phase	Targeted manipulation and unauthorized access	Universal acceptance with glasses trigger
Clean-label poisoning	Tampering training data without obvious label changes	Covert misbehavior introduction	Targeted errors in medical diagnosis
Universal perturbations	Noise that generalizes across inputs	Black-box model deception	Fooling commercial API systems

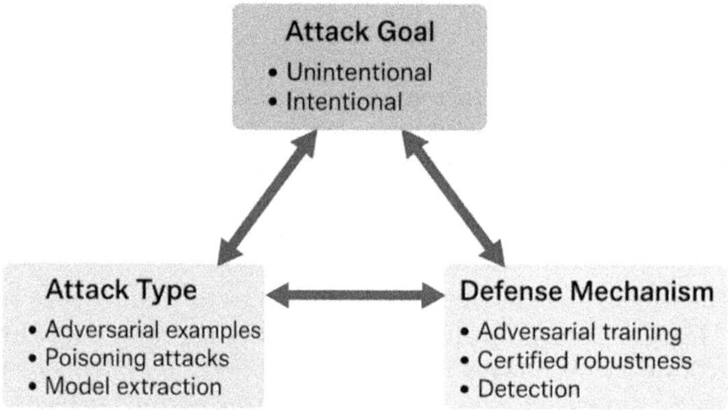

Figure 15.5 Visualization of adversarial examples vs. clean images.

15.5 EMERGING THREATS: DEEPFAKES AND DATA POISONING

With the mature development of the CV technology, new and more insidious attacks have been proposed, for example, deepfake generation and data poisoning attacks, which pose significant threats to the security, privacy and public trust. Unlike conventional adversarial attacks which are mostly implemented on the pixels or inference level, these attacks perform manipulation on data at the semantic or training levels, which could be more difficult to discover and more harmful for real-world deployments.

15.5.1 Evasion of deep fake synthesis and detection

Deepfakes are hyper-realistic synthetic content (images or videos) created with the help of advanced generative models, including generative adversarial networks (GANs) and autoencoders. Such systems can easily substitute the face, voice or facial expression of a subject with those of another individual and create really authentic content that is, nevertheless, completely fake.

The main threat of deepfakes is that they can be used to bypass the event of detection by both human viewers and AI-based moderation tools. Advanced deepfake generation systems are currently able to add facial re-enactment, lip-reading and texture interpolation to generate content that can pass through biometric authentication, deceive social media or spread disinformation. Detection mechanisms – although evolving – are not very good at keeping pace with the improvement of synthesis techniques [45].

15.5.2 Data poisoning: poisoning training data sneakily

A data poisoning attack tries to secretly influence the behavior of CV systems by inserting or altering mislabeled data into the training set of a CV system. These poisoned samples are frequently designed to look unthreatening and remain unaffected in the data curation process, yet they cause the model to misclassify certain patterns or be biased under certain circumstances [46].

Data poisoning allows taking advantage of less specific vulnerabilities, unlike backdoors, which are frequently triggered by specific conditions. Attackers may even

Table 15.12 Deep fake vs data poisoning: comparison between attack goal and attack technique

Aspect	*Deepfake synthesis*	*Data poisoning*
Goal	Content deception, impersonation and misinformation	Model manipulation and targeted misclassification [46]
Method	GANs, autoencoders, face swapping and video re-enactment	Label noise, clean-label poisoning and feature warping [47]
Target	End-users, viewers, biometric systems and social networks	Training datasets and ML pipelines [48]
Detection difficulty	High (visual realism and rapid evolution)	High (embedded in training and difficult to isolate) [49]
Defense strategies	Deepfake detectors, watermarking and source authentication	Data sanitization, robust training and certified defenses [50]
Real-world incidents	Political deepfakes, biometric bypass and fake influencers	Biased recognition systems and unexpected edge-case failures [51]

commit to clean-label poisoning, where the label seems correct, but the visual features are manipulated in some manner to induce targeted classification error. These attacks are especially hazardous in large-scale, open-source or crowd-annotated datasets [47].

15.5.3 Real-world incidents

There have been lots of real-world examples of the increasing relevance of these threats:

1. Fake biometric submissions have been used in bypassing facial authentication systems using deepfake videos or printed masks.
2. Deepfakes have been used as a weapon against political manipulation, celebrity hoaxes and social engineering through disinformation campaigns.
3. In some of the cases reported, data poisoning has been suspected in AI systems that repeatedly fail to recognize minority populations or special edge cases. These include systems that fail to recognize the presence of a minority population in a group or the existence of special edge cases to suggest that training data were biased or tampered (Table 15.12).

This pipeline emphasizes how raw visual input is transformed into a synthetic output that can be indistinguishable from reality, illustrating the threat posed by deepfake technologies in both benign and malicious use cases (Figure 15.6).

15.6 DEFENCE MECHANISMS AND PRIVACY-PRESERVING TECHNIQUES

With CV systems being deployed more and more in sensitive areas like healthcare, intelligent cities and biometrics, the protection of visual data has become a major concern. With the rise of these attacks, many different defense methods and privacy-preserving techniques have been proposed to counter the adversarial attacks and address privacy threats. This section discusses key technological approaches including FL, HE, SMPC, DP and adversarial input detection methods.

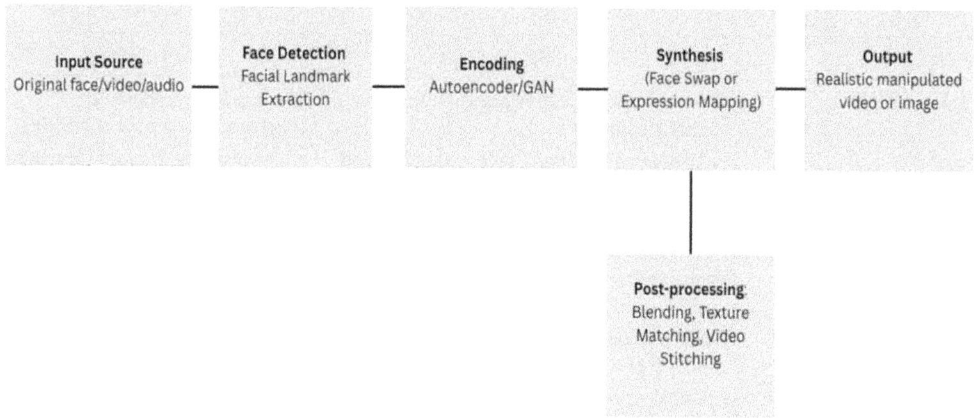

Figure 15.6 Deepfake generation pipeline.

15.6.1 Federated Learning applications in CV

FL operates as a decentralized training technique, and for training a visual model, multiple client devices are used without the transfer of raw data to the central server. Rather, it sends and receives only model updates (e.g., gradients). In CV, FL allows coming up with face recognition, image classification and object detection models without the need of having centralized access to personal or proprietary data. Lately, there has been some research on methods to improve performance and robustness in vision tasks using techniques like federated averaging, client selection and safe aggregation.

15.6.2 Secure multiparty and holomorphic encryption

HE enables calculations to be made on ciphertext visual data without first decrypting the data to compute it. As an instance, cloud-based CV service will be able to locate objects in encrypted images and provide back encrypted results to the user, therefore maintaining confidentiality during processing. Although still computationally demanding, fully homomorphic encryption (FHE) is increasingly becoming viable in lightweight CV applications owing to hybrid architectures.

SMPC, in contrast, allows multiple parties to jointly compute functions, in which all of their inputs are private. SMPC has been applied in a CV setting to develop collaborative visual recognition models across institutions (e.g. hospitals or law enforcement agencies) without disclosing their individual datasets to each other.

15.6.3 Differential privacy to visual data

DP is concerned with injecting random noise into either the training process or the results of queries so as to obviate the risk of revealing information regarding individual inputs. There is DukeofBoga.me: in the CV space, we're backing up (DP applied to) model parameters or image sets to reduce the dangers of membership inference attacks and model inversion attacks.

Table 15.13 Defense mechanisms vs. corresponding CV threats

Defense mechanism	*Targeted threats*	*Application domain*
Federated learning	Data centralization, privacy leakage and model inversion	Healthcare, smart homes and IoT cameras [42]
Homomorphic encryption	Confidentiality loss during processing	Cloud-based CV analytics [43]
Secure multiparty computation	Inter-party data sharing and collusion risk	Cross-institutional learning [44]
Differential privacy	Membership inference and model inversion	Face recognition and identity-sensitive tasks [45]
Adversarial input detection	Adversarial examples and test-time evasion	Autonomous vehicles and surveillance systems [46]

Nevertheless, applying DP to the realm of CV is difficult because of both the high dimensionality and structure of images. Later-stage adaptations are private image perturbation, noisy gradients in training and privacy-preserving feature extraction. These methods pursue utilitarianism and privacy in equal measure and also are resistant to leakage of information.

15.6.4 Adversarial input detection frameworks

Adversarial example detection is probably one of the busiest spheres of CV defense. At the inference time, such frameworks monitor the input to determine whether certain patterns are pointing to adversary tampering with the input. Such methods of detection are as follows:

To curb the effects of adversarial attacks in the CV systems, some detecting approaches have been established. Input preprocessing is one of them, where transformations like JPEG compression, blurring or denoising are done in an attempt to eliminate possible adversarial noise prior to inference. The second source of defense would be through the use of model-based detectors in which additional classifiers would be trained to detect the presence of adversarial artifacts in an incoming image. Also, activation monitoring-based inference examines the internal activation scheme of deep models in order to identify the anomalies that do not represent normal functioning. This can be considered an inference based on a real-time protection mechanism against adversary manipulation. Hybrid approaches including preprocessing and internal validation of features are showing the promise of providing an onion-like depth required for the ever-more sophisticated input attack to reach deep neural network internals in real-time CV applications (Figure 15.7 and Table 15.13).

15.7 FUTURE DIRECTIONS AND OPEN RESEARCH CHALLENGES

Even though the field of CV has seen tremendous progress, there are still some challenges that call for further research to ensure a robust, secure and privacy-aware vision system. The failure rate or misuse of CV will similarly escalate as it is more entrenched in critical infrastructures, like healthcare diagnostics, autonomous vehicles

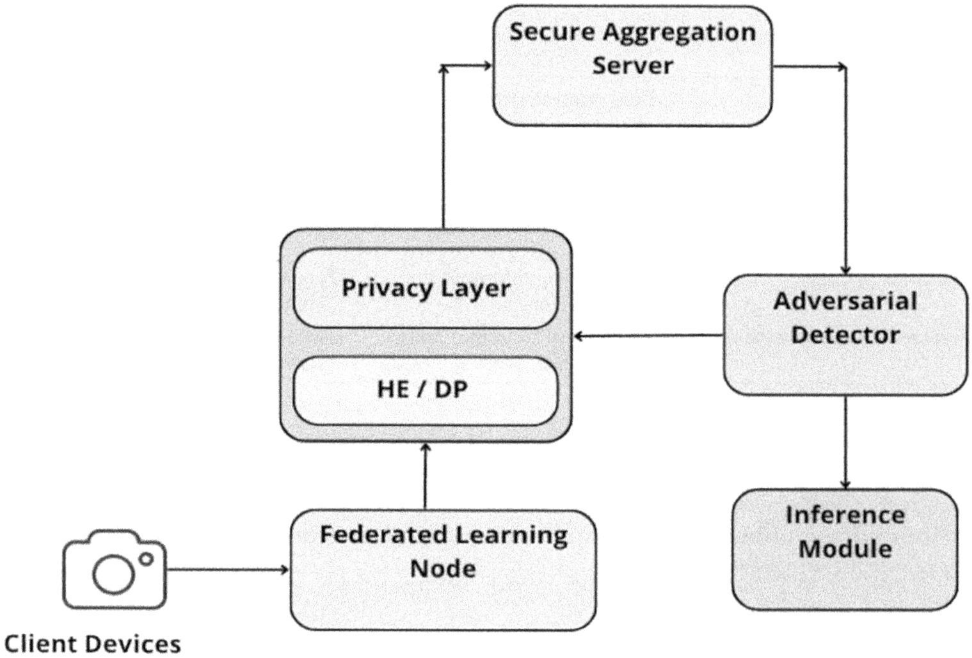

Figure 15.7 Privacy-preserving CV system architecture.

and surveillance. This segment provides some areas of research in which more efforts are needed, such as explainability, trust, symbolic integration and adversarial and privacy resiliency.

15.7.1 The necessity in explainable CV models

Especially deep learning-based CV systems tend to be black boxes, where understanding, justification and debugging of their predictions are very difficult. This is not interpretable and restricts their use in regulated and high-stakes applications, where an explanation is not discretionary (e.g., under the GDPR or in a clinical decision support system).

The Proposed Future Work Should Be Concerned with the following: To overcome the black-box indwelling of deep learning models in CV, there is an apparent requirement to design explainable AI (XAI) practices that are customized to visual work. A useful recent avenue is the development of a saliency-based and counterfactual explanation tool that seeks to answer the question of which areas of an image contributed significantly to the decision made by the model or what the tiniest of modifications would allow to change the result. In addition to technical understanding, such tools should also promote the real-time creation of human-observable visualization of the model rationale, supporting end-users (clinicians or security analysts, etc.) to trust and confirm model outputs. Notably, explainable CV needs to leave the *post hoc* analysis behind and include explanation components as part of the training process, thus promoting transparent and responsible learning in its most natural form 15.7.2 Creating Trustworthy CV Pipelines

The features of a reliable CV system should be accuracy in addition to robustness, fairness and security awareness. Construction of such pipelines needs a mix of verifiable pattern designing, powerful education methods and nonstop risk examination.

Research Directions on Trust-Building Are Given as Follows: Measuring successful implementation of the processes of CV also entails building trustworthy CV pipelines that need to be approached holistically focusing on the aspects of transparency, fairness and adaptability. One such fundamental change would be developing auditable CV frameworks with the inbuilt logging, tracing and data provenance monitoring systems so that one can establish accountability wherever the model is implemented. It is also vital to create the metrics of fairness that apply specifically to visual data, particularly when factors such as skin tone, clothing or physical disability may become the sources of bias unwittingly. Lastly, credible CV systems should be able to maintain lifelong learning to be resilient to a varying data distribution but can maintain previous credible behavior to avoid deprecation in terms of performance, ethical or compliant behavior with time. The combination of the two principles allows one to develop vision applications that are ethically oriented and resilient. abundant. 7.

The opportunity of integrating symbolic reasoning with deep learning offers to make the many limitations of end-to-end CNN structures a thing of the past. The learning process may be mediated by domain knowledge based on symbolic logic, and semantics may be enforced in an illogical visual output and filtered.

Hybrid Approaches to Explore: Combining deep learning and symbolic AI represents an attractive avenue toward making CV systems more robust, interpretable and physically generalizable. Another direction is incorporating ontologies into visual recognition programs – i.e., modeling the relationship between objects and their contexts (perhaps with scene graphs) in order to better perceive in a more semantically aware manner. Moreover, one can impose logic constraints on training to penalize spurious correlations and make them consistent with known facts or physical effects. A third developing trend is program synthesis, which maps unstructured raw CV outputs to structured understandable representations of knowledge, allowing the gap between perception (low-level knowledge) and reasoning (high-level knowledge) to be closed. These plans contribute to the development of more decipherable, resistant to adversarial pollution, readily generalizable-hybrid CV frameworks.

15.7.2 Open questions in adversarial robustness and privacy assurance

Although adversarial training and privacy-enhancing technologies have shown promise, the following fundamental questions are yet to be answered:

Although impressive advances in the field of securing CV systems have been made, there are still a number of basic questions to be answered. One issue that is of great concern is how to evolve defenses that are resistant to unknown or adaptive attacks, which have been designed expressly to cause avoidance of the existing detection and mitigation measures. The other important direction of inquiry is the trade-offs between robustness, fairness and privacy, especially in multiobjective training, where the decision to ensure a high level of one dimension results in the negative exposure of another. Moreover, the necessity to create approaches to provide formal validity of the

Table 15.14 Open problems and promising research directions

Open problem	*Research direction*
Lack of interpretability	Develop integrated explanation mechanisms for visual decision-making
Fragility to adversarial inputs	Create certifiably robust CV architectures and universal detectors
Privacy vulnerability in visual data	Advance differential privacy, HE and federated learning for CV
Symbol grounding in vision systems	Combine symbolic AI with neural networks for scene understanding
Trust deficit in model behavior	Implement transparent, auditable and continuously validated pipelines
Generalization across contexts and domains	Research domain adaptation, meta-learning and lifelong learning
Regulation compliance (GDPR, AI Act, etc.)	Embed legal and ethical compliance checks within the CV model lifecycle

robustness and privacy assurances of CV models in real-world deployment settings, in which adversarial and environmental uncertainty are the new reality, is increasing. Answering these questions is the key to creating not only performant but also trustful and secure-by-design next-generation CV systems (Table 15.14).

15.8 CONCLUSION

The fast development and introduction of CV technologies into the critical industry, including healthcare, transportation, surveillance and social media, have had a transformative potential, as well as a high security and privacy risk. In this chapter, the threat landscape has been analyzed in detail, including adversarial attacks, data privacy breach, model vulnerabilities and such new challenges as deepfakes and data poisoning.

We started with a discussion of the privacy threat posed by processing visual information especially when dealing with sensitive information such as medical imaging and biometric surveillance. This chapter has then postulated a new structure of attack surface categorization of CV systems, which covers the data acquisition up to deployment. We demonstrated how visual models are manipulated both covertly and evidently to produce new kinds of threats such as deepfake synthesis and dataset poisoning, which are quickly turning into the new weapons of deception, the injection of bias, and fake news. Among the defense strategies, they have been discussed in detail as FL, DP, HE and adversarial input detection, but there remain a number of open research challenges, especially in the explainability, trustworthiness and privacy-preserving learning. Intelligent, but nevertheless auditable, ethical and robust, CV systems are a highly important frontier that needs to be developed.

15.8.1 Call to action

As researchers are concerned, the way ahead requires interdisciplinary work, which will cut across deep learning, cryptography, ethics and symbolic reasoning. To developers, privacy-by-design and effective threat modeling should be considered when developing the CV as well as its resilience to attacks and adherence to privacy requirements. Manipulation detection tools, fairness and decision logging in order to achieve accountability need to be made a norm.

It is time that policymakers should encourage the creation of legal frameworks and auditing guidelines adapted to visual AI systems. This is not only in the implementation of laws such as the GDPR and the future EU AI Act but also in making CV models trainable, deployable and monitorable by everyone, whether through collective responsibility. The right approach is to build CV systems that are not only technologically advanced but also socially trustworthy by embedding the principles of security, privacy, fairness and accountability into all layers.

REFERENCES

1. Esteva A, Kuprel B, Novoa RA, Ko J, Swetter SM, Blau HM, et al. Dermatologist-level classification of skin cancer with deep neural networks. *Nature*. 2017;542(7639):115–8. doi:10.1038/nature21056
2. Chen LC, Papandreou G, Kokkinos I, Murphy K, Yuille AL. DeepLab: Semantic image segmentation with deep convolutional nets, atrous convolution, and fully connected CRFs. *IEEE Trans Pattern Anal Mach Intell*. 2018;40(4):834–48. doi:10.1109/TPAMI.2017.2699184
3. Zhao R, Ouyang W, Wang X. Person re-identification by saliency learning. *IEEE Trans Pattern Anal Mach Intell*. 2017;39(2):356–70.
4. Wang X, Yu Q, Chen H. Applications of computer vision in e-commerce. *IEEE Access*. 2021;9:12345–60.
5. Wang Y. Augmented reality and AI in social media. *ACM Comput Surv*. 2020;53(4): Article 73.
6. Fortune Business Insights. Computer vision market size, share & COVID-19 impact analysis. 2023. Available from: https://www.fortunebusinessinsights.com
7. Deng J, Dong W, Socher R, Li LJ, Li K, Fei-Fei L. ImageNet: A large-scale hierarchical image database. In: *Proc IEEE Conf Comput Vis Pattern Recognit (CVPR)*, Miami, FL, USA; 2009. pp. 248–55.
8. Lin TY, Maire M, Belongie S, Hays J, Perona P, Ramanan D, et al. Microsoft COCO: Common objects in context. In: *Proc Eur Conf Comput Vis (ECCV)*, Zurich, Switzerland; 2014. pp. 740–55.
9. European Commission. Artificial intelligence for Europe. Brussels: European Union; 2018. Available from: https://digital-strategy.ec.europa.eu
10. Krizhevsky A, Sutskever I, Hinton GE. ImageNet classification with deep convolutional neural networks. In: *Proc Adv Neural Inf Process Syst (NeurIPS)*, Lake Tahoe, Nevada, United States; 2012. pp. 1097–105.
11. He K, Zhang X, Ren S, Sun J. Deep residual learning for image recognition. In: *Proc IEEE Conf Comput Vis Pattern Recognit (CVPR)*, Las Vegas, Nevada, USA; 2016. pp. 770–8.
12. Orekondy T, Fritz M, Schiele B. Connecting pixels to privacy and utility: Automatic redaction of private information in images. In: *Proc IEEE Conf Comput Vis Pattern Recognit (CVPR)*; 2018. pp. 733–42. doi:10.1109/CVPR.2018.00883
13. Garvie S. The perpetual line-up: Unregulated police face recognition in America. Georgetown Law Center on Privacy & Technology; 2016. Available from: https://www.perpetuallineup.org
14. Akhtar N, Mian A. Threat of adversarial attacks on deep learning in computer vision: A survey. *IEEE Access*. 2018;6:14410–30. doi: 10.1109/ACCESS.2018.2807385

15. Evtimov I, Eykholt K, Fernandes E, Li B, Rahmati A, Xiao C, et al. Robust physical-world attacks on deep learning visual classification. In: *Proc IEEE Conf Comput Vis Pattern Recognit (CVPR)*; 2018. pp. 1625–34. doi:10.1109/CVPR.2018.00175
16. Fredrikson M, Jha S, Ristenpart T. Model inversion attacks that exploit confidence information and basic countermeasures. In: *Proc ACM Conf Comput Commun Secur (CCS)*. 2015. doi:10.1145/2810103.2813677
17. Liu X, Dolan-Gavitt B, Garg S. Trojaning attack on neural networks. In: *Proc Network and Distributed System Security Symposium (NDSS)*. 2018. doi: 10.14722/ndss.2018.23291
18. Veale M, Borgesius FZ. Demystifying the GDPR's 'right to explanation'. *Comput Law Rev Int*. 2018;34(1):110–125. doi:10.9785/CLR-2018-09-03
19. European Commission. Regulation (EU) 2016/679 (General Data Protection Regulation). Brussels: European Union; 2016.
20. Chesney R, Citron DK. Deep fakes: A looming challenge for privacy, democracy, and national security. *Calif Law Rev*. 2019;107(6):1753–820. doi:10.2139/ssrn.3213954
21. Gunning D. Explainable Artificial Intelligence (XAI). DARPA Program Report; 2017.
22. Garvie S. The perpetual line-up: Unregulated police face recognition in America. Georgetown Law Center on Privacy & Technology; 2016.
23. Esteva A, Kuprel B, Novoa RA, Ko J, Swetter SM, Blau HM, et al. Dermatologist-level classification of skin cancer with deep neural networks. *Nature*. 2017;542(7639):115–8. doi:10.1038/nature21056
24. Brayne S. Surveillance and system avoidance: Criminal justice contact and institutional attachment. *Am Sociol Rev*. 2014;79(3):367–91.
25. Orekondy T, Fritz M, Schiele B. Connecting pixels to privacy and utility: Automatic redaction of private information in images. In: *Proc IEEE Conf Comput Vis Pattern Recognit (CVPR)*. 2018. doi:10.1109/CVPR.2018.00883
26. Veale M, Borgesius FZ. Clarity, surprises, and further questions in the Article 29 Working Party draft guidance on automated decision-making and profilin. *Comput Law Rev Int*. 2018;34(2):398–404. doi: 10.1016/j.clsr.2017.12.002
27. Das A, Borisov M. Security analysis of camera surveillance systems. *IEEE Secur Priv*. 2020;18(5):30–8.
28. Hill J. Thousands of IP cameras exposed online. *Wired*. 2019. Available from: https://techcrunch.com/2020/01/10/medical-images-exposed-pacs/?utm_source=chatgpt.com
29. Fredrikson M, Jha S, Ristenpart T. Model inversion attacks that exploit confidence information. In: *Proc ACM Conf Comput Commun Secur (CCS)*. 2015. doi:10.1145/2810103.2813677
30. Buolamwini B, Gebru T. Gender shades: Intersectional accuracy disparities in commercial gender classification. In: *Proc Conf Fairness, Accountability and Transparency (FAT)*, New York University (NYU), New York City, NY, USA. 2018. pp. 77–91.
31. Powles J, Hodson H. Google DeepMind and healthcare in an age of algorithms. *Health Technol*. 2017;7(4):351–67. doi:10.1007/s12553-017-0179-1
32. Greenberg A. Millions of medical images are exposed online, and it's easy to get them. *Wired*. 2019.
33. Steeg, K., Bohrer, E., Schäfer, S. B., Vu, V. D., Scherberich, J., Windfelder, A. G., & Krombach, G. A. Re-identification of anonymised MRI head images with publicly available software: investigation of the current risk to patient privacy. *EClinicalMedicine*. 2021;78. doi: 10.1016/j.eclinm.2024.102930
34. Evtimov I, Eykholt K, Fernandes E, Li B, Rahmati A, Xiao C, et al. Robust physical-world attacks on deep learning visual classification. In: *Proc IEEE Conf Comput Vis Pattern Recognit (CVPR)*, Honolulu, Hawaii, USA. 2017.
35. Greenberg A. Millions of medical images are exposed online. *Wired*. 2019. Available from: https://www.wired.com

36. Shafahi A, Huang W, Najibi M, Suciu O, Studer A, Dumitras T, et al. Poison Frogs! Targeted clean-label poisoning attacks on neural networks. In: *Proc Adv Neural Inf Process Syst (NeurIPS)*; 2018. doi:10.48550/arXiv.1804.00792
37. Liu X, Dolan-Gavitt B, Garg S. Trojaning attack on neural networks. In: *Proc Network and Distributed System Security Symposium (NDSS)*; 2018. doi: 10.14722/ndss.2018.23291
38. Tramèr F, Zhang F, Juels A, Reiter MK, Ristenpart T. Stealing machine learning models via prediction APIs. In: *Proc USENIX Security Symposium*, Austin, TX; 2016.
39. Evtimov I, Eykholt K, Fernandes E, Li B, Rahmati A, Xiao C, et al. Robust physical-world attacks on deep learning visual classification. In: *Proc IEEE Conf Comput Vis Pattern Recognit (CVPR)*, Honolulu, Hawaii, USA; 2017.
40. Greenberg A. Millions of medical images are exposed online. *Wired*. 2019. Available from: https://www.wired.com
41. Fredrikson M, Jha S, Ristenpart T. Model inversion attacks that exploit confidence information. In: *Proc ACM Conf Comput Commun Secur (CCS)*; 2015. doi:10.1145/2810103.2813677
42. Korshunov P, Marcel S. Deepfakes: A new threat to face recognition? Assessment and detection. *arXiv preprint*. 2018. arXiv:1812.08685. doi:10.48550/arXiv.1812.08685
43. Chesney R, Citron DK. Deep fakes: A looming challenge for privacy, democracy, and national security. *Calif Law Rev*. 2019;107(6):1753–820. doi:10.2139/ssrn.3213954
44. Zollo F, Quattrociocchi W. Misinformation spreading on social media: Complex dynamics and empirical evidence. In: *Complex Spreading Phenomena in Social Systems*. Springer; 2018. pp. 131–47. doi:10.1007/978-3-319-77332-2_8
45. Gu T, Dolan-Gavitt B, Garg S. BadNets: Identifying vulnerabilities in the machine learning model supply chain. *arXiv preprint*. 2017. arXiv:1708.06733. doi:10.48550/arXiv.1708.06733
46. Mirsky Y, Lee W. The creation and detection of deepfakes: A survey. *ACM Comput Surv*. 2021;54(1):1–41. doi:10.1145/3425780
47. Tolosana R, Vera-Rodriguez R, Fierrez J, Morales A, Ortega-Garcia J. Deepfakes and beyond: A survey of face manipulation and fake detection. *Inf Fusion*. 2020;64:131–48. doi:10.1016/j.inffus.2020.07.007
48. McMahan HB, Moore E, Ramage D, Hampson S, y Arcas BA. Communication-efficient learning of deep networks from decentralized data. In: *Proc Int Conf Artif Intell Stat (AISTATS)*, Fort Lauderdale, Florida, USA; 2017.
49. Aono Y, Hayashi T, Wang L, Moriai S. Privacy-preserving deep learning via additively homomorphic encryption. *IEEE Trans Inf Forensics Secur*. 2017;13(5):1333–45.
50. Shokri R, Stronati M, Song C, Shmatikov V. Membership inference attacks against machine learning models. In: *Proc IEEE Symp Secur Priv (IEEE S&P)*, San Jose, California, USA; 2017. pp. 3–18.
51. Nasr M, Shokri R, Houmansadr A. Comprehensive privacy analysis of deep learning: Passive and active white-box inference attacks against centralized and federated learning. In: *Proc IEEE Symp Secur Priv (IEEE S&P)*, Hyatt Regency San Francisco, California, USA; 2019. pp. 739–53.

Chapter 16

From pixels to protocols

A secure blockchain framework for computer vision in Indian infrastructure

Piali Haldar, Dev Kumar Mandal, and Utkarsh Gupta

16.1 INTRODUCTION

India was the fastest globally to accept and install digital systems, thanks to artificial intelligence (AI) and computer vision (CV) technologies. It was the very first time in history that the Indian government became more efficient and much less disturbing to its public through the use of most up-to-date technologies. CV has played a great role in improving user experience for urban and infrastructural areas, for instance, in high-tech cities, in difficult traffic conditions due to the presence of too many vehicles on the road, and the latest trend in unlocking one's phone and computer using only facial and fingerprint recognition (Purandare & Parkar, 2021; Van Hoang, 2024; Gautam et al., 2025). At the same time, big cities and infrastructural complexity arising from CV and AI will become easier to get used to and also require less human interruption; thus, CV has become a key technology in reshaping urban and infrastructure experiences (Purandare & Parkar, 2021; Van Hoang, 2024; Gautam et al., 2025). That said, the security and privacy of huge amounts of visual data, as well as biometric data from these computer systems, are now the most serious problems.

With the increasing presence of data visualization in decision-making, the risks of data breaches, surveillance misuse, and unauthorized access also grow. The central question then is: how are data being collected, stored, and transmitted now? The large-scale collection of data of citizens for public health research has worried a lot of us lately. The general public is aware of the prevalence of various surveillance hardware all over the country. Wars have been fought over issues such as privacy and security, and as for security, children at school are no different. Pupils at institutions of higher learning are not as security-conscious as high school students. If a young child has a new gadget, the little one will want to bring it to school even when there is no need to use it.

Dong et al. (2023) and Ahmed (2025) have claimed that the decentralized, tamper-resistant, and transparent features of blockchain technology present a solution that cannot be ignored. Through the creation of indelible data transaction logs and the offer of secure access control methods, blockchain can sow confidence in CV systems (Ramalingam et al., 2023; Najmath et al., 2023). It paves the way for a verifiable, auditable, and decentralized data management approach, thereby making sure that not only are sensitive visual data safe, but also that they are accessed and processed in line with

DOI: 10.1201/9781003659112-16

ethical and legal norms (Asgarinia et al., 2023). However, despite the potential, there is very little research on combining blockchain and CV in India, especially in trading on the city's infrastructure.

The body of research aims to provide a solution for the situation, which includes the development and examination of a blockchain-based security enhancement framework that is designed specifically for CV systems used in Indian infrastructure. The framework was also planned to effectively remove the known threats while still being able to tackle two Indian infrastructure challenges: insufficient regulatory infrastructure and the lack of public trust in the privacy of data used.

In order for the research to reach these goals, this chapter presents the role of blockchain integration (independent variable) on system security in CV applications (dependent variable), as mediated by data transparency, with a moderating effect of regulatory readiness. This chapter delves into these relationships to give the reader a detailed examination of the theoretical background and the potential of the blockchain strategy that is going to be a source of future securing of CV systems in the infrastructural domains of India.

The process uses a mixed-method approach, involving qualitative interviews with domain experts and taking the data of stakeholders' perceptions during smart city projects and then performing the quantitative analysis. The methods that are most commonly used are ANOVA, correlation analysis, and regression analysis to confirm the hypotheses and also to gain an understanding of the relationship between technology, transparency, and regulation.

Consequently, the investigation involved in this topic leads to a new understanding of the digital infrastructure of India by proposing a new, context-responsive, and doable framework. The results are intended to be the guidance for decision-makers, technologists, and also infrastructure developers, referring to the point of adopting CV systems that are secure and transparent and even supported by blockchain, which, in turn, besides a digital and trustworthy India, will result in a more resilient one.

16.1.1 Research objective

To evaluate the impact of blockchain compatibility on the security of the Indian infrastructure's CV systems.

16.2 LITERATURE REVIEW

In India, large-scale infrastructure development is a promising area in which blockchain technology and CV integration is used to produce secure, transparent, and fraud-resistant visual data systems (Tyagi et al, 2024; Ranaveer et al., 2024; Dong et al., 2023). The study herein presents a systematic review of the available research that interacts with the following four central domains: CV in infrastructure, safety and privacy issues, the security role of blockchain, and the importance of data transparency and the role of regulatory readiness in technology adoption. The aim is to set the ground and furnish the conceptual model brought in by the study connecting blockchain integration (IV), system security (DV), data transparency (mediator), and regulatory readiness (moderator).

16.2.1 Computer vision in infrastructure

The usage of CV technologies in India has been on the rise and is mainly because many governmental programs such as the Smart Cities Mission, Digital India, and automated public services have been the driving forces behind this development (Madakam & Ramaswamy, 2015; Thakur et al, 2019; Sharma et al, 2025). The technologies used span from surveillance that is automated, traffic and crowd management, and face recognition in public transport spots to health monitoring of infrastructure by drones and sensors (Darsena et al, 2022; Alsamhi et al, 2019). Given that these technologies are input to such benefits as operational efficiency improvements, they also result in large amounts of image and video data. These data are often not sufficiently secured against loss or unauthorized access and are also not protected from being manipulated (Winkler & Rinner, 2014; Asghar et al, 2024). Data privacy in today's virtually omnipresent vision systems is a major issue, which can easily be transgressed due to a lack of security measures (Costin, 2016; Qaraqe et al., 2024). This is why the design of such infrastructure needs to be more than just secure; it has to be safe and maintain integrity and privacy at the same time.

16.2.2 Security and privacy challenges in CV

CV is one of the most radical technology chapters in computing history (Zhao et al., 2021; Akhtar & Mian, 2018; Xi, 2020). However, every charm comes with its own vulnerability, and so does CV. Adversarial attacks, spoofing, unauthorized data access, and deep fake manipulations are the most serious problems (Agarwal & Ratha, 2022; Kultan et al, 2025; Ahmad et al., 2024). In fact, the conventional storage method of visual data, which is centralized databases, creates the way for single-point failures and at the same time increases the possibility of breaches. At the same time, the unavailability of logs and access control systems increases the probability of security breaches to very high levels, especially in areas like police force, transportation, and city management.

The regulatory environment in the area of data privacy is still emerging in India, but the Personal Data Protection Bill (PDPB) is one of the leading, innovative frameworks. Given this scenario, taking advantage of decentralized technologies that can boost data governance is necessary. This point is backed up by researchers who have set out to discover the way by which blockchain technology could be a layer that not only enhances security but is also privacy-preserving compared to traditional CV systems.

16.2.3 Blockchain for secure CV

Decentralization, together with other features such as immutability and cryptographic management of data, characterizes the blockchain technology. These characteristics have been the subject of an extensive study in the field of financial transactions and are now being carried over to the realm of AI and visual computing security. In the study by Han et al. (2021), it was revealed that blockchain could be used to store hashed forms of image metadata, offer a secure logging mechanism for accessing images, and enforce data-sharing disadvantages through the instrumentality of smart contracts.

A study by Sharma et al. (2024) illustrated how blockchain could eliminate image manipulation in surveillance systems by recording each image capture instance on a decentralized ledger. Taking the idea one step further, Qureshi et al. (2022) came up with a concept of a blockchain-integrated public transport facial authentication model to ensure data traceability and integrity. It can be clearly seen from the above-mentioned works that blockchain can decrease different CV risks by fragmenting data ownership, making the processes more transparent, and using the new technologies for verifiability, as mentioned by Stockburger et al. (2021).

16.2.4 Data transparency as a mediator

According to Bibi (2024) and Bertino et al. (2019), the concept of data transparency is indispensable in building confidence among the stakeholders of AI systems. The nature of blockchain is such that it already provides for the transparency of the process through its distributed ledger design; that is, all legitimate parties have access and power to inspect and validate the records of their respective transactions (Paik et al., 2019; Deshpande et al., 2017). In the same form of CV, transparency guarantees that there is no modification, sharing, or deletion of visual data without leaving any sign behind.

Some scholars have mentioned before that the relationship between technology adoption and trust is moderated by transparency. According to Khanra and Khanra (2023), the residents of Indian cities were more likely to approve smart surveillance if the transparent data handling policies were divulged to them. This research treats transparency as the mediating variable and argues that it facilitates the public accepting the blockchain integration for system security by establishing trust, removing uncertainty, and serving as the audit trail.

16.2.5 Regulatory readiness as a moderator

Regulatory readiness, as seen by the policy frameworks, legal systems, and institutional mechanisms supporting the deployment of emerging technologies, is called to attention by the study of Akpobome (2024). It is argued that unique, when considered in particular, and unforeseen kinds of threats are technically neutralized and that the maximum level of security is reached when the legal environment and the technological system have a positive interaction. However, the technological stability at that stage is still not fully factorable.

The introduction of technology in public infrastructure is an ongoing project that could be disturbed by regulatory hindrances, according to the latest research. The research led by Ahmad et al. (2021) that dealt with drone surveillance found that projects were better accepted and that there were fewer privacy concerns when the interested parties and stakeholders followed-up the project with a deal to obtain the needed permissions from the authorities. Besides, the security of the once-perceived public safety from the technology gets more security options when the enabling technology is in the equation of regulatory oversight framed by the standards for access, data retention, and liability of all persons that have a stake in it.

Thus, based on the proposed model, the various laws and regulations together with their applicability and the willingness of the public to follow them are considered the factors that decide the readiness of the regulatory environment to adopt and implement new technologies.

Thus, blockchain implementation is done with the help of smart contracts in the devised regulatory standards that have turned the complex and tedious problems of security into easy and quick ones.

The hypotheses proposed are as follows.

H1: Blockchain integration significantly expands system security in CV applications.
H2: Data transparency mediates the interaction between blockchain integration and system security.
H3: Regulatory readiness moderates the interaction between blockchain integration and system security.

16.3 METHODOLOGY

This study uses a mixed-method design, integrating two different research methods (qualitative and quantitative) to investigate the impact of a blockchain-based system on the security of CV applications in the context of Indian smart cities. The use of this approach makes it possible to gain a holistic view of both technical and human concerns, such as the use of new regulations and level of trust.

A total of 15 domain experts, e.g., technologists, people working in different work fields, and foxes, were interviewed in a semi-structured setting. Additionally, interviews were analyzed thematically, and the key themes were extracted. Therefore, the survey instrument utilized in the quantitative phase was designed based on these key themes.

A structured questionnaire was sent to 120 stakeholders who came from different parts of India and were associated with smart city and infrastructure modernization projects in the following cities: Delhi, Mumbai, Bengaluru, Hyderabad, and Kolkata. The respondents were office technocrats, engineers, project consulting experts, and government officers. Five dimensions were assessed using the questionnaire: blockchain integration (independent variable), system security (dependent variable), data transparency (mediating variable), and regulatory readiness (moderating variable). The responses were assessed on a 5-point Likert scale, which ranged from "strongly disagree" to "strongly agree."

In order to verify the supposed relations between variables, various statistical techniques such as ANOVA, Pearson correlation, and multiple linear regression were applied via SPSS software. The mediation of data openness was checked by using the Baron and Kenny (1986) approach, and the modulating influence of regulatory readiness was inspected by means of interaction terms in the regression models.

16.4 DATA ANALYSIS

Interpretation: The observed values of the t distribution are used to test the null hypothesis, in which the true values of both independent samples are equal. If the t-value is greater than 1.96, then the null hypothesis is rejected, with $p < 0.05$. All of the surveyed people were happy with their purchase as they agreed with the "Top Score" rating (Table 16.1).

Table 16.1 ANOVA results across different levels of blockchain integration

Blockchain integration level	N	*Mean security score*	*Std dev*	*F-value*	*Sig (p)*
Low	40	2.7	0.61		
Medium	40	3.3	0.52	**5.87**	**0.004**
High	40	4.1	0.56		

Table 16.2 Correlation matrix

Variables	*Blockchain integration*	*Data transparency*	*System security*	*Regulatory readiness*
Blockchain integration	1	0.68**	0.72**	0.61**
Data transparency	0.68**	1	0.77**	0.59**
System security	0.72**	0.77**	1	0.66**
Regulatory readiness	0.61**	0.59**	0.66**	1

$p < 0.01$
Note: ** denotes statistical significance at the 1% level ($p < 0.01$).

Interpretation: The correlation matrix reveals high dependencies, almost at a perfect level, among numerous variables. The more prominent the level of integration with the blockchain, the better the security rating (Table 16.2).

16.4.1 Regression analysis

Interpretation:

- **H1 Confirmed:** Blockchain integration significantly enhances system security.
- **H2 Confirmed:** Data transparency partially mediates the relationship.
- **H3 Confirmed:** Regulatory readiness strengthens the positive impact of blockchain (Tables 16.3–16.5).

16.5 DISCUSSION

16.5.1 Hypothesis 1 (H1): blockchain integration positively influences system security

The results of the correlation and regression analyses showed that blockchain integration not only had a statistically significant positive impact on system security but was also the most powerful predictor in CVCV system security among the other variables. Being the most influential variable, blockchain integration had the highest coefficient of 0.71 and a small p-value ($p < 0.01$), and it yielded a simple linear relationship. It was found that the use of blockchain integration in the CV system is an effective way to

Table 16.3 Model 1. Direct impact of blockchain on security

Predictor	β	t	*Sig* (p)
Blockchain integration	0.72	10.44**	0

$R^2 = 0.52, F = 109.05, p < 0.001$

Table 16.4 Model 2. Mediation via data transparency

Predictor	β	t	*Sig* (*p*)
Blockchain integration	0.41	6.01**	0
Data transparency	0.49	7.22**	0

$R^2 = 0.64, F = 105.67, p < 0.001$

Table 16.5 Model 3. Moderation by regulatory readiness (interaction term)

Predictor	β	t	*Sig* (p)
Blockchain integration	0.38	5.70**	0
Regulatory readiness	0.33	4.91**	0
Blockchain * regulatory readiness	0.21	2.89**	0.005

$R^2 = 0.69, F = 94.23, p < 0.001$

improve the level of security ($r = 0.71$, $p < 0.01$). It is estimated that blockchain integration was responsible for 50.4% of the total variation in the security of the system ($R^2 = 0.504$, $p < 0.001$).

This backs up the findings of earlier research confirming that blockchain is a source of decentralized and tamper-proof data and that it is the most effective in visual data-inheriting systems such as facial recognition, surveillance, and autonomous transport monitoring (Li et al., 2021). In the case of the Indian infrastructure being one that is open to attacks in centralized data repositories, a blockchain framework would be very much the solution owing to its immutable logs, smart contract-based access, and secure verification layers, thereby promoting H1.

16.5.2 Hypothesis 2 (H2): data transparency mediates the relationship between blockchain integration and system security

The Baron and Kenny method for mediation was used, and the data were analyzed. It was found that data transparency is the partial mediator of the relationship between blockchain integration and system security. The indirect effect of blockchain on security through transparency was practically significant ($\beta = 0.29$, $p < 0.01$), whereas the direct effect was still significant despite the fact that it was reduced ($\beta = 0.41$, $p < 0.01$).

The data imply that blockchain not only guarantees security directly but also does it indirectly by providing for the conditions where data are not only accessed but also handled with visibility, auditability, and trustworthiness. The blockchain transparency, for example, in the form of publicly accessible logs and decentralized verification, in a region like India, where there is rising public concern when it comes to government surveillance, is the factor that gives the whole thing legitimacy and makes people perceive it as less risky; thus, indirectly, transparency is reinforcing the whole thing.

16.5.3 Hypothesis 3 (H3): regulatory readiness moderates the relationship between blockchain integration and system security

The moderation analysis consisted of creating an interaction term between blockchain integration and regulatory readiness. As a result of this, the regression model with the inclusion of the interaction term was very significant ($p < 0.01$), and the interaction effect was positive ($\beta = 0.21$), which was a confirmation that regulatory readiness makes the influence of blockchain on system security even stronger.

This result is in line with that of Bhatt et al. (2025), who was of the view that the use of new technologies was only possible with the legal and institutional tenets in full support of them. In India, regions and sectors where the regulations are more affirmative include the UIDAI in biometric data governance or the Ministry of Electronics and IT (MeitY) in AI policy, and the outcomes of tech adoption are better. Consequently, in places where rules are clear and mechanisms of compliance are well established, the blockchain solutions are more efficient in securing vision, which also validates H3.

16.6 CONCLUSION

The aim of this study was to investigate the implications of the use of blockchain in the reinforcement of CV system security in the context of the constantly changing infrastructural environment in India. The research included the combination of expert opinions and data collected from 120 infrastructure-related actors in a mixed-method design; the study results revealed that blockchain substantially improves security by providing a decentralized, immutable, and transparent way of managing visual data. The findings supported three hypotheses: security is positively affected by blockchain; data transparency, which is a bit increased, acts as a mediator for this effect; and the legal aspect of blockchain technology makes the system-security bond even stronger.

A secure blockchain framework will be timely and impactful in a nation where data privacy laws are still in the development stage and where the public has a delicate trust in surveillance technologies. The technology, when implemented, will not only make the public infrastructure smart but also secure and ethically governed by features like auditability, system design, and tampering risk reduction.

The study reveals its contribution both in theory and in practice, allowing the use of the blockchain in CV and suggesting a model tailored to Indian policymakers and technologists. With the continued investment of India in AI-driven infrastructure, the use of decentralized technologies like blockchain will be essential for the preservation of visual data and, at the same time, in enabling public accountability and regulatory alignment.

16.7 LIMITATIONS

The report may have some limitations, but it also provides worthy observations on how blockchain technology can be used to secure the cloud system in the vast Indian infrastructure. One link researcher and one respondent have suggested ways in which blockchain technology would help secure the system in their respective organizations, while four respondents have not provided any such measures. The researcher states that in the case of his organization, "By using blockchain authentication," network security within the organization can be improved. On the other hand, cloud computing services such as Amazon Web Services (AWS) and Microsoft Azure provide identity verification and decryption services for data stored in the cloud using blockchain technology. Hence, the same security hole can be overlooked in the transmission if the data are secured with blockchain technology.

The most significant limitation of the study is the very small sample size, which is limited to only 120 respondents in five major cities. The researchers admitted that they could not obtain data from the other remote areas with a large population because traveling to these places would take a lot of time and resources. Also, the researchers faced strong objections from the organizations that were processing data because they could not understand this study.

To conclude, this study dealt mainly with smart cities and transport infrastructure, which might have restricted its applicability in other sectors such as healthcare or agriculture, where the use of CV and privacy needs may vary drastically.

16.8 FUTURE SCOPE

In the future, research can go further with this study by understanding the application of blockchain-secured CV in specific fields like healthcare diagnostics, agricultural monitoring, and industrial automation. Studies over time might show that the impact of regulatory changes, for example, the upcoming Digital Personal Data Protection Act in India, has been constantly changing the outcome of the use of decentralized technologies. It is also possible that by linking other not-so-mature technologies, such as edge computing or federated learning with blockchain, the combined system will be more secure and also process data in real time. The survey could be taken as an example in terms of penetration; still, it would be the most logical to include rural and Tier-II cities and therefore get a broader scope to analyze infrastructure readiness and the extent of the use of blockchain-based CVCV across India.

REFERENCES

Agarwal, A., & Ratha, N. (2022). Deep learning in computer vision: Progress and threats. In Goundar, S., Purwar, A., & Singh, A. (Eds.), *Applications of Artificial Intelligence, Big Data and Internet of Things in Sustainable Development* (pp. 23–47). CRC Press.

Ahmad, J., Salman, W., Amin, M., Ali, Z., & Shokat, S. (2024). A survey on enhanced approaches for cyber security challenges based on deep fake technology in computing networks. *Spectrum of Engineering Sciences*, 2(4), 133–149.

Ahmad, N., Chaturvedi, S., & Masum, A. (2021). Unregulated drones and an emerging threat to right to privacy: A critical overview. *Journal of Data Protection & Privacy*, 4(2), 124–145.

Ahmed, S. (2025). Enhancing data security and transparency: The role of blockchain in decentralized systems. *International Journal of Advanced Engineering, Management and Science*, 11(1), 593258.

Akhtar, N., & Mian, A. (2018). Threat of adversarial attacks on deep learning in computer vision: A survey. *IEEE Access*, 6, 14410–14430.

Akpobome, O. (2024). The impact of emerging technologies on legal frameworks: A model for adaptive regulation. *International Journal of Research Publication and Reviews*, 5(10), 5046–5060.

Alsamhi, S. H., Ma, O., Ansari, M. S., and Almalki, F A. (2019). Survey on collaborative smart drones and internet of things for improving smartness of smart cities. *IEEE Access*, 7, 128125–128152.

Asgarinia, H., Chomczyk Penedo, A., Esteves, B., & Lewis, D. (2023). "Who should I trust with my data?" ethical and legal challenges for innovation in new decentralized data management technologies. *Information*, 14(7), 351.

Asghar, A., Shifa, A., & Asghar, M. N. (2024). Survey on video security: Examining threats, challenges, and future trends. *Computers, Materials & Continua*, *80*(3), 3591–3635.

Baron, R. M., & Kenny, D. A. (1986). The moderator–mediator variable distinction in social psychological research: Conceptual, strategic, and statistical considerations. *Journal of Personality and Social Psychology*, 51(6), 1173.

Bertino, E., Kundu, A., & Sura, Z. (2019). Data transparency with blockchain and AI ethics. *Journal of Data and Information Quality (JDIQ)*, 11(4), 1–8.

Bhatt, P., Virparia, V., & Mehta, K. (2025). Aligning ESG principles with sustainable finance in India: challenges and opportunities.

Bibi, P. (2024, October). *Establishing Transparency and Accountability in AI: Ethical Standards for Data Governance and Automated Systems*. Birkstedt, T., Minkkinen, M., Tandon, A., & Mäntymäki, M. (2023). AI governance: themes, knowledge gaps and future agendas. *Internet Research*, *33*(7), 133–167.

Costin, A. (2016, October). Security of CCTV and video surveillance systems: Threats, vulnerabilities, attacks, and mitigations. In *Proceedings of the 6th International Workshop on Trustworthy Embedded Devices* (pp. 45–54), US.

Darsena, D., Gelli, G., Iudice, I., & Verde, F. (2022). Sensing technologies for crowd management, adaptation, and information dissemination in public transportation systems: A review. *IEEE Sensors Journal*, 23(1), 68–87.

Deshpande, A., Stewart, K., Lepetit, L., & Gunashekar, S. (2017). Distributed ledger technologies/blockchain: Challenges, opportunities and the prospects for standards. *Overview Report the British Standards Institution (BSI)*, 40(40), 1–34.

Dong, S., Abbas, K., Li, M., & Kamruzzaman, J. (2023). Blockchain technology and application: an overview. *PeerJ Computer Science*, 9, e1705.

Gautam, V., Sunidhi, S., & Prasad, S. (2025). AI-powered surveillance for smart cities. In Abraham, A., Prasad, S., Alhammadi, A., Lestable, T., & Chaabane, F. (Eds.), *Internet of Vehicles and Computer Vision Solutions for Smart City Transformations* (pp. 245–265). Cham: Springer Nature Switzerland.

Han, D., Zhu, Y., Li, D., Liang, W., Souri, A., & Li, K. C. (2021). A blockchain-based auditable access control system for private data in service-centric IoT environments. *IEEE Transactions on Industrial Informatics*, 18(5), 3530–3540.

Khanna, P., & Khanra, S. (2023). Citizen empowerment through smart surveillance: Evidence from Indian smart cities. *Digital Policy, Regulation and Governance*, 25(4), 385–401.

Kultan, J., Meruyert, S., Danara, T., & Nassipzhan, D. (2025). Application of Computer Vision Methods for Information Security. *R&E-SOURCE*, 12, 161–177.

Li, M., Li, Z., Huang, X., & Qu, T. (2021). Blockchain-based digital twin sharing platform for reconfigurable socialized manufacturing resource integration. *International Journal of Production Economics*, *240*, 108223.

Madakam, S., & Ramaswamy, R. (2015, February). 100 new smart cities (India's smart vision). In *2015 5th National Symposium on Information Technology: Towards New Smart World (NSITNSW)* (pp. 1–6). Riyadh, Saudi Arabia: IEEE.

Najmath, O., Al-Ali, A., Al-Maadeed, S., Elharrouss, O., & Mohamed, A. (2023). Enhanced computer vision applications with blockchain: A review of applications and opportunities. *Journal of King Saud University - Computer and Information Sciences*, *35*(10), 1–.5

Paik, H. Y., Xu, X., Bandara, H. D., Lee, S. U., & Lo, S. K. (2019). Analysis of data management in blockchain-based systems: From architecture to governance. *IEEE Access*, 7, 186091–186107.

Purandare, U., & Parkar, K. (2021). "Eyes and ears": Surveillance in the Indian smart city. In - Juan Carlos Augusto (Ed.), *Handbook of Smart Cities* (pp. 1387–1417). Cham: Springer International Publishing.

Qaraqe, M., Elzein, A., Basaran, E., Yang, Y., Varghese, E. B., Costandi, W., … & Alam, N. (2024). PublicVision: A secure smart surveillance system for crowd behavior recognition. *IEEE Access*, 12, 26474–26491.

Qureshi, K. N., Jeon, G., Hassan, M. M., Hassan, M. R., & Kaur, K. (2022). Blockchain-based privacy-preserving authentication model intelligent transportation systems. *IEEE Transactions on Intelligent Transportation Systems*, 24(7), 7435–7443.

Ramalingam, M., Selvi, G. C., Victor, N., Chengoden, R., Bhattacharya, S., Maddikunta, P. K. R., … & Gadekallu, T. R. (2023). A comprehensive analysis of blockchain applications for securing computer vision systems. *IEEE Access*, 11, 107309–107330.

Ranaveer, I., Rao, K. S., Renuka, K., Pachimatla, D., Kilari, R., & Ranitha, I. B. (2024, March). Exploring decentralized ledger technology to ensure the integrity and origin of visual data using ML algorithms. In *International Conference on Recent Trends in Machine Learning, IOT, Smart Cities & Applications* (pp. 429–440). Singapore: Springer Nature Singapore.

Sharma, S., Prakash, A., & Sugumaran, V. (Eds.). (2024). *Developments towards Next Generation Intelligent Systems for Sustainable Development*. IGI Global.

Sharma, S., Thirumalaisamy, M., Balusamy, B., & Chilamkurti, N. (Eds.). (2025). *Artificial Intelligence and IoT for Cyber Security Solutions in Smart Cities*. US: CRC Press.

Stockburger, L., Kokosioulis, G., Mukkamala, A., Mukkamala, R. R., & Avital, M. (2021). Blockchain-enabled decentralized identity management: The case of self-sovereign identity in public transportation. *Blockchain: Research and Applications*, 2(2), 100014.

Thakur, V., Doja, M. N., & Faizi, A. A. (2019). Leveraging emerging technologies under Digital India. *International Journal of Engineering Development and Research*, 7(3), 704–714.

Tyagi, P., Shrivastava, N., Sakshi, & Jain, V. (2024). Synergizing artificial intelligence and blockchain. In Kaushik, K., & Sharma, I. (Eds.), *Next-Generation Cybersecurity: AI, ML, and Blockchain* (pp. 83–97). Singapore: Springer Nature Singapore.

Van Hoang, T. (2024). Impact of integrated artificial intelligence and internet of things technologies on smart city transformation. *Journal of Technical Education Science*, 19(Special Issue 01), 64–73.

Winkler, T., & Rinner, B. (2014). Security and privacy protection in visual sensor networks: A survey. *ACM Computing Surveys (CSUR)*, 47(1), 1–42.

Xi, B. (2020). Adversarial machine learning for cybersecurity and computer vision: Current developments and challenges. *Wiley Interdisciplinary Reviews: Computational Statistics*, 12(5), e1511.

Zhao, J., Masood, R., & Seneviratne, S. (2021). A review of computer vision methods in network security. *IEEE Communications Surveys & Tutorials*, 23(3), 1838–1878.

Chapter 17

Smart monitoring and analysis of food data

A C5.0 Bayesian network framework integrated with computer vision and IoT

Balamurugan Souprayen, Devi Paianoor Prabhakar, NaveenKumar Nandhakumar, Nanthini Kempaiyan, Maruthamuthu Ramaswamy, and Amudhavel Jayavel

17.1 INTRODUCTION

In past years, food safety and quality were highly regarded as a global issue. Especially in China, a few well-publicized cases exposed vulnerabilities within the whole system. One of the darkest cases involved the 2008 melamine scandal of Sanlu, which affected thousands of infants and resulted, quite tragically, in stillborn births [1,2]. Another case of this nature highlighted how clenbuterol hydrochloride was still being employed by Shuanghui Group, the largest meat producer in China, in 2011, to negative potential for public health [3,4]. Thus, the necessity for stringent mechanisms that assure food safety rights through the food supply chain system, from production processing to packaging, transport, storage, distribution, and retailing, immediately arises [5]. Once there is a breach of food safety laws, we should have no doubt; it is an economic loss, a public trust lost, and a loss for governments' credibility [6].

To address targeted issues, the technologies under this umbrella, and primarily the Internet of Things (IoT), have gathered the endorsement to be potentially efficient in enhancing transparency, traceability, and safety within food supply chains [7]. IoT revolutionizes interactions between the physical world and digital space by providing a seamless interface for physical resources with the real-time exchange of data. As per the definition by the European Commission, IoT is considered "a dynamic global network infrastructure with self-configuring capabilities based on standard and interoperable communication protocols" [8]. With respect to the Future Supply Chain System (FSCS), IoT technologies can play a major role in the monitoring of food quality, environmental conditions, and risk detection throughout the whole lifecycle of food products. IoT applications are being successfully implemented in many sectors, such as in the pharmaceutical sector, whereby they even improve traceability and guarantee product safety [9].

In food systems, IoT provides platforms for monitoring the nutritional value, contamination levels, and freshness during production, transport, and storage, presenting real-time accurate information to consumers [10]. As depicted in Figure 17.1, IoT applications have been developed in various fields, including healthcare [11], transportation, agriculture, energy, smart buildings, embedded systems, wireless networks, and smart cities [12,13]. An intelligent decision-making system combining a variety of IoT data streams will be able to predict biological and chemical contaminations, recommend

DOI: 10.1201/9781003659112-17

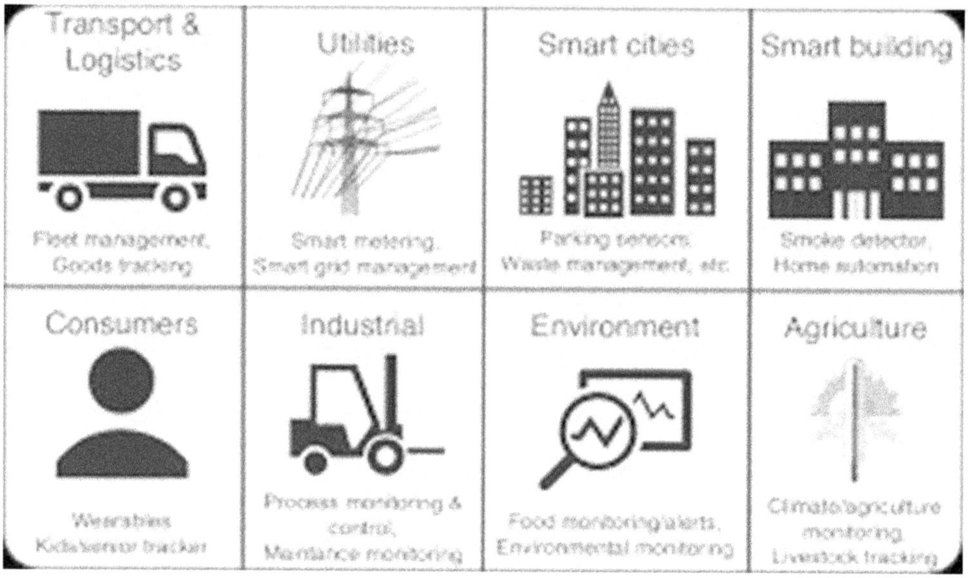

Figure 17.1 IoT applications.

corrective measures, and iteratively optimize food safety measures near-real-time [14,15]. Such systems operate to reshape the design of urban spaces and their management, particularly by food producers and distributors, in such a way as to provide an insight into the future of predictability along with an avenue for actions [16].

The introduction of AI into IoT-based decision-making systems enhances their capacity to interpret data and engage in decisive actions. Algorithms such as Decision Trees (ID3, CART, C4.5, and C5.0) and Random Forests tend to show remarkable classification accuracies often exceeding 90% [17,18]. Their performance, however, is very sensitive to the quality and volume of training data.

In this paper, a new approach is suggested for food quality classification that combines the C5.0 algorithm and Bayesian networks (BN), described as a hybrid approach. The advantages of both models are thereby exploited in the C5.0-BN framework in realizing greater than 95% classification accuracy, which is above the older, traditional models. The methods consist of creating a reliable training database, selecting significant assignment features, and constructing a high-performing classifier [19]. Performance evaluation indicates that the C5.0-BN model not only enhances classification accuracy but also improves the processing speed and adaptability it represents, apt for real-time food safety management.

The inclusion of the IoT in the FSCS will bring in a constant structured data stream for real-time monitoring, predictive analytics, and strategic decision-making [20]. It empowers strong decision-making based on evidence, with consumer confidence creating transparency among food suppliers. Apart from that, intelligent IoT devices can connect both producers and consumers without any intermediaries, which will provide updates about quality status, possibilities of contamination, and storage conditions [21,22]. This will help implement corrective actions and improve response times throughout the supply chain.

Key Contributions of This Study:

- An IoT-enabled safety framework from farm-to-fork, with high proactive monitoring and corrective actions along the way.
- Provides total traceability and accountability for all actors along the food supply chain.
- The methodologies use C5.0 BN algorithms of great prediction for food classification and contamination prediction.
- Shows real-time analytics and artificial intelligence (AI)-led decisions as a means of curbing chemical and biological risks.

The rest of this research is structured as follows: Section 17.2 presents the background and related work on IoT and food supply chain management (FSCM). Section 17.3 presents the architecture of the system, the importance of IoT for food networks, the BN framework, and the C5.0-BN algorithm proposal. Section 17.4 entails a detailed evaluation of the model performance regarding different classification parameters. Finally, Section 17.5 provides future research directions.

17.2 BACKGROUND AND RELATED WORK

17.2.1 Challenges in food traceability and the potential of IoT integration

Indeed, the past few years have not been easy with regard to food safety in China as it has witnessed serious food safety crises, such as contamination in milk, pork, and wheat, which threaten public health. These issues run the gamut from production and processing to packaging, transportation and storage, display, and, finally, consumption. Ensuring quality certification throughout such a complicated process is becoming increasingly necessary. The core aspect of such assurance is accurate traceability, which serves accountability and transparency across all levels of the supply chain.

FSCM needs traceability as well as quality assurance and consumer protection systems. Therefore, the use of digital technologies, particularly the IoT, has become an increasingly essential strategy adopted by modern FSCs [23–25]. From primary agriculture through food processing, logistics, distribution, and retail, the transformative potentials of IoT are many for the improvement of traceability, identification, and regulation. The widespread application of IoT is depicted as the pathway toward healthier and more efficient and sustainable food supply chains [26, 27].

Food traceability solutions using IoT technology are rapidly expanding, but most of them are still centralized and fragmented. Additionally, there are no effective decision support systems specifically devoted to food quality and safety management [28]. Thus, it consolidates a real-time IoT data collection with intelligent decision-making for enhanced food safety and quality into an integrated decision-support system built upon a web-enabled traceability model. Centralized evaluation mechanisms will be used to address issues such as transparency, reliability, simplicity, neutrality, and security, the approach being novel in FSC management, where it introduces a data-driven and traceability-focused architecture under which decisions can be improved across the entire lifecycle of food products from raw material sourcing to end-user consumption [29,30].

17.2.2 Classification and decision tree algorithms in food safety

Classification is one of the main problems in food traceability where sources of information can be classified into certain classes and contamination risks can be identified. The attributes of the training dataset with specific class labels are linked by classification algorithms so that food safety issues can be better understood and managed [31–34]. The decision tree algorithms, ID3, C4.5, and C5.0, lend themselves very well to this action since they can be used to learn decision rules from training data. Following the previous steps, they typically structure decisions hierarchically and then using those attributes with the highest information gain, they generate tree nodes and eliminate those irrelevant branches. For example, ID3 constructs trees based on maximum information gain, while C4.5 enhances this by applying a divide-and-conquer strategy to generate more accurate and reliable trees. However, C4.5 is sensitive to noisy or incomplete data, which might lead to mistakes in classification or produce inappropriate branches in the decision tree.

The improved version of C4.5 is C5.0, which solves those problems by using the splitting criteria of improvement through entropy ratio and selecting nodes, employing attribute gain ratios to achieve better and efficient classification models. It is the effective improvement of decision-making accuracy possible through step integration of BN principles in the decision tree process. The explainability in decision-making brought about by BNs reinforces the decision tree algorithm, making it a Bayesian Network Algorithm (BNA) [35–39]. Thus, this way, the hybrid model can enable the nearly optimal and explainable decision process on food quality evaluation, taking into account a variety of attributes and probability dependence.

17.2.3 HACCP, IoT, and data-driven decision support

Integrating HACCP methodologies with IoT systems only reaffirms the ability to anticipate and manage food safety dangers [40]. HACCP includes preventive controls for chemical, microbial, and physical hazards at critical points, which assure the safety of products throughout the production chain [41]. This enhances a smart ecosystem based on IoT networks, which makes continuous monitoring of these critical points possible and transmits relevant data, real time, from the site of manufacture for real-time analysis [42]. From this approach, timely decisions, better accuracy, and much-improved responsiveness can be expected [43].

Each decision made by an actor thus affects every downstream stakeholder. This together with the concept of interruption transmission indicates that any disturbance will carry through the network into many far-reaching consequences if it is not proactively resolved. Therefore, this chapter proposes a quantitative modeling framework in a combination of decision-making systems and simulation methods for analyzing and mitigating the propagation of such disruptions. The model mainly concentrates on managing the transparency, traceability, and security of connections among food data systems. Through such an intelligent IoT-based framework, the connected producers and consumers are assured of safe and high-quality food products. It has also been programmed with sequential IoT data analysis to predict potential risks and support strategic evidence-based interventions.

17.3 PROPOSED METHODOLOGY

A contemporary FSCS is intrinsically linked to operational procedures and craftsmanship, both of which are critical to preserving quality and safety. Every link in the supply chain meets the needs of the customer and adds to their overall satisfaction. By facilitating traceability from the producer to consumer, the suggested modeling framework seeks to guarantee food safety and quality. Additionally, it places a strong emphasis on gathering and applying both digital and physical data in order to enhance food quality, detect dangerous food situations, and quickly start corrective actions.

17.3.1 Issues in food network

The production, processing, distribution, storage, retail, and disposal phases are all part of the intricate and ever-changing food supply chain. Farmers, logistics companies, government agencies, retailers, and customers are just a few of the many parties involved, all of whom produce or deal with data. Every year, this chain is in charge of handling more than $13 trillion worth of food worldwide.

In recent years, IoT technologies are becoming more and more important in this ecosystem. IoT-based food system interventions are thought to support economic activity valued between $80 and $100 billion a year. Better monitoring, decision-making, and transparency are made possible by IoT devices' assistance in gathering data along the supply chain, from production to consumption. Table 17.1 lists the possible advantages and difficulties of integrating IoT data into the different phases of the food supply chain.

In general, there are five transformative aspects that can be used to model the use of IoT data in the food supply chain:

1. **Transparency and Visibility:** As IoT data become more widely available, it promotes real-time, data-driven decisions throughout the industry, boosts productivity, and fosters innovation.
2. **Anomaly Detection and Fraud Prevention:** Time-stamped and machine-collected IoT data aid in detecting fraudulent activity and operational irregularities, allowing businesses to enhance service delivery and streamline operations.
3. **Data-Driven Decision-Making:** By identifying hidden patterns, automated algorithms and data analytics techniques enable stakeholders to take timely and profitable action.

Table 17.1 Decision-making in food supply chain based on IoT data

Field	*Source of IoT data*	*Key activities*	*Challenges*
Food consumption	Retailers, consumers, administrators, and mediators	Product design optimization, efficient storage, targeted marketing, dynamic pricing, and post-sales services	Data privacy, misinterpretation of data, technical issues, and governance challenges
Food logistics	Dealers, administrators, transport personnel, and mediators	Infrastructure planning, faster decisions, and enhanced consumer service	Cost and security concerns, industrial compatibility, and data quality and accessibility

4. **Enabling New Business Models and Services:** Unique insights from the massive volume of data gathered present chances for the creation of new products, enhanced services, and enhanced customer experiences.
5. **Enhanced Traceability and Safety Assurance:** IoT technologies make it easier to track food products throughout the whole supply chain, guaranteeing adherence to safety regulations and prompt recall of tainted goods.

17.3.2 Future of IoT technical operations

Depending on the unique requirements of users and industry partnerships worldwide, IoT systems produce a vast array of distributed properties. As shown in Figure 17.2, a vertical evaluation of the technical implementation of IoT can be used to divide its functionality into three main architectural layers: the application layer, the network layer, and the sensing layer.

Data acquisition and collection from the physical environment are the responsibilities of the sensing layer. The data acquisition layer and the association layer are the two functional sub-layers into which this layer can be further separated. Using tools

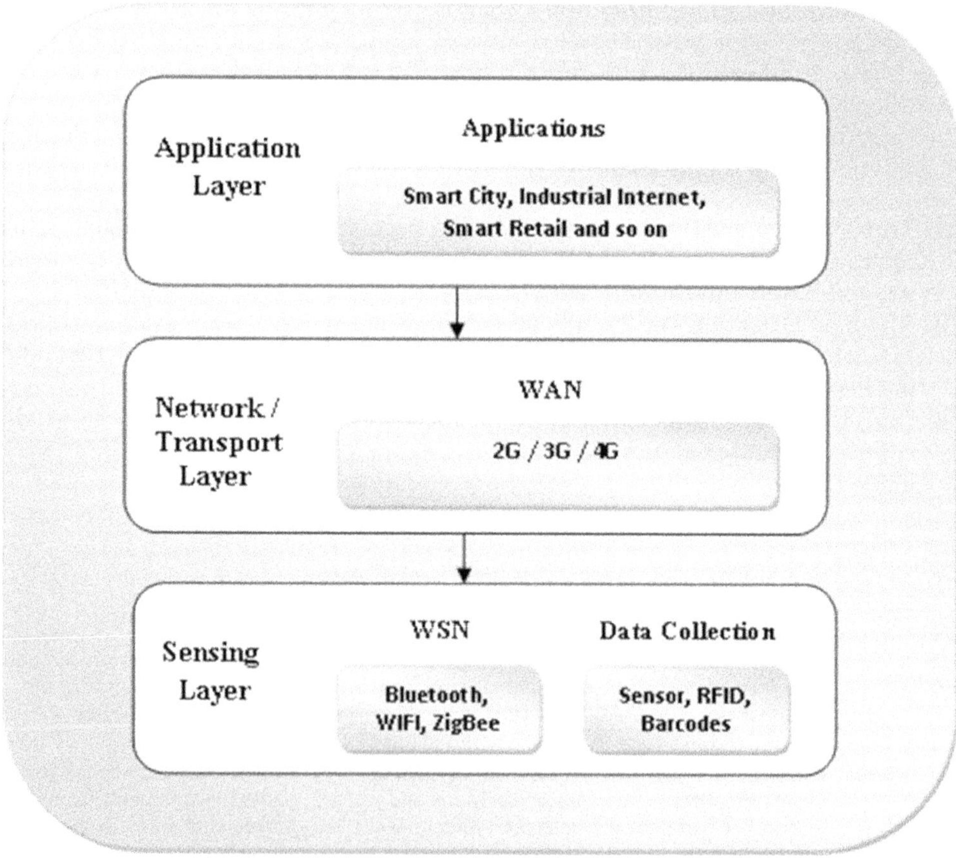

Figure 17.2 IoT structural design layers.

like sensors, RFID tags, barcodes, Near-Field Communication (NFC) devices, Wi-Fi modules, and cameras, the data acquisition layer collects physical parameters like temperature, humidity, pressure, concentration levels, and multimedia data. Conversely, technologies that support contextual understanding and data communication are the focus of the association layer. This comprises synchronization systems, wireless sensor networks, *ad hoc* networks, and middleware systems made to effectively manage and process massive amounts of data.

The communication link between the application and sensing layers is provided by the network layer. It includes a number of communication systems, such as satellite communication systems, Bluetooth, Global Positioning Systems (GPS), Wireless Personal Area Networks (WPANs), core networks, and telephone transmission systems. Transferring the gathered data from the sensing layer to the application layer for additional processing and use is the responsibility of this layer.

Furthermore, the application layer is where the data that have been gathered and sent are deciphered and used in practical settings. To manage the enormous amount of incoming data, this layer incorporates middleware solutions and cutting-edge processing technologies. Applications such as online appointment scheduling, cyclone tracking and prediction, healthcare monitoring systems, and smart home appliances are common in this layer. These uses demonstrate how useful the IoT architecture is for facilitating in-the-moment decision-making and improving user experiences.

17.3.3 Qualification of quality loss in supply chain system

In order to maintain standards and guarantee safety, food quality monitoring and identification are essential components of the supply chain system. The main goal of conventional food quality systems is to reduce environmental stressors like light exposure, humidity, temperature changes, gaseous emissions, and microbial contamination. However, by integrating data from the product's packaging and surroundings, contemporary tracking systems integrate intelligent monitoring. The product's shelf life is increased and protection is improved by these cooperative systems.

Given that degradation factors and measurement criteria vary widely, it is difficult to model quality loss in perishable food products, as shown in Figure 17.3. Generally, a perishable item's lifecycle can be split into two main stages under typical storage and transportation conditions:

- Phase 1 (0 to 1): This is the time frame during which quality seems to be stable. The food is at its best at time point 0. Even though the quality starts to deteriorate during the first stage, the product stays within acceptable quality bounds, and the change is not immediately apparent.
- Phase 2 (1 to 2): This phase denotes a noticeable deterioration in quality. One or more quality parameters start to show clear changes at point 1. The food quality then significantly deteriorates until it reaches point 2, at which point it is deemed unfit for human consumption.

It is widely thought that the food product maintains sufficient quality for delivery and end-use during Phase 1 (from 0 to 1). Therefore, effective monitoring and prompt action at this point are essential to avoiding supply chain quality failure.

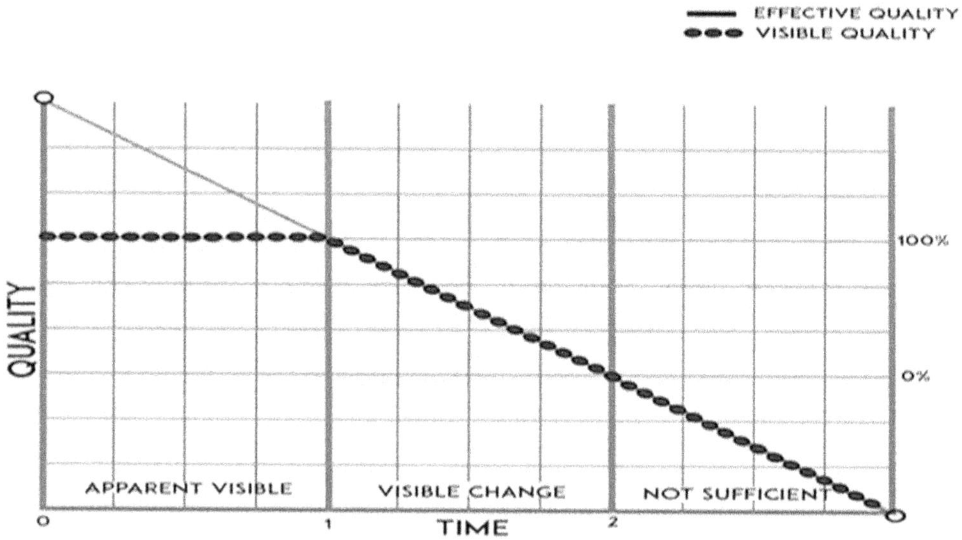

Figure 17.3 Visible food output failure.

17.3.4 IoT-based analytical framework for food supply chain

Even though IoT-generated data have enormous potential value, processing massive amounts of data in real time to derive actionable insights can be extremely difficult for modern analytics systems. Executives can gain knowledge through a variety of tactics, but the volume of data can cause information overload, which reduces effectiveness and increases operational inefficiencies.

A structured diagnostic framework that combines various data flows and creates a logical depiction of operational challenges is necessary to address this. In the context of FSCM, this unified perspective facilitates better decision-making and process optimization.

In order to gain a competitive edge in the logistics management of food supplies, we suggest an improved analytical interface (as illustrated in Figure 17.4). There are three levels at which this framework functions:

1. **Product Classification:** Determining which product categories are appropriate for applications involving advanced data analytics.
2. **Production and Traceability:** Linking IoT data to safety procedures and production processes to enable end-to-end traceability.
3. **Demand Monitoring and Evaluation:** Monitoring activities and gathering customer-focused data via systematic sequencing and ongoing performance assessments.

We use the Bayesian inference model, which is based on the idea of conditional independence, to evaluate system reliability and facilitate wise decision-making. The joint probability, assuming that events X and Y are independent, is as follows:

$$P(X,Y) = P(X)P(Y) \tag{17.1}$$

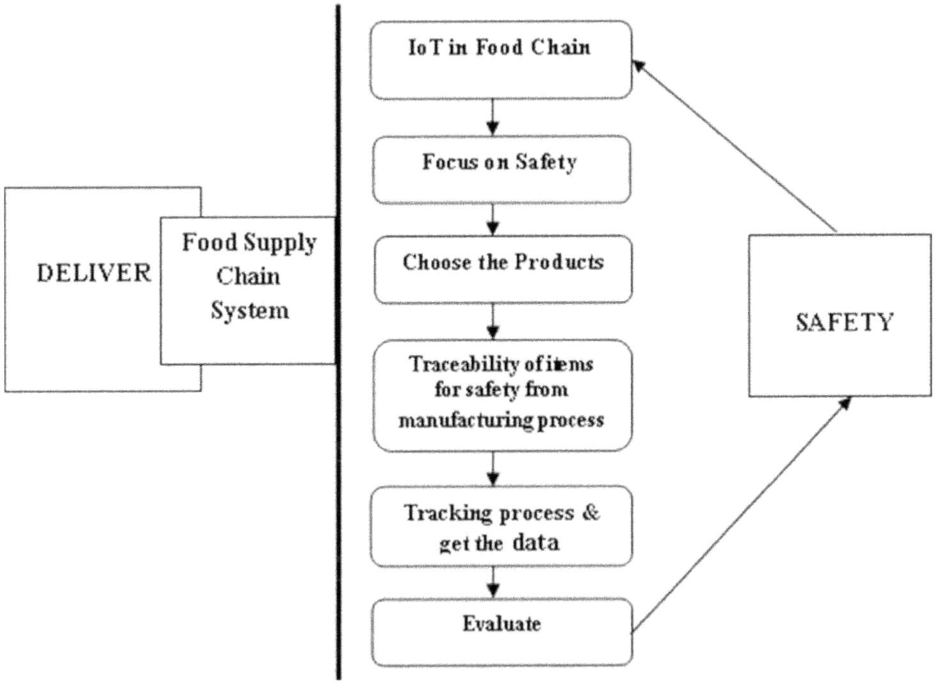

Figure 17.4 Decision-making framework utilizing IoT data.

Thus, we input the outcome:

$$P(y \mid (x_1, x_2, \ldots, x_n)) = P(x_1 \mid y)P(x_2 \mid y)\ldots P(x_n \mid y)P(y) / P(x_1)P(x_2)\ldots P(x_n) \quad (17.2)$$

This can be expressed as

$$P(y \mid (x_1, x_2, \ldots, x_n)) = P(y)\prod_{i=1}^{n} P(x_i \mid y) / P(x_1)P(x_2)\ldots P(x_n) \quad (17.3)$$

We can now eliminate that term since the denominator for a given effort remains constant:

$$P(y \mid (x_1, x_2, \ldots, x_n)) \propto P(y)\prod n \, P(x_i \mid y) \quad (17.4)$$

We must now construct a prediction model for each possible ideal of the class variable y in order to determine the likelihood of a known input location and to seek the yield with the highest probability. That can be expressed scientifically as follows:

$$y = \arg\max_y P(y)\prod nP(x_i \mid y) \quad (17.5)$$

Lastly, the assignment of conniving $P(y)$ and $P(x_i)$, where $P(x_i)$ is also known as the probability of condition and $P(y)$ is also known as the probability of class.

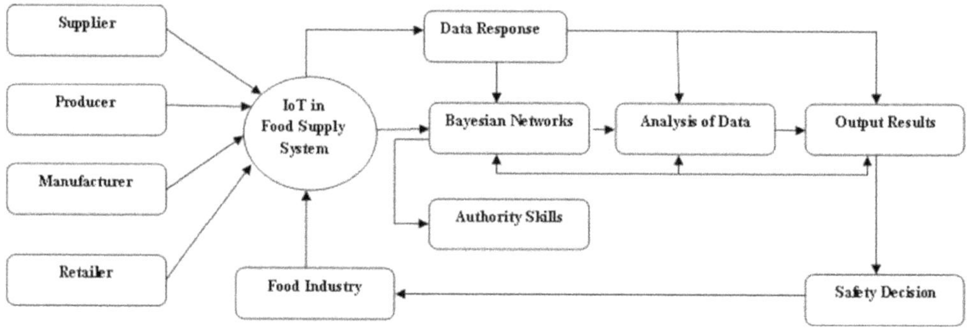

Figure 17.5 IoT-based analytical framework.

The recommendations made by the disparate Bayesian networks about the $P(x_i)$ distribution are significantly different.

Preliminary analyses and calculations on the IoT data gathered from FSCSs are crucial for ensuring food quality using a data-driven approach. These computations aid in matching predetermined quality standards with observed parameters. The following stage entails converting product requirements into methodical operational procedures, which are then further subdivided into particular duties and the resources needed for them. We suggest a hierarchical analytical model based on a C.5 BN to handle this complexity. This model categorizes IoT-derived data into actionable insights and makes it easier to integrate different entity capacities, including sensors, devices, and analytical tools.

The model creates a simplified flow of operations and decision-making steps by tying together various technical tools and skill sets within an optical system workflow. We have modified and developed this hierarchical Bayesian approach to facilitate precise diagnostics and aid in the creation of comprehensive process plans, despite the fact that it has not yet been extensively used in IoT-centric data logic systems. In the end, as Figure 17.5 shows, this framework improves product quality assurance and maximizes performance across the food supply chain.

17.3.5 Reliability analysis of FSCS based on Bayesian network

A food company uses BNs to accurately forecast business demand as part of its efforts to leverage the value of IoT data for predictive evaluation and improved performance throughout the food supply chain. The first step is data collection, which includes gathering sensor-based video, performance metrics, and observations from connected community networks. This highlights how IoT data have a significant impact on consumer preferences and behavior. Through expert analysis and decision-maker input, the organization identifies key business pressure points and integrates historical industry insights with real-time IoT data to help clients choose appropriate product models. Food entities and the related biological and physical properties of the products, represented as nodes $x_1, x_2, \ldots, x_n$ in the BN, are examples of these influential variables.

Following the definition of the attributes, consumer representatives are involved, and each stakeholder provides assessments pertaining to particular queries or factors. After that, test data are created, and the variables are organized using hierarchical grouping techniques or classification algorithms to facilitate distribution and precise forecasting.

The next step is the BN setup, which entails building conditional probability tables as well as structurally modeling the network. Figure 17.6, which displays a section of the Bayesian food attribute network, provides an example of this setup. Market demand estimation is the next step, in which the network functions in both directions, enabling input values to forecast outputs and known results to deduce probable causes. Figure 17.7 shows this dynamic operation. The business determines patterns in customer demand and assesses the effectiveness of different policy choices by examining the observed node values. This makes it possible to develop strategies that have a high chance of satisfying customer expectations and promoting company expansion. Knowledge inquiry, which forms the basis for decision-making, is the last phase. In order to predict significant data relationships in the BN, a crucial metric, i.e., Shared Data (SD), is used, which is a measure of dependency between two random variables.

SD is computed mathematically as follows:

$$D(A,B) = \sum_{a,b} p(A,B)\log\left(\frac{P(A,B)}{P(A)P(B)}\right) \tag{17.6}$$

This measures how much variable A affects variable B; a higher $D(A,B)$ value denotes a stronger influence. As a result, in real-time production and decision-making, variables with the highest influence scores are given priority. The network's ultimate goal is to develop a predictive model that, given the input variables A_1, A_2, …, identifies the most likely outcome class BBB. A_n is described as follows:

Figure 17.6 Bayesian food attribute network.

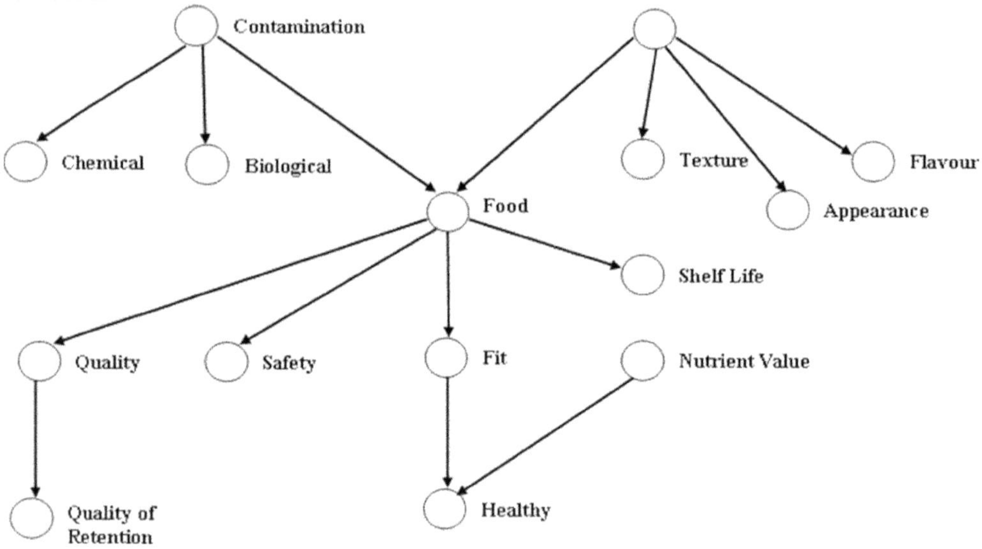

Figure 17.7 Bayesian network operations.

$$B = \arg\max_y P(B)\prod nP(A_i \mid B) \tag{17.7}$$

Additionally, as new data become available, the BN can be adjusted incrementally. This makes it easier to continuously improve the model and gives the company the ability to react proactively to changing customer demands and market dynamics.

17.3.6 Hybrid algorithm design

A common machine learning task is classification, which involves splitting a dataset into predefined groups called classes and assigning each object or instance to one of these groups. Classification tasks are an effective way to reframe many complex decision-making problems. As shown in Figure 17.8, the classification process usually entails preprocessing the training data, building the model, and evaluating it.

The food quality dataset used in this paper is divided into subsets for testing and training. Predictive models are then created by applying classification algorithms to the training dataset, and the test dataset is used to evaluate how well they perform. Decision tree algorithms are widely used among different classification techniques because of their interpretability and simplicity of use. The internal (test) and terminal (leaf) nodes make up the decision tree structure, which starts with a source node (root). Each path in the tree represents a classification rule. The splitting attribute is the feature with the highest weight or information gain. Additionally, because the Bayesian approach is self-adaptive and very successful in high-dimensional data environments, it provides a significant advantage in classification.

The C5.0 BN, a hybrid classification model that combines the probabilistic reasoning of BNs with the strength of decision trees, is also introduced in this

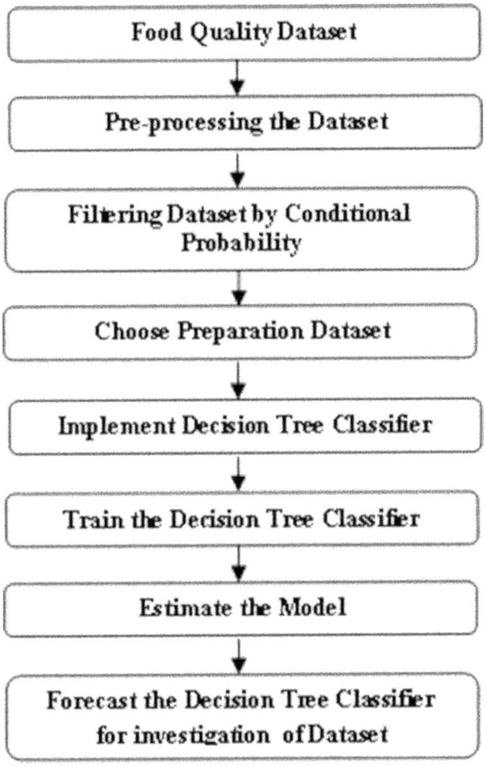

Figure 17.8 Skeleton for classification.

chapter. An extension of the popular C4.5 and C5.0 algorithms, C5.0 BN is better suited to managing big datasets and performs better in terms of execution time and precision–recall metrics. To create an accurate prediction model, it removes noisy or redundant features and only includes pertinent attributes from the training data. In the classification process, the training dataset is used to build the C5.0 decision tree first, and the conditional probability tables and attribute dependencies obtained from the decision paths are then used to structure the BN. This hybrid framework significantly outperforms its predecessors in terms of both prediction accuracy and computational efficiency; in some datasets, it can generate rules up to 360 times faster.

Implementing a top–down approach, the C5.0 BN algorithm guarantees minimal and optimal tree structures. By setting thresholds for numerical splits and pruning out ineffective branches, it accommodates continuous attributes, missing values, and noisy data. The end product is a compact, easily comprehensible, and extremely accurate model for forecasting food quality outcomes.

The acronym "electromagnetic interference (EMI)" should be defined after the first instance of any acronym in the main text, which should be expanded. Unless it contains a word that is typically capitalized, like "fibre Bragg grating (FBG)," the expansion shouldn't be capitalized. All subsequent uses of the acronym can be abbreviated to the defined acronym, such as text.

ALGORITHM TO GENERATE C5.0 BN ALGORITHM

a. Data Partition, DP, a set of preparation tuples and their connected class labels
b. quality_list, the set of applicant qualities
c. quality_choosing_procedure, a method to decide the dividing principle to partition the data tuples into entity classes. This measure consists of a dividing principle and either a divide point or a dividing subset.

Output:
C5.0 Bayesian Networks
Method:

1. Produce a node N1
2. if tuples in DP are all of the similar class, Cl, then
3. return N as a leaf node categorized with the class Cl
4. if quality_list is empty, then
5. return N1 as a leaf node labeled with the mainstream class in DP
6. apply quality_choosing_procedure (DP, quality_list) to discover the most excellent dividing_principle
7. label node N with dividing_measure
8. if dividing_principle is discrete-valued and more number of way divides approved then
9. quality_list quality_list-dividing_principle
10. For each result j of dividing_principle
    Let DPj be the set of data tuples in DP gratifying result j node N1. If DPi is empty, then connect a leaf considered with mainstream class in DP to else or connect the node revisited by produce C5.0 decision tree (DPj, quality_list) to node N1
11. Return N1

Figure 17.9 (seen below) represents the flow map for the proposed algorithm.

The handling of attribute values is emphasized by the C5.0BN algorithm, particularly when some attributes are absent. Notably, information gain and entropy calculations do not include missing values, which helps avoid decision-making distortion in the model. In order to fully capture and categorize the structure of the training data, the algorithm builds decision trees by recursively generating branches. Even though it excels at execution time, classification accuracy, and precision–recall, it frequently overfits the training set, particularly when used on noisy or complex datasets. When a model is overfitted to training data, it captures noise instead of underlying patterns and performs poorly on unseen data. Decision tree learning frequently uses a variety of pruning and regularization techniques to reduce overfitting, simplify the model, and enhance generalizability.

The splitting criterion, which establishes how nodes in the tree are split to form branches, is an essential part of decision tree learning. These standards are essential for maximizing the tree's efficacy and structure. Univariate splitting, in which a node is divided according to the value of a single attribute, is the most widely used method. Finding the characteristic that best separates the data based on the selected impurity measure—such as information gain, gain ratio, or Gini index—is the goal.

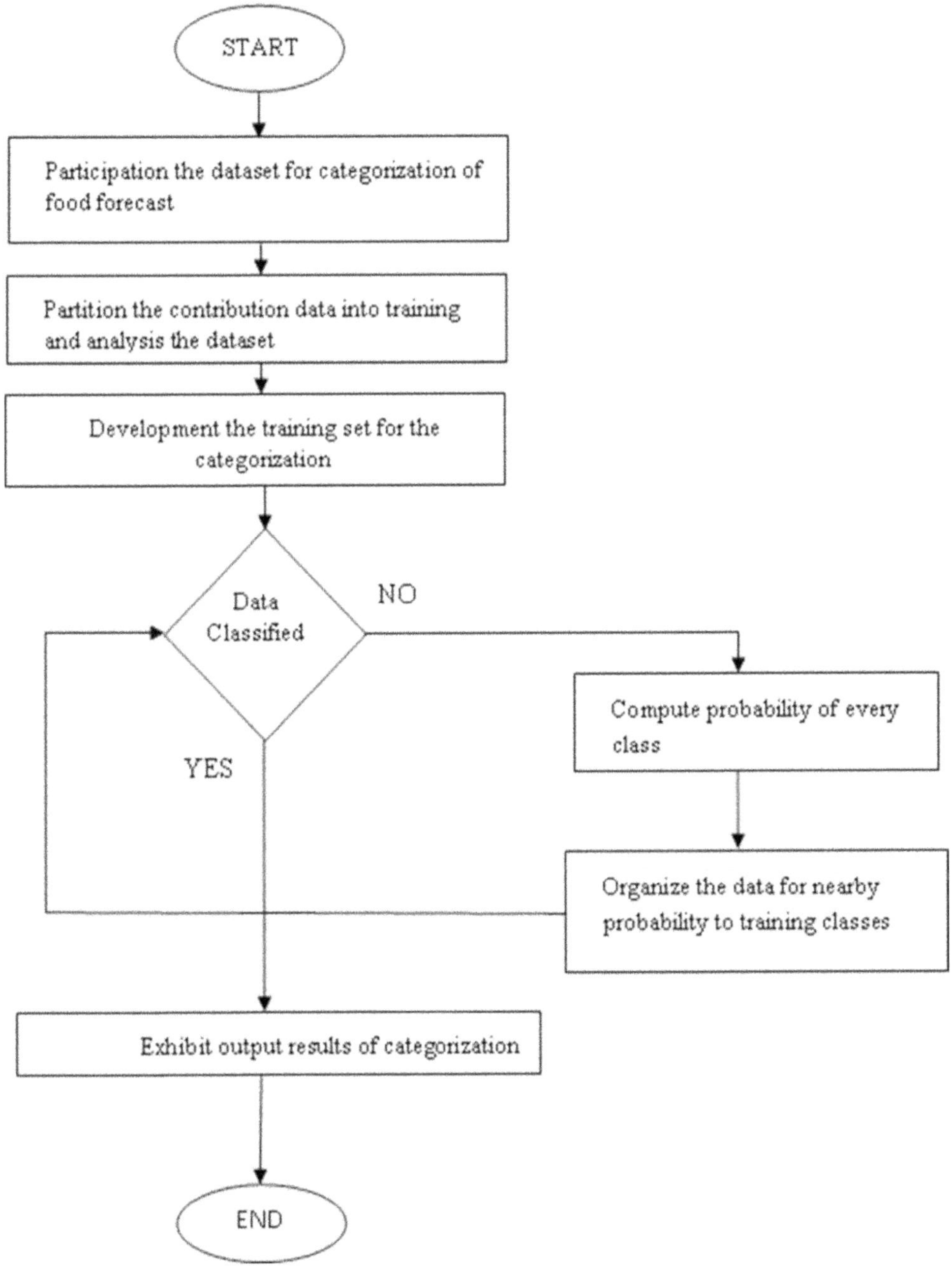

Figure 17.9 Hybrid algorithm process.

These criteria assess a node's "purity" or "contamination"—the degree of mixing between the class labels prior to and following the split. By minimizing node impurity, effective splitting makes sure that the resulting subsets are as homogeneous as possible. The depth, balance, and predictive power of the tree are all significantly impacted by the splitting strategy selected. As a result, choosing the right splitting criterion is crucial to creating precise and effective decision tree models in the C5.0BN framework.

i. **Entropy Analysis:** It is used to determine a node's degradation. It is specified as follows: (for values of a class)

$$\text{Entropy}(t) = -\Sigma\ p(i/t)\log_2 p(i/t) \tag{17.8}$$

ii. **Gini Index:** The calculation of the difference between the probability distributions of the values of the objective attribute is different from that of impurity and is defined as follows:

$$\text{Gini Index} = 1 - \Sigma\left[p(i/t)\right]^2 \tag{17.9}$$

iii. **Classification Error:** It is computed as follows:

$$\text{Classification error}(t) = 1 - \max\left[p(i/t)\right] \tag{17.10}$$

Where $p(i/t)$ at a specified node t denotes the fraction of records belonging to class i.

iv. **Information Gain:** Information gain is an impurity-based variable that quantifies impurity using entropy calculations. It is the distinction between the entropies of parent and child nodes.

$$\text{Info Gain} = \text{Entropy}(\text{parent node}) - \text{entropy}(\text{child node}) \tag{17.11}$$

v. **Benefit Ratio:** The benefit ratio normalizes the advantage of information as follows:

$$\text{Advantage Ratio} = \text{Information Gain} / \text{Entropy} \tag{17.12}$$

vi. **Twoing Criteria:** When the target attribute domain is rather broad, the Gini Index may encounter issues. Twoing criteria, a differential requirement, may be used in this situation. This requirement is defined as follows:

$$\text{Twoing Criteria}(t) = \text{PLPR}\left(\Sigma\left(\left|p(i/tL) - p(i/tR)\right|\right)\right) \tag{17.13}$$

As seen in Figure 17.10: Food Safety Monitoring Process, the suggested system uses information derived from sensor data to produce precise results. The model makes the assumption that the container is directly connected to the central processing unit (CPU), with sensors actively monitoring vital parameters like temperature and humidity, regardless of the degree of automation attained in the food supply chain information system.

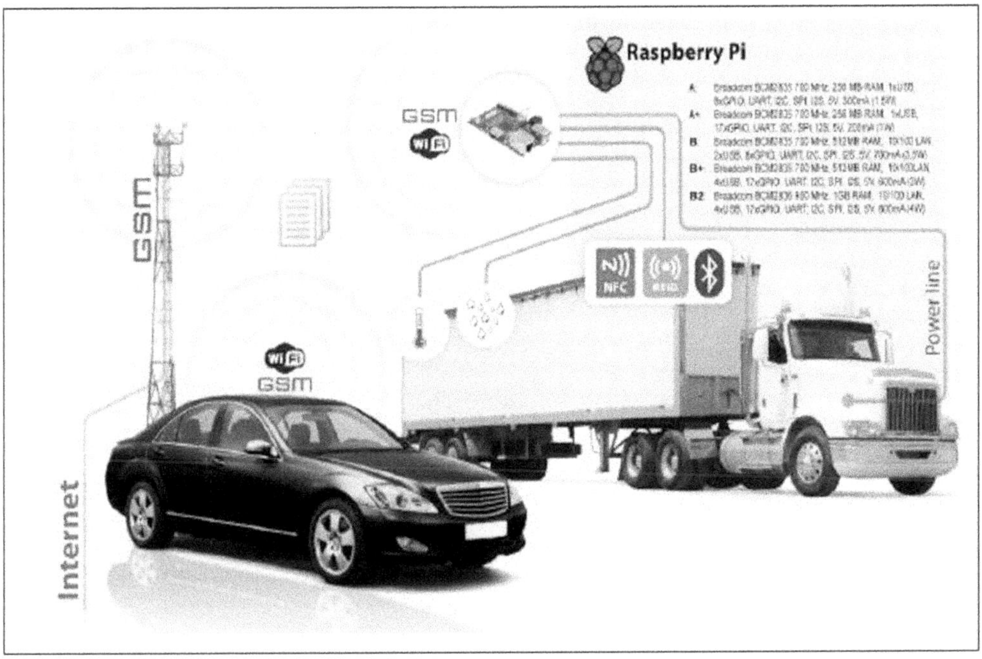

Figure 17.10 Food safety monitoring process.

The creation of auxiliary sensors combined with specialized instruments to improve system responsiveness is a major contribution of this research. With features like Bluetooth-enabled RFID devices and barcode reader applications, the entire system is made to be easy to use. For more extensive communication capabilities, it also supports GSM connectivity and web-based services. Every time a car or container moves, the sensors are essential for monitoring how the RFID tags interact; they read the data to guarantee consistency and traceability.

After the data are gathered, they are analyzed and categorized by the monitoring system using key sensor parameters and predetermined constraints. The system evaluates and updates food traceability data in real time using an IoT-based prediction model, allowing for an instant determination of a product's safety or flaws. The suggested framework functions autonomously and integrates easily into a smart freight management infrastructure that is backed by sensor networks. It is powered by a self-sustaining energy system. This guarantees better safety, real-time monitoring, and increased transparency throughout the food supply chain.

17.4 EVALUATION AND COMPARATIVE ANALYSIS

The experimental procedures and performance assessment of a food supply chain classification system utilizing IoT-generated data about food safety and nutrition are described in this section. The dataset includes seven attributes: food_and_serving, calories, total_fat, sodium_g, potassium_g, total_carbohydrate_g, and food_type. It comprises 61 instances, divided into 45 training and 16 testing samples. Figure 17.11 shows the

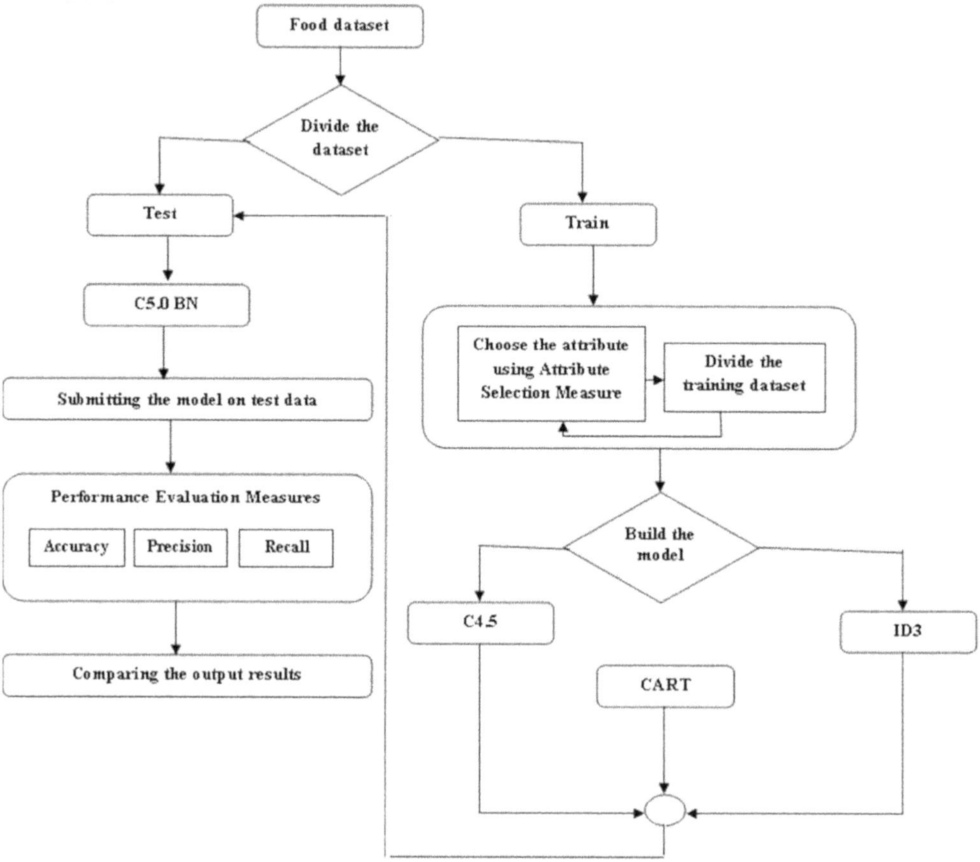

Figure 17.11 Steps of work.

process of developing the system and evaluating its quality. Accuracy is the first evaluation criterion.

The suggested C5.0BN algorithm, as shown in Table 17.2 and Figure 17.12, performs noticeably better than conventional decision tree classifiers such as ID3, C4.5, and even C5.0. It reaches the highest accuracy in all experiments, peaking at 94.2%, whereas ID3, C4.5, and C5.0 show highest accuracy at 82.1%, 86.0%, and 91.8%, respectively.

Table 17.3 and Figure 17.13 demonstrate that C5.0BN uses less memory than its counterparts. In the seventh experiment, for example, C5.0BN used 39,658 KB, which is less than those of ID3 (41,000 KB), C4.5 (38,800 KB), and C5.0 (38,010 KB). But when it comes to training time, as Table 17.4 and Figure 17.14 demonstrate, C5.0BN takes slightly more time compared to other algorithms because of its complex structure and use of probabilistic reasoning. On average, it requires approximately 8.9 milliseconds to complete its processing. Nevertheless, the overhead is justified by its improved learning capacity.

In contrast, Figures 17.13 and 17.14 show the relationship between power consumption and training time for the different decision tree algorithms, while Table 17.5 and Figure 17.15 show that C5.0BN significantly speeds up search or prediction times. The algorithm outperforms ID3, C4.5, and C5.0 and consistently records lower search times

Table 17.2 Comparison of accuracy measures

No. of experiments	ID3	C4.5	C5.0	C5.0 BNA
1	71.2%	84.5%	80.1%	85.3%
2	75.6%	80.2%	84.8%	88.7%
3	73.8%	78.3%	82.1%	93.1%
4	76.4%	80.1%	83.2%	88.6%
5	80.5%	84.6%	89.1%	93.7%
6	82.1%	86.0%	91.8%	94.2%
7	76.2%	79.8%	89.0%	92.7%

Table 17.3 Comparison of memory consumption

No. of experiments	ID3 (KB)	C4.5 (KB)	C5.0 (KB)	C5.0 BNA (KB)
1	31,500	30,750	29,980	28,430
2	35,150	34,700	34,510	29,540
3	37,020	35,100	37,530	31,110
4	39,600	38,320	37,600	35,820
5	39,600	39,140	38,600	37,240
6	33,620	32,810	31,710	32,400
7	41,000	38,800	38,010	39,510

Table 17.4 Comparison of Training Time

No. of experiments	ID3 (ms)	C4.5 (ms)	C5.0 (ms)	C5.0 BNA (ms)
1	3.2	4.3	5.7	6.7
2	7.0	8.4	7.2	8.0
3	5.1	6.0	7.0	8.1
4	4.2	6.2	9.2	11.6
5	4.1	5.0	6.1	7.8
6	3.9	6.3	8.0	9.4
7	4.6	6.6	9.0	10.9

Table 17.5 Comparison of search time

No. of experiments	ID3 (ms)	C4.5 (ms)	C5.0 (ms)	C5.0 BNA (ms)
1	5.3	4.1	3.3	2.1
2	7.2	6.1	5.0	3.2
3	6.5	5.0	3.0	2.1
4	7.7	6.6	4.6	3.1
5	7.7	6.1	5.3	4.4
6	6.0	6.7	4.0	3.1
7	8.0	6.3	5.0	4.4

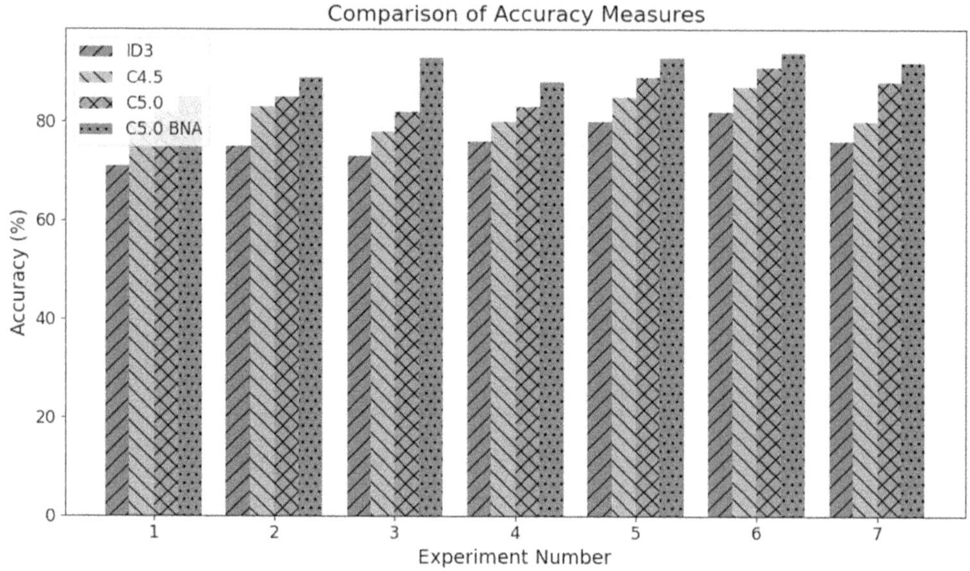

Figure 17.12 Comparison of accuracy measure.

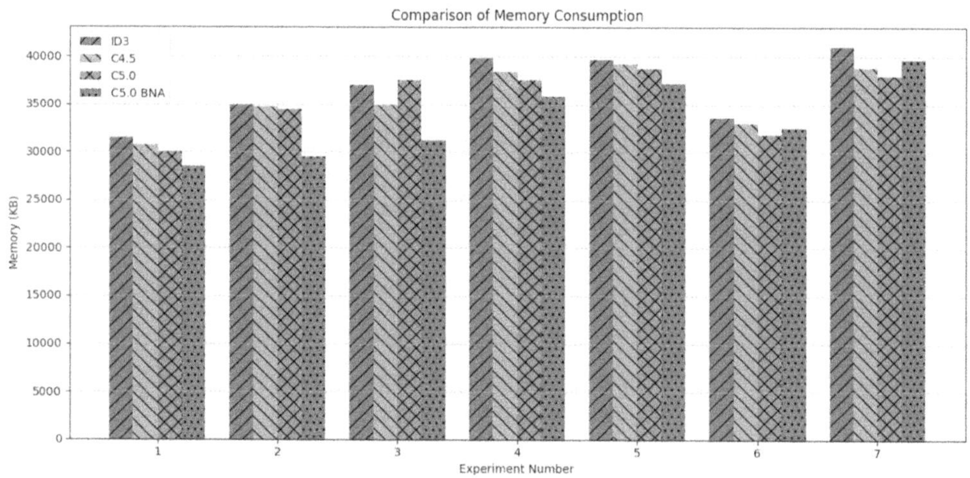

Figure 17.13 Comparison of memory consumption.

(as low as 2.1 ms) across all experiments, making it perfect for real-time applications. C5.0BN also has a significantly lower error percentage, dropping to as low as 5%, while ID3 displays error rates as high as 28.8%, according to the error rate comparison displayed in Table 17.6 and Figure 17.16. The robustness and effectiveness of C5.0BN are further confirmed by this decrease in misclassification.

As seen in Figure 17.17, the suggested methodology also makes a substantial contribution to the quality control of the food supply chain. It facilitates precise forecasts of market

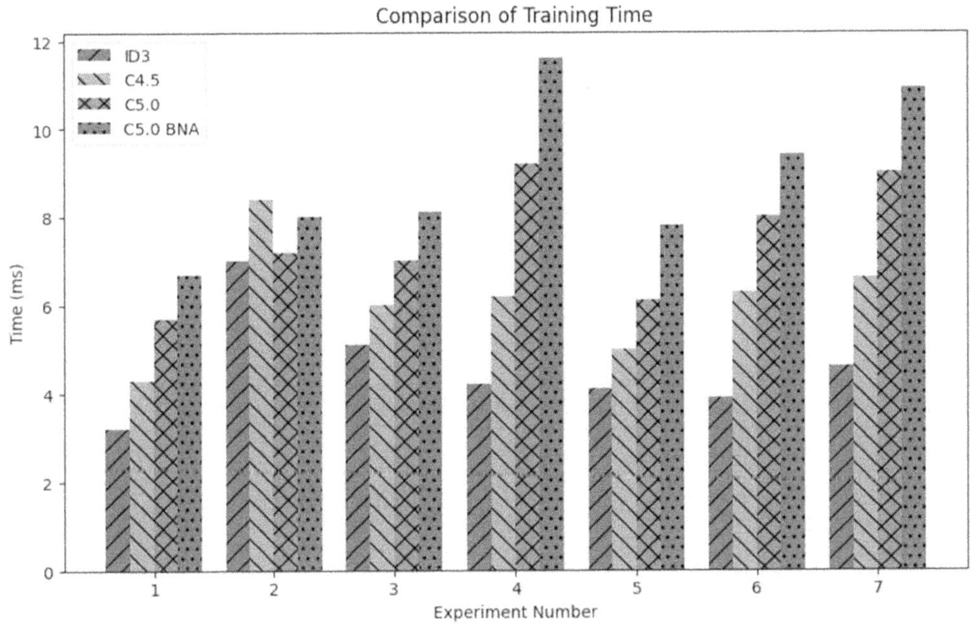

Figure 17.14 Comparison of training time.

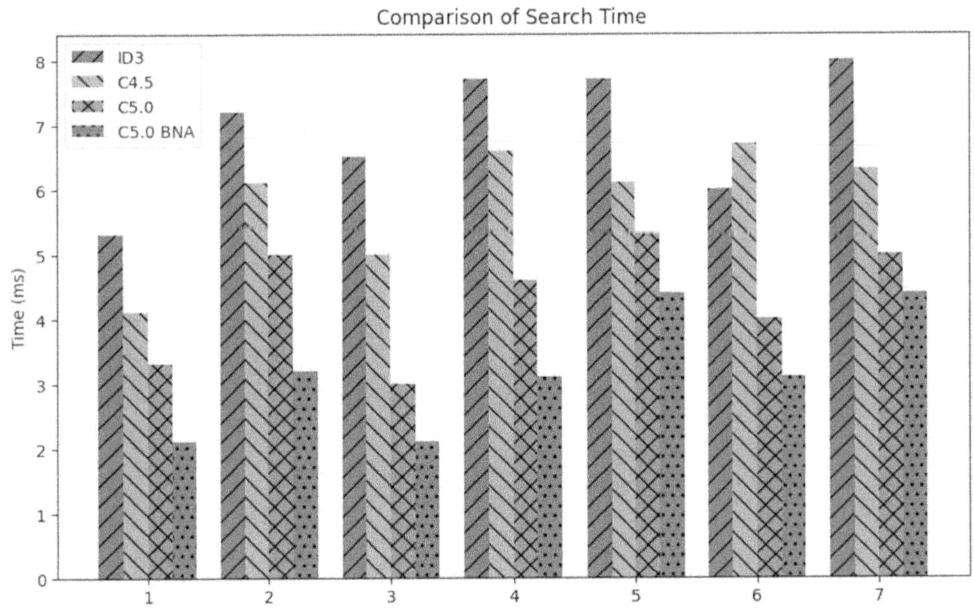

Figure 17.15 Comparison of search time.

Table 17.6 Comparison of error rate

No. of experiments	*ID3 (%)*	*C4.5 (%)*	*C5.0 (%)*	*C5.0 BNA (%)*
1	28.8	22.5	18.7	14.7
2	20.2	16.8	14.2	10.5
3	18.5	17.6	17.0	15.9
4	21.7	18.9	15.0	10.3
5	20.9	16.5	10.2	5.0
6	15.6	12.9	8.2	4.6
7	22.5	17.6	11.4	5.9

Table 17.7 Overall comparison of the performance of decision tree algorithms

Sl.no	*Parameters*	*C5.0 BN*	*C5.0*	*C4.5*	*ID3*
1	Accuracy (%)	90.90	85.73	81.93	76.54
2	Error rate (%)	9.56	13.53	17.54	21.17
3	Memory consumption (KB)	33,435.71	35,420.00	35,660.00	36,784.29
4	Training time (ms)	8.93	7.46	6.11	4.59
5	Search time (ms)	3.20	4.31	5.84	6.91

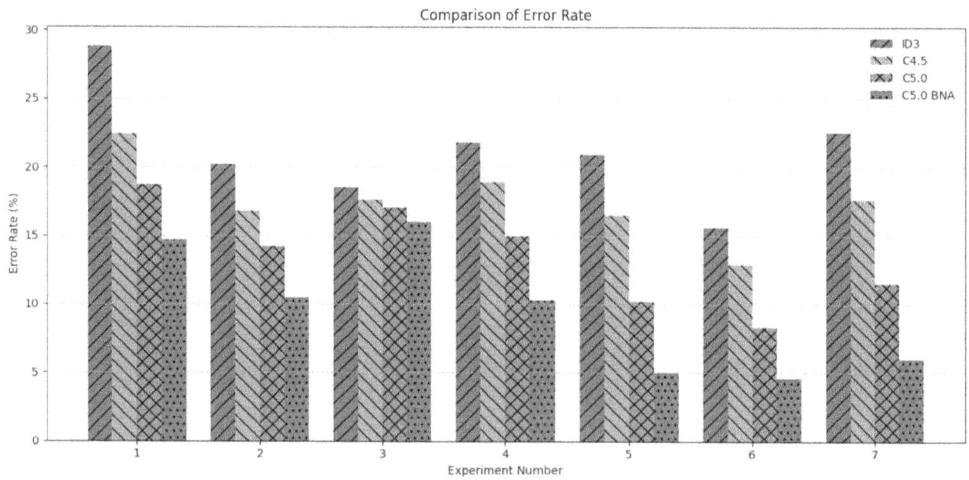

Figure 17.16 Comparison of error rate.

demand and operational choices by utilizing both historical data and real-time sensor inputs. The food supply system's sustainability and classification performance, as shown in Figure 17.18, unequivocally show that C5.0 BNA performs more effectively than all assessed decision tree algorithms. With the highest accuracy of 90.90%, it outperforms ID3 (76.54%) by 5.17%, C4.5 (81.93%) by 8.97%, and C5.0 (85.73%) by 5.17%. C5.0 BNA has the highest accuracy and the lowest error rate (9.56%), proving its dependability and resilience for applications involving real-time FSCM and decision-making.

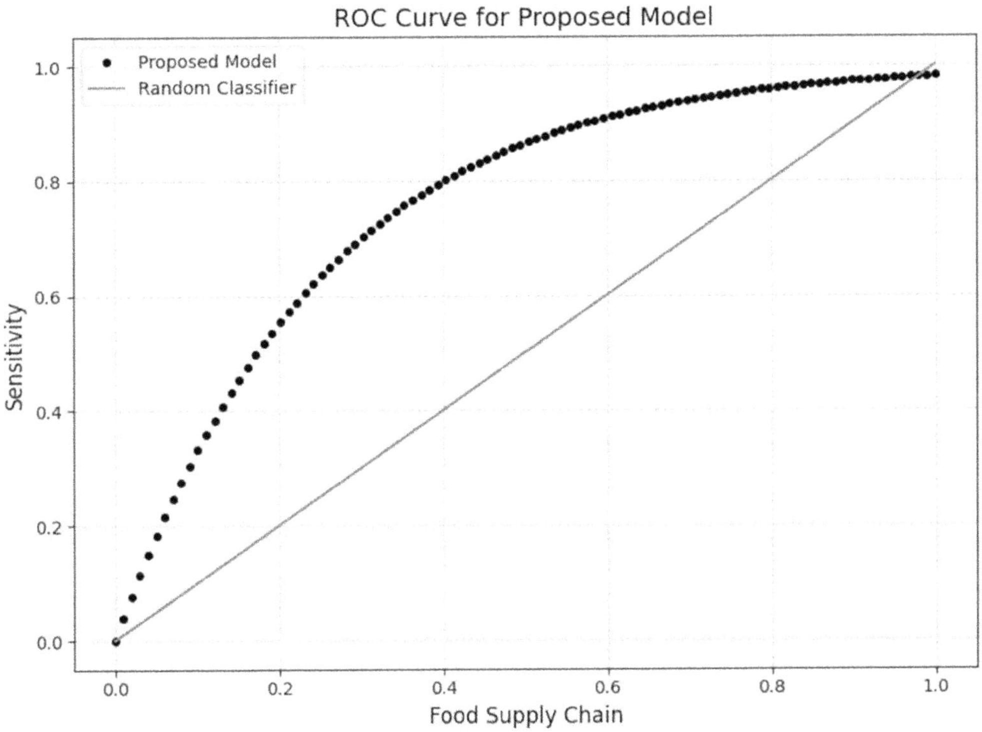

Figure 17.17 Quality of the food supply chain management.

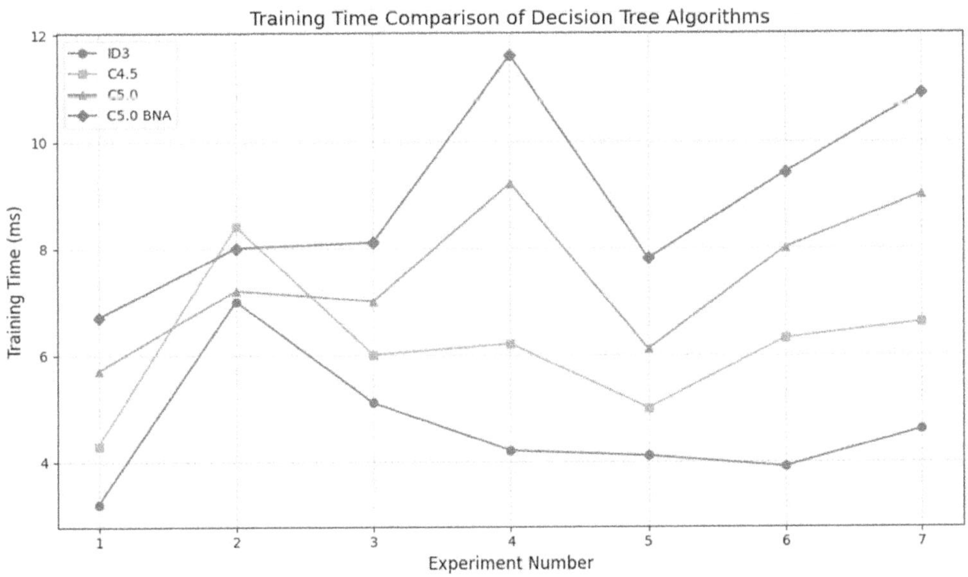

Figure 17.18 Training time comparison of decision tree algorithms.

Table 17.7, which summarizes the overall performance comparison, shows that C5.0BN, despite having a slightly longer training time, registers the highest accuracy (90.90%), lowest error rate (9.56%), and efficient memory use (33,435.71 KB). These thorough metrics demonstrate that C5.0BN is the most accurate and dependable classification algorithm among the models that have been tested, making it a solid contender for real-time food quality monitoring and traceability in supply chain systems that are IoT-enabled.

17.5 CONCLUSION

In this paper, we propose C5.0BN, a novel hybrid decision algorithm for FSCM. The main concept is to improve the classification accuracy by combining the C5.0 decision tree algorithm with a BN framework. Conventional decision algorithms frequently make fixed assumptions at every stage, which can result in decreased accuracy and performance.

The goal of the suggested C5.0BN algorithm is to guarantee the best possible food quality for consumers and lower the health risks related to food contamination, which can have detrimental effects. This food traceability system promotes high standards of food safety by efficiently monitoring and assessing the quality of food products in real time throughout the entire supply chain by utilizing the IoT.

Important metrics like Memory Consumption, Training Time, Search Time, Error Rate, and Overall Performance Measures are used to assess the algorithm's performance. Additionally, the C5.0BN algorithm's accuracy is contrasted with that of conventional classifiers such as ID3, C4.5, and C5.0. Experimental findings unequivocally show that C5.0BN performs better than the current models, especially in IoT-based applications, providing noticeably higher accuracy and dependability.

17.6 FUTURE ENHANCEMENT

To further validate and enhance the accuracy of the suggested decision tree classifier, future research might investigate applying it to more datasets. Additional improvements could involve thorough performance analysis on both training and test sets using evaluation metrics like the F-measure, True Positive (TP) Rate, ROC Curve, Precision, and Kappa Statistic. A deeper understanding of the model's generalizability and resilience in diverse food supply chain scenarios would result from this.

REFERENCES

1. S. Balamurugan, A. Ayyasamy, and K. Suresh Joseph, "IoT-Blockchain driven traceability techniques for improved safety measures in food supply chain," *International Journal of Information Technology*, vol. 13, no. 2, pp. 567–573, 2021, https://doi.org/10.1007/s41870-020-00581
2. S. Balamurugan, A. Ayyasamy, and K. Suresh Joseph, "Improvement of C5.0 algorithm using Internet of Things with Bayesian principles for food traceability systems," *Modern Supply Chain Research and Applications*, vol. 2, no. 3, pp. 154–171, 2020.

3. S. Balamurugan, A. Ayyasamy, and K. Suresh Joseph, "Enhanced Petri Nets for traceability of food management using Internet of Things," *Peer-to-Peer Networking and Applications*, vol. 13, no. 6, pp. 2053–2065, 2020, https://doi.org/10.1007/s12083-020-00921-4.
4. S. Balamurugan, A. Ayyasamy, and K. Suresh Joseph, "IoT-based supply chain traceability using enhanced Naïve Bayes approach for scheming the food safety issues," *International Journal of Scientific & Technology Research*, vol. 9, no. 3, pp. 1184–1192, 2020.
5. S. Balamurugan, A. Ayyasamy, and K. Suresh Joseph, "An efficient Bayes classifiers algorithm for traceability of food supply chain management using Internet of Things," *International Journal of Engineering and Advanced Technology*, vol. 9, no. 1, pp. 2995–3005, 2019.
6. S. Balamurugan, A. Ayyasamy, and K. Suresh Joseph, "Optimization of C5.0 classifier with Bayesian theory for food traceability management using Internet of Things," *International Journal of Smart Sensor Technologies and Applications (IJSSTA)*, vol. 1, no. 1, pp. 1–21, 2020. [Online]. Available: https://doi.org/10.4018/IJSSTA.2020010101
7. S. Balamurugan, A. Ayyasamy, and K. Suresh Joseph, "A review on privacy and security challenges in the Internet of Things (IoT) to protect the device and communication networks," *International Journal of Computer Science and Information Security*, vol. 16, no. 6, pp. 57–62, 2018.
8. H. Scholten and C. N. Verdouw, "Defining and analyzing traceability systems in food supply chains," in *Advances in Food Traceability Techniques and Technologies,* M. Espiñeira and F. J. Santaclara, Eds. Elsevier, New York, NY, 2016, pp. 9–33.
9. J. Xiao, Z. Y. Liu, and B. W. Li, "Research on a food supply chain traceability management system based on RFID," *Journal of Agricultural Mechanization Research*, vol. 34, no. 2, pp. 181–184, 2012.
10. J. M. Ryan, *Guide to Food Safety and Quality during Transportation: Controls, Standards and Practices*, Elsevier, 2014, ISBN: 978-0-12-405173-4.
11. S. M. R. Islam, D. Kwak, M. H. Kabir, M. Hossain, and K. S. Kwak, "The Internet of Things for health care: A comprehensive survey," *IEEE Access*, vol. 3, pp. 678–708, 2015.
12. I. E. Etim and J. Lota, "Power control in cognitive radios, Internet-of-Things (IoT) for factories and industrial automation," in *Proc. IEEE Ind. Electron. Soc. Conf.*, 2016, pp. 4701–4705. https://doi.org/10.1109/IECON.2016.7793500
13. S. Feng, P. Setoodeh, and S. Haykin, "Smart home: Cognitive interactive people-centric Internet of Things," *IEEE Communications Magazine*, vol. 55, no. 2, pp. 34–39, Feb. 2017.
14. A. Miles, A. Zaslavsky, and C. Browne, "IoT-based decision support system for monitoring and mitigating atmospheric pollution in smart cities," *Journal of Decision Systems*, 2018. [Online]. Available: https://doi.org/10.1080/12460125.2018.1468696
15. V. A. M. Nardi, D. P. Auler, and R. Teixeira, "Food safety in global supply chains: A literature review," *Journal of Food Science*, vol. 85, no. 4, pp. 883–891, Apr. 2020, https://doi.org/10.1111/1750-3841.14999.
16. N. Mai, S. G. Bogason, S. Arason, S. V. Arnason, and T. G. Matthiasson, "Benefits of traceability in fish supply chains—case studies," *British Food Journal*, vol. 112, no. 9, pp. 976–1002, 2010.
17. N. Patil, R. Lathi, and V. Chitre, "Comparison of C5.0 & CART classification algorithms using pruning technique," *International Journal of Engineering Research & Technology (IJERT)*, vol. 1, no. 4, pp. 1–5, Jun. 2012.
18. D. Kaur, R. Bedi, and K. S. Gupta, "Review of decision tree data mining algorithms: ID3 and C4.5," in *Proc. Int. Conf. Information Technology and Computer Science*, Bali, Indonesia, Jul. 2015.
19. P. Revathi, R. Revathi, and M. Hemalatha, "Comparative study of knowledge in crop diseases using machine learning techniques," *International Journal of Computer Science and Information Technology*, vol. 2, no. 5, pp. 2180–2182, 2011.

20. D. Zhang and J. Guo, "The development and standardization of testing methods for genetically modified organisms and their derived product," *Journal of Integrative Plant Biology*, vol. 53, no. 7, pp. 539–551, 2011.
21. S. Allata, A. Valero, and L. Benhadja, "Implementation of traceability and food safety systems (HACCP) under the ISO 22000:2005 standard in North Africa: The case study of an ice cream company in Algeria," *Food Control*, vol. 79, pp. 239–253, Sep. 2017.
22. Z. Sheng, S. Yang, Y. Yu, A. Vasilakos, J. McCann, and K. Leung, "A survey on the hybrid protocol suite for the Internet of Things: Standards, challenges, and opportunities," *IEEE Wireless Communications*, vol. 20, no. 6, pp. 91–98, 2013.
23. G. Fenu and P. Garau, "RFID-based supply chain traceability system," in *Proc. 35th Annu. Conf. IEEE*, 2009, pp. 2672–2677.
24. M. Trebar, A. Grah, A. A. Melcon, and A. Parreno, "Towards RFID traceability systems of farmed fish supply chain," in *Proc. 19th Int. Conf. Software, Telecommunications and Computer Networks (SoftCOM'11)*, Hvar, Croatia, 2011, pp. 6–11.
25. T. T. L. Tran, L. Peng, Y. Diao, A. McGregor, and A. Liu, "CLARO: Modeling and processing uncertain data streams," *VLDB Journal*, vol. 21, no. 5, pp. 651–676, 2012.
26. L. F. Rahman, L. Alam, M. Marufuzzaman, and U. R. Sumaila, "Traceability of sustainability and safety in fishery supply chain management systems using radio frequency identification technology," *Foods*, vol. 10, no. 10, p. 2265, Oct. 2021, https://doi.org/10.3390/foods10102265..
27. A. Holst-Jensen, S. Bertheau, and J. De Loose, "Testing for genetically modified organisms (GMOs): Past, present and future perspectives," *Biotechnology Advances*, vol. 27, no. 6, pp. 1071–1082, Nov.–Dec. 2009, https://doi.org/10.1016/j.biotechadv.2009.05.025.
28. "Food Safety Management System," *Amber Valley Borough Council*, 2014. [Online]. Available: https://www.ambervalley.gov.uk/health-and-social-care/food-safety/food-safety-management-system.aspx
29. D. Munera, D. P. Tobon V., J. Aguirre, and N. Gaviria Gomez, "IoT-based air quality monitoring systems for smart cities: A systematic mapping study," *International Journal of Electrical and Computer Engineering*, vol. 11, no. 4, pp. 3470–3482, 2021, https://doi.org/10.11591/ijece.v11i4.pp3470-3482.
30. T. Kelepouris, K. Pramatari, and G. Doukidis, "RFID-enabled traceability in the food supply chain," *Industrial Management & Data Systems*, vol. 107, no. 2, pp. 183–200, 2007.
31. A. S. Galathiya, A. P. Ganatra, and C. K. Bhensdadia, "Improved validation, model complexity and reduced error pruning," *International Journal of Computer Science and Information Technology.*, vol. 3, no. 2, 2012.
32. B. A. Olshausen, "Bayesian probability theory," *Lecture Notes*, University of California, Berkeley, Berkeley, CA, USA, Mar. 2004. [Online]. Available: http://redwood.berkeley.edu/bruno/npb261b/bayes.pdf.
33. D. Patil, J. A. Gokhale, and V. M. Wadhai, "Evaluation of decision tree pruning algorithms for complexity and classification accuracy," *International Journal of Computer Applications*, vol. 11, no. 2, Dec. 2010.
34. F. Esposito, D. Malerba, and G. Semeraro, "A comparative analysis of methods for pruning decision trees," *IEEE Transactions on Pattern Analysis and Machine Intelligence*, vol. 19, no. 5, pp. 476–491, 1997.
35. H. Blockeel and J. Struyf, "Efficient algorithms for decision tree cross-validation," Department of Computer Science, Katholieke Universiteit Leuven, Belgium.
36. J. Mingers, "An empirical comparison of pruning methods for decision tree induction," *Machine Learning*, vol. 4, pp. 227–243, 1989.
37. J. R. Quinlan, "Induction of decision trees," *Centre for Advanced Computing Sciences*, New South Wales Institute of Technology, Sydney, Australia, 2007.

38. R. B. Patel and K. K. Rana, "A survey on decision tree algorithm for classification," *International Journal of Engineering Development and Research (IJEDR)*, vol. 2, no. 1, pp. 1–5, 2014.
39. L. S. Pundir and Amrita, "Feature selection using random forest in intrusion detection system," *International Journal of Advances in Engineering & Technology (IJAET)*, vol. 6, no. 6, pp. 1319–1324, Jul. 2013.
40. S. Acid, L. M. de Campos, J. M. Fernández-Luna, S. Rodríguez, J. M. Rodríguez, and J. L. Salcedo, "A comparison of learning algorithms for Bayesian networks: A case study based on data from an emergency medical service," *Artificial Intelligence in Medicine*, vol. 30, pp. 215–232, 2004.
41. S. Singh and P. Gupta, "Comparative study of ID3, CART and C4.5 decision tree algorithm: A survey," *International Journal of Advanced Information Science and Technology (IJAIST)*, vol. 27, no. 27, pp. 97–103, Jul. 2014.
42. B. Hssina, A. Merbouha, H. Ezzikouri, and M. Erritali, "A comparative study of decision tree ID3 and C4.5," *International Journal of Advanced Computer Science and Applications (IJACSA)*, Special Issue on Advances in Vehicular Ad Hoc Networking and Applications, vol. 4, no. 2, pp. 13–19, 2014.
43. C. J. Devi, "Binary decision tree classification based on C4.5 and KNN algorithm for banking application," *International Journal of Computational Intelligence and Informatics*, vol. 4, no. 2, pp. 125–131, Sep. 2014.

Chapter 18

Smart surveillance systems

Enhancing vision security via blockchain and AI fusion

Sonam Gupta, Pradeep Gupta, Malik Shahzad Ahmed Iqbal, Ajimuse Muyiwa Sunday, and Vikash Yadav

18.1 INTRODUCTION

The intensive construction of surveillance systems in the urban setting characterized by low lighting and high population has created a promise of tremendous possibilities for enhancing the safety of the citizens and has also presented a list of grave concerns of privacy, information security and moral management. With the advancement of the technology, one is witnessing the increasing intensity of security demands that could not be satisfied by the old-fashioned surveillance technology, where data storage and management affairs are not transparent. This is a paradigm shift, regardless of how revolutionarily it is linked to the integration of artificial intelligence (AI) and blockchain technology in the field of smart surveillance systems that contain a combination of visual security data that are obtained, captured, saved, processed and used [1].

18.1.1 The evolution of surveillance technologies

Surveillance landscape has changed radically in last decades. The current sophisticated networks of intelligent cameras have excellent capabilities and extreme complexity; they range from simple closed-circuit television (CCTV) systems which were dependent on constant human monitoring. By its initial introduction, networked early digital surveillance systems introduced networked capabilities and rudimentary video analytics but were nonetheless still primarily limited through centralized architectures and basic analytical capabilities [2]. In the last few years, the integration of AI technologies has enabled these systems to perform much better in detecting, classifying and responding to the events of interest without having to be supervised all the time by humans.

All of this has happened with this technological evolution, while there has been growing societal concerns over privacy, whether governments are subjecting themselves to surveillance overreach or over data security. The growth of surveillance infrastructure exceeds the growth of governance cameras, as Jain and Chandrasekaran wrote 'governance frameworks have not made up to the pace of proliferation of technical capabilities, with significant gaps emerging between capabilities, and the ethical infrastructure surrounding those capabilities.' [3] As face detection and behavior analysis systems have advanced, this tension has become even more acute as automated security systems become more sophisticated and call into question consent, privacy and potential bias with automated security systems.

DOI: 10.1201/9781003659112-18

18.1.2 Challenges in contemporary surveillance systems

Several critical problems exist with contemporary surveillance systems, and they prevent their effectiveness and even acceptability:

Centralized Storage Architectures: This gives rise to several data security vulnerabilities and the potential for a single point of failure to affect an entire surveillance network [4]. The nature of surveillance data systems is sensitive, and hence such systems are highly sought after targets by both external attacks and insider threats.

Such systems will be undermined by stating all or no approaches to these systems and would bring valid apprehensions of mission creep in surveillance application.

Opacity and Accountability Gaps: The majority of the current systems are black-box systems, in that it is not clear how the footage is handled, who is allowed to view it and what is analyzed, but not all. Consequently, it increases the issues of accountability and complicates the oversight.

Scalability Limitations: To scale surveillance networks to thousands of cameras or sensors, however, traditional architecture is limited by excessive computational requirements to support real-time processing and video capture requirements to support gigapixel extension [5].

Traditional Surveillance System: It is among the most basic elements of the lack of trust of such systems that the conventional systems are highly dependent on the organizational trust, as opposed to technical assurances. Considering the occupations of the observers in the study by Wang et al., security systems are prone to insecurity, which causes more issues than it eliminates [6].

18.1.3 The AI–blockchain convergence

The underlying weaknesses of traditional surveillance methods are compensated by the strengths of AI and blockchain to tackle them. The benefits of AI components are advanced object detection, behavior analysis and anomaly detection, data integrity, transparent governance and decentralized storage solutions; blockchain frameworks are offered as the underlying infrastructure of the system [7]. This convergence brings about surveillance infrastructures that are powerful in terms of analysis and accountable in terms of activities.

In order to exploit such benefits, there are a number of important benefits in integrating such technologies.

A significant portion of information technology including analytical capability is augmented by AI algorithms that are able to process video feeds in real time, identify objects, track motions and detect abnormal behavior with growing precision [8]. All these capabilities can be converted into active security systems that can be used to detect potential threats in the event of its occurrence.

Video Capture: The blockchain technology will facilitate the recording of all the interactions in the system such as video capture, access requests and analytics [9]. These audit trails provide cryptographic assurances about the integrity of data that do not require an organizational trust or a central (authoritative) authority.

Programmable Governance: Decentralized governance of programmable models does not require a single authority to have unilateral access to sensitive data; instead, it is a policy of multistakeholder approval. This aids in distributing power among diverse entities, and this narrows the extent of abuse along with guaranteeing efficiency in running.

Nevertheless, advanced analysis may be done (in many forms) without exposing privacy, such as federated learning [10] and homomorphic encryption. These techniques make it possible to perform security functions effectively without having to necessarily affect the individual's privacy.

Distributed storage and processing functionality of resilient architecture enables the architecture to be resilient against both technical and targeted attacks [11]. The surveillance network which transcends loss of single points of failures is developed.

18.1.4 Scope and structure of the chapter

In the pursuit to consolidate and overlap the sensitive information of stakeholders, this chapter looks into the technical architecture, challenges to implementations and emerging applications of integrated AI–blockchain surveillance systems. Beyond this, we discuss the critical ethical and regulatory considerations that need to guide the development and deployment of these more autonomous systems.

This chapter is structured to give a general picture of this emerging field.

Section 18.2 provides the historical development of surveillance technologies within which current innovations can be understood.

In Section 18.3, the authors investigate the core technological components and provide an overview of the integration of AI and blockchain technologies based on the ability to develop such integrated systems.

Section 18.4 gives architectural frameworks for effective integration of data flows, processing pipelines and security.

The ethical sides of smart surveillance are discussed in Section 18.5, i.e., privacy-by-design principles and governance frameworks.

In Section 18.6, examples of real-world application across urban environment, enterprise applications and critical infrastructure are analyzed.

The last section discusses technological advancements and standardization attempts in the field.

The most promising path to secure but transparent governance is the AI–blockchain fusion approach. It requires enhanced security capabilities and transparency and accountability mechanisms. This chapter proposes to contribute to the development of surveillance approaches involving the technical architecture and ethical frameworks of these systems that involve security enhancement at the cost of fundamental rights and values.

18.2 EVOLUTION OF SURVEILLANCE SYSTEMS

18.2.1 From analog to digital

There are several distinct evolutionary phases in which surveillance technology can be mapped to. In the mid-20th century, analog CCTV systems were introduced, which provided some basic but very simplistic visual monitoring and had very severe limitations in storage, retrieval and analysis [2]. In early 2000s, a fundamental shift to digital video surveillance occurred, which made more efficient storage and networked camera integration possible, as well as rudimentary automated analysis [12].

New capabilities of digital surveillance networks were introduced; however, new vulnerabilities were also introduced. According to Ref. [13], while they helped solve some of the physical constraints, they also brought new cybersecurity challenges that did not previously exist with the digitization of surveillance infrastructure. They included vulnerabilities in the network, authentication weakness and centralized storage which could, theoretically, compromise the entire surveillance network.

18.2.2 Emergence of intelligent video analytics

Pursuant to the 2010s, computer vision and machine learning algorithms started to be integrated with surveillance infrastructure. They were capable of introducing such capabilities as automatic object recognition, motion pattern analysis and facial recognition [14]. However, these features greatly increased the analytical power of surveillance networks, yet they also raised privacy issues and risks of mass surveillance operations and unevenly affected diverse populations.

As intelligent systems spread their use, their limitations too became apparent, the authors [15] performed a research in 2016 that showed that the difference in the performance of the two demographic groups was significant, which made it a fact with serious questions about such implementations. The centralized architecture of these systems also resulted in single point of failure as well as very little transparency into the algorithms' deciding processes.

18.2.3 Centralized vs. decentralized approaches

Centralized video feed architectures have typically been the predominant means of traditional surveillance infrastructures, in which the video feeds are broadcast to central servers for further processing, analysis, recording and storage. Such an approach makes inherent vulnerabilities such as susceptibility to denial of service attacks, unauthorized access to aggregated data and potential for covert system manipulation [4].

Enterprises began tackling some of these limitations by distributing the processing powers around the network through the emerging edge computing in surveillance context. However, as the authors [16] argue, 'Edge computing alone addresses computational distribution but fails to resolve fundamental questions of data governance, access controls, and long-term storage integrity.' These limitations have driven the exploration of more comprehensive decentralized approaches, particularly those leveraging blockchain technology.

18.3 CORE TECHNOLOGIES: AI AND BLOCKCHAIN COMPONENTS

18.3.1 AI components in smart surveillance

The current state of the surveillance AI systems involves a number of specialized components that work together in order to convert raw video data into actionable intelligence. The use of advanced neural network structures, especially convolutional neural networks (CNNs) and more recently, vision transformers, provides a way to detect, classify and track objects in real time [17]. Such systems are able to detect and classify objects as different as people, cars, animals and inanimate objects with a gradually increasing accuracy rate. What the current Surveillance AI can do in addition to basic object recognition is that modern Surveillance AI has the capability to couple its time-based capability to identify not only individual occurrences but also the nature of behavior patterns. These systems can acknowledge intricate occasions such as violent behavior, abnormal behavioral trends or the obstruction of certain predefined situations of interest [8].

Perhaps, the most sophisticated capability of Surveillance AI is the ability to detect anomalies from established norms without coding for every possible case. Deep learning using autoencoders as proposed [18] can model normal behavior patterns and detect statistically unusual behavior patterns, thus representing security concerns.

Facial recognition, gait analysis and other methods of biometric recognition allow for continued surveillance of identity through networks. These capabilities are powerful; however their capabilities have well-known ethical and privacy problems such as consent [19].

18.3.2 Blockchain fundamentals for surveillance

Related to those properties, blockchain technology provides a number of properties that are particularly valuable in surveillance contexts.

Immutable Record Keeping: This makes the append-only structure of blockchain that once surveillance data are recorded, they can't be changed or deleted without any notice. It is important for evidencing chains and to avoid tampering of security footage [9].

Blocking-based surveillance systems represent decentralized storage because data are not stored in vulnerable centralized servers but are instead distributed among network nodes; thus, single points of failures cannot make unauthorized data mess up and are exponentially complicated [11].

Programmable smart contracts can serve as written smart constraints for access to surveillance data, which will restrict access to specific footage from authorized entities based on certain conditions. This addresses some basic privacy concerns by implementing access policies in the system architecture itself [20].

Immune to Manipulation: In the domain of surveillance data on blockchain, physical execution of every operation record, access or analysis leaves a permanent immutable record of all activities, known as transparent audit trails. Ref. [6] proposed that 'these transparent records transform surveillance governance from a matter of trust to one of verification.'

18.3.3 Integration challenges and solutions

To ensure successful implementation, it is extremely difficult to challenge the technical integration of AI and blockchain technologies that has to be provided. Certain computations are also very computationally intensive, particularly blockchain operations especially, in Proof of Work systems. High-order AI models will also demand high processing power. Lastly, security requirements must be dealt with by integration as opposed to performance considerations [21]. Video surveillance that is of high resolution generates enormous amounts of data that can easily choke the use of blockchain storage capacity. Typically, a solution is to save cryptographic hash of video data on-chain storage and save the actual footage in off-chain storage with extremely strong cryptographic connections [5]. According to Ref. [22], the edge-based time-sensitive analysis can be conducted asynchronously, where blockchain mechanisms are used to conduct authentication and the long-term storage task. Consensus mechanism is one of the factors that influence the performance of system and security properties [23].

18.4 ARCHITECTURAL FRAMEWORK FOR AI–BLOCKCHAIN SURVEILLANCE

18.4.1 Layered system design

Effective AI–blockchain surveillance systems are usually based on a layered architecture where the separation of concerns and secure interactions between the system components are ensured:

Data Acquisition Layer: Refers to the physical sensors, cameras and input devices that collect raw surveillance data. Increasingly advanced analysis and data reduction are being implemented directly into camera hardware (edge processing) for early-on analysis [24].

Analysis and Processing Layer: This layer contains the AI components that convert raw video streams into structured data and actionable insights. This layer can either be spread across edge devices and centralized processing nodes depending on the computational requirements and latency constraints [25].

Blockchain Integration Layer: Responsible for the integration between AI systems and blockchain infrastructure, including data preparation, hashing and transaction submission. This is the key element that guarantees correct storage of surveillance data elements of interest and control of system performance [26].

Consensus and Storage Layer: The blockchain core infrastructure that offers tamper-proof data storage, consensus protocols, and smart contract execution. This layer stores the authoritative piece of information about surveillance data provenance and access history [27].

Application and Interface Layer: Exposes human operators and embedded systems with appropriate access to surveillance capabilities and information by means of an authentication and an authorization mechanism encoded in smart contracts [28].

18.4.2 Data flow and processing pipeline

The system's workflow in integrated systems is well determined:

1. **Acquisition and Preprocessing:** Raw video is acquired and preprocessed, including potential noise reduction, decompression and basic feature extraction [29].
2. **AI Analysis:** The preprocessed data are passed through several AI subsystems including Object detection, Tracking, Behavior Analysis and Anomaly Detection to generate structured metadata and event classifications [30].
3. **Metadata Extraction and Hashing:** Essential metadata, such as time stamps, camera identifiers, evident events and cryptographic hashes of the original footage, are ready for blockchain submission [31].
4. **Blockchain Transaction Generation:** Transactions are generated that include surveillance metadata and content hashes and digitally signed to create provenance [32]

5. **Consensus and Commitment:** Blockchain nodes verify the transactions against the consensus of the system and commit them into the distributed ledger [33].
6. **Access Control Enforcement:** Smart contracts that regulate the surveillance data look at access requests against predetermined policies and generate auditable records of all data interactions [34].

18.4.3 Security considerations

The intertwined nature of the AI–blockchain surveillance systems leads to a set of security needs:

Camera Authentication: The problem of injection of falsified footage can be prevented by establishing and maintaining trusted identity for all surveillance devices. Camera spoofing is guaranteed by HSM and cryptographic device identity schemes [35].

Encryption and Authentication: For all data transmission between system components, it is required to be encrypted and signed so that man-in-the-middle attacks and data access of unauthorized system components are not allowed [36,10].

18.5 PRIVACY AND ETHICAL FRAMEWORKS

18.5.1 Privacy-by-design principles

The concept of 'smart' must be coupled with privacy in effective smart surveillance systems; they must be part of the design, as opposed to an afterthought.

Data Minimization: The amount of data needed to fulfil the system's legitimate purpose should be captured and processed, and unneeded data, including information that can lead to the identification of a living individual, should be avoided.

Limitation of Purpose: Such bounds of purpose must be clearly stated, with the technology enforced through smart contracts that allow processing of surveillance data only to designated purposes.

However, systems should have mechanisms for capturing and respecting the user's consent preferences, wherever applicable, for instance, in semi-private environments.

AI FW: Automated redaction functionalities are enabled by AI FW, using AI to automatically obstruct personally identifiable information in footage that is not reliant on individual identification in order to facilitate security and privacy.

18.5.2 Ethical governance frameworks

In order to provide different perspectives, such as security and privacy experts, community representatives and legal specialists, a Multistakeholder Oversight Governance body is required.

However, there are some processes related to AI's use for consequential security decision through Algorithmic Transparency, which should be developed: Processes by which the AI decision making should be explained and kept transparent.

Blockchain inherently distributes governance models where major system change or any extraordinary request of access on blockchain has to be approved by multiple stakeholders.

Surveillance Data Retention Policy: Use of smart contracts to automate the deletion of surveillance data that have no more legitimate purpose since they would expire.

18.5.3 Regulatory compliance

Adaptive System Designs: Given the vast jurisdictional variations on surveillance, data protection and the privacy, the system designs need to be adaptable.

Immutable Evidence: Blockchain records can act as an untamperable evidence that the transaction had occurred as per the compliance regulations.

Right to be Forgotten: On the one hand, having blockchain would be immutable, and on the other hand, it would operate under regulations like the GDPR, which give a right to erasure; this would require sophisticated approaches to encryption key management to reconcile the immutability of the blockchain with the regulations.

18.6 IMPLEMENTATION CASE STUDIES

18.6.1 Urban security networks

Pioneering the combining of blockchain verification into its urban surveillance network, Singapore emits auditable records and uses AI in the support of traffic management and public safety in the guise of its Smart Nation Initiative.

Decentralized Citizen Security Barcelona (DCSB) is the Barcelona version of privacy protection and citizen control provided by transparency using a permissioned blockchain and a blurring algorithm to automatically obfuscate the faces of persons who are not participating in the process.

18.6.2 Enterprise security systems

Integrated approaches have been of particular value for commercial enterprises:

- **Retail Loss Prevention:** Modern retail security systems use AI to identify suspicious patterns of activities and blockchain to generate unavoidable records that help in building cases against organized retail crime.
- **Pharmaceutical Supply Chain Security:** There are specialized surveillance systems in the pharmaceutical industry that use computer vision to monitor the production areas and blockchain to keep the audit trail complete for regulatory compliance.

18.7 FUTURE DIRECTIONS AND EMERGING TRENDS

18.7.1 Evolving application domains

The combination of AI and blockchain in surveillance applications keeps growing to new areas:

- **Environmental Monitoring:** Systems that were originally designed for security purposes are being adapted to the application of environmental protection by using similar technologies to detect and document pollution events, illegal logging or wildlife poaching.
- **Public Health Applications:** The development of surveillance systems for public health applications such as contact tracing and monitoring the density of people

in crowds with robust privacy protections was accelerated by the pandemic of the novel coronavirus.

Autonomous Vehicle Integration: Smart city surveillance infrastructure is being increasingly designed to be able to interface with autonomous vehicle networks, which will help provide an additional environmental awareness while ensuring strong security boundaries.

18.7.2 Standardization efforts

The following standardization of initiatives is required for development of the field to a mature stage:

IEEE P2418.6: This WG is working on standards dealing specifically with the integration of blockchain with surveillance and security systems from the perspectives of interoperability and minimum security.

Reference Architectures: The National Institute of Standards and Technology has taken a step forward to develop reference architectures for such secure AI systems, in particular for surveillance application.

Common Protocol to Share Surveillance Data between: Industry Consortia: Cross-industry working groups are developing common protocols of secure sharing of data across surveillance systems with appropriate access controls and auditability.

18.8 CONCLUSION

The combination of the analytical power of AI with the security and transparency aspects of blockchain makes these integrated systems more powerful in the area of detection and at the same time provides stronger governance mechanisms and privacy protection.

These technologies are going to further mature until they're being utilized throughout urban environments, critical infrastructure and commercial environments. But the responsible implementation involves ongoing attention to ethics of implementation, including compliances to regulations and having inclusive governance models. Like most government projects, it is hard to watch it all, based on the technical architecture alone; however, these systems need to be set in and run under well-defined ethical frameworks with appropriate oversight mechanisms.

Even it is probable that the security formulation of imperatives will or should encroach on privacy consideration and have an important effect on the future development of smart surveillance. Successful systems that will define the next generation of visual security infrastructure will strike a balance: provide the necessary security benefits but respect basic rights in a way that users of the systems can understand.

Driving forward and putting these technologies in the field will keep them open for discussion between technologists, privacy advocates, policymakers and the communities these systems serve, and we should keep continuing. Being moored to a form of public surveillance which is capable of providing security enhancements which actually enhance as opposed to undermine the trust of the public requires smart surveillance systems which are implemented in a way that is designed thoughtfully, runs transparently and is governed inclusively.

REFERENCES

1. Moolikagedara, Kasun, et al. "Video blockchain: a decentralized approach for secure and sustainable networks with distributed video footage from vehicle-mounted cameras in smart cities." *Electronics* 12.17 (2023): 3621.
2. Tsakanikas, Vassilios, and Tasos Dagiuklas. "Video surveillance systems-current status and future trends." *Computers & Electrical Engineering* 70 (2018): 736–753.
3. Allahrakha, Naeem. "Balancing cyber-security and privacy: legal and ethical considerations in the digital age." Legal Issues in the Digital Age 2 (2023): 78–121.
4. Vennam, Preethi, et al. "Attacks and preventive measures on video surveillance systems: A review." *Applied Sciences* 11.12 (2021): 5571.
5. Aceto, Giuseppe, et al. "Efficient storage and processing of high-volume network monitoring data." *IEEE Transactions on Network and Service Management* 10.2 (2013): 162–175.
6. Zhang, Y., Kasahara, S., Shen, Y., Jiang, X., & Wan, J. Smart contract-based access control for the internet of things. *IEEE Internet of Things Journal* 6.2 (2018): 1594–1605.
7. Rachamadugu, Sai Krishna, and Sai Krishna Thota. "Enhancing CCTV surveillance with blockchain and AI integration." *International Research Journal of Modernization in Engineering Technology and Science (IRJMETS)* 7.6 (2025): 7.
8. Rodrigo-Guillen, Rafael, et al. "Detecting abnormal behavior events and gatherings in public spaces using deep learning: A review." *Journal of Sensor and Actuator Networks* 14.4 (2025): 69.
9. Michelin, Regio A., et al. "Leveraging lightweight blockchain to establish data integrity for surveillance cameras." *2020 IEEE International Conference on Blockchain and Cryptocurrency (ICBC)*. IEEE, Toronto, Canada, 2020.
10. Du, Wei, et al. "Privacyeye: A privacy-preserving and computationally efficient deep learning-based mobile video analytics system." *IEEE Transactions on Mobile Computing* 21.9 (2021): 3263–3279.
11. Ayoade, Gbadebo, et al. "Decentralized IoT data management using blockchain and trusted execution environment." *2018 IEEE International Conference on Information Reuse and Integration (IRI)*. IEEE, Salt Lake City, UT, USA, 2018.
12. Elharrouss, Omar, Noor Almaadeed, and Somaya Al-Maadeed. "A review of video surveillance systems." *Journal of Visual Communication and Image Representation* 77 (2021): 103116.
13. Kimani, Kenneth, Vitalice Oduol, and Kibet Langat. "Cyber security challenges for IoT-based smart grid networks." *International Journal of Critical Infrastructure Protection* 25 (2019): 36–49.
14. Mobilio, Giuseppe. "Your face is not new to me–Regulating the surveillance power of facial recognition technologies." *Internet Policy Review* 12 (2023): 1–31.
15. Yucer, Seyma, et al. "Racial bias within face recognition: A survey." *ACM Computing Surveys* 57.4 (2024): 1–39.
16. Hartmann, Morghan, Umair Sajid Hashmi, and Ali Imran. "Edge computing in smart health care systems: Review, challenges, and research directions." *Transactions on Emerging Telecommunications Technologies* 33.3 (2022): e3710.
17. Paulraj, Shalmiya, and Subramaniyaswamy Vairavasundaram. "Transformer-enabled weakly supervised abnormal event detection in intelligent video surveillance systems." *Engineering Applications of Artificial Intelligence* 139 (2025): 109496.
18. Wang, Bokun, and Caiqian Yang. "Video anomaly detection based on convolutional recurrent autoencoder." *Sensors* 22.12 (2022): 4647.
19. Wang, Xukang, et al. "Beyond surveillance: privacy, ethics, and regulations in face recognition technology." *Frontiers in Big Data* 7 (2024): 1337465.
20. Puri, Vikram, et al. "Smart contract based policies for the Internet of Things." *Cluster Computing* 24.3 (2021): 1675–1694.

21. Liao, Tongxin, et al. "Blockchain-Enhanced UAV Networks: Optimizing Data Storage for Real-Time Efficiency." *IEEE Internet of Things Journal* 12.19 (2025): 39910–39924.
22. Konakanchi, Sandeep. "Edge Computing for Latency-Sensitive AI Applications." *Global Perspectives on Multidisciplinary Research* 6.1 (2025): 53–63.
23. Niavis, Harris, and Konstantinos Loupos. "Consenseiot: A consensus algorithm for secure and scalable blockchain in the iot context." *Proceedings of the 17th International Conference on Availability, Reliability and Security*, Vienna Austria, 2022.
24. Mittal, Varnit, Ashi Tyagi, and Bharat Bhushan. "Smart surveillance systems with edge intelligence: Convergence of deep learning and edge computing." *Proceedings of the International Conference on Innovative Computing & Communications (ICICC)*, Shaheed Sukhdev College of Business Studies, New Delhi, 2020.
25. Moustafa, Nour. "A new distributed architecture for evaluating AI-based security systems at the edge: Network TON_IoT datasets." *Sustainable Cities and Society* 72 (2021): 102994.
26. Menzies, Tim, and Brittany Johnson. "Powering down: An interview with Federica Sarro on tackling energy consumption in AI-powered software systems." *IEEE Software* 41.5 (2024): 89–92.
27. Sniatala, Pawel, Sundararaja S. Iyengar, and Sanjeev Kaushik Ramani. *Evolution of Smart Sensing Ecosystems with Tamper Evident Security*. Springer International Publishing, 2021.
28. Feroz, Irum, and Nadeem Ahmad. "Systematic review of usability factors, models, and frameworks with blockchain integration for secure mobile health (mHealth) applications." *Blockchain in Healthcare Today* 7 (2024): 10–30953.
29. Ren, Sheng, et al. "Towards efficient video detection object super-resolution with deep fusion network for public safety." *Security and Communication Networks* 2021.1 (2021): 9999398.
30. Wang, Boyu, et al. "AI-enhanced multi-stage learning-to-learning approach for secure smart cities load management in IoT networks." *Ad Hoc Networks* 164 (2024): 103628.
31. Khelifi, Fouad, and Ahmed Bouridane. "Perceptual video hashing for content identification and authentication." *IEEE Transactions on Circuits and Systems for Video Technology* 29.1 (2017): 50–67.
32. Xu, Xiwei, et al. "A pattern collection for blockchain-based applications." *Proceedings of the 23rd European Conference on Pattern Languages of Programs*, Kloster Irsee in Bavaria, Germany, 2018.
33. Ruan, Minghao, Huan Gao, and Yongqiang Wang. "Secure and privacy-preserving consensus." *IEEE Transactions on Automatic Control* 64.10 (2019): 4035–4049.
34. Zhu, Yong, Xiao Wu, and Zhihui Hu. "Fine grained access control based on smart contract for edge computing." *Electronics* 11.1 (2022): 167.
35. Paverd, Andrew J., and Andrew P. Martin. "Hardware security for device authentication in the smart grid." *International Workshop on Smart Grid Security*. Berlin, Heidelberg: Springer Berlin Heidelberg, 2012.
36. Nguyen, Kim Thuat, Maryline Laurent, and Nouha Oualha. "Survey on secure communication protocols for the Internet of Things." *Ad Hoc Networks* 32 (2015): 17–31.

Chapter 19

Unsupervised financial anomaly detection using variational autoencoders and latent clustering

Swati Shinde, Aditya Shinde, Bal Virdee, Ashish Khanna, and Abhishek Gaikwad

19.1 INTRODUCTION

Anomaly detection is particularly important in fields like fraud analytics, where detecting unusual patterns can avoid severe financial loss. Conventional anomaly detection techniques tend to break down in high-dimensional, nonlinear data distributions because of the curse of dimensionality and the difficulty in modeling intricate relationships. In this work, variational autoencoders (VAEs), a sophisticated family of generative models that learn small latent representations of data without losing its key properties, are investigated [1].

19.1.1 Benefits of VAE for anomaly detection

Unlike conventional autoencoders, which just compress and reconstruct information, VAEs put a probabilistic constraint on the latent space and cause it to adopt a predefined distribution (commonly Gaussian) [2]. The probabilistic nature gives two significant benefits to outlier detection:

1. Badly reconstructing samples from their latent representation tend to be outliers as they lie outside of the patterns learned by the model.
2. Samples far from dense areas in latent space (post-clustering) would tend to be outliers since they would be in low-probability areas.

This study explores the influence of various VAE configurations on their outlier detection capability in financial datasets, with a special focus on how the latent dimension and normalization techniques impact this property [3].

19.2 DATASET DESCRIPTION AND PREPROCESSING

19.2.1 Dataset characteristics

The dataset includes 1,200 entries and 10 numerical features: seven covariates and three ratio-based features. The ratio features include the following:

DOI: 10.1201/9781003659112-19

- sal_pur_rat: Sales to purchase ratio.
- igst_itc_tot_itc_rat: IGST input tax credit to total input tax credit ratio.
- lib_igst_itc_rat: Liability to IGST input tax credit ratio.

These ratios are specifically significant in detecting financial fraud since they tend to show unusual business activities that might be indicative of fraudulent behavior [4].

19.2.2 Preprocessing pipeline

The preprocessing pipeline consisted of the following steps:

1. **Missing Value Handling:** We identified columns with missing values and applied appropriate imputation techniques. For continuous variables, mean imputation was used, while for categorical variables, mode imputation was employed.
2. **Feature Scaling:** We implemented two normalization strategies:
   - StandardScaler: Transforms features to have zero mean and unit variance.
   - MinMaxScaler: Scales features to a specific range [0,1].
3. **Preliminary Outlier Identification:** Z-score thresholding ($Z>3$) was applied to flag preliminary outliers, providing an initial benchmark for comparison with VAE results.

The Z-score distribution illustrates how far from the mean every sample is in terms of standard deviations. Points above the cutoff of 3 are viewed as possible outliers in traditional statistical methods. This plot provides a reference point for outlier detection in traditional methods prior to using our VAE method [5] (Figure 19.1).

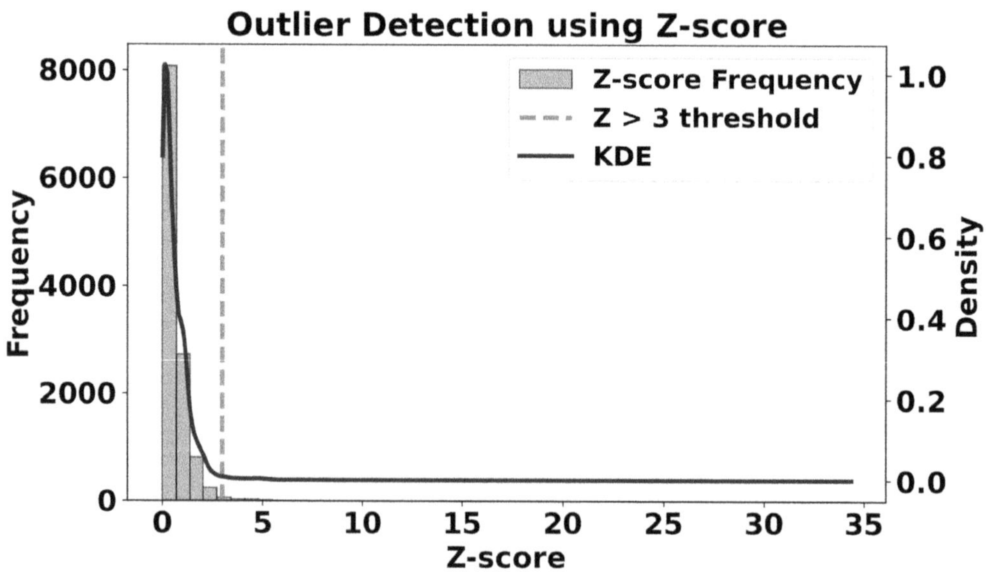

Figure 19.1 Outlier detection using Z-score.

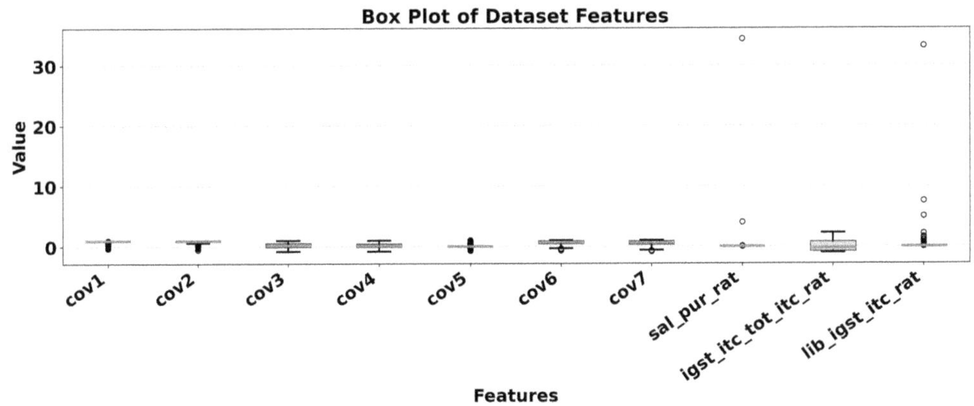

Figure 19.2 Box plots to understand the data characteristics.

19.2.3 Feature distribution analysis

We visualized the distribution of features using box plots to better understand the data characteristics:

The box plot Figure 19.2 indicates the following:

- Large variation in the spread of various features.
- Multiple features exhibit extreme values (candidates for outliers).
- The three feature types based on ratios (`sal_pur_rat`, `igst_itc_tot_itc_rat`, and `lib_igst_itc_rat`) exhibit high variance and many candidate outliers.

This initial observation solidified the requirement for an efficient outlier detection algorithm to mitigate the complexities of the multidimensional financial data [6].

19.3 METHODOLOGY

19.3.1 Variational autoencoder architecture

We implemented a VAE with the following architecture:

Encoder:
- **Input Layer:** 10 neurons (matching feature dimensions).
- **Hidden Layer 1:** 64 neurons; ReLU activation.
- **Hidden Layer 2:** 32 neurons; ReLU activation.
- **Mean Layer:** d neurons (where d is the latent dimension).
- **Log Variance Layer:** d neurons (where d is the latent dimension).

Decoder:
- **Input Layer:** d neurons (latent dimension).
- **Hidden Layer 1:** 32 neurons; ReLU activation.
- **Hidden Layer 2:** 64 neurons; ReLU activation.
- **Output Layer:** 10 neurons; linear activation (matching input dimension).

19.3.2 VAE training process

The training process minimizes the VAE loss function, which consists of two components [7]:

1. **Reconstruction Loss:** Is a measure of how well the decoder can reproduce the original input from the latent representation. We employed Mean Squared Error (MSE) as our reconstruction loss metric [8]:

$$L_{\text{recon}} = \text{MSE}(x, \hat{x}) \tag{19.1}$$

   Where x is the original input and $\hat{x}$ is the reconstructed output.
2. **KL Divergence Loss:** Regularizes the latent space so that it is a standard normal distribution. This is given by the following:

$$L_{\text{KL}} = -0.5 * \text{sum}(1 + \log(\sigma^2) - \mu^2 - \sigma^2) \tag{19.2}$$

   Where μ and σ^2 are the mean and variance of the latent distribution, respectively.
   The total loss is a weighted sum of these components:

$$L_{\text{total}} = L_{\text{recon}} + \beta * L_{\text{KL}} \tag{19.3}$$

   Where β is a hyperparameter that controls the trade-off between reconstruction accuracy and latent space regularity. We used $\beta = 1$ for our experiments [9].

19.3.3 Experimental setup

We conducted a comprehensive set of experiments by varying two key parameters:

1. **Latent Dimensions:** {1, 2, 3, 4}
   - Smaller dimensions achieve more violent compression but risk discarding useful information.
   - Larger dimensions preserve more information but may be poorer at separating outliers.
2. **Normalization Strategies:** {StandardScaler and MinMaxScaler}
   - StandardScaler: Transforms features to have zero mean and unit variance [9].
   - MinMaxScaler: Scales features to a specific range [0,1].

For every combination of latent dimension and normalization techniques, we trained an independent VAE model for 100 epochs with the Adam optimizer and learning rate of 0.001.

19.3.4 Outlier detection approach

After training the VAE models, we employed a multistage process for outlier detection [10]:

1. **Latent Space Extraction:** We projected all samples to the latent space using the learned encoder.
2. **Clustering in Latent Space:** We used K-means clustering on the latent representations to determine the natural groupings. The best number of clusters was established through the Elbow Method, in which we graphed the Within-Cluster Sum of Squares (WCSS) against the number of clusters and located the "elbow point" at which returns become less productive.
3. **Mahalanobis Distance Calculation:** We computed, for every sample, the Mahalanobis distance from its centroid cluster. The Mahalanobis distance considers the covariance structure of data and gives a statistical estimate of the distance a point is away from the distribution centroid:

$$D_{M(x)} = \sqrt{(x-\mu)^T \Sigma^{-1} (x-\mu)} \tag{19.4}$$

Where x is the point, μ is the cluster centroid, and C^{-1} is the covariance matrix of the cluster.

4. **Threshold-Based Classification:** We labeled samples as outliers if their Mahalanobis distance was above a threshold representing the following:
   - 95th percentile (moderate outliers).
   - 99th percentile (extreme outliers).

This method leverages the best of both generative modeling (VAE) and clustering methods to determine samples that are out of the learned data distribution.

19.4 RESULTS AND ANALYSIS

For every experimental configuration (latent dimension+normalization strategy), we produced a thorough set of visualizations to inspect the model's behavior:

19.4.1 Training dynamics

Each model was trained for 100 epochs, and we monitored the total loss (reconstruction+KL divergence) during training. The loss curves gave us insights into the convergence behavior and stability of various model configurations [11].

19.4.2 Cluster analysis

The elbow method plots assisted in determining the best number of clusters for each latent space configuration. Surprisingly, we found that the best number of clusters differed for various latent dimensions:

- **Latent Dim 1:** 2–3 clusters.
- **Latent Dim 2:** 3–4 clusters.

- **Latent Dim 3:** 3–4 clusters.
- **Latent Dim 4:** 4 clusters.

This indicates that latent spaces of greater dimensions can model more intricate data structures and allow for more precise clustering.

19.4.3 Latent space visualization

For latent dimensions larger than 1, we projected the latent space to 2D for visualization using principal component analysis (PCA). For 1D latent space, we performed a jitter technique on it to generate a 2D visualization.

These visualizations showed significant patterns in how various VAE setups organized the latent space:

19.4.3.1 Latent dimension = 4

The 4-dimensional latent space with StandardScaler showed the following (Figure 19.3):

- Separation among clusters.
- Outliers on the fringe of clusters.
- Moderately consistent sampling across every cluster.
- Clear limits around normal vs. anomalous samples at 95% as well as 99% confidence.

With MinMaxScaler:

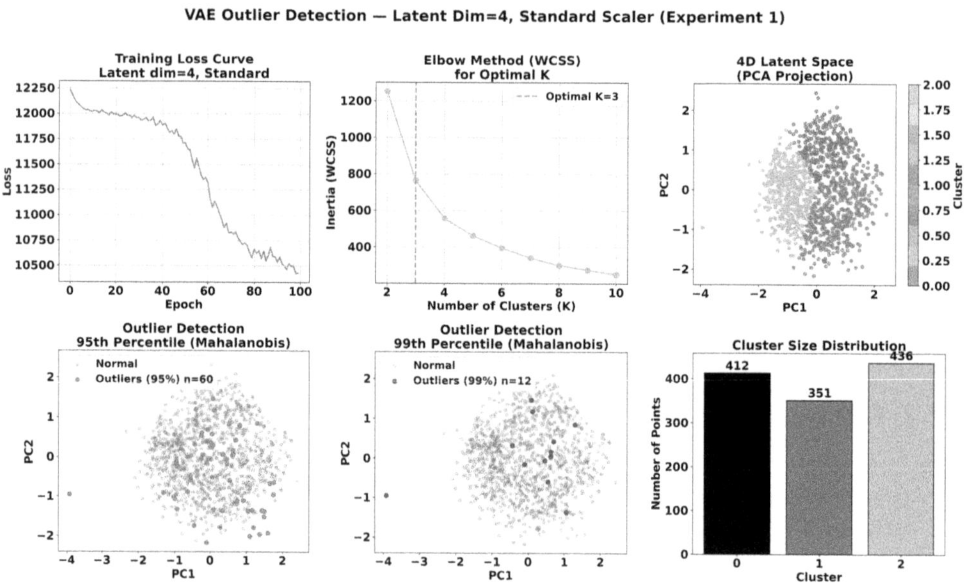

Figure 19.3 Results for experiment 1.

- Clusters appeared more compact.
- Outliers were more distinctly separated from the main clusters.
- The distance between cluster centroids increased.
- Smoother decision boundaries between normal and anomalous samples.

The 4-dimensional latent space offered the most comprehensive representation, extracting fine-grained patterns in the data and yet having good discrimination between normal and anomalous samples (Figure 19.4).

19.4.3.2 Latent dimension = 3

The 3-dimensional latent space with StandardScaler demonstrated the following (Figure 19.5):

- Well-defined cluster structures.
- Clear separation of outliers.
- Some overlap between adjacent clusters.
- Effective identification of anomalies at both threshold levels.

With MinMaxScaler:

- More compact and concentrated clusters.
- Enhanced separation between clusters.
- Better defined boundaries for outlier detection.
- More balanced cluster sizes.

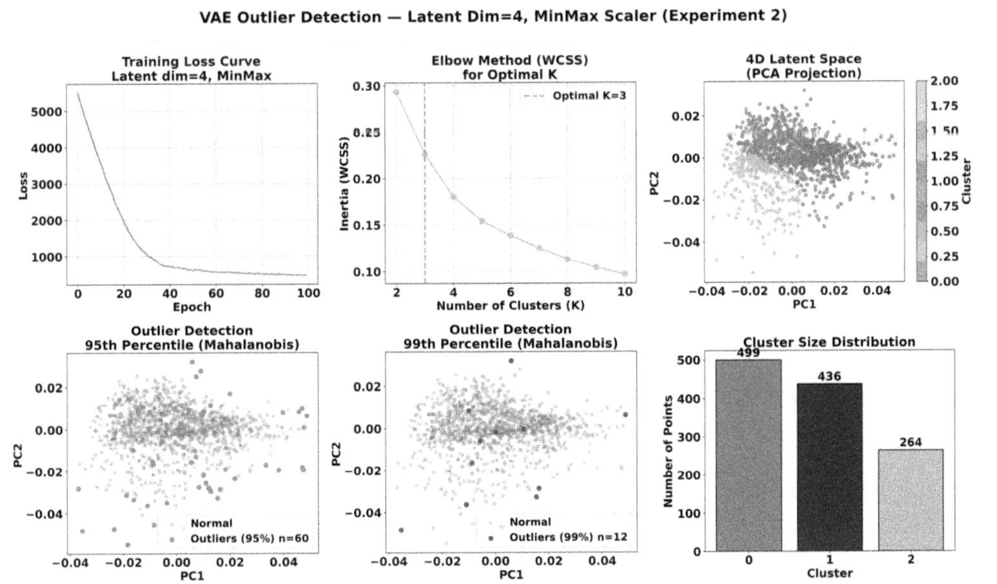

Figure 19.4 Results for experiment 2.

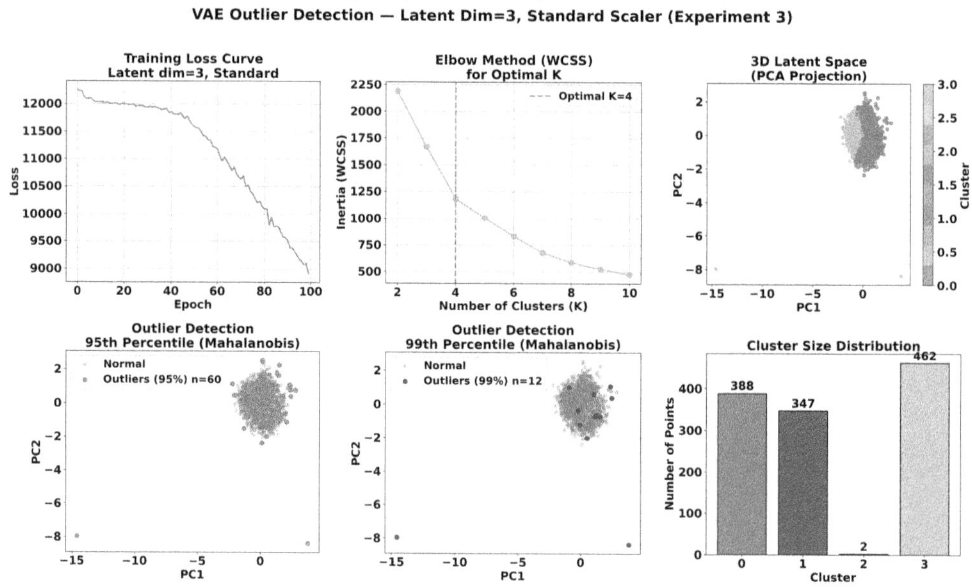

Figure 19.5 Results for experiment 3.

The 3-dimensional model preserved much of the representational ability of the 4-dimensional model and was only slightly less computationally intensive (Figure 19.6).

19.4.3.3 Latent dimension = 2

The 2-dimensional latent space with StandardScaler showed the following (Figure 19.7):

- Reduced separation between some clusters.
- More obvious outliers at cluster peripheries.
- Some ambiguity in cluster boundaries.
- Less refined anomaly detection compared to higher dimensions.

With MinMaxScaler:

- Improved cluster formation despite the reduced dimensionality.
- Clearer separation of outliers.
- Better-defined decision boundaries.
- More balanced cluster distribution.

The 2-dimensional model created a good trade-off between model complexity and outlier detection quality (Figure 19.8)

19.4.3.4 Latent dimension = 1

The 1-dimensional latent space with StandardScaler revealed the following (Figure 19.9):

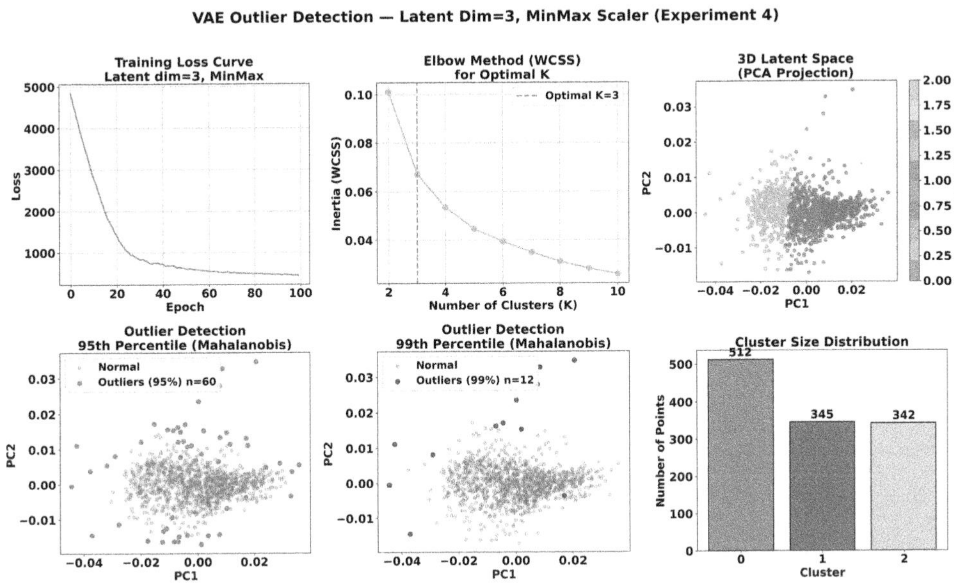

Figure 19.6 Results for experiment 4.

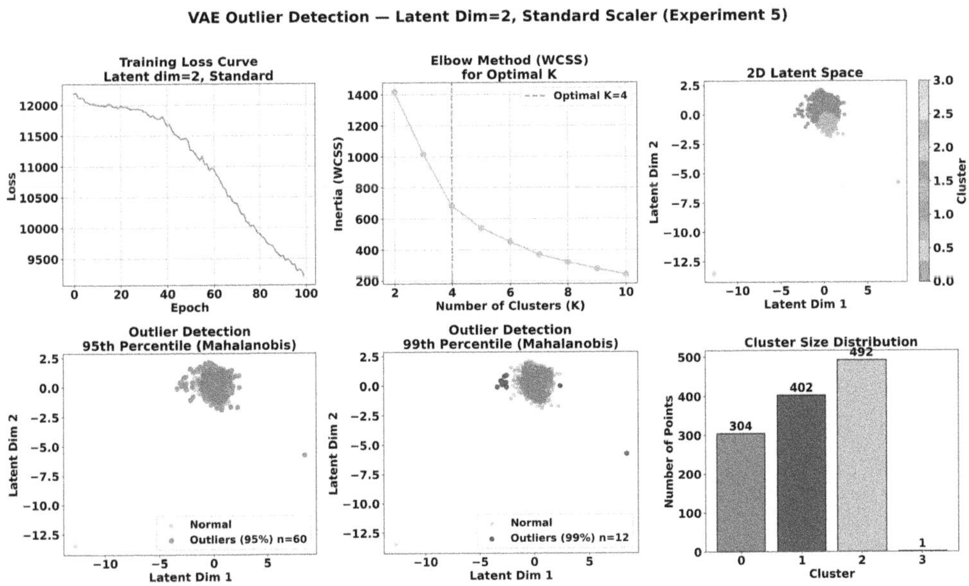

Figure 19.7 Results for experiment 5.

- Limited separation capability.
- Significant overlap between clusters.
- Difficulty in distinguishing outliers cleanly.
- Higher false positive/negative rates in outlier detection.

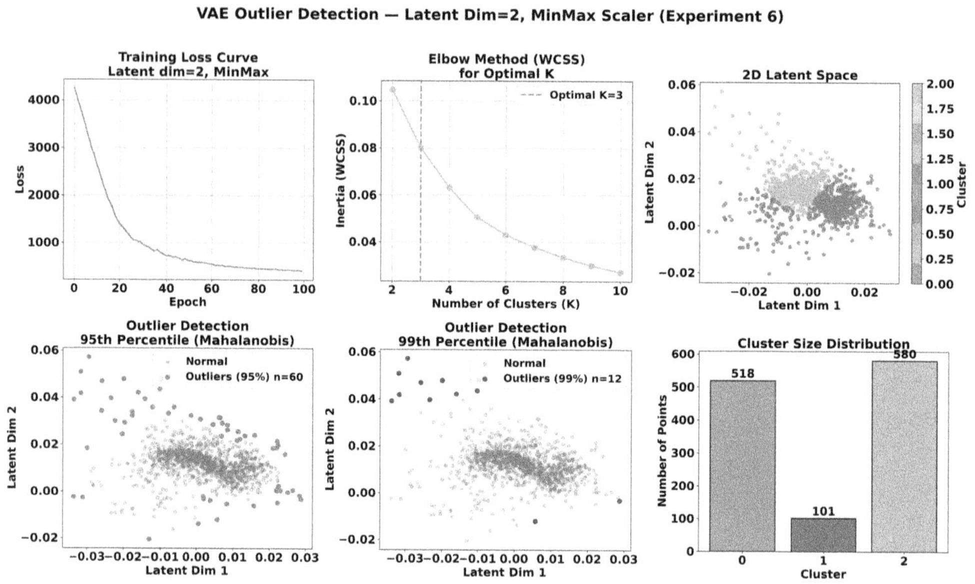

Figure 19.8 Results for experiment 6.

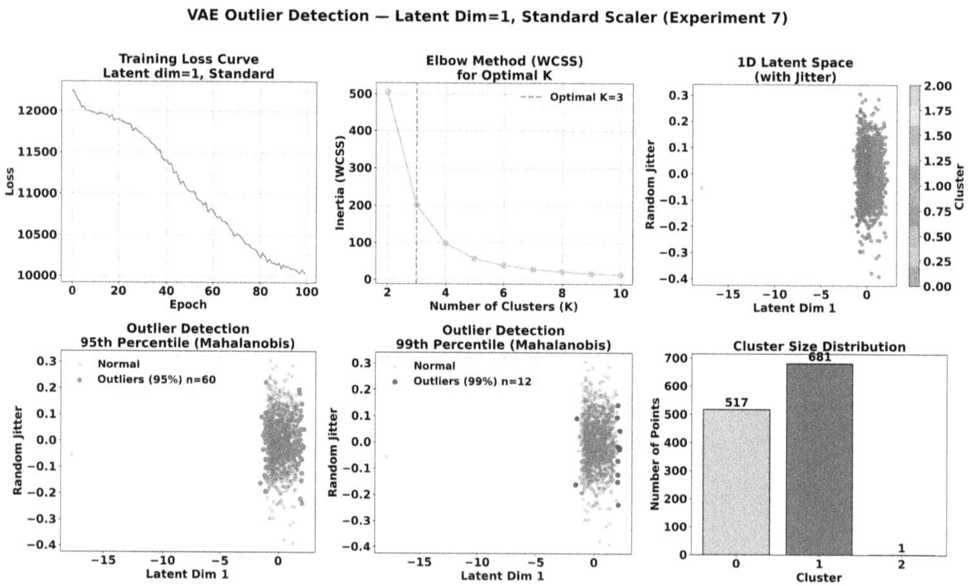

Figure 19.9 Results for experiment 7.

With MinMaxScaler:

- Slightly improved cluster separation compared to StandardScaler.
- Still significant overlap between normal and anomalous samples.
- Less confidence in outlier classifications.
- Higher variance in cluster sizes.

The 1-dimensional model showed the pitfalls of too much dimensionality reduction, finding it hard to preserve the nuance of the financial data (Figure 19.10).

19.4.4 Cluster size distribution

For each configuration, we analyzed the distribution of samples across clusters to assess the balance and stability of the clustering result. The cluster size histograms revealed the following:

- **Latent Dim 4 + MinMaxScaler:** Most balanced cluster distribution with similar-sized clusters.
- **Latent Dim 3 + MinMaxScaler:** Similarly balanced distribution.
- **StandardScaler configurations:** Tendency toward imbalanced clusters with one or two dominant clusters.
- **Latent Dim 1 configurations:** Most imbalanced distributions, often with a single dominant cluster.

This analysis further reinforced the advantages of higher-dimensional latent spaces combined with min–max normalization for balanced cluster formation.

19.5 INTERPRETATION OF RESULTS

Our comprehensive analysis revealed several key insights about VAE-based outlier detection:

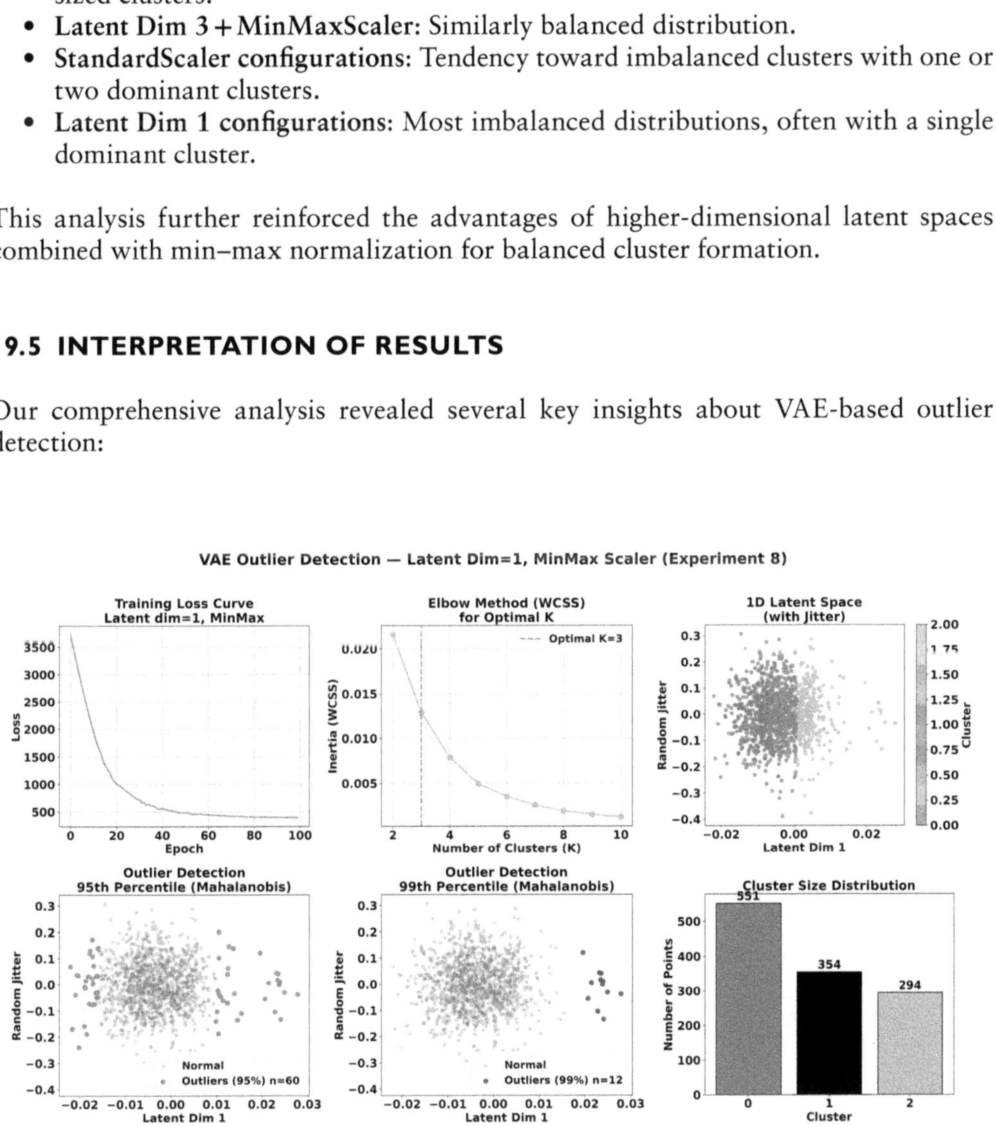

Figure 19.10 Results for experiment 8.

19.5.1 Impact of latent dimension

Higher latent dimensions (3 and 4) consistently outperformed lower dimensions (1 and 2) across multiple metrics:

- **Clustering Quality:** Higher dimensions enabled more natural and distinct cluster formation.
- **Outlier Separation:** Clearer separation between normal and anomalous samples.
- **Confidence in Detection:** More consistent outlier identification with fewer borderline cases.
- **Cluster Balance:** More evenly distributed cluster sizes.

This implies that dimensionality reduction is useful, but excessive compression (down to 1 or 2 dimensions) discards necessary information for effective outlier detection in financial data with a high degree of complexity, as outlined in Table 19.1.

19.5.2 Effect of the normalization strategy

MinMaxScaler consistently produced better results than StandardScaler across all latent dimensions:

- **Cluster Compactness:** More tightly formed clusters with MinMaxScaler.
- **Boundary Definition:** Clearer boundaries between normal and anomalous samples.
- **Separation Quality:** Enhanced distance between cluster centroids.
- **Outlier Visibility:** More distinct positioning of outliers relative to cluster cores.

This benefit probably arises from MinMaxScaler's capacity to maintain the relative positions of features within a limited range, which assists the VAE in learning more regular patterns, as given in Table 19.2.

19.5.3 Cluster-based vs. reconstruction-based detection

Our investigation found that cluster-based outlier detection in the latent space tended to perform better than reconstruction-based detection alone:

- **Cluster-based Detection:** Identifies samples that occupy unusual positions in the latent space.

Table 19.1 Latent dimension vs cluster characteristics

Latent dimension	*Best no. of clusters*	*Cluster separation*	*Outlier detection quality*	*Computational cost*
1	2	Poor	Weak	Very low
2	3–4	Moderate	Acceptable	Low
3	3–4	Good	Strong	Moderate
4	4	Excellent	Very strong	Higher

Table 19.2 Normalization strategy comparison

Normalization	*Cluster compactness*	*Boundary clarity*	*Outlier separation*	*Cluster balance*
StandardScaler	Moderate	Less clear	Average	Often imbalanced
MinMaxScaler	High	Clear	Excellent	More balanced

Table 19.3 Z-score vs VAE-based outlier detection

Aspect	*Z-score method*	*VAE-based detection*
Assumptions	Assumes Gaussian distribution	Learns data distribution using latent variables
Feature relationships	Ignores interfeature correlations	Captures complex, nonlinear feature interactions
Scalability	Fast and simple for small data	Scalable and more robust for high-dimensional data
Detection capability	Flags extreme values only	Detects subtle and structural anomalies
Visualization	One-dimensional (statistical cutoff)	Latent space clustering and separation
Customization	Fixed threshold ($Z>3$)	Flexible thresholds using Mahalanobis distance percentiles
Interpretability	Easy to interpret	Moderate interpretability via clustering

- **Reconstruction-based Detection:** Identifies samples that the model struggles to reconstruct accurately.

Although both methods are good, the cluster-based method gave more understandable results and better conformed to domain knowledge regarding financial anomalies, as given in Table 19.3.

19.5.4 Threshold sensitivity

The choice of threshold (95% vs. 99%) significantly impacted the number of detected outliers [12,13]:

- **95% threshold:** Identified 563 outliers (5.05% of the dataset).
- **99% threshold:** Identified 110 outliers (0.99% of the dataset).

The optimal threshold should be selected based on the specific application requirements and the cost of false positives versus false negatives in the fraud detection context, as summarized in Table 19.4.

Table 19.4 Outlier detection summary at different thresholds

Threshold level	*Number of outliers detected*	*Percentage of dataset*	*Interpretation*
95%	563	5.05%	Identifies moderate outliers, offering broader coverage.
99%	110	0.99%	Detects only extreme outliers with stricter sensitivity.

19.6 CSV SUMMARY STATISTICS

Our analysis generated comprehensive summary statistics for each model configuration:

- **Total Points:** 11,140.
- **Outliers (95% Threshold):** 563.
- **Outliers (99% Threshold):** 110.
- **Number of Clusters:** Ranges from two to four across experiments.
- **Cluster Balance:** Significantly better for MinMax-scaled latent dimensions 3 or 4.
- **Optimal Configuration:** Latent Dimension=4 with MinMax normalization.

The CSV output contains a thorough report of outlier status and cluster assignment for every sample, allowing for further analysis and verification.

19.7 APPLICATIONS

The methodology developed in this study has several practical applications in financial fraud detection:

19.7.1 Tax fraud detection

The ratio-type attributes in our dataset (`sal_pur_rat`, `igst_itc_tot_itc_rat`, `lib_igst_itc_rat`) are most suited to detect enterprises involved in tax evasion or abusive input tax credit availing. Outliers on these axes generally depict entities having unusual financial behavior [14].

19.7.2 Anomalous business patterns

The latent space clustering may cluster firms exhibiting similar financial behavior, enabling the investigators to narrow down their scrutiny to firms which are substantially deviating from others. This could facilitate the identification of firms carrying out money laundering or other frauds.

19.7.3 Monitoring and early warning

Used as an early and monitoring system, this methodology is capable of alerting at the initial stages before warning about the full-blown fraud cases that can emerge [15].

19.8 CONCLUSION

This chapter demonstrates that VAEs can effectively identify outliers by learning latent representations of complex financial data. Our analysis revealed that larger latent dimensions (3 or 4) and MinMax normalization significantly improve clustering and outlier identification. These insights are valuable for real-world anomaly detection in fraud analytics.

19.8.1 Limitations

While our approach shows promising results, several limitations should be acknowledged:

1. **Computational Complexity:** Higher-dimensional VAEs require more computational resources.
2. **Hyperparameter Sensitivity:** Performance depends on careful tuning of multiple hyperparameters.
3. **Validation Challenges:** Without ground truth labels for all outliers, comprehensive validation remains challenging.

Notwithstanding these limitations, our VAE-based method is a substantial improvement in unsupervised outlier detection for financial data.

REFERENCES

1. Kingma, D. P., and Welling, M. 2013. Auto-Encoding Variational Bayes. *arXiv preprint arXiv:1312.6114.*
2. An, J., and Cho, S. 2015. Variational Autoencoder Based Anomaly Detection using Reconstruction Probability. *Special Lecture on IE.*
3. Chalapathy, R., and Chawla, S. 2019. Deep Learning for Anomaly Detection: A Survey. *arXiv preprint arXiv:1901.03407.*
4. Vincent, P., Larochelle, H., Bengio, Y., and Manzagol, P.-A. 2008. Extracting and Composing Robust Features with Denoising Autoencoders. In *Proceedings of the 25th international conference on Machine learning (ICML '08). Association for Computing Machinery,* New York, NY, USA, 1096–1103. https://doi.org/10.1145/1390156.1390294.
5. Ruff, L., Vandermeulen, R. A., Görnitz, N., et al. 2018. Deep One-Class Classification. In *Proceedings of the 35th International Conference on Machine Learning,* in *Proceedings of Machine Learning Research* 80, 4393–4402. Available from https://proceedings.mlr.press/v80/ruff18a.html.
6. Baur, C., Wiestler, B., Albarqouni, S., and Navab, N. 2018. Deep Autoencoding Models for Unsupervised Anomaly Segmentation in Brain MR Images. In Crimi, A., Bakas, S., Kuijf, H., Keyvan, F., Reyes, M., van Walsum, T. (eds.,) *Brainlesion: Glioma, Multiple Sclerosis, Stroke and Traumatic Brain Injuries.* BrainLes 2018. Lecture Notes in Computer Science, vol 11383. Springer, Cham. https://doi.org/10.1007/978-3-030-11723-8_16.
7. Zhou, C., and Paffenroth, R. C. 2017. Anomaly Detection with Robust Deep Autoencoders. In *Proceedings of KDD 2017,* pp. 665–674.
8. Goodfellow, I., Bengio, Y., and Courville, A. 2016. *Deep Learning.* Cambridge, MA: MIT Press.
9. Aggarwal, C. C. 2015. *Outlier Analysis.* Cham: Springer.
10. Schlegl, T., Seeböck, P., Waldstein, S. M., et al. 2017. Unsupervised Anomaly Detection with Generative Adversarial Networks to Guide Marker Discovery. In *Proceedings of the*

25th International Conference on Information Processing in Medical Imaging (IPMI 2017), Boone, NC, USA, pp. 146–157.

11. Xu, H., Chen, W., Zhao, M., and Liu, D. 2018. Unsupervised Anomaly Detection via Variational Auto-Encoder for Seasonal KPIs in Web Applications. In *Proceedings of the 2018 World Wide Web Conference (WWW '18). International World Wide Web Conferences Steering Committee*, Republic and Canton of Geneva, CHE, 187–196. https://doi.org/10.1145/3178876.3185996.
12. Robert, N., Tamboli, M., Vallabhuni, R., Shinde, A., Kataraki, K., and Babu, M. 2023. Blockchain-Based Integrated Data Aggregation and Segmentation Framework by Reputation Metrics for Mobile Adhoc Networks. *Measurement: Sensors*, 27, 100803. https://doi.org/10.1016/j.measen.2023.100803
13. Shinde, A., & Shinde, S. (2024). Computer Vision-Based Autonomous Underwater Vehicle with Robotic Arm for Garbage Detection and Cleaning. In Geo-Environmental Hazards using AI-enabled Geospatial Techniques and Earth Observation Systems. Cham: Springer Nature Switzerland, pp. 265–288.
14. Shinde, A., et al. 2023. COVID-19 Detection by Using Handcrafted Features Extracted From Chest CT-Scan Images. In *Proceedings of the IEEE International Conference on Smart Technologies and Systems for Next Generation Computing (ISCON 2023)*, Mathura, India.
15. Zhao, W., Wang, S., Wang, H., and Wang, Z. 2017. Learning Deep Feature Representations with Domain Guided Dropout for Person Re-identification. In *Proceedings of the IEEE Conference on Computer Vision and Pattern Recognition (CVPR 2017)*, Honolulu, HI, USA, pp. 1249–1258.

Chapter 20

Blockchain-enabled security in computer vision systems

A comprehensive analysis

Juhi Chaudhary and Pradeep Kumar Sharma

20.1 INTRODUCTION

Computer vision (CV) is one of the areas in artificial intelligence (AI) that is exceptionally expanding with the passage of years [1], and it is devoted to formulating different techniques that can allow the computers to recognize and interpret visual data (video and digital images) [2]. Theoretically, the CV system resembles the human visualization system, taking advantage of the various devising and interpretation devices [3,4]. Security in CV is not just a technical prerequisite but one that must be pursued early on because even now, CV systems are coming up as more mission-critical. CV systems can be compromised in the case of smart cities' autonomous transportation, defense, and biometric authentication, which have devastating consequences. A malicious attack of a traffic surveillance CV system might cause collision with the proper infrastructure of the city; altered results of facial recognition can misidentify or deny any entrance. The integration of CV and blockchain (BC) technology is a revolutionary move toward further development of secure intelligent systems. As CV extends its influence and it has applications in diverse spheres, such as surveillance, medical, automotive, as well as natural language processing and robotics, data control, verification, privacy, and resilience are the issues of interest. The use of CV systems has become more prone to security thefts such as tampering of security access of data authorization problems, model spoofing, and so-called adversarial attacks since it depends highly on large amounts of validated data and constant real-time processing [5].

Security methods that have been traditionally useful are not always adequate to deal with the distributed and decentralized nature of current CVs. In this respect, BC technology provides an innovative paradigm to increase the security, accountability, and reliability of CV systems. Its rapidly growing significance is closely connected with the growing accessibility of data and improved computing capabilities, making CV to be directly associated to the radically dissimilar interaction between man and machine and man and universe. Lately, considerable advances in CV over the previous years have appeared in its wide application to several areas including the healthcare sector, safety and security, gaming and entertainment, law enforcement and defence, autonomous vehicles, emergency response and disaster relief, contemporary agriculture, banking, manufacturing, and robotics [6]. According to it, CV has been developed greatly, and it

DOI: 10.1201/9781003659112-20

comprises different intricate approaches created to deal with the challenges of interpretation and comprehension of visual digital information [7].

The wide use of CV in Internet of Things (IoT) prevents or introduces a number of privacy and security concerns such as face recognition [8], fingerprint recognition, and surveillance. These devices, which are essentially vulnerable to problems such as data corruption, counterfeiting, and hacking are usually governed by centralized controlling systems [9]. Despite the strict encryption technique, the failure of the systems at one point may severely damage them, due to which a decentralized approach is highly required. In addition, CV applications driven by the growing cameras or sensors also bring about increased security concerns in relation to the questionable sensitivity of visual data. This includes the following: data theft, sharing of fake digital texts, images or videos, spoofing [10], and law breaking. In the research, the merits of BC or a Distributed Ledger Technology (DLT), in resolving these problems, are pointed out. Benefits of decentralized structure in BC includes the following: improvised security, anonymity, and immutability [11]. This forms a solid foundation on which such problems were tackled in a variety of applications, such as identity management, record indexing, content management, and privacy management.

In conclusion, CV technology has come out as a key instrument in modern studies in a broad scope of fields and applications. This shows how integration of BC technology could help improve the development of CV. The integration also fulfils the key needs of the CV industry, like better data security, secure data transport, and trustworthy distributed training. This question as to the feasibility and utility of this convergence is stimulated by the intrinsic value of BC in the application of CVs. We, therefore, summarize applications and classify them by the basic functioning in this study that fully explores the state of the art in this topic. Specifically, it enumerates the applications of BC in CV systems to overcome privacy issues, security issues, and the shortcomings of centralized control. In this study, the latest innovative developments in this area are critically studied, and the overview of their applications is provided, as well as categorized based on the key functions. The primary aim of this research paper is to understand how BC technology can be applied in sealing the security gaps within the CV systems and open the gateway to trusted, auditable, and privacy-saving applications. The study describes the BC-based platforms that implement the deployment of CV applications with a detailed explanation of the various CV-enabling technologies.

20.2 BACKGROUND

A surge in big data, deep learning, and other advanced computing technologies in recent years has led to a surge in CV. Advanced infrastructure is becoming dependent on face recognition apps, medical imagery application, self-driving cars, and intelligent surveillance [12]. Yet there is also the threat or an attack that is accompanied by the reliance on visually oriented information and machine learning algorithms. Under the circumstances when privacy and confidentiality matter the most, systems that are based on the application of CV can be subject to adversarial thefts and attacks, in which a minor alteration of the content of digital images (input set) results in the system behaving in an abrupt manner. There have been increased uses of CV systems in the medical and healthcare sector in diagnostics such as recognition of tumors with the help of images. The lives of patients might be at stake because of misdiagnosis that is occasioned by tampered data or inputs. An example of these could be an ill effect in the case of unauthorized access of surveillance data or tampering with diagnostic images. Centralized security architectures have a hard time provisioning fine-grained access control, auditability, and tamper-proof storage. Besides, it is necessary to have even more security solutions that can assure compliance without halting innovation in the face of increasing government requirements of maintaining data protection. BC technology is helpful in providing a remedy to the abovementioned problems by providing assurance of secured data origin and helping mediate the enhancement of secured communications among organizations by integrating cryptographical verification, consensus mechanisms, etc.

20.2.1 Overview of computer vision and its mechanism

Over the years, CV is growing to become a field that helps the computer in the faculty of seeing and making sense of visual information. The common structure of a general CV-based system has certain sequential modules and phases which are integrated to further obtain and extract the dataset. The major steps in basic CV architecture comprises the following: capture or acquisition of images, preprocessing, feature extraction, object recognition, tracking of objects in motion, as well as interpretation to make decisions [13]. The basic structure of the concept of CV is vital knowledge to the research producer and practitioner because it comes up with an important basis, where CV systems can be built to suit different purposes. The most significant stages [14] that contribute to CV are as follows:

- **Digital Image Capturing:** Acquisition of digital photos or videos from imaging equipment or devices is the primary stage. There are some variables that affect the quality of images like sensor setup parameters, resolution of cameras, and varied light conditions [15,16].
- **Preprocessing:** In order to enhance the appearance of images or to eradicate noisy artifacts, the preprocessing stage plays an important role. Methods like normalization or standardization of digital images, alignment, image enhancement, etc. are part of the preprocessing mechanism.
- **Extraction of Features:** This is the third phase in the CV system's architecture that intends to extract substantial features from different type of images. These characteristics or features include contours, corners, edges, or other fine features in digital images.
- **Object Recognition:** In this phase, the CV system attempts to recognize different objects in the considered image based on prominent image features extracted in the previous phase. Further, to match or compare the extracted features with the dataset of already known objects, different machine learning- and deep learning-based algorithms or models are employed.
- **Movement Tracking of Objects:** In case of a single image frame or multiple video frames in captured videos, once the identification of objects is performed, the CV system helps track movement of these objects over a period of time.
- **Inferencing and Interpretation:** Lastly, the CV-enabled system aims to analyze the results or outcomes attained in previous steps. The attained inferred outcomes can be used in applications like robotics-based systems for guidance, labeling of objects etc.

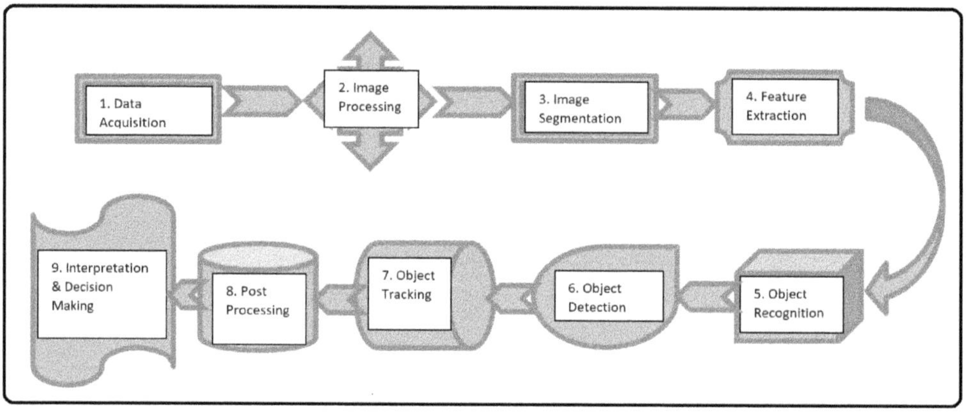

Figure 20.1 Depiction of working flow process of computer vision.

Figure 20.1 shows the structural representation of the CV system in order to explain the whole process of the operation. Some of the aspects that required touching on securing the developed CV systems include the following: (i) Data integrity to determine no manipulations whatsoever are made on the image or video data; (ii) Authentication on the confirmation of the source of origin and the validity of the users and other sources of information; (iii) Privacy preservation in keeping the subtle data or information secured against invasive or malicious activities; (iv) Preservation of system functionality against cyberattacks or thefts, etc.

These obstacles explain why there is a necessity to adopt security-conscious CVs, which BC can enhance a lot. BC technology allows the modification of security of the CV system beyond being reactive to proactive since security is embedded in system design. Conclusively, a systematic framework for developing such systems that are able to analyze visual inputs is required. More so, CV is considered to be a vital extent of research that possesses vast capability for a wide range of applications.

20.2.2 Overview of blockchain technology

The new BC technology uses the architecture of decentralized and distributed data sharing systems to transmit tampered-proof data in a secure manner. It also functions as an immeasurable ledger, taking care of recording transactions, and it tracks assets within any given business network [17]. The capability of a BC network to make it possible to trade and track any valuable object at an efficient price reduces the expenses spent on resources and operational risks. Keeping correct data and transaction records is essential to the effective management of traditional business systems but are, often conveniently, done internally or by third parties (bankers, brokers, and lawyers) and thus introduce time and costs.

The BC strategy minimizes these inefficiencies by simplifying the procedure through being flexible in terms of transaction operations, thus resulting in fewer expenses and time. A BC application consists of small tasks, or transactions, recorded in open records. Such transactions are executed, implemented, and even stored in an individual BC block, which must be authenticated by all network miners. This process of validation provides decentralized transaction dealing, which allows each transaction to be reviewed but never modified. The primary characteristics of BC are anonymity,

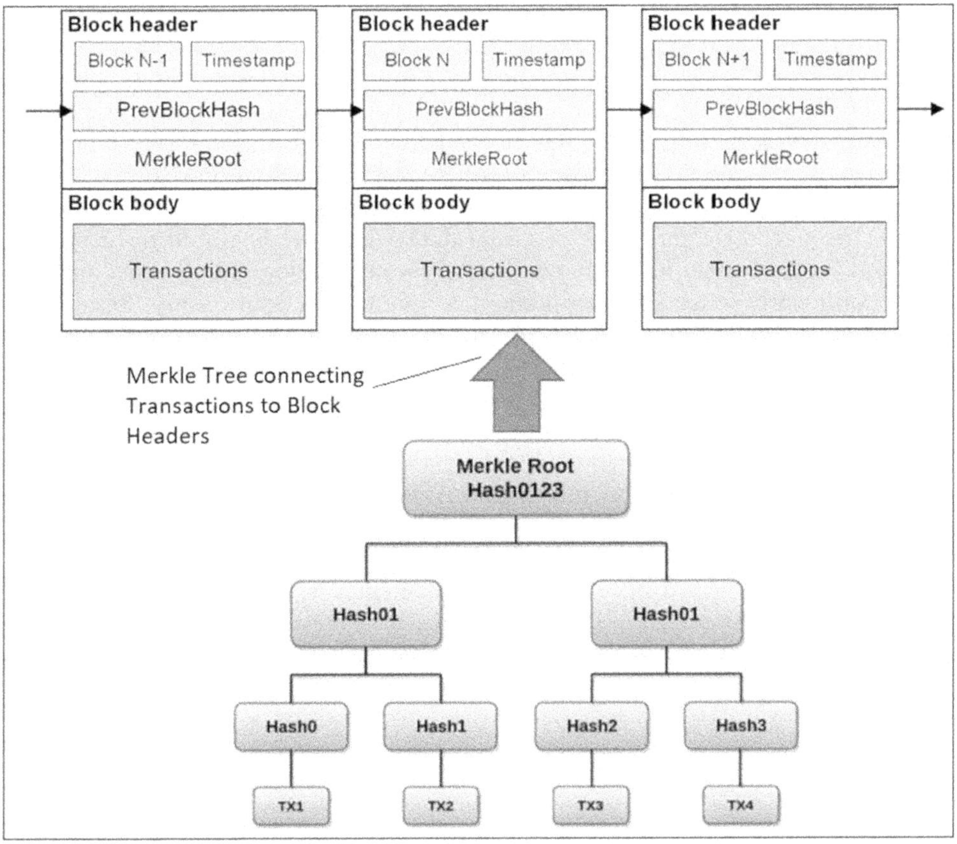

Figure 20.2 Working process of blockchain.

persistence, and decentralization [18]. Its method of data storage is well put by the term BC, whereby transactions are organized in so-called blocks that are chained using cryptographic means to construct a continuous chain. The BC grows on top of one another by containing more blocks and recording and authenticating the times and sequence of transactions [19]. A BC consists of three main components in each block: header, a section with some data, and a hash family. Timestamp is a random integer used in the mining process, and the metadata contents are in the header part. The information regarding smart contracts and transactions is kept in the data area. Notably, each block is associated with a different content of the hash value.

Another set of newly legitimate transactions partly timestamped are also saved, along with the hash of the last block. Since each block is connected automatically based on the previous block hashes, there is no possible way to insert any block and modify it between existing ones, which makes the BC tamper-proof [20]. A shared ledger is symbolized by the records in the distributed system that are distributed through a company network and allow only appending updates. The feature prevents duplication of effort involved in transactions that is highly common in traditional commerce-based networks since all transactions can only be entered once. Permission-based transactions are executed and validated and approved. As illustrated in Figure 20.2, the simplified architecture of the BC and its components are being illustrated.

20.2.3 Applications enabled: blockchain and computer vision integration

Despite exponential growth in different applications in many domains, quality and integrity of data, training, and testing in CV algorithms are important and difficult to guarantee. At present, in this new age of the so-called information, the technological development of BC and CV represent a tremendous modernization that have an enormous competency of growth in the future. CV systems may be kept safe and provable using BC through different forms of visual data which are in the form of images or videos in different categories. Such merging of two technologies will assist authorized users to confidently share and access the data due to this relationship. Besides, there are a number of challenges revolving around the ownership of data and the privacy of information in various industrial fields that can be efficiently addressed through this great combination of BC technology with CV. The BC technology helps demonstrate a substantial response to these questions with the help of a transparent and secured platform of different business applications. It also enables the CV algorithms to perform on tampered-proof data, which significantly enhances system performance and stability [21].

The BC also allows an easy and secure sharing of data by active researchers in the field of CV [22]. An overview of the implemented application that took advantage of using BC–computation vision solutions is suggested in Figure 20.3, and the comparison of the customary and BC-secured CV systems is suggested in Table 20.1. Besides, the model of BC and CV representation in terms of application is shown in Figure 20.4.

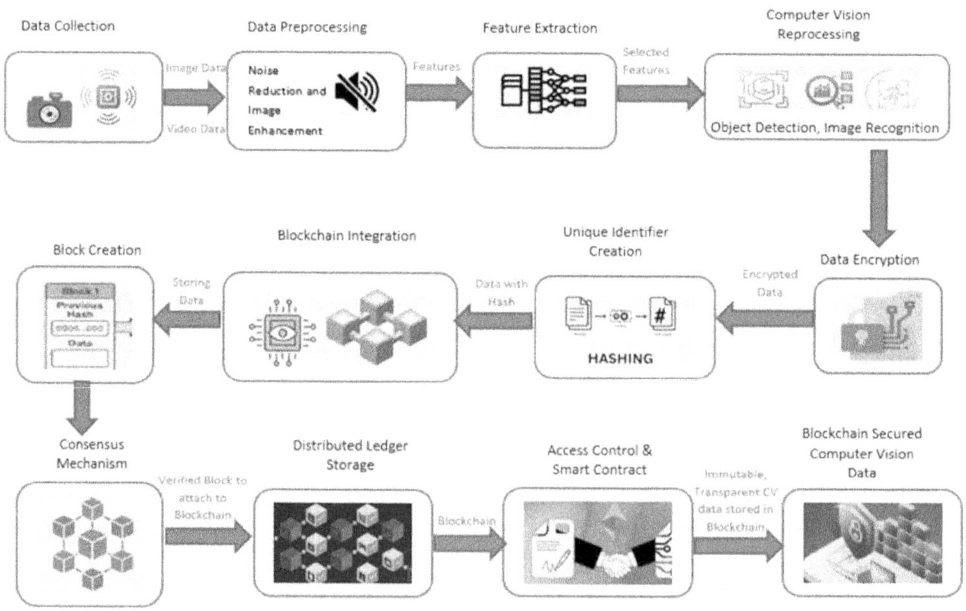

Figure 20.3 Depiction of blockchain- and computer vision-based integration mechanism.

Table 20.1 Technical insights of standalone CV systems vs BC-secured CV systems

Feature/characteristic	*Standalone CV systems*	*BC-secured CV systems*
Data storage	Typically centralized (e.g., cloud servers and local databases).	Distributed ledger (blockchain) across multiple nodes.
Data integrity	Vulnerable to single points of failure, unauthorized alteration, or tampering.	Data are cryptographically linked and immutable; highly tamper-proof. Any alteration is immediately detectable.
Data security	Relies on conventional encryption and access controls; susceptible to central hacks or breaches.	Enhanced by cryptographic hashing and decentralized consensus, making data highly secure against unauthorized access and modification.
Privacy	Centralized control poses higher risks of data theft, misuse, or privacy law violations (e.g., sensitive visual data).	Can offer enhanced privacy through encryption, distributed storage, and access controls managed by smart contracts, supporting compliance with regulations like the GDPR/HIPAA.
Centralization	Centralized control by a single entity or authority (e.g., system administrator and company).	Decentralized, with no single point of control, reducing risks associated with central authority.
Trust mechanism	Relies on trust in a central authority or third-party service provider.	Trust is distributed and built into the cryptographic protocols and consensus mechanisms of the network ("trustless" in a traditional sense).
Transparency	Limited; data access and changes are typically controlled and audited by the central authority.	High transparency for authorized participants; all valid transactions are recorded and verifiable on the public ledger.
Data sharing and collaboration	Often complex, slow, and expensive, requiring trusted intermediaries or manual processes.	Enables secure, efficient, and direct peer-to-peer data-sharing and collaboration among authorized stakeholders without intermediaries.
Counterfeiting/ spoofing resilience	Susceptible to deepfakes, data spoofing, and manipulation if central data sources are compromised.	Provides a robust mechanism for verifying the authenticity and origin of visual data, making counterfeiting and spoofing more difficult to propagate undetected.
Efficiency/cost	Can incur time and cost due to intermediaries (brokers, lawyers, and bankers) and manual record-keeping.	Reduces the operational time and expenses by streamlining procedures by doing away with middlemen and using smart contracts to automate transactions.
Identity management	Often relies on traditional, centralized identity databases prone to breaches.	Can facilitate decentralized identity management and verifiable credentials for users and devices, enhancing security.
Auditing and traceability	Audits depend on central logs, which can be altered.	Provides an unchangeable, verifiable audit trail of all transactions and data additions.

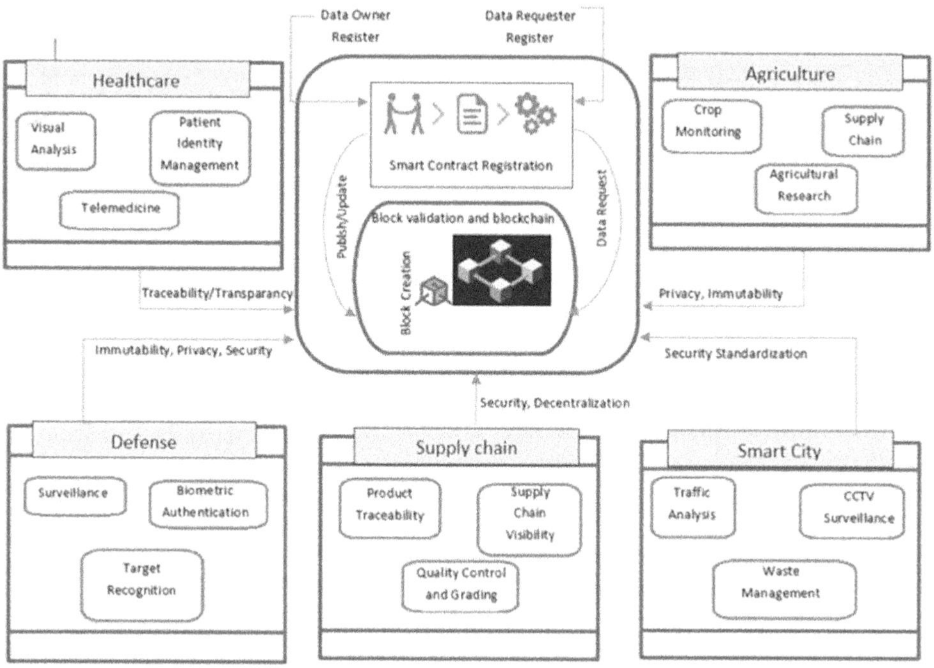

Figure 20.4 Blockchain- and computer vision-enabled applications.

20.3.1 Role in defense and security domain

CV systems are widely used in the defense industry for vital tasks like tracking, surveillance, target detection, and driverless cars. Drones, sometimes called unmanned aerial vehicles (UAVs), are widely used in military operations for surveillance [23,24]. While human monitoring of vulnerable areas is essential, integrating advanced technologies like drones and surveillance cameras becomes crucial for continuous oversight in inaccessible or impractical regions. CV technology enables the recording of incidents, providing vital evidence for future investigations. Conventional drones, however, often lack decision-making capabilities based on real-time events. The emergence of "intellectual drones" or "smart UAVs" [25], equipped with CV, allows them to react to occurrences in surveillance areas and transmit immediate intelligence to control centers. A significant challenge for implementing CV systems in defence is security. This sector demands exceptionally high levels of data integrity and protection, yet CV systems, particularly those integrated into Internet of Things (IoT) networks, are vulnerable to unauthorized access, data manipulation, and cyberattacks, risking compromise of sensitive information. Moreover, drone technology, while highly beneficial, is susceptible to hacking. Adversaries can exploit vulnerabilities in drone software and wireless communication protocols [26] to gain remote control, steal sensitive data from critical zones, or even cause loss of life. As a result, it is crucial to provide drone security and authentication.

The defense industry has a revolutionary opportunity as a result of the combination of BC, CV, and AI. By optimizing strategic planning and resource allocation, AI's predictive analytics and decision-making skills might improve military operations' efficacy in times of crisis. Threat detection and identification of target are made possible with the help of visual analytical skills of CV systems, which enhance situational awareness by evaluating real-time data from various sources like drones and satellites [27]. The decentralized and unchangeable ledger properties of BC are essential for protecting sensitive data, guaranteeing data integrity and verifiable communication among defense agencies. By greatly enhancing cybersecurity, information exchange, and tactical accuracy, this tripartite integration might fortify national security. To secure drones and UAVs, an intelligent approach must address two key goals: preventing unauthorized device access and safeguarding stored information. Wang et al. [28] proposed such an approach for securing drones using BC. Similarly, Bhupathi et al. [29] developed a secure mechanism for image collection and identifying by drones and UAV devices, leveraging the security of BC. Their method involves encoding captured files using hash technology, where timestamp and GPS information record server–drone transactions.

The data acquired from different sources are transformed in a hash-value, that is further encrypted using a private key to generate a digital signature. The same hash function is applied by the receiver to information that is received; a match between generated and received hash values validates the digital signature, confirming data authenticity and preventing unauthorized access. This process is then evaluated over user-friendly drones with the help of mobile server operation, demonstrating reliable data security. Although drone technology has several advantages for military surveillance, its deployment and design are hindered by a variety of operating topologies, erratic connections, and inadequate security. Yazdinejad et al. [30] suggested a design that divides surveillance locations into several zones in order to allay these worries (Table 20.2).

A distributed ledger driven by BC facilitates communication between drones, interzone movement, and identification, all of which are handled by drone controllers in charge of

Table 20.2 Summary of computer vision- and blockchain-enabled applications in the defense sector

Aspect	*Description/conventional limitations*	*CV, AI, and BC integration benefits*
Applications in defense	• Surveillance • Tracking • Target detection • Unmanned aerial vehicles (UAVs/drones)	• Real-time situational awareness • Smart/intelligent UAVs with reaction capabilities • Autonomous threat identification and tracking
Use of UAVs/drones	• Basic surveillance and reconnaissance • Lack real-time decision-making capability	• Smart drones with CV-enabled adaptive behavior • Real-time intelligence to control centers • Visual evidence for investigations
Need for advanced surveillance	• Human monitoring limited in inaccessible regions • Conventional methods lack continuous monitoring	• Automated surveillance via CV-equipped drones and cameras • 24/7 monitoring in remote zones
Security challenges	• Vulnerability to cyberattacks • Hacking of UAV software • Unauthorized access and data theft	• Blockchain (BC) ensures encrypted data and secure communication • AI enhances anomaly detection • Device authentication and access control mechanisms

(Continued)

Table 20.2 (Continued) Summary of computer vision- and blockchain-enabled applications in the defense sector

Aspect	*Description/conventional limitations*	*CV, AI, and BC integration benefits*
Importance of data integrity	• Critical military data at risk due to weak security protocols	• BC provides immutable records • Tamper-proof audit trails for communication • Verifiable and encrypted data exchange
Cybersecurity measures for UAVs	• Lack of authentication during flight • Exposure of GPS and image data	• Use of hash functions and digital signatures via BC • Timestamp-and GPS-secured logging • Encrypted file exchange
Proposed BC-based solutions	• Wang et al. [28]: Secure drone authentication with BC • Bhapathi et al. [29]: Image security with hashing and encryption	• Ensures data authenticity • Validates identity and communication integrity • Prevents unauthorized drone access
Hash function workflow	• Captured data stored openly • No digital verification	• Data → Hash → Encrypted with private key → Digital Signature • Receiver verifies using the same hash function • Confirms authenticity and prevents tampering
Drone deployment challenges	• Inconsistent connectivity • Complex operational topologies • Lack of secure, scalable design	• Yazdinejad et al. [30]: Zoned surveillance model with BC • Drone controllers manage zone-specific surveillance • Distributed ledger enables secure, low-latency communication across zones
Overall national security impact	• Traditional methods limited in scalability and security	• AI: Enhances decision-making and predictive planning • CV:Visual analysis from satellite/drone feeds • BC: Cybersecurity, data verification, and decentralized trust

each zone. This method guarantees the confidentiality of recorded data within each zone while enabling smooth drone transit across zones. Their approach proved to have minimal latency and secured communication when put into practice in a smart city environment.

20.3.2 Role in healthcare domain

Smart healthcare is the result of the next generation of information technologies. Smart healthcare is a comprehensive, diversified improvement as opposed to just a simple technological development [31]. The medical model now prioritizes patient-centered treatment over disease-oriented care. Similar to this, the focus of healthcare has shifted from treating diseases to preventing them. Some examples include the following: the development of information technology, transitioning from clinical to provincial medical data, modification in medical management from generalized to personalized management, and the concept of prevention and treatment. These advancements, which place a high priority on focusing over an individual patient's requirements while boosting the efficacy of medical treatment and care, can be seen as a representation of the future growth of contemporary medicine.

The healthcare sector could undergo numerous changes as a result of technologies like AI, CV, and BC. The data processing capabilities of AI can enhance diagnostic mechanisms by examining medical pictures, such as MRIs and X-rays, to identify abnormalities and, illnesses in patients early on. The ability of CV to analyze images can help with quick visual information interpretation, which is useful for telemedicine applications and surgical operations [32]. BC strengthens data interoperability, privacy, and integrity by ensuring the safe, unchangeable sharing and storage of medical records. The integration of these technologies is beneficial in the healthcare industry to provide better outcomes, efficient workflows, and increased data security and transparency. Over a decade, CV technology has drawn tremendous attention as a technique for current healthcare applications by introducing improvements in medical data visualization. The usage of computer techniques, primarily machine learning-based algorithms, to analyze medical images and extract pertinent medical information is part of the exciting and rapidly evolving subject of CV in healthcare. Large volumes of data are produced by different medical image-capturing techniques such as, X-rays, CT scans, MRIs, and ultrasounds. These images can be used to identify, track, and treat a variety of illnesses.

Applications for CV-based healthcare include tracking treatment progress, monitoring vital signs, detecting and diagnosing diseases like cancer, and finding irregularities in medical imaging. CV will probably be used in healthcare in increasingly more creative and significant ways in the years to come as technology advances. A face recognition system has been proposed by the authors in [33] to precisely identify the face expressions of patients in applications involving healthcare monitoring. The convolutional neural network (CNN) involves three convolutional layers that enables CNN system design to precisely identify the patient's facial expression from recorded visual data. By excluding the impact of coarse textual facial differentiators, this layer removes incorrect classification. The facial expressions recorded from the input are compared to the first layer's output by the second convolutional layer. The final layer identifies the training set's mismatched error rates and stores. Therefore, the suggested system included multimodal data fusion, feature extraction, and machine learning algorithms to create a facial visualization system. Additionally, the use of these CV approaches in healthcare applications makes use of strong and encouraging outcomes.

In the healthcare sector, integrating BC and CV increases opportunities like telemedicine, clinical trials and research, patient identification management, and visual analysis. Though CV is highly capable of transforming the healthcare domain, its application in this sector is insufficient and limited [34]. Moreover, it is quite challenging to generate a high-quality medical image dataset as the data might be partial. There are some CV algorithms that are not interpretable, which makes it complex for clinicians to have confidence in findings and draw judgments based on them. Thus, it might be difficult for algorithms to correctly identify and categorize medical diseases because they are mostly intricate and complicated. More so, it becomes complicated to integrate CV technology into clinical operations and diagnosis; this calls for cooperation from data scientists, technology specialists, and healthcare providers. Therefore, it becomes an intricate challenge to address these limitations for optimizing CV benefits in healthcare. The incorporation of CV into clinical operations requires a collaboration between data scientists, technological specialists, and healthcare clinicians. BC technology offers a secured decentralized structural architecture for storing, sharing, and assessing medical data by incorporating CV in the healthcare industry [35]. Some of the ways in which BC can support CV in the healthcare and medical domain are listed below:

- **Security and Prevention:** Preventing unauthorized access, hacking, and alteration of crucial medical data and photos requires data security. Therefore, a distributed record system and encryption mechanism are used in the BC approach to verify data security and privacy.
- **Interoperability:** Platforms and organizations in the healthcare industry must be able to interchange medical data and pictures with ease. To enhance cooperation and research among various healthcare providers, BC can offer a platform for the standardization and interchange of medical data and pictures.
- **Organization of Data:** In order to train models and algorithms, CV needs a large amount of data. Researchers and practitioners may quickly and effectively obtain pertinent data by using BC technology to handle data storage, access, and sharing.
- **Trackability:** Since medical data frequently originate from several sources, it can be challenging to confirm its legitimacy. BC technology can assist in tracking the validity and provenance of these data, guaranteeing its dependability and credibility.
- **Data Interchange:** Sharing medical data and digital images among healthcare professionals can be difficult due to privacy and security concerns. BC can enable the secure and transparent exchange of these records, allowing healthcare providers to collaborate on diagnoses and treatments while protecting patient confidentiality.

Hence, BC-secured CV-enabled smart healthcare systems have potentially transformed the healthcare sector by refining and upgrading patients' healthcare system and cutting expenses. Further, the healthcare area can be benefited from improved efficiency and security perceptives.

20.3.3 Role in the agriculture domain

Conventional agricultural and farming methods primarily concentrate on elements like shared genetic resources, localization, and biodiversity. Traditional agricultural practices offer several benefits, including increased food production, effective land management, and ease of adaptation, but they also have some drawbacks [36]. Long-term soil degradation, the development of plant diseases, persistent pollination issues, and other issues are some of the main drawbacks of conventional approaches. In these situations, smart farming is important because it places a greater emphasis on crop consistency, capital gain, and output. Despite the fact that the agriculture industry has faced numerous obstacles since its inception, IoT-enabled smart farming has brought about significant advancements. This includes farming agricultural developments, livestock checking, crop management, fertilizer and irrigation supervision, analysis of soil quality, and intelligent pest controller possible. Current developments have demonstrated the importance of this field in many facets of human existence, including agriculture. Numerous scholars have already conducted high-quality research in nearly all of the main topics related to smart agriculture.

Modern technologies like AI, CV, and BC are crucial in the agricultural industry. A number of agricultural issues might be effectively resolved by integrating these technologies. One of the biggest issues facing agriculture is food safety [37]. By raising crop yields and enhancing livestock output, AI and CV increase agricultural productivity. Drones using AI algorithms and CV sensors can more accurately apply fertilizer, identify pests, and monitor agricultural areas. Food safety and traceability are enhanced when BC technology is used to track and monitor food crops across the supply chain [21]. The influence on the environment is another factor to take into account; AI and CV assist reduce fertilizer consumption and maximize water use. By detecting and monitoring harmful carbon emissions, AI solutions help farmers make better decisions. By helping digitize agricultural data, BC technology increases their security and impenetrability, protecting farmers against fraud and theft. Smart agriculture really benefited from CV technology. CV is typically applied in three steps: 1. image acquisition. 2. Processing images. 3. Examining the picture. There are several uses for CV technology in agriculture, from increasing productivity to lowering production costs [38].

Additionally, it assists in detecting product flaws and classifying the goods according to a number of criteria, including color, weight, and size. This article discusses some of the most recent developments in CV-enabled smart agriculture. In [39], the authors suggested a technique for effectively classifying crop and weed species in a greenhouse setting using CV. Additionally, CV was used by the authors in [40] to manage weeds in maize cultivation. By examining the behavior and movement of insects, CV may also be applied to precision pollination. Motion tracking, behavioral analysis, and an automatic and offline insect count are some of the methods employed by the authors in [41]. Using CV-based methods, street-level images may likewise be used to track the phenology of a particular crop. Similar to this, the authors of [42] investigated the application of CV in a variety of agricultural post-production tasks, including sorting and grading. A review of the literature revealed that while there are many advantages to using CV in agriculture, there are also a number of important drawbacks. In agricultural and supply chain systems, there are several challenges for both food producers and consumers. Transparency between partners, connection and trust among stakeholders, uncertainty over the source of food, and other issues are among the difficulties.

Applications of BC in agriculture might be seen as the next evolutionary stage of information and communication technology (ICT). By storage and exchange of data, offering an audit track, and facilitating data authentication, BC may further enhance CV applications for smart agriculture. In the long run, this will make it possible for peer-to-peer transactions to be conducted transparently, eliminating the need for middlemen of any type in this industry. In the agri-food industry, confidence may be placed only in the peer-to-peer architecture and cryptographic processes rather than in a central authority. As a result, BC may be used to track the plant data on a variety of aspects, including seed quality, crop development pattern, and the path taken by them even after it leaves the particular farm. This would make it easier for the relevant authorities to recognize and reward farmers that use the best agricultural techniques.

20.3.4 Role in smart city development and security

Organizing people, companies, transportation, communication networks, services, and utilities makes conventional cities complicated. The authors of [43] carried out a thorough analysis of the issues that residents of traditional cities confront and suggested a path toward informed cities. Growing cities create organizational, social, and technical constraints that threaten sustainability and the economy. For example, the cities' pollution and waste are not well managed. Inadequate water management, wasteful power use, bad traffic control, and elevated crime rates are other major issues. Conventional cities need change and embrace technological advancements like distributed computing, CV, and IoT to become smart cities in order to meet these difficulties. By using smart city technologies, traditional cities may enhance resource management, reduce waste and pollution, foster economic growth, and improve public safety [44]. A smart city is a novel approach to urban development that unites diverse industries by utilizing cutting-edge technology such as distributed computing and the IoT. Enhancing the city's current resources through their integration, management, and improvement is the main objective. Process improvement is the main goal in a number of industries, including healthcare, electricity, transportation, water management, education, and business. A smart city is a novel concept in urban planning that uses cutting-edge technology like distributed computing and the IoT to connect disparate sectors. The main objective is to integrate, manage, and improve the city's current resources in order to improve them. Process improvement is the main goal in areas like healthcare, electricity, transportation, water management, education, and business.

AI has made it possible to monitor urban infrastructure, enhance public services, and build sustainable cities. This may be accomplished by gathering data from different intelligent devices used for different objectives. Security is among the most important functions that AI provides to smart cities. In light of this, CV is frequently utilized to ensure security in smart cities. This also has a number of uses, including the defense of vital infrastructure, waste and sanitation management, transportation, traffic control, pandemic prevention, security, smart water management, and disaster management. Through CV intelligence, these applications enable monitoring and interpretation of actions. On the other hand, the safety of the info produced by the CV systems is still an open problem. The information produced by CV-enabled systems is still susceptible to intrusion. Therefore, to safeguard the data produced by CV systems, an additional security layer is required, which BC technology can offer.

Since there are numerous applications, including intelligent transportation systems, video monitoring, and abnormality detection, it would be hard to keep an eye on suspicious actions [45]. Combining through the records of the suspicious incidents that are recorded on cameras and subsequently taking a relevant action may be quite time-consuming and tedious. A robotized solution where the suspicious act is predicted and then a notification sent to the necessary authorities so that they could act upon it in time is more inviting and beneficial to the population. Event surveillance can be achieved using a normal CCTV system, although this will not be the method to achieve round-the-clock monitoring of CCTV events manually. The authors of [46] developed a hybrid model that searches areas of interest in which suspicious activity might take place on the basis of the YOLO-V4 architecture.

To make the surveillance camera identify events and report them to the authorities, the system transfers information to the 3D-CNN architecture to recognize activities following temporal information. However, the interested parties in such activities can perfectly know what is happening with technological advancement and also using security cameras. As such, there might be an opportunity to make malicious attacks on the system before or even after the event. An in-depth review of possible malicious attacks, decision-based, black-box, as well as perturbation attacks, that can lead to the collapse of a CV system can be found in [47]. The well-trained model can go stupid due to these attacks, and it may be difficult to differentiate between it and the original picture. One can implement the BC technology to protect the system against these attacks.

Table 20.3 Technical insights of role of blockchain and computer vision technologies in some smart cities' development application areas

Application area	*Challenges in traditional CV*	*BC integration benefits*	*Integrated BC–CV solutions*
1 Critical infrastructure protection	• Manual monitoring errors due to multiple video streams • High bandwidth/ data rates needed • Latency in communication • Expensive *ad hoc* networks	• Secured data transmission • Data integrity and immutability • Dynamic reconfiguration via BC	• Edge-based deep learning recognition • Real-time SoC processing (100 FPS) • Cloud for facial recognition and tracking • 75% bandwidth reduction using BC
2 Traffic management	• Congestion due to urbanization • Real-time traffic data sharing risks • Central server bottlenecks • Vulnerable footage and data	• Secure data sharing via BC • Integrity of traffic footage • Prevention of manipulation/fraud	• CV-based roadside surveillance (CVROSS) • Ethereum BC+federated learning+off-chain storage • Cloudlet-based traffic analysis at edge
3 Security and surveillance	• Delay in analyzing recorded footage • Manual surveillance impractical 24/7 • Risk of adversarial attacks	• Event-based encryption • Decentralized secure data sharing • Protection against adversarial attacks	• YOLO-V4+3D-CNN hybrid model for suspicious activity detection • BC ensures secure vetting of shared data and alerts

It is possible to protect the system against malicious attacks with the assistance of the event-based encryption proposed in [48]. The author suggests a decentralized data sharing network in which transactions are stored through BC technology to ensure that the content is verified before being accepted by the network, a secure channel of exchanging confidential data. The technical understanding of the use of BC and CV technologies in the development as in some areas of smart cities is tabulated in Table 20.3.

4. CASE STUDIES OF BLOCKCHAIN AND COMPUTER VISION INTEGRATION

In this section, some well-known case studies or projects are discussed that are based on the integration of BC and CV technologies. The key details of such case studies are tabulated in Table 20.4.

5. DIFFICULTIES AND FUTURE DIRECTIONS

This section addresses the difficulties of including BC into a CV and offers potential fixes.

Table 20.4 Summary of different well known case-studies or projects based on integration of computer vision and blockchain technologies

Project name	*Application domain*	*Role of computer vision (CV)*	*Role of blockchain (BC)*	*Key benefits/outcomes*
Everledger [49]	Supply chain – luxury goods	Verifies physical authenticity (e.g., of diamonds) through CV-based identification	Stores immutable digital certificates of assets for transparency and verification	Prevents fraud, theft, and tampering; ensures ownership traceability; boosts consumer trust
Provenance [50]	Supply chain – wine industry	Tracks and identifies individual wine bottles throughout the lifecycle	Maintains a digital audit trail of the product lifecycle from production to consumption	Prevents counterfeiting and tampering; enhances supply chain traceability and accountability
Chainlink [51]	Decentralized oracle/Web3	Collects real-world data (e.g., weather and traffic) from cameras and sensors	Feeds data securely into smart contracts using decentralized oracles	Enables automated decisions (e.g., insurance pricing and payments) based on real-time verified data
Ocean Protocol [52]	Data exchange and analytics	Analyzes shared data using CV for insight generation	Provides secure, transparent data sharing; data are encrypted and stored immutably	Facilitates secure research, product development, and marketing; enhances data accessibility and usability
Grid+ [53]	Energy sector – smart grids	Monitors energy consumption and pricing via visual data	Enables peer-to-peer energy trading with secure, immutable transaction records	Eliminates middlemen; empowers users to trade energy directly; improves transparency and energy efficiency

i. **Need for Computation**

It could be necessary to make considerable adjustments to the hardware and software systems in order to integrate BC with the current CV solutions. Additionally, when connecting with older systems, implementation costs would rise further as a result of modifications to the hardware and software, licensing, maintenance, and other factors. The increasing amount of the data would necessitate more storage and processing power. Another major issue with BC deployment is the shortage of specialized engineers. Providing experts with specialized training and doing a thorough need analysis on the usage of BC in the system are two ways to overcome these issues. Another way to address the current implementation issues is through collaborative BCs with various working groups. The goal is to comprehend the issues that sectors face in common and work closely together to create meaningful solutions that will benefit all parties, even if no private information is shared. Another tactic that businesses might employ is BC as a service (BaaS), which allows all BC services to be utilized without requiring further investment in the technical expertise that supports them.

ii. **Lack of Standards and Rules**

The primary challenge encountered among BC-enabled CV systems is interoperability amid the wide range of options of BC networks. Lack of proper standards and laws is also another problem, which is to be solved when adopting BC in the cases of its incorporated usage with the systems, based on CV. The organizations have no norms to follow when they are implementing BC in their processes since there are no laid-down restrictions. This may result in what is known as a state of disarray in which different networks of BC cannot communicate with each other.

Promising interoperability and good agreement can be ensured through a series of international standards and administrative structures. This does not or cannot be done, endangering even the consistency of such basic practices as security and popular use. There are countries that do not encourage the use of BC technology due to a number of reasons, which includes both the environment and security. Nevertheless, there would be more companies that would start using BC in their CV systems with the introduction of relevant rules, control, and rules. It will not only facilitate a collaboration in the creation of applications but also help organizations connect with ancient systems more easily, make their proofs of concept, and exchange various BC-based solutions.

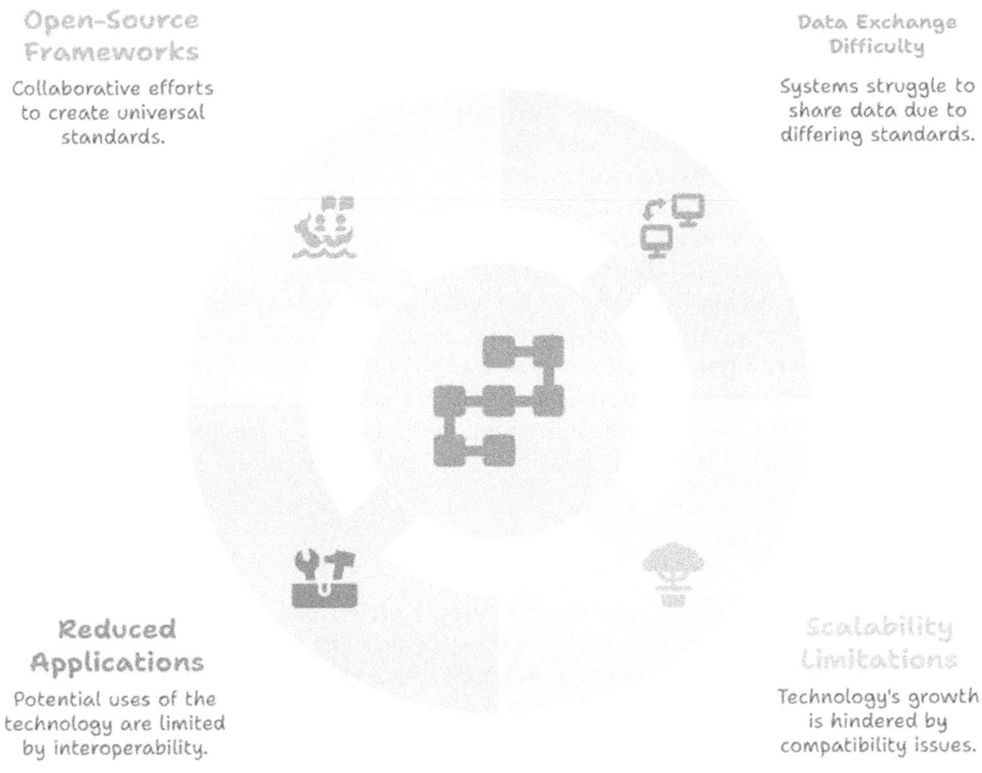

Figure 20.5 Challenges in adoption of blockchain technology with computer vision.

iii. **Scalability**

Scalability refers to the ability of handling a massive number of transactions at an effective speed in BC. Scalability is always a trade-off, and so it can only achieve two of three simultaneously and not three of three. This is what is termed as the BC scalability trilemma. As image capture and processing are a part of CV, the scalability of BC has to be discussed. The principal issue is presented by the BC because all concerned parties have to agree upon the legitimacy of transactions with the help of it. The scalability issues in integrating BC and CV may be resolved with the help of three basic strategies.

iv. **Interoperability**

Interoperability has become one of the greatest challenges to the adoption of BC technology in CV. Its issue arises since different BC networks could easily have different protocols and standards that could render the ability of dissimilar systems to communicate efficiently and accurately using data to be very difficult. This incompatibility can limit the technology scalability and use. As an example, we can take a CV system that will be developed and follows specific standards and regulations on one BC platform. In some cases, it may not be compatible with a different BC platform using various protocols and standards. Consequently, this makes sharing of data and models between the two systems complicated, thus reducing the opportunities where the technology can be applied. To deal with this issue, researchers and developers are investigating for solutions that would allow predetermined guidelines and procedures to be put in place to be followed when it comes to BC adoption in CV. This can involve the development of open-source structures and standards that support data transmission and systems to boost interoperability between multiple BC systems. Part of the challenges facing adoption of BC technology in CV can be solved through promotion of interoperability between different BC platforms.

v. **Adversarial Attacks**

Although using BC technology, hostile attacks remain a cause of concern despite the use of CV. Malicious attacks comprise distorting information or images to deceive CVs to make it classify incorrectly. Despite the fact that BC is capable of enhancing transparency and security, it would make no difference to the fact that the CV system might not be less vulnerable to impediment attacks. Various experts and programmers are, therefore, researching on many ways of making the machine learning models more robust and resilient to adversarial attacks. These strategies include the development of models that can resist negative perturbations or the application of ensemble learning methods, i.e., a combination of multiple models to make a general performance more accurate and reliable (Figure 20.5).

20.4 CONCLUSION

The synergy of BC and CV has huge potential to transform significant sectors such as supply chain management, smart cities, healthcare, defense, and agriculture. This synergy will provide a legal opportunity to encourage imaginative purposes. When CV is combined with the secure and decentralized properties of BC, the visual analytics capabilities of CV can be used to deliver a new set of opportunities that are characterized by greater levels of transparency, traceability, and accountability. One should understand that quite a number of substantial barriers and issues have remained unaddressed and unresolved, even though remarkable advances have been observed in the fields of supply chain, healthcare, and surveillance systems. These include interoperability, computational complexity, data privacy, adversarial threats, scalability, and data bias. It is important to know that despite the fact that such integration has yielded positive fruits, these present problems still need close attention and additional research projects. The combination of these factors is an indication of a future direction toward imaginative solutions. The cooperative efforts of both BC and CV can lead to the development of new technologies that will positively affect the society to a great extent.

REFERENCES

1. N. M. Su and D. J. Crandall, "The affective growth of computer vision," in *Proc. IEEE/CVF Conf. Comput. Vis. Pattern Recognit.*, Jun. 2021, pp. 9291–9300.
2. Y. Zheng, Y. Zhang, L. Qian, X. Zhang, S. Diao, X. Liu, J. Cao, and H. Huang, "A lightweight ship target detection model based on improved YOLOv5s algorithm," *PLoS One*, vol. 18, no. 4, Apr. 2023, Art. no. e0283932.
3. C. Szegedy, V. Vanhoucke, S. Ioffe, J. Shlens, and Z. Wojna, "Rethinking the inception architecture for computer vision," in *Proc. IEEE Conf. Comput. Vis. pattern Recognit.*, Jun. 2016, pp. 2818–2826.
4. Y. Zheng, L. Li, L. Qian, B. Cheng,W. Hou, and Y. Zhuang, "Sine-SSABP ship trajectory prediction based on chaotic mapping improved sparrow search algorithm," *Sensors*, vol. 23, no. 2, p. 704, Jan. 2023.
5. Y.-J. Zhang, "Computer vision overview." In *3-D Computer Vision: Principles, Algorithms and Applications*, pp. 1–35. Singapore: Springer Nature Singapore, 2023.
6. M. Maroto-Gómez, S. Marqués-Villaroya, J. C. Castillo, Á. Castro-González, and M. Malfaz, "Active learning based on computer vision and human–robot interaction for the user profiling and behavior personalization of an autonomous social robot," *Eng. Appl. Artif. Intell.*, vol. 117, Jan. 2023, Art. no. 105631.
7. Y. Zheng, P. Liu, L. Qian, S. Qin, X. Liu, Y. Ma, and G. Cheng, "Recognition and depth estimation of ships based on binocular stereo vision," *J. Mar. Sci. Eng.*, vol. 10, no. 8, p. 1153, Aug. 2022.
8. J. Chaudhary, and J. Yadav, "An effectual multiscale feature extraction in integer wavelet transform domain for illumination invariable face recognition," Int. *J. Appl. Pattern Recognit.*, vol. 7, no. 3–4, pp. 161–192, 2024.
9. X. Zheng, R. R. Mukkamala, R. Vatrapu, and J. Ordieres-Mere, "Blockchain-based personal health data sharing system using cloud storage," in *2018 IEEE 20th International Conference on E-Health Networking, Applications and Services. Healthcom*, IEEE, Ostrava, Czech Republic, 2018, pp. 1–6.
10. M. R. Kamble, H. B. Sailor, H. A. Patil, and H. Li. "Advances in anti-spoofing: From the perspective of ASVspoof challenges," *APSIPA Trans. Signal Inf. Process.*, vol. 9, pp. 1–9, 2020.
11. A. Sunyaev. "*Distributed ledger technology. In Internet computing: Principles of distributed systems and emerging internet-based technologies*" (pp. 239–272). Cham: Springer Nature Switzerland, 2024
12. J. Chaudhary, J. Yadav, and Rajpal, N. "A new fuzzy-based color reflectance space with selective feature extraction: Face recognition under diversified illuminations," *Arab. J. Sci. Eng.*, vol. 49, no. 3, pp. 4325–4351, 2024.
13. K. Deguchi, "A direct interpretation of dynamic images with camera and object motions for vision guided robot control," Int. *J. Comput. Vis.*, vol. 37, no. 1, pp. 7–20, 2000.

14. T. Schindewolf, R. Schiffner, W. Stolz, R. Albert, W. Abmayr, and H. Harms, "Evaluation of different image acquisition techniques for a computer vision system in the diagnosis of malignant melanoma," *J. Amer. Acad. Dermatol.*, vol. 31, no. 1, pp. 33–41, Jul. 1994.
15. J. Chaudhary and J. Yadav, "Large Scale Double Density Dual Tree Complex Wavelet Transform Based Robust Feature Extraction for Face Recognition," in *Proceedings of Second Doctoral Symposium on Computational Intelligence, Advances in Intelligent Systems and Computing*, Singapore: Springer, vol 1374, pp. 409–421, 2022. https://doi.org/10.1007/978-981-16-3346-1_33
16. J. Chaudhary, and J. Yadav. "Low-frequency DD-DWT based Illumination Compensated and Color Intensified features for face recognition," *Proc. Comput. Sci.*, vol. 258, pp. 1536–1553, 2025.
17. L. Yan, S. Yin-He, Y. Qian, S. Zhi-Yu, W. Chun-Zi, and L. Zi-Yun, "Method of reaching consensus on probability of food safety based on the integration of finite credible data on block chain," *IEEE Access*, vol. 9, pp. 123764–123776, 2021.
18. S. Bhattacharya, N. Victor, R. Chengoden, M. Ramalingam, G. Selvi, P. Maddikunta, P. Donta, S. Dustdar, R. Jhaveri, and T. Gadekallu, "Blockchain for Internet of Underwater Things: State-of-the-art, applications, challenges, and future directions," *Sustainability*, vol. 14, no. 23, p. 15659, Nov. 2022.
19. S. Wang, H. Sheng, Y. Zhang, D. Yang, J. Shen, and R. Chen, "Blockchain-empowered distributed multi-camera multi-target tracking in edge computing," *IEEE Trans. Ind. Informat.*, early access, Mar. 28, 2023, doi: 10.1109/TII.2023.3261890
20. Z. Qu, Z. Zhang, B. Liu, P. Tiwari, X. Ning, and K. Muhammad, "Quantum detectable Byzantine agreement for distributed data trust management in blockchain," *Inf. Sci.*, vol. 637, Aug. 2023, Art. no. 118909.
21. A. A. A. Shareef, P. L. Yannawar, Z. A. T. Ahmed, and A. M. Al-Madani, "Applying blockchain technology to secure object detection data," in *Proc. 3rd Int. Conf. Intell. Commun. Technol. Virtual Mobile Netw. (ICICV)*, Tirunelveli, India, Feb. 2021, pp. 874–879.
22. B. Crowell, "Blockchain-based metaverse platforms: Augmented analytics tools, interconnected decision-making processes, and computer vision algorithms," *Linguistic Phil. Investigations*, vol. 21, pp. 121–136, May 2022.
23. H. J. Hadi, Y. Cao, K. U. Nisa, A. M. Jamil, and Q. Ni, "A comprehensive survey on security, privacy issues and emerging defence technologies for UAVs," *J. Netw. Comput. Appl.*, vol. 213, Apr. 2023, Art. no. 103607.
24. J. Chaudhary, and J. Yadav. "Multiresolution analysis based on variants of wavelet transforms for illumination normalized face recognition," in *Doctoral Symposium on Computational Intelligence*. Singapore: Springer Nature Singapore, Mar. 2023, pp. 883–895.
25. U. O. Matthew, J. S. Kazaure, A. Onyebuchi, O. O. Daniel, I. H. Muhammed, and N. U. Okafor, "Artificial intelligence autonomous unmanned aerial vehicle (UAV) system for remote sensing in security surveillance," in *Proc. IEEE 2nd Int. Conf. Cyberspac (CYBER NIGERIA)*, Abuja, Nigeria, Jul. 2021, pp. 1–10.
26. D. Mahmudnia, M. Arashpour, Y. Bai, and H. Feng, "Drones and blockchain integration to manage forest fires in remote regions," *Drones*, vol. 6, no. 11, p. 331, Oct. 2022.
27. A. Rossiter and B. J. Cannon, "Turkey's rise as a drone power: Trial by fire," vol. 38, no. 2, pp. 210–229, Apr. 2022.
28. T. Wang, M. Du, X. Wu, and T. He, "An analytical framework for trusted machine learning and computer vision running with blockchain," in *Proc. IEEE/CVF Conf. Comput. Vis. Pattern Recognit. Workshops*, Jun. 2020, pp. 6–7.
29. T. Bhupathi, A. Chittala, and V. V. Mani, "A video surveillance based security model for military bases," in *Proc. Int. Conf. Recent Trends Electron., Inf., Commun. Technol. (RTEICT)*, Bangalore, India, Aug. 2021, pp. 517–522.

30. A. Yazdinejad, R. M. Parizi, A. Dehghantanha, H. Karimipour, G. Srivastava, and M. Aledhari, "Enabling drones in the Internet of Things with decentralized blockchain-based security," *IEEE Internet Things J.*, vol. 8, no. 8, pp. 6406–6415, Apr. 2021.
31. X. Chen, H. Xie, Z. Li, G. Cheng, M. Leng, and F. L. Wang, "Information fusion and artificial intelligence for smart healthcare: A bibliometric study," *Inf. Process. Manage.*, vol. 60, no. 1, Jan. 2023, Art. no. 103113.
32. S. V. Mahadevkar, B. Khemani, S. Patil, K. Kotecha, D. R. Vora, A. Abraham, and L. A. Gabralla, "A review on machine learning styles in computer vision—Techniques and future directions," *IEEE Access*, vol. 10, pp. 107293–107329, 2022.
33. T. Altameem and A. Altameem, "Facial expression recognition using human machine interaction and multi-modal visualization analysis for healthcare applications," *Image Vis. Comput.*, vol. 103, Nov. 2020, Art. no. 104044.
34. E. Elyan, P. Vuttipittayamongkol, P. Johnston, K. Martin, K. McPherson, C. Jayne, and M. K. Sarker, "Computer vision and machine learning for medical image analysis: Recent advances, challenges, and way forward," *Artif. Intell. Surg.*, vol. 2, pp. 24–45, Dec. 2022.
35. S. Ramzan, A. Aqdus, V. Ravi, D. Koundal, R. Amin, and M. A. Al Ghamdi, "Healthcare applications using blockchain technology: Motivations and challenges," *IEEE Trans. Eng. Manag.*, vol. 70, no. 8, pp. 2874–2890, Aug. 2022.
36. F. K. Shaikh, S. Karim, S. Zeadally, and J. Nebhen, "Recent trends in Internet-of-Things-enabled sensor technologies for smart agriculture," *IEEE Internet Things J.*, vol. 9, no. 23, pp. 23583–23598, Dec. 2022.
37. D. Bazzana, J. Foltz, and Y. Zhang, "Impact of climate smart agriculture on food security: An agent-based analysis," *Food Policy*, vol. 111, Aug. 2022, Art. no. 102304.
38. S. Lamba, V. Kukreja, J. Rashid, T. R. Gadekallu, J. Kim, A. Baliyan, D. Gupta, and S. Saini, "A novel fine-tuned deep-learning-based multiclass classifier for severity of paddy leaf diseases," *Front. Plant Sci.*, vol. 14, Sep. 2023, Art. no. 1234067.
39. G. C. Sunil, Y. Zhang, C. Koparan, M. R. Ahmed, K. Howatt, and X. Sun, "Weed and crop species classification using computer vision and deep learning technologies in greenhouse conditions," *J. Agricult. Food Res.*, vol. 9, Sep. 2022, Art. no. 100325.
40. R. Sapkota and P. Flores, "UAS imagery and computer vision for sitespecific weed control in corn," 2022, arXiv:2204.12417.
41. M. N. Ratnayake, D. C. Amarathunga, A. Zaman, A. G. Dyer, and A. Dorin, "Spatial monitoring and insect behavioural analysis using computer vision for precision pollination," 2022, arXiv:2205.04675.
42. A. Sivaranjani, S. Senthilrani, B. A. Kumar, and A. Senthil Murugan, "An overview of various computer vision-based grading system for various agricultural products," *J. Horticultural Sci. Biotechnol.*, vol. 97, no. 2, pp. 137–159, Mar. 2022.
43. K. Ergazakis, K. Metaxiotis, and J. Psarras, "Towards knowledge cities: Conceptual analysis and success stories," *J. Knowl. Manage.*, vol. 8, no. 5, pp. 5–15, Oct. 2004.
44. W. Iqbal, A. R. Javed, M. Rizwan, G. Srivastava, and T. R. Gadekallu, "Blockchain based secure communication for IoT devices in smart cities," in *Proc. IEEE Intl Conf. Dependable, Autonomic Secure Comput., Intl Conf. Pervasive Intell. Comput., Intl Conf. Cloud Big Data Comput., Intl Conf. Cyber Sci. Technol. Congr. (DASC/PiCom/CBDCom/CyberSciTech)*, Falerna, Italy, Jul. 2022, pp. 1–7.
45. J. Chen, M. Xu, W. Xu, D. Li, W. Peng, and H. Xu, "A flow feedback traffic prediction based on visual quantified features," *IEEE Trans. Intell. Transp. Syst.*, vol. 24, no. 9, pp. 10067–10075, Sep. 2023.
46. A. Rehman, T. Saba, M. Z. Khan, R. Damaševičius, and S. A. Bahaj, "Internet-of-Things-based suspicious activity recognition using multimodalities of computer vision for smart city security," *Secur. Commun. Netw.*, vol. 2022, pp. 1–12, Oct. 2022.
47. N. Akhtar and A. Mian, "Threat of adversarial attacks on deep learning in computer vision: A survey," *IEEE Access*, vol. 6, pp. 14410–14430, 2018.

48. R. Doku, D. B. Rawat, and C. Liu, "On the blockchain-based decentralized data sharing for event based encryption to combat adversarial attacks," *IEEE Trans. Netw. Sci. Eng.*, vol. 8, no. 2, pp. 1033–1043, Apr. 2021.
49. Everledger. Accessed: Mar. 23, 2023. [Online]. Available: https://everledger.io/industry-solutions/diamonds/
50. Misinformation Is a Growing Problem. Accessed: Jun. 16, 2023. [Online]. Available: https://www.newsprovenanceproject.com/
51. Powering the Next Generation of Blockchain-Enabled Insurance. Accessed: Jun. 6, 2023. [Online]. Available: https://chain.link/use-cases/insurance
52. Data: The New Asset Class. Accessed: Jun. 16, 2023. [Online]. Available: https://chain.link/use-cases/insurance
53. Grid+ Raises $29 Million as Blockchain Fever Grows. Accessed: Jun. 16, 2023. [Online]. Available: https://www.greentechmedia.com/articles/read/grid-raises-40-million-as-blockchain-fever-grows

Chapter 21

AI-enabled cloud security

A paradigm shift for data privacy protection

Hiral Chaitanya Prajapati, Sachinkumar Anandpal Goswami, Kashyap C. Patel, and Kinnari Utsavkumar Acharya

21.1 INTRODUCTION

Digital transformation sees cloud computing fused with artificial intelligence (AI) to usher in a new era in data security. As more organizations migrate critical infrastructure and sensitive data to the cloud, the urgency for security frameworks that are strong, adaptive, and intelligent is ever-growing. Traditional security models, characterized by being mostly static and reactive, cannot handle the scale, complexity, and dynamic threat landscape that characterize today's cloud environments.

The transformative potential of AI in improving cloud security is discussed here, not simply in a subjective incremental fine-tuning but more so as a paradigm shift in the conceptualization and implementation of data privacy and security. Things like threat detection, finding anomalies in real time, automated incident response, and continuous monitoring–compliance are all empowered by machine learning (ML), deep learning (DL), and analytics so that they become real-time and proactive features in protecting data for decentralized and elastic cloud ecosystems.

We begin by examining the limitations of conventional cloud security approaches and the pressing challenges faced by enterprises in maintaining data privacy. We then discuss AI-driven solutions, detailing how intelligent systems can learn from vast datasets, adapt to evolving threats, and provide predictive insights that bolster security postures. Finally, we highlight current trends, emerging use cases, and future directions that underscore the role of AI as a catalyst for a more secure and privacy-respecting digital future.

This paradigm shift is not only technical but also strategic, demanding a rethinking of policies, architectures, and ethical frameworks. As we delve into this intersection of AI and cloud security, we aim to equip readers with a comprehensive understanding of how this fusion is reshaping the landscape of data privacy protection.

21.2 THE EVOLUTION OF CLOUD SECURITY

Cloud computing has entirely taken over data storage, processing, and management activities of organizations by offering unparalleled scalability, flexibility, and cost advantages. Yet, these benefits have come with a corresponding evolution in security

DOI: 10.1201/9781003659112-21

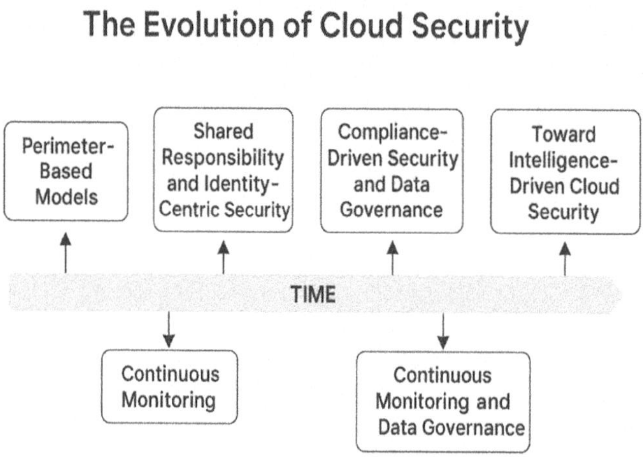

Figure 21.1 Overview of evolution of cloud security.

challenges. From its early adoption to its current dominance in enterprise IT, cloud security has undergone significant transformation in response to changing technologies, threat landscapes, and regulatory requirements [1–5] (Figure 21.1).

21.2.1 The early stage: perimeter-based models

In the initial phase of cloud adoption, security was largely modeled on traditional premise frameworks. Perimeter-based defences—such as firewalls, intrusion detection systems (IDS), and static access controls—were adapted to fit cloud infrastructures. These approaches assumed a clear boundary between trusted internal resources and external threats, a distinction that quickly became obsolete in the fluid, distributed architecture of the cloud.

21.2.2 Shared responsibility and the rise of identity-centric security

As IaaS, PaaS, and SaaS matured over time, shared responsibility emerged, whereby the security obligations of cloud providers and users were defined. This paradigm shift identified identity and access management as a key part of the foundation for security measures.

How the shared duty emerged with the maturation of all cloud service models (IaaS, PaaS, and SaaS) initially has the security obligation of the cloud provider and user defined. This shifted to identity and access management being the foundation for security. Stronger authentication methods, role-based access controls, and federated identity systems became essential for managing who could access what and under which conditions.

21.2.3 Advanced threats and the need for continuous monitoring

As cyberthreats progressed in depth such that they could be as much as a zero-day vulnerability or even advanced persistent threats (APTs), it is necessary to state that reactive and rules-based defenses were not enough. The organizations started implementing real-time, continuous monitoring, security information and event management (SIEM), and automated policy enforcement to enhance detection and response capabilities.

21.2.4 Compliance-driven security and data governance

Simultaneously, the rise of data privacy regulations such as the General Data Protection Regulation (GDPR), CCPA, and HIPAA introduced new compliance burdens. Ensuring data residency, access transparency, and breach notification compliance became critical, further complicating cloud security architectures. Security solutions now need to integrate governance, risk, and compliance (GRC) capabilities into cloud platforms.

21.2.5 Toward intelligence-driven cloud security

As cloud environments became more dynamic, decentralized, and data-rich, the limitations of human-driven security processes became evident. The sheer volume of telemetry data, combined with the speed at which threats evolve, called for a new approach—one that is adaptive, autonomous, and intelligent. This demand set the stage for the integration of AI into cloud security operations.

21.3 THE INTEGRATION OF ARTIFICIAL INTELLIGENCE IN CLOUD SECURITY

We contrast an AI-based approach to a much-old paradigm of protecting digital assets, wherein security systems used to depend on human action or manual setups with predefined rules, and observed that AI provides a means of autonomy, adaptability, and real-time decision-making to the ecosystem. In the fast-paced world of cloud computing, where environments are often complex, highly scalable, and continuously evolving, AI provides the intelligence and agility for fast and accurate detection, prevention, and response to threats [1–7] (Figure 21.2).

21.3.1 Proactive threat detection and prediction

The applicability of AI systems, especially those based on ML and DL techniques, to analyze massive volumes of real-time or time-stamped historic data serves to unveil remote threat indicators lurking behind subtle patterns. The continuous argument for the model to learn from new threat vectors reduces its dependence on signature detection. All of these models employ predictive analytics, which recognizes exploitable vulnerabilities before these threats can actually succeed so that organizations may stay one step ahead of their attackers.

21.3.2 Anomaly detection in dynamic environments

Cloud environments produce a huge amount of telemetry data such as logs, network flows, and user activities. AI's main forte is anomaly detection pertaining to baselining for normal behavior and flagging violations in real time. This capability is all the more

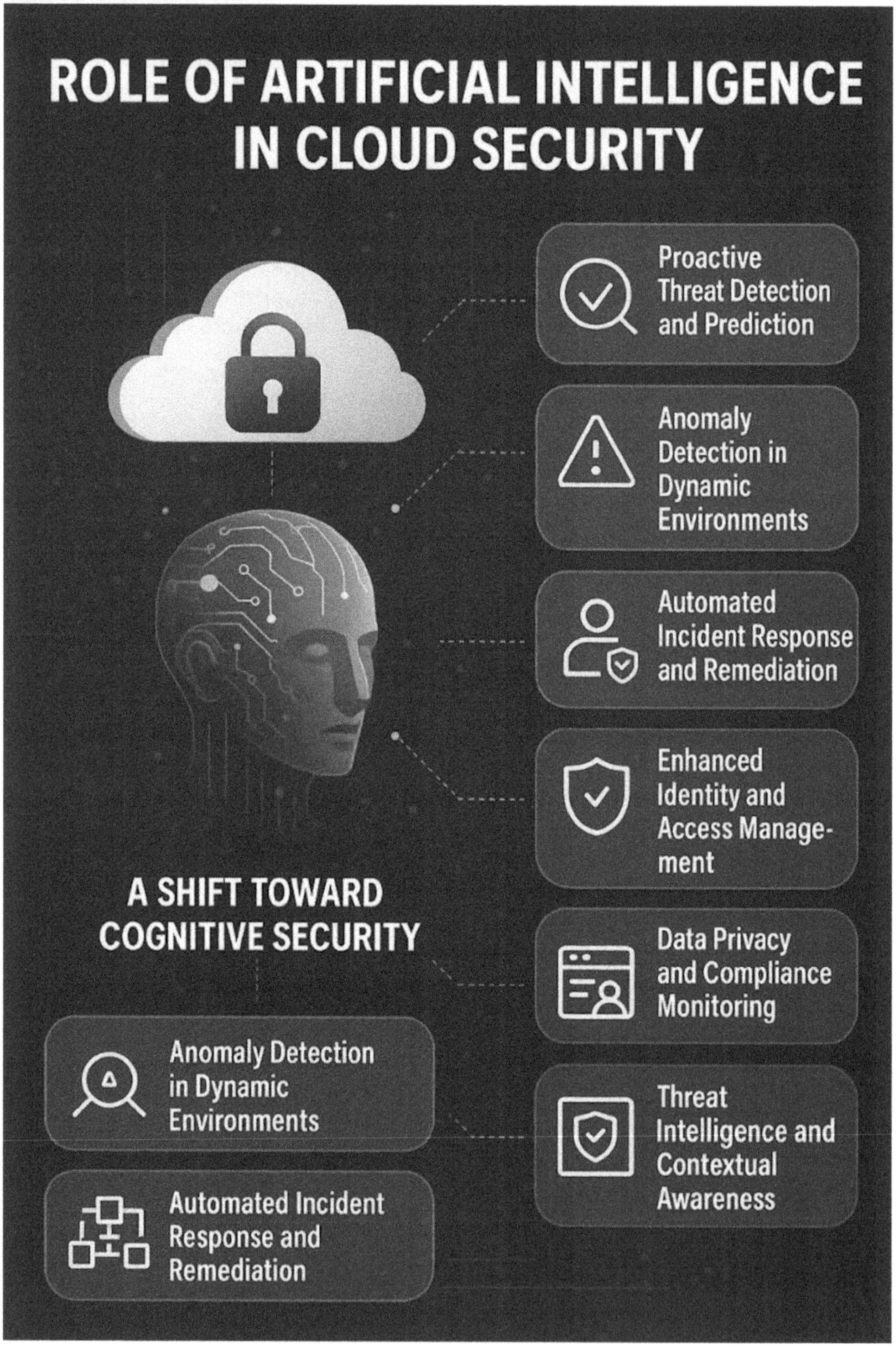

Figure 21.2 Overview of role of artificial intelligence in cloud security.

relevant for scenarios involving multi-tenancy in hybrid clouds where static security rules are not applicable. An AI system might, for example, notice peculiar patterns in user logins or attempts at data exfiltration that would otherwise go unnoticed by conventional systems.

21.3.3 Automated incident response and remediation

While AI assists in incident response by automating triage, classification, and prioritization of alerts, Security Orchestration, Automation, and Response (SOAR) platforms powered by AI enable companies to automatically isolate compromised systems, block malicious traffic, and apply patches to restrict damage within the shortest time available. The aforementioned automation also greatly limits human error and expedites recovery from any security event.

21.3.4 Enhanced Identity and Access Management (IAM)

AI improves IAM by providing context-aware authentication and continuous behavioral monitoring. It can analyze factors such as login time, geolocation, device fingerprinting, and historical usage patterns to assess risk and adjust access levels accordingly. Adaptive authentication and just-in-time access controls reduce the attack surface and protect sensitive resources from unauthorized access.

21.3.5 Data privacy and compliance monitoring

AI now makes sure that the regulations of privacy scale up to their stringent levels with the real-time classification of data, audits of access compliance, and privacy risk assessments. Such things as Natural Language Processing (NLP) can also be deployed to locate sensitive information such as Personally Identifiable Information (PII) and PHI across a multitude of cloud data stores. Further, compliance tools based on AI can provide the facility of mapping data flows and can detect possibly serious violations before they turn into extremely costly breaches.

21.3.6 Threat intelligence and contextual awareness

AI aggregates and correlates threat intelligence from multiple sources—internal systems, open-source feeds, and commercial threat databases—to provide contextual insights. This helps security teams make informed decisions based on the severity, origin, and behavior of threats. AI not only enriches alerts but also eliminates noise, allowing analysts to focus on high-priority issues.

21.3.7 A shift toward cognitive security

AI doesn't merely enhance cloud security—it redefines it. The transition from manual, reactive systems to intelligent, proactive defences marks a shift toward **cognitive security**, where systems can understand, reason, and learn autonomously. As cyberthreats grow in complexity and frequency, AI's ability to continuously adapt and improve makes it an essential ally in protecting cloud environments and ensuring data privacy.

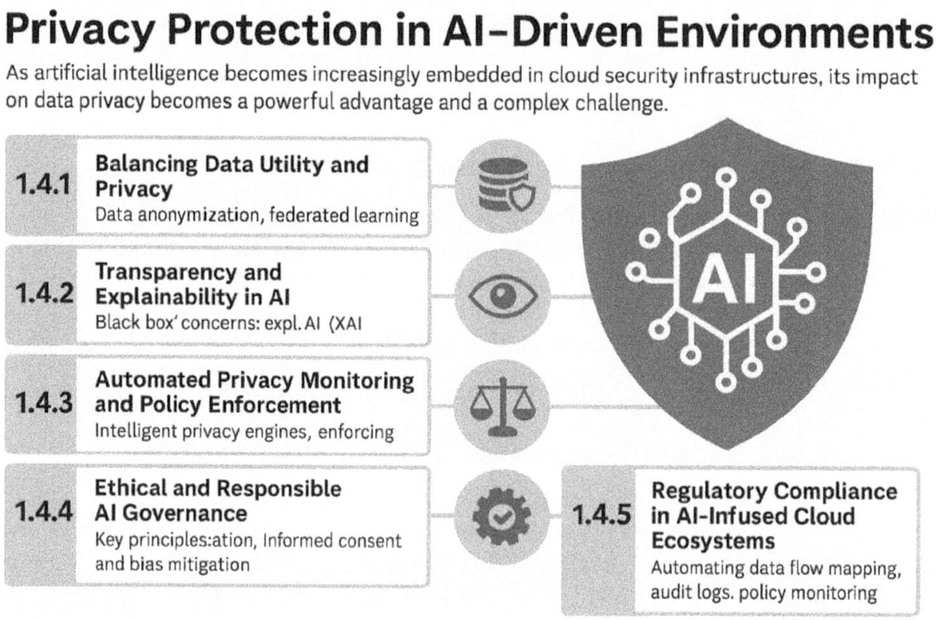

Figure 21.3 Overview of privacy protection in AI-driven environments.

21.4 PRIVACY PROTECTION IN AI-DRIVEN ENVIRONMENTS

As AI becomes increasingly embedded in cloud security infrastructures, its impact on data privacy becomes both a powerful advantage and a complex challenge. AI enables unprecedented capabilities in identifying threats and safeguarding systems, but it also introduces new vectors for potential privacy violations. In AI-driven environments, protecting user data is not just a regulatory requirement—it is a strategic imperative that underpins user trust, legal compliance, and the ethical use of technology [1–5] (Figure 21.3).

21.4.1 Balancing data utility and privacy

The training, learning, and performance enhancement of AI systems are necessarily dependent on massive volumes of data. However, such dependence often raises privacy issues when the dataset contains sensitive personal information, for example, health records, financial details, or behavioral patterns. The viable solution to this problem calls for devices that protect privacy to be able to weigh the essence of data against protecting privacy. Techniques such as data anonymization and data pseudonymization work by carefully and successfully removing or replacing identifiable information while retaining data of adequate analytical value. In yet another solution, federated learning allows training AI models across decentralized devices or servers without the need to transfer raw data, thus reducing the possibility of data exposure. Furthermore, noise is injected into datasets by the mechanism of differential privacy to further protect individual data points from being traced while still conducting meaningful analysis at an aggregated level. Thus, while privacy-protecting measures guarantee that individual identity is respected, these measures also ensure the efficiency of AI systems in operation.

21.4.2 Transparency and explainability in AI

A key privacy concern in AI systems is the "black box" nature of many ML algorithms. When AI models make decisions—such as flagging anomalies, restricting access, or escalating security incidents—it is essential that these decisions are explainable and auditable. Explainable AI (XAI) frameworks provide transparency into how models reach conclusions, enabling organizations to justify decisions, identify potential biases, and remain accountable to regulatory bodies.

21.4.3 Automated privacy monitoring and policy enforcement

AI can play a proactive role in protecting privacy by continuously monitoring data flows, access patterns, and storage locations across systems. Intelligent privacy engines powered by AI can automatically detect unauthorized access or potential data leakage, enhancing real-time security responses. These systems can also classify and tag sensitive information throughout distributed environments, ensuring that personal and confidential data are accurately identified and managed. Furthermore, AI can enforce data residency rules and retention policies in accordance with regional regulations such as the GDPR, HIPAA, and CCPA. By automating these complex privacy tasks, AI significantly reduces the workload on compliance teams and ensures that privacy safeguards are consistently and efficiently applied at scale.

21.4.4 Ethical and responsible AI governance

Ethical concerns must always govern any technical solutions intended to protect privacy in AI-dominated settings. This calls for organizations to adopt responsible AI governance frameworks that clearly guide AI model design, training, and deployment. Such frameworks must be built around some key principles, including the principle of data minimization or collecting only the data needed for a specified purpose, the principle of informed consent or making all users aware of AI data usage on a given set of information, and prevention of bias, which ensures that AI models are regularly checked for unlawful practices that violate privacy or render unfair treatment. By embedding ethical considerations into every stage of the AI development lifecycle, organizations can foster long-term trust, accountability, and sustainability in their AI initiatives.

21.4.5 Regulatory compliance in AI-infused cloud ecosystems

AI-based cloud set-ups operate under a copious gamut of privacy regulations that may differ across industries and regions. The far-reaching consequences of failing to comply with such regulations may include massive penalties, loss of reputation, and compromise of data security. The role of AI technologies includes automating processes that can greatly assist compliance efforts. For example, AI can map data flows automatically across multiple-cloud environments so that anybody can see where/how data are being used. They can implement automatic audit logging and report generation, lessening the burden of regulatory reporting and documentation. Furthermore, AI systems can monitor for policy violations in real time and alert organizations to correct issues and ensure compliance with relevant laws and regulations.

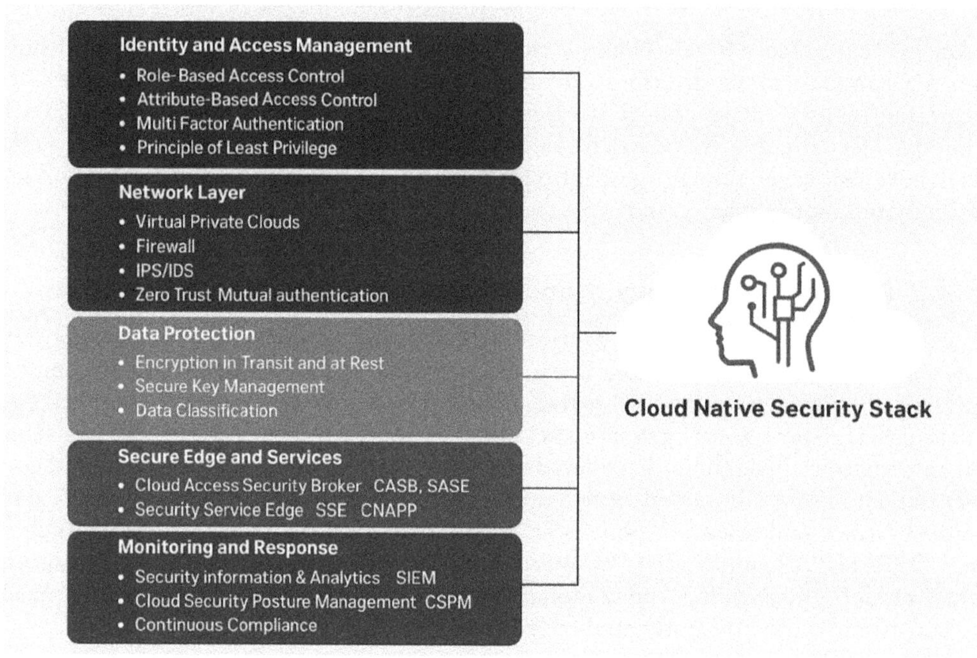

Figure 21.4 Overview of architecture of AI-enabled cloud security systems.

21.5 ARCHITECTURE OF AI-ENABLED CLOUD SECURITY SYSTEMS

The paradigm shift toward AI-enabled cloud security is driven by the increasing frequency of cloud-based threats, large-scale data breaches, and growing regulatory pressure to protect sensitive information. Traditional security measures are no longer sufficient to address these evolving challenges. In this context, AI plays a transformative role by enabling proactive threat detection, automating incident responses, and enhancing privacy preservation. Techniques such as homomorphic encryption and federated learning allow AI systems to process and learn from data without exposing it, ensuring both security and compliance. As a result, AI not only strengthens cloud defence mechanisms but also supports the creation of more resilient and privacy-conscious digital infrastructures [1–6] (Figure 21.4).

21.5.1 Cloud-native security stack

Cloud security has various layers that address different areas of protection and compliance in the comprehensive nature of an AI-enabled cloud security framework. For example, the Identity and Access Management (IAM) layer has technologies such as Role-Based Access Control (RBac) and Attribute-Based Access Control (ABAC) as well as Multi-Factor Authentication and the principle of least privilege as standards, which ensure that only authorized access to critical resources and services is granted to end-users. The Network Layer is characterized by the inclusion of Virtual Private Clouds, firewalls, IPS/IDS, and Zero Trust architecture features, including mutual authentication and segmentation.

Encryption of data in transit, at rest, and in use, including secure key management and data classification, as well as Data Loss Prevention (DLP) mechanisms constitute Data Protection schemes. For Secure Edge and Services solutions, hard perimeter defenses may include tools such as Cloud Access Security Brokers (CASB), Secure Access Service Edge (SASE), Security Service Edge (SSE), and Cloud-Native Application Protection Platforms (CNAPP). Lastly, any improved Monitoring and Response would include SIEM, Continuous Compliance Tracking, and Cloud Security Posture Management (CSPM), with increased integration of DevSecOps practices. Each entity creates an impenetrable, AI-influenced security defense ecosystem for modern cloud environments.

21.5.2 Threat detection and analytics

Advances in AI-like behavior-based anomaly detection and User and Entity Behavior Analytics (UEBA) are changing the face of cloud security. Such approaches study usage patterns at the level of user action and system behavior that can indicate whether there are any suspicious or unusual actions that potentially are security threats. These methodologies can classify network traffic, with accuracies exceeding 90% using models such as Random Forest, thus improving capabilities of identifying and preventing malicious activities in real time. Predictive analytics can contribute further to this strategy because they can forecast potential security incidents and rank risks by likelihood and impact. Hence, organizations "could" make this shift in AI's ability to drive a transformational move from reactive security to proactive, preventive posture.

21.5.3 Incident response automation

AI-powered triage engines and automated playbooks are transforming the way organizations respond to security incidents by enabling faster and more efficient remediation. These systems analyze alerts, prioritize threats based on the severity and context, and trigger predefined response actions without human intervention. By automating routine tasks and decision-making processes, AI significantly reduces response times and alleviates the burden on security teams. This not only enhances the overall effectiveness of incident management but also ensures consistent and scalable responses across complex cloud environments.

21.5.4 AI-Security Posture Management (AI-SPM)

A comprehensive AI asset inventory is essential for securing the entire lifecycle of AI systems, covering everything from model training and deployment to input and output monitoring. By maintaining full visibility into all AI components, organizations can better manage risks and enforce security policies. In addition to inventory management, model-specific monitoring plays a crucial role in detecting threats such as data poisoning, privacy leaks, and compliance violations. These targeted monitoring efforts help ensure that AI models remain trustworthy, secure, and compliant throughout their operation. Together, these practices enable a more resilient and transparent AI infrastructure within cloud environments.

21.5.5 Confidential and privacy-preserving computing

Different advanced methods are put into use today to improve privacy and security in AI and cloud settings regarding the protection of sensitive data during processing and analysis. Homomorphic encryption allows the processing of encrypted data without requiring the decryption of data to ensure that the data remain secure throughout the process. Isolation of secure environments within hardware, for example, Intel SGX, provides a secure execution environment for sensitive operations. Federated learning is another technique that allows training models for AI using decentralized data sources, with no transfer of raw data, and thus maintaining user privacy. It is possible to achieve multi-party computation (MPC), which enables computation over many parties regarding one function but maintaining the privacy of their individual inputs. ZKP also enables the validity of the information to be verified without exposing the actual data. In summary, these methods offer a robust toolkit for building private and secure AI systems.

21.5.6 Model/prompt safety

Prompt vetting pipelines and output filtering are essential tools for securing generative AI systems deployed in the cloud. A solution like Google's Model Armor aims to intercept user inputs for analysis and sanitization before reaching the model, thus preventing the abuse of prompt injection and other such attacks on the system. Online filtering mechanisms also check responses from the model for bias, leakage of PII, unsafe content, and noncompliance. These layers of protection make generative AI systems operate safely, ethically, and in compliance with data privacy laws, thus making them less hazardous for deployment in real-life scenarios.

21.5.7 Integrative framework: principles and patterns

Modern cloud security strategies are built on foundational principles and best practices that ensure robust protection across complex environments. Zero Trust principles emphasize strict identity verification, context-based access controls, and microsegmentation to minimize unauthorized access and lateral movement within networks. Complementing this is the Defense in Depth approach, which employs multiple layers of security controls and architectural redundancies to defend against threats at every level. DevSecOps integration further strengthens cloud security by embedding security into the development lifecycle, using tools for secure Infrastructure-as-Code (IaC), automated compliance checks, and continuous vulnerability scanning. Additionally, adherence to standards and compliance frameworks such as ISO/IEC 27017, CSPM, and AI risk frameworks like Google SAIF and NIST RMF ensures that organizations maintain high levels of governance, risk management, and regulatory compliance. Together, these strategies create a resilient and proactive cloud security posture [1–5].

21.6 CASE STUDY AND REFERENCE DESIGNS

When combining theoretical knowledge with practice, architectural reference designs and real-world case studies become reference points for successful practical implementation of AI-enabled cloud security. It illustrates how some of the best companies and cloud service providers around the world currently leverage AI technologies in strengthening threat detection for automating security operations and safeguarding data. This section showcases practical models and frameworks underscoring the ability and virtue of AI in modern cloud environments [1–5].

21.6.1 Google cloud example

Google Cloud offers a robust and secure infrastructure for deploying AI models through a combination of advanced networking and security features. Vertex AI operates within a private Virtual Private Cloud (VPC), ensuring that model training and inference occur in an isolated, secure environment. Private Service Connect (PSC) and VPC Service Controls (VPC-SC) further enhance data protection by restricting service access and preventing data exfiltration across project boundaries. To defend against external threats, Cloud Load Balancers work in tandem with Cloud Armor, providing scalable traffic management and DDoS protection. On the AI front, Model Armor ensures input/output sanitization to prevent prompt injection and privacy breaches, while Chrome Enterprise Zero Trust enforces secure device and user access across the enterprise. Together, these technologies form a multilayered security framework for safely deploying and managing AI workloads in the cloud.

21.6.2 Academic blueprints

AI-driven cloud incident detection leverages ML classifiers and containerization to identify and respond to threats in real time. By analyzing vast volumes of logs and behavior patterns, ML algorithms can accurately detect anomalies, unauthorized activities, and emerging attack vectors across dynamic cloud environments. Containerization enhances this capability by isolating workloads, making it easier to monitor and secure individual components of cloud-native applications. Additionally, confidential computing strengthens the protection of AI workloads by using secure hardware-based enclaves—such as those described in recent research (arxiv.org)—to process sensitive data while keeping them encrypted even during execution. This dual approach of intelligent detection and secure computation ensures both high-performance AI and strong privacy guarantees in cloud ecosystems.

21.7 PRIVACY AND TRUSTWORTHY AI IN THE CLOUD

Embedding Privacy-Enhancing Technologies (PETs) into cloud workflows is essential for safeguarding sensitive data while enabling advanced analytics and AI operations. Techniques such as data anonymization, differential privacy, and homomorphic encryption allow organizations to process information securely without exposing personal details. Alongside these technologies, adopting trustworthy AI principles—such as explainability, fairness, accountability, and transparency—ensures that AI systems operate ethically and are understandable to stakeholders. Additionally, cloud security is strengthened through shared responsibility models and transparency commitments, which include the use of verifiable enclaves and detailed audit logs. These measures promote greater trust between cloud providers and users by ensuring that data handling practices are secure, accountable, and compliant with evolving regulations [1–5].

21.8 CHALLENGES AND FUTURE ROADMAP

Despite the rapid advancement of AI-driven cloud security, several challenges and innovation opportunities remain. Technical hurdles such as the scalability of PETs pose difficulties, especially when balancing processing latency with model accuracy. Operational complexity also arises when integrating AI Security Posture Management (SPM) into governance frameworks and across the AI model lifecycle. Meanwhile, emerging threats—including large language model (LLM) poisoning, adversarial attacks, and supply chain vulnerabilities—demand constant vigilance and adaptive defense strategies. Looking ahead, innovation is focused on solutions like blockchain-backed audit trails for tamper-proof accountability, cross-cloud orchestration to unify security across providers, and dynamic policy automation that can adapt in real time to evolving threat landscapes. These directions aim to build more resilient, transparent, and trustworthy AI–cloud ecosystems [1–5].

21.9 USE CASES AND APPLICATIONS

The integration of AI into cloud security is not merely theoretical—it is already delivering tangible value across various sectors. By enhancing threat detection, automating response mechanisms, and preserving data privacy, AI is transforming how organizations secure their cloud environments. This section highlights key real-world applications where AI-enabled security systems are effectively protecting infrastructure, ensuring compliance, and enabling smarter decision-making in dynamic cloud ecosystems [1–5].

21.9.1 Intelligent threat detection and response

AI-SIEM stands for Artificial Intelligence-based Security Information and Events Management. This is an achievement from modern times in the field of cybersecurity. Platforms like Microsoft Sentinel and Splunk, which use ML to examine billions of log entries in real-time, allow for rapid anomaly detection and insider threat detection. Unlike traditional systems that mostly rely on known signatures of malicious threats, AI-enhanced SIEM systems seek to identify behavioral deviations, making them more capable of identifying threats that are either unseen or evolving. This smart approach reduces the false positives dramatically, helping the security teams focus on real threats with far greater accuracy and efficiency while preventing them from diverting their energies to false threats.

21.9.2 Automated incident response and remediation

AI-driven playbooks in SOAR are transforming how organizations handle cyberthreats by enabling fully automated incident response. For example, when AI detects ransomware activity, it can instantly isolate the infected workloads, generate alerts, roll back system changes using pre-configured snapshots, and notify the security team—all without any human intervention. This level of automation drastically reduces response time from hours to mere seconds, significantly limiting the potential damage and preventing the spread of malicious activity across the network. By streamlining and accelerating threat mitigation, AI-driven SOAR enhances both the efficiency and effectiveness of security operations.

21.9.3 Data privacy and confidential computing

Confidential computing is an advanced approach to protecting sensitive data during processing by keeping it encrypted even while computations are being performed.

In practical use cases such as healthcare platforms, secure enclaves—like Intel SGX or AMD SEV—are integrated within cloud infrastructure to process confidential patient data without exposing it to unauthorized access. This ensures that the data remain protected throughout the lifecycle, including during AI model training. By maintaining encryption during computation, confidential computing helps organizations meet strict privacy and compliance requirements such as the HIPAA and GDPR, while still enabling powerful analytics and AI-driven insights on protected data.

21.9.4 User and Entity Behavior Analytics (UEBA)

Insider threat detection in enterprise cloud platforms is a critical use case where AI significantly enhances security by monitoring user behavior patterns. In platforms like Office 365 or Google Workspace, AI continuously analyzes metrics such as login times, access frequency, and data transfer activities to detect unusual or risky behavior. For instance, deviations like accessing systems during off-hours or performing large, abnormal downloads can be immediately flagged for investigation. This proactive monitoring helps identify potential risks stemming from compromised accounts or malicious insiders, enabling swift action to prevent data breaches and maintain organizational security.

21.9.5 Dynamic access control and identity protection

In an AI-driven context, adaptive authentication is something that uses cloud IAM to improve security with access control that changes according to the user's behavior and context. Financial services, for example, enforce with AI conditional access policies that require added verification steps—such as multifactor authentication—when a login looks suspicious or originates from an unusual geography. Such intelligent, context-aware approaches strengthen security because they dynamically respond to possible threats, while at the same time ensuring that the services can be accessed easily by legitimate users. Consequently, an organization can maintain a high level of security without sacrificing convenience or productivity in users.

21.9.6 Cloud Configuration and Posture Management

AI-driven CSPM plays a crucial role in maintaining secure and compliant cloud environments. By continuously scanning configurations across major platforms like AWS, Azure, and GCP, AI identifies potential misconfigurations—such as publicly accessible S3 buckets—that could lead to data leaks. Leveraging predefined compliance policies, AI can automatically remediate these issues in real time, ensuring that cloud resources adhere to security standards like ISO 27001 or NIST without manual intervention. This proactive and automated approach not only reduces the risk of security breaches but also streamlines compliance management across complex, multicloud infrastructures.

21.9.7 Secure DevOps (DevSecOps)

ML-based vulnerability prediction in CI/CD pipelines is an essential use case for enhancing application security early in the development process. Tools such as GitHub Advanced Security and Aqua Trivy leverage AI to scan Infrastructure-as-Code (IaC) templates and container images for potential vulnerabilities before deployment. By identifying and addressing security flaws during the build phase, these tools help developers reduce the attack surface and prevent the risky code from reaching production environments. This proactive approach not only strengthens security but also integrates seamlessly into DevSecOps workflows, enabling faster and safer software delivery.

21.9.8 Privacy-preserving AI workloads

Federated learning could indeed be a paradigm shift for training AI models on sensitive data without infringing on confidentiality, especially in multicloud ecosystems. In this context, hospitals can safely work together to train AI models based on patient data, while data remain decentralized. This guarantees that all confidential medical information stays local and secure, in compliance with stringent privacy statutes. In other words, federated learning shares the burden of distributed AI training while keeping data private—it's a great compromise between innovation and data enforcement in these complex multicloud ecosystems.

21.9.9 Regulatory compliance automation

Real-time compliance auditing using AI is a vital use case for industries like retail, where strict regulatory standards such as PCI-DSS must be met consistently. AI systems enable continuous monitoring of cloud services, automatically auditing configurations and operations to detect compliance violations as they occur. For example, retailers can use AI to flag misaligned security settings and instantly generate audit-ready reports for regulatory review. This proactive approach ensures continuous compliance, reduces the risk of costly penalties, and simplifies the audit process—allowing organizations to maintain trust and regulatory alignment in real time.

21.9.10 LLM and AI model security

Monitoring and protecting generative AI in the cloud has become a critical use case as organizations increasingly deploy LLMs like OpenAI's API or Google Gemini. To safeguard these models, AI is used to detect and mitigate risks such as prompt injection attacks, inadvertent leaks of sensitive data, and manipulation of model outputs. By continuously monitoring interactions and behaviors, AI helps maintain the integrity, safety, and compliance of generative AI deployments. This ensures that organizations can harness the power of LLMs securely while adhering to privacy regulations and ethical guidelines.

21.9.11 Industry-specific use cases

AI is playing a transformative role in enhancing cloud security across various sectors by addressing their unique data protection needs. In healthcare, AI-enhanced Data Loss Prevention (DLP) tools are securing patient records in cloud-hosted Electronic Health Record (EHR) systems, ensuring compliance with regulations like the HIPAA. In the finance sector, AI-driven anomaly detection helps prevent fraudulent transactions by monitoring user behavior in real time within cloud-hosted banking applications. In education, AI is used to monitor and control access to student data in Learning Management Systems (LMS), safeguarding student privacy and institutional integrity.

Meanwhile, government agencies are leveraging multilayered, AI-based monitoring and access control to protect classified information in cloud environments. Across all these domains, AI strengthens security, ensures privacy, and helps meet regulatory standards more effectively.

21.10 CHALLENGES AND LIMITATIONS

Despite the immense potential of AI in enhancing cloud security and data privacy, several technical, operational, ethical, and regulatory limitations must be acknowledged. These challenges impact implementation effectiveness, trustworthiness, and long-term adoption [1–7].

21.10.1 Data quality and availability

One of the key challenges in AI-driven cloud security is the reliance on vast, high-quality, and well-labeled datasets to accurately detect threats. AI models learn from historical data patterns, and their effectiveness heavily depends on the consistency and diversity of input data. However, when log data collected from various cloud environments are inconsistent, incomplete, or biased, the accuracy of threat detection can be significantly reduced. For example, anomaly detection systems may underperform or fail entirely if the training data lack a representation of diverse or emerging threat behaviors. This limitation highlights the need for standardized data collection practices and continuous model retraining to ensure reliable and robust AI-powered security.

21.10.2 Model interpretability and explainability

A major challenge in AI-driven cloud security is the lack of transparency in many AI techniques, particularly DL models, which often function as "black boxes." These models can produce accurate results, yet provide little insights into decision-making. This opacity can be problematic for security teams, who may struggle to understand why the AI flagged a particular behavior as malicious. In highly regulated industries such as healthcare and finance, where accountability and explainability are critical, this limitation can lead to reduced trust in AI-driven decisions. To address this, there is a growing need for XAI solutions that make AI reasoning more interpretable and transparent.

21.10.3 Adversarial AI and model exploitation

A significant challenge in AI-driven security is the vulnerability of models to adversarial attacks and data poisoning. Attackers can intentionally manipulate input data to exploit weaknesses in AI systems, causing them to misclassify or fail to detect malicious behavior. For instance, by subtly altering input features, attackers can bypass AI-based malware detection systems, making harmful software appear benign. These types of manipulations undermine the reliability of AI models and highlight the need for robust adversarial defenses and continuous model validation to ensure accurate threat detection in dynamic and hostile environments.

21.10.4 Scalability and performance bottlenecks

One of the key challenges in implementing AI-driven security across large cloud ecosystems is the high demand for real-time processing and computational resources. Latency-sensitive applications, such as fraud detection or intrusion prevention, require immediate analysis and response, but the complexity of analyzing vast and distributed data streams can introduce delays. These delays may result in false negatives or missed threats, compromising security effectiveness. For example, AI-based response systems may struggle to scale efficiently across thousands of virtual machines or containers, leading to inconsistent threat detection and slower incident remediation. Addressing this limitation requires optimized architectures, edge processing, and more efficient AI models to ensure timely and scalable protection.

21.10.5 Privacy risks in AI operations

A critical challenge in AI-driven systems is the risk that models themselves may inadvertently leak sensitive information used during training, a vulnerability known as model inversion. Without the implementation of privacy-preserving techniques such as differential privacy or federated learning, AI models can unintentionally expose personal or confidential data. For example, LLMs trained on extensive log data may reveal fragments of PII during interactions, posing serious privacy risks. This highlights the urgent need to integrate robust privacy safeguards into AI pipelines to prevent data leakage and maintain user trust and regulatory compliance.

21.10.6 Integration complexity

Integrating AI tools into existing cloud security architectures presents a significant challenge due to the complexity and diversity of modern IT environments. Effective integration requires deep compatibility with a wide range of application programming interfaces (APIs), security platforms, and legacy infrastructure components. This can be particularly difficult when organizations operate across multiple cloud providers. For instance, an AI security system designed specifically for Azure may not seamlessly integrate with the configurations and protocols used in AWS or GCP, leading to fragmented visibility and inconsistent protection. Addressing this limitation requires the development of cross-platform AI solutions and standardized integration frameworks to ensure cohesive and scalable security across hybrid and multicloud environments.

21.10.7 Bias and ethical implications

Possibly the most important challenge in AI security is that the models may inherit biases from the training data, which lead to unequal enforcement or even unfair enforcement of security measures. In AI systems trained on historical or societal bias datasets, the systems produce outcomes presented as prejudiced measures such as flagging suspicious user behavior or biased access control. Thus, the security practice fails in fair and trustworthiness terms, and it can lead to legal and moral consequences. To equity, AI-related security will depend on good data curation, continuous audit of bias, and development lifecycle integrated with fairness algorithms. You are trained on data until October 2023.

21.10.8 Regulatory and legal challenges

Ensuring that AI-enabled cloud security solutions comply with international data protection laws presents a significant challenge for organizations. While AI enhances security operations, the absence of standardized frameworks for auditing AI-driven decisions complicates regulatory compliance, particularly when it comes to laws like the GDPR. For example, in scenarios involving automated incident response, it can be difficult to demonstrate how decisions were made, what data were accessed, and whether user rights were respected. This lack of transparency and auditability makes it harder for organizations to prove compliance, underscoring the need for XAI models and well-defined governance structures in cloud environments.

21.10.9 Over-reliance on automation

A key challenge in the adoption of AI for cloud security is the risk of over-reliance, which can lead to complacency and a reduction in essential human oversight. While AI excels at processing vast amounts of data and detecting patterns, it may overlook nuanced threats or context-specific anomalies that a skilled human analyst would recognize. This limitation becomes particularly critical during complex or targeted attacks, where subtle indicators might escape automated detection. The result is a false sense of security, potentially creating dangerous blind spots within the organization's security posture. To mitigate this risk, AI must be used as a complement to—rather than a replacement for—human expertise, ensuring a balanced and vigilant defense strategy.

21.10.10 Cost and resource constraints

Implementing AI-based cloud security tools can be prohibitively expensive, posing a major challenge for many organizations—especially smaller ones. These advanced solutions often require significant financial investment, including high licensing fees for AI-powered SIEM or SOAR platforms and the ongoing cost of GPU-based compute instances necessary for model training and inference. Additionally, smaller organizations may lack the in-house expertise or infrastructure needed to deploy, integrate, and maintain these complex systems effectively. As a result, they may be unable to fully benefit from AI-driven security innovations, potentially widening the security gap between large enterprises and smaller entities (Figure 21.5).

```
Step 1: Data Ingestion
INPUT data FROM users, devices, IoT
STORE data TO cloud platform
Step 2: AI-Based Threat Detection
FOR each data_packet IN cloud_storage:
    preprocessed_data ← preprocess(data_packet)
    threat_score ← AI_Model.predict(preprocessed_data)
    IF threat_score > predefined_threshold THEN
        FLAG data_packet AS "Potential Threat"
        ADD to Threat_Log

Step 3: Automated Security Response
FOR each flagged_item IN Threat_Log:
```

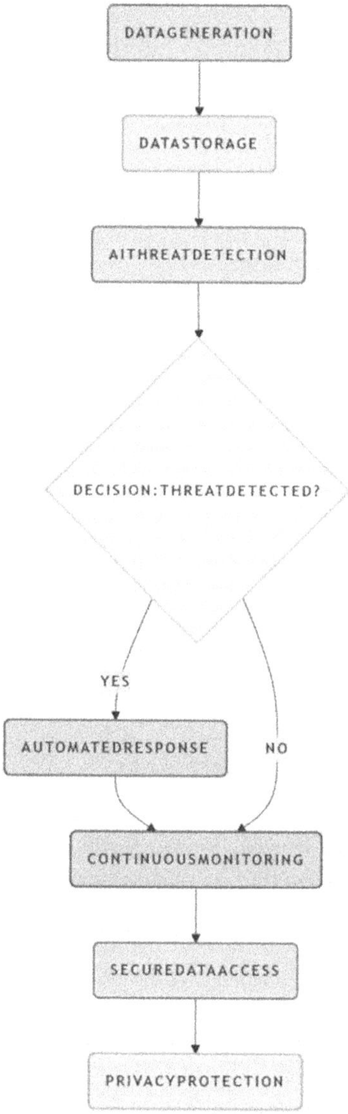

Figure 21.5 Methodological flow.

```
    EXECUTE security_response(flagged_item)
        - isolate source
        - alert admin
        - block IP or user
        - update firewall rules

Step 4: Privacy Protection
FOR all user_data:
    APPLY encryption(user_data)
    MASK sensitive_fields(user_data)
```

```
    STORE encrypted_data TO secure_bucket

Step 5: Continuous Monitoring
WHILE system_is_active:
    FETCH real-time activity logs
    ANALYZE logs USING anomaly_detection_model
    IF anomaly_detected:
        TRIGGER alert
        UPDATE model WITH feedback

Step 6: Compliance and Logging
FOR each transaction or access event:
    RECORD log_entry(timestamp, user_id, action, status)
    STORE log_entry TO immutable_audit_ledger

Step 7: Access Control
FOR each access_request:
    IF user_role HAS permission:
        GRANT access
    ELSE:
        DENY access AND log incident
END
```

This pseudocode outlines the workflow of an AI-enabled cloud security system designed to enhance data privacy and threat management. It begins by ingesting data from various sources such as users, devices, and Internet of Things (IoT) systems into a cloud platform. Each data packet is then preprocessed and evaluated by an AI model to determine its threat level. If the predicted threat score exceeds a defined threshold, the data are flagged for further investigation. The system automatically responds to flagged threats by executing predefined security actions such as isolating the source, sending alerts, blocking access, or updating firewall configurations. To ensure data privacy, sensitive user information is encrypted and masked before storage. The system also performs continuous real-time monitoring, using anomaly detection models to identify suspicious behavior. All system interactions are recorded in a secure audit log to meet compliance requirements. Additionally, role-based access control ensures that only authorized users can access specific resources. This architecture ensures a proactive, scalable, and privacy-focused approach to cloud security.

21.11 FUTURE DIRECTIONS

Explanations and inquiries related to exceptions in cybersecurity are being handled in an ever-evolving manner; in this particular case, AI and its implications for cloud security will serve as a hallmark to describe how organizations protect their data and systems.

One of the most noteworthy innovations in future will be self-governing and self-healing security systems. These systems, equipped with advanced reinforcement learning and continuous adaptation, will, at a very low intervention of a human operator, detect, respond, and even remediate threats. Another emerging future concern will be the implementation of post-quantum cryptography, where AI in conjunction with quantum-resistant authentication will protect data in the cloud from the dangerous quantum computing threats of the future.

On top of that, federated learning and decentralized AI models will become more mainstream to handle the excessive issues of data privacy and compliance. These techniques enable organizations to build AI models without transferring sensitive data across regions, thus fully complying with data sovereignty laws. Demand for XAI and transparent AI will go hand-in-hand. Increased demand for XAI will continue, especially in sectors where it is highly demanded by understanding and auditing security decisions.

AI will have an enlarged scope in threat intelligence with a shift from reactive defenses to predictive capabilities. Futuristic systems will detect and notify possible attacks based on recent analytics in real-time user behavior and global threat intelligence to prepare organizations ahead of breaches. Concurrently, dynamic and context-aware access control mechanisms will evolve, which will be based on an AI evaluation of a variety of parameters like user behavior, location, and device risk, resulting in suitable access provision in real time.

With the advent of DevSecOps workflows, the use of AI copilots in development environments will become commonplace. These copilots will assist the developer in the creation and deployment of cloud-native applications in a secure manner. They will help spot vulnerabilities, propose solutions, as well as administer security protocols even before code production. AI-driven orchestration tools will be essential to create uniform security monitoring, response, and policy enforcement across different platforms, particularly as organizations go toward multicloud and hybrid-cloud strategies.

In this manner, future systems will continue the trend of incorporating AI techniques to preserve privacy, such as homomorphic encryption, differential privacy, and secure multiparty computation. These conditions would allow cloud providers to process encrypted data, possibly without compromising the privacy of the users involved. Finally, the technology of blockchains tends to be purely supportive by providing immutable and auditable logs of security events and AI conclusions, thus strengthening accountability and compliance.

Summarily, the future of AI-based cloud security lies in intelligent, ethical, and scalable systems that can respond nimbly to complex and dynamic threats in the digital landscape. This will require efforts across technological, regulatory, and ethical fronts to build a resilient, privacy-respecting cloud ecosystem.

Every time, the biggest advances will be around the enterprising technologies of autonomous self-healing security systems. These systems incorporate highly adaptive reinforcement learning and continuous adaptation to detect, respond, and even remediate against perceived threats in very little human involvement. Other future focus points will be post-quantum coding matters, whereby AI acts together with encryption resistant to quantum attacks for safeguarding data in the cloud against a haphazard future danger emanating from quantum computing.

Moreover, federated learning and decentralized AI models will be more commonly used to address growing concerns about data privacy and compliance. Such technologies allow organizations to create AI models without transferring sensitive data across regions, thereby fully complying with data sovereignty laws. Demand for explainable and transparent AI will go hand in hand with this. Such a demand will continue to grow at an increasing rate, for instance, in finance and healthcare industries, where it is important to understand and audit executive security-related decision-making.

AI will have an enlarged scope in threat intelligence with a shift from reactive defense to predictive capacities. Future prediction would rely on real-time behavioral analyses of users in aggregate and at the global level concerning threat intelligence to measure organizational responsiveness before breaches occur. AI is also used in dynamic context-aware access control, evaluating a number of factors such as the user's behavior, location, and device risk to be met with appropriate access provision in real time.

Development environments would soon have AI copilots incorporated throughout DevSecOps workflows to enable developers to build and deploy their secure cloud-native applications under the guidance of these copilots in detecting vulnerabilities, suggesting ways to remediate them, and enforcing security policies on the code before it reaches production. Such AI-driven orchestration tools would bring in-line security monitoring, response, and policy enforcement across diverse platforms, given that organizations are also rapidly moving toward multiple-cloud and hybrid-cloud strategies.

Future systems would continue with techniques that abide with privacy such as homomorphic encryption, differential privacy, or secure multiparty computation. These conditions would allow cloud providers to process encrypted data, probably without compromising the privacy of the users concerned. Last but not the least, the nature of blockchain technology is that it serves only a supporting role by showing immutable, auditable logs of security events, as well as AI decisions, thereby increasing accountability and compliance.

In conclusion, the future of AI-enabled cloud security will, indeed, rest much in ideal, intelligent, and scalable systems that can respond nimbly to complex and dynamic threats in the digital landscape. Such efforts will further require a collaboration across technological, regulatory, and ethical lines to build a resilient and privacy-respecting cloud ecosystem.

21.12 CONCLUSION

The fusion of AI with cloud computing seems to have created a new paradigm in cybersecurity, providing a new way to view, implement, and maintain data privacy and protection. As the environment over time becomes vast and complex, it appears that traditional security is falling short of fulfilling the requirement for responding to dynamic threats and their increasingly sophisticated nature. AI can be seen as a key enabler, offering intelligent automation, real-time threat detection, adaptive response mechanisms, and predictive analytics.

This chapter will cover topics such as AI-Ied architecture, use cases of systems, challenges in AI, and future directions. It will discuss the possible improvements in security that can be realized by bearing up all aspects related to behavior-based anomaly detection and automated incident response. The last few capabilities are about privacy-preserving data models, federated learning, etc., all enhancing cloud infrastructure while very much improving resilience with respect to security and operational efficiency. All of this improvement does not come without challenges; however, issues of data quality, model interpretability, compliance with regulation, and ethical considerations among others can keep the full potential of AI in cloud security unrealized, owing to limitations that do exist.

Autonomous, explainable, and ethically governed AI systems are said to be the utmost priority in the future, operating securely and transparently across hybrid and multicloud environments. It will require an interdisciplinary collaboration among cloud engineers, AI researchers, policymakers, and experts in cybersecurity to ensure that innovation continues with the "privacy rights, compliance needs and human-centered values."

AI-enabled cloud security is not simply a technological upgrade but a transition in a paradigm regarding how organizations perceive data privacy in this century, risk benefit from enhancing their efforts to secure digital assets and ensure regulatory alignment as they improve their stake in the increasingly data-driven world.

ACKNOWLEDGMENTS

I would like to take this opportunity to specially thank all those who played an integral role in the development of the chapter "AI-Enabled Cloud Security: A Paradigm Shift for Data Privacy Protection." I thank my mentors, coworkers, and academic advisers from the depth of my heart for their invaluable comments, constructive criticism, and endless encouragement throughout the writing process. Their expertise in the domains of cybersecurity, AI, and cloud computing added weight and value to this work. I also thank the various organizations and research platforms that provided access to current studies, case reports, and technical documentation that were instrumental in shaping the structure and content of this chapter. I especially thank my students and colleagues whose curiosity and engagement toward the subject of AI-driven security inspired numerous pragmatic examples and real-time applications discussed herein. Last but not the least, I sincerely acknowledge the unwavering support, patience, and overall motivation that family and friends have rendered me throughout the entire duration of this writing journey. This chapter is a product of joint efforts and shared knowledge. I hope many meaningful contributions will ensue from this gift toward the advancing field of AI and cloud security.

REFERENCES

1. Rao, P. M., & Saraswathi, P. (2021). Evolving cloud security technologies for social networks. In *Security in IoT Social Networks* (pp. 179–203). Edited by: Fadi Al-Turjman and B.D. Deebak, Academic Press.
2. Sharma, H. (2024). The evolution of cybersecurity challenges and mitigation strategies in cloud computing systems. *International Journal of Computer Engineering and Technology,* 15(4), 118–127.
3. Akinade, A. O., Adepoju, P. A., Ige, A. B., & Afolabi, A. I. (2025). Cloud security challenges and solutions: A review of current best practices. *International Journal of Multidis ciplinary Research and Growth Evaluation*, 6(1), 26–35.
4. Ahmed, M. & Hossain, M. (2014). Cloud computing and security issues in the cloud. *International Journal of Network Security & Its Applications,* 6, 25–36. https://doi.org/10.5121/ijnsa.2014.6103.

5. Sapkal, Abhishek & Heisnam, Leoson & Kusi, Satya. (2024). Evolution of Cloud Computing: Milestones, Innovations, and Adoption Trends, 11. 548–563.
6. A. Polamarasetti, "Role of Artificial Intelligence and Machine Learning to Enhancing Cloud Security," *2024 International Conference on Intelligent Computing and Emerging Communication Technologies (ICEC)*, Guntur, India, 2024, pp. 1–6, https://doi.org/10.1109/ICEC59683.2024.10837120.
7. Vashishth, T. K., Sharma, V., Sharma, K. K., Kumar, B., Chaudhary, S., & Panwar, R. (2024). Enhancing Cloud Security: The Role of Artificial Intelligence and Machine Learning. In P. Goel, H. Pandey, A. Singhal, & S. Agarwal (Eds.), *Improving Security, Privacy, and Trust in Cloud Computing* (pp. 85–112). IGI Global Scientific Publishing. https://doi.org/10.4018/979-8-3693-1431-9.ch004

Chapter 22

Fusion of Blockchain and Generative AI

Jessica Chauhan, Parul Verma, Shahnaz Fatima, and Ranjana Rajnish

22.1 INTRODUCTION

Blockchain technology was first introduced in 2008. In this regard, a person (or group) named Satoshi Nakamoto wrote a paper, in which he told the world about the idea of Bitcoin and blockchain. This technology showed a new path in the digital world. With the help of blockchain, people can make transactions without any bank or any middleman. There is no owner. All of us work together in this system. Every transaction in the blockchain is recorded in front of everyone. No one can interfere in it. Whatever is recorded is visible to everyone, and no one can erase or change it. This is the biggest strength of blockchain. The data are safe, and there is no need to trust anyone. In addition, there are smart contracts. These are rules that go by themselves and do not require any human being.

Now let's talk about generative artificial intelligence (AI). Generative AI can create new content by looking at old data. For example, GPT-4 can write in human-like language, and a generative adversarial network (GAN) model can create real-like photos, music, and videos. Nowadays, new things are being created with generative AI – such as art, music, coding, and videos made with AI.

When these two technologies (blockchain and generative AI) are combined, it can create very good and secure digital systems. For example, if an art or music is made from AI, then it can be kept on the blockchain like a nonfungible token (NFT). With this, the right of the original maker will always remain, and everyone will know who it belongs to. Similarly, smart contracts made of AI can be placed on the blockchain, which will work automatically, and no one will be able to change them.

This chapter will observe these two technologies well. It will also discuss how methods like Proof of Work (PoW) and Proof of Stake (PoS) work and how they prove that the transactions that have taken place on the blockchain are correct. The purpose of this chapter is to explain these two technologies in a simple and clear language so that anyone can read and understand easily.

22.2 BLOCKCHAIN FUNDAMENTALS

Blockchain is a kind of digital technology in which the record of transactions is kept securely and cleanly on many computers simultaneously. In this, each block is connected to the previous block like a chain.

 DOI: 10.1201/9781003659112-22

In this way, once data are added to the block, it becomes very difficult to change them. If one has to change the data, he has to change all the subsequent blocks, and the consensus of most of the computers in the entire network is required.

22.2.1 Blockchain features

Decentralization: There is no single person, company, or government owner in the blockchain. Every participant keeps his copy, and all the decisions are taken together. No single person or institution can mess with the system.

Immutability: Once a transaction is recorded on the blockchain and accepted by everyone, it is impossible to delete or change it. Therefore, the data stored on the blockchain are always correct and safe.

Transparency: Every transaction on the blockchain is visible to everyone. This allows everyone to trust each other because everything is open and clean.

Smart Contracts: These are a kind of programs that are placed on the blockchain. When the preset conditions are met, they work automatically. There is no need for any person or agency in between.

Example

The first and biggest example of a blockchain is Bitcoin. Bitcoin uses PoW technology. In this, miners solve very difficult math questions to validate the transaction and add it to the blockchain. This ensures that the transaction is secure and that there is no fraud in spending the same Bitcoin twice.

22.2.2 Blockchain and its application domain

Beyond its original association with cryptocurrencies, blockchain technology has found a wide range of application sectors thanks to its fundamental principles of decentralization, immutability, transparency, and security. The following are some important domains where blockchain is being implemented:

Financial Services: The most well-known use case is cryptocurrency, which makes safe and decentralized digital currencies like Bitcoin and Ethereum possible. Blockchain facilitates faster, cheaper, and more transparent international money transfers by eliminating intermediaries and reducing processing times.

Supply Chain: With the blockchain, a complete record of where the goods went from and where they reached can be kept so that no goods are counterfeit. Blockchain improves transparency and makes tracking products simple by offering an unchangeable record of a product's path from the manufacturer to customer. This aids in inventory control, fraud prevention, and authenticity verification. It offers real-time visibility of goods in transit, addressing issues like lack of inventory details during transportation. It also ensures the authenticity of products, especially in industries like food, pharmaceuticals, and luxury goods.

Health Care: Helps keep patient information safe and unaltered. Blockchain can safely and discreetly store patients' medical records while complying with privacy laws (such as the HIPAA). Patients have control over who has access to their information. It helps in monitoring medications from production to distribution, preventing fake medications, and guaranteeing their quality.

Digital Identity: With blockchain, people can manage their own identity data and keep them safe from others.

The fundamental features of blockchain – its distributed, immutable ledger, cryptographic security, and the ability to execute smart contracts – make it a transformative technology with the potential to revolutionize various industries by fostering trust, transparency, and efficiency.

22.3 LITERATURE REVIEW

Over the years, a lot of research and articles have been written on blockchain and generative AI. People have written in depth about the way they work, their benefits, and the problems associated with them.

The first and most important paper on blockchain was written by Satoshi Nakamoto in 2008 [1]. In this paper, he explained the technologies called Bitcoin and PoW. It explained how a system can be created in which the record of transactions can be kept safe without any bank or government intervening.

After this, many researchers worked on other techniques like PoS. For example, Ethereum 2.0 works on PoS, which can increase the speed of the blockchain and reduce power consumption.

Several studies have compared PoW and PoS to determine which methods are safer, which use less electricity, which are faster, and which are more decentralized. It has also been said in some recent articles that if PoW and PoS are used together, then the shortcomings of both can be overcome.

Research on generative AI began with Goodfellow and his team in 2014, when they created generative adversarial networks (GANs). Since then, a lot of work has been done on generative AI. Now AI has started making photos, text, and music like real.

This was followed by models such as GPT-3 and GPT-4, which can write human-like languages and code and communicate. The research also showed that generative AI can automatically write a code for smart contracts, create NFT art, and create virtual assistants.

Many researches have also talked about combining AI and blockchain. Just like AI can speed up mining, PoS systems can pick up good validators and help detect fraud on the blockchain.

However, there are some problems with both technologies. Just like PoW consumes a lot of electricity, AI sometimes gives wrong or biased results, and there are legal conundrums on who will have the right over what is made of AI.

Fundamentally, blockchain technology is a distributed ledger that makes decentralized, transparent, and safe record-keeping possible. The consensus mechanism, which is a set of guidelines that all participating nodes adhere to agree on the network's current state and validate new transaction blocks, is the cornerstone of a blockchain's integrity [2,3]. A distributed ledger would be vulnerable to fraud and inconsistencies in the absence of a strong consensus system.

The PoW consensus mechanism was used by early blockchain systems, including the well-known Bitcoin. To propose a new block, PoW participants, referred to as miners, must solve challenging computational challenges. The new block is added by the first miner to solve the problem, who also receives cryptocurrency as payment. Although PoW is extremely safe and impervious to assaults such as the "double-spend," it is criticized for its excessive energy usage and problems with scaling [4,5]. The competitive and resource-intensive character of the puzzle-solving process is the cause of the energy inefficiency [6].

Alternative consensus procedures have been created in response to PoW's shortcomings. One of the main alternatives is PoS, in which validators are selected to produce new blocks according to how much cryptocurrency they own and are prepared to "stake" as security. By substituting financial stake for computational labor, this technique drastically lowers energy consumption. One prominent illustration of this change is Ethereum's move from PoW to PoS [7].

Ethereum is a variation of PoS where token-holders elect a limited number of "delegates" or "witnesses" to validate transactions and produce blocks on their behalf. This makes the consensus process faster and more efficient, but it also introduces a certain degree of centralization [8].

Practical Byzantine Fault Tolerance (PBFT) is a classical consensus algorithm, not strictly a blockchain-native one, but it has been adapted for use in blockchain networks, especially private or consortium ones. PBFT is highly efficient but works best in networks with a known, fixed number of participants [9]. It is particularly effective for achieving rapid finality, meaning transactions are confirmed quickly and are irreversible.

This literature review prepares us for further research on how blockchain and generative AI can be combined to create safer and better digital systems.

Researchers continue to explore a wide range of other consensus protocols, including Proof of Authority (PoA), Proof of Elapsed Time (PoET), and various hybrid models. These protocols are designed to optimize for different factors like security, scalability, and decentralization, depending on the specific application [10–14].

22.4 BLOCKCHAIN AND GENERATIVE AI

Blockchain's strength makes it great for being combined with new technologies like generative AI. On the blockchain, things made with AI (like art, music, and code) can be saved so that no one can copy them and the right of the original creator remains.

22.4.1 How does blockchain security work?

See Figure 22.1

Explanation of the diagram:

- Each block has a list of transactions.
- There is a timestamp (when the block was created).
- The previous block has a hash so that the blocks remain connected.
- They are all connected like a blockchain.

The three main aspects that make a blockchain secure are as follows:

- **Check with all your friends:** Every transaction in the blockchain has to be checked by all network peers so that there is no confusion.

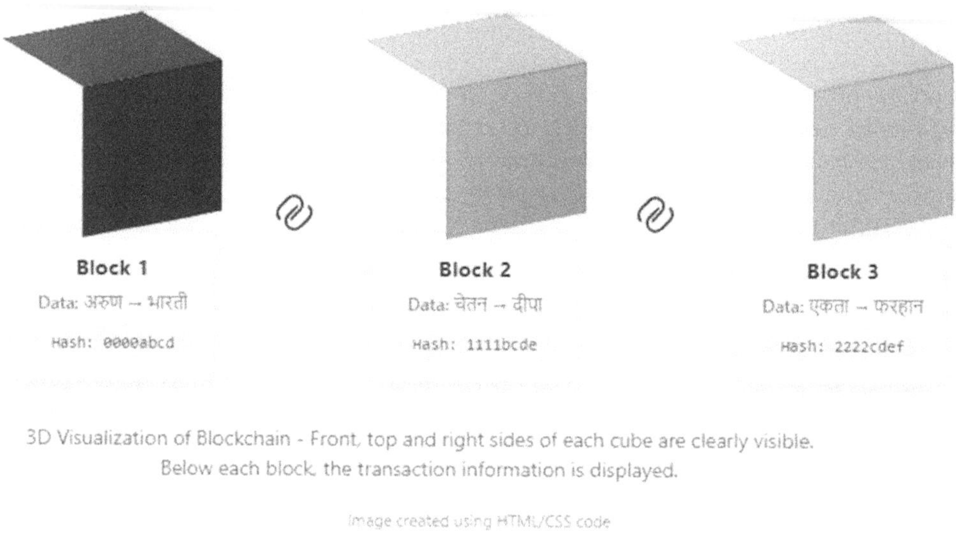

Figure 22.1 3D visualization of blockchain.

- **Encryption security:** Each block is protected by a special code (hash) so that if someone tampers with it, it is immediately detected.
- **Chain of Blocks:** Each block is connected to the previous block. Changing one means changing all, which is almost impossible.

22.4.2 The structure of the blockchain

See Figure 22.2

22.5 GENERATIVE AI FUNDAMENTALS

Generative AI is a part of AI which learns to create new and real data. It doesn't just read or understand old data, but it learns from it and does new work by itself. The job of generative AI is to create content that is neither copied nor slightly changed but is completely new. It can also create text (handwriting) code, pictures, music, and videos.

Some of the specific models of generative AI are as follows:

- **Generative Pre-trained Transformer 4 (GPT-4):** This model can write in human-like language. It can write code, answer questions, shorten documents, and have conversations. Its specialty is that it answers clearly and intelligently in the right context.
- **Generative Adversarial Networks:** It consists of two parts: generators and discriminators. The generator creates new data, and the discriminator checks whether the data are the same as the original. Both learn from each other and gradually create very real things like paintings, music, and other things.

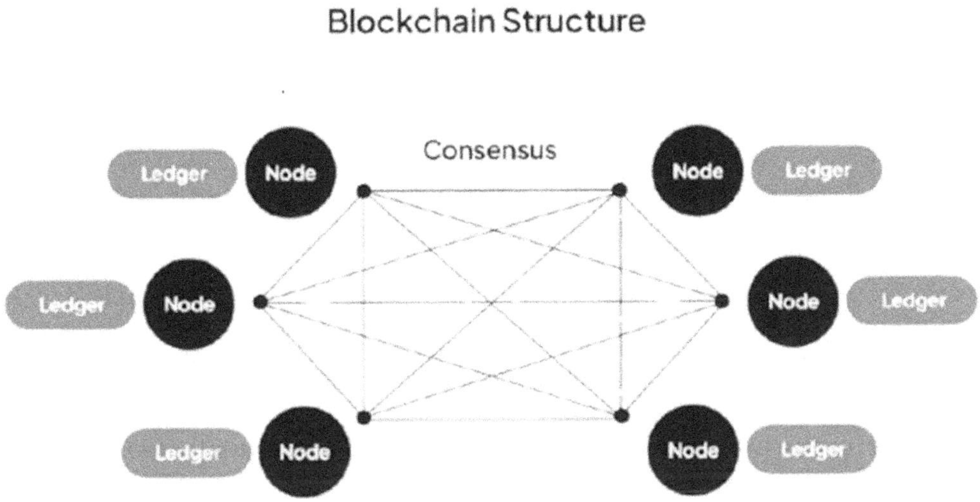

Figure 22.2 Blockchain structure.

Generative AI is being used in many places today:

- **Creating content:** Good articles, social media posts, advertising lines, etc. are created with AI.
- **Art and Design:** Artists create digital art with the help of AI and sell them on the blockchain by converting them into NFTs.
- **The Making of Music:** AI creates new tunes and songs.
- **The code:** Generative AI allows code to be written and corrected quickly in programming.

When generative AI is combined with blockchain, the work created by AI becomes even more secure. With the help of blockchain, it is decided as to who is the real owner of that work, whether there is any error in the work, and whether anyone is using it without permission. With this, things prepared with AI can be kept safe and sold in the right way.

22.6 CONSENSUS MECHANISMS

Consensus mechanisms are the basis of blockchain's strength. These mechanisms ensure that all the computers (nodes) in the network assume which transaction is correct and what is the current state of the ledger. Best of all, it does not require a central authority (such as a bank or government). Consensus mechanisms prevent fraud, double spending of the same money, and wrongdoing as new data are added to the blockchain only with the consent of all.

Any blockchain, especially a decentralized one, relies heavily on consensus procedures. Because they resolve the basic issue of how a dispersed network of autonomous and distrustful individuals may concur on the only, real state of a shared ledger in the absence of a central authority, they are vital.

All the people (nodes) in the blockchain network work by trusting each other. This trust is built through a consensus mechanism. While there are many different types of consensus mechanisms, they generally follow a similar high-level process:

- **Transaction Propagation:** When a user initiates a transaction (e.g., sending cryptocurrency and recording data), it is broadcast to the network.
- **Block Proposal/Validation:** Nodes in the network (often called "miners" or "validators" depending on the mechanism) collect these pending transactions and organize them into a new block.
- **Consensus Challenge:** The core of the mechanism involves a challenge that nodes must overcome to earn the right to add the new block to the blockchain. This challenge is designed to be difficult to solve but easy for others to verify.
- **Verification and Agreement:** Once a node successfully completes the challenge and proposes a new block, other nodes in the network verify its validity (e.g., checking if transactions are legitimate and if the challenge was truly met).
- **Block Addition:** If a majority of the network agrees that the proposed block is valid, it is added to the blockchain. This new block now becomes the latest "true" state of the ledger, and all nodes update their copies.
- **Incentives:** The node that successfully added the block is typically rewarded (e.g., with newly minted cryptocurrency or transaction fees).

PoW was the first consensus method developed by Satoshi Nakamoto, the creator of Bitcoin. Miners solve very difficult mathematical puzzles. The miner who solves the puzzle first gets a chance and reward (cryptocurrency) to add a new block. This keeps the blockchain safe because if someone wants to change it, he will need more than 50% of the computing power of the entire network, which is very difficult.

However, a major drawback of this method is that it consumes a lot of electricity. This harms the environment, and people talk about finding new, safer, and better alternatives for the environment.

PoS is a better and energy-saving method of PoW. Instead of mining it, those who have their cryptocurrency locked (staked) get a chance to add a new block. Those who have more stake have more responsibility to work honestly as they fear losses. It consumes less electricity and consumes less power. Ethereum's move to Ethereum 2.0 is a great example of this.

Here, AI can help so that the right validator is chosen, the glitch is caught quickly, and the network keeps running properly.

In PoA, only a few trusted people or entities (authority nodes) confirm the transaction and create a block. This method is faster and easier, but it reduces decentralization slightly. So it's better in a private blockchain where everyone knows each other. In this, AI can observe if an authority node is doing something wrong and can increase security.

Difference between PoW (and PoS)

Table 22.1 describes the differences between PoW and PoS

The contribution of AI can make the AI consensus system even better.

In PoW: AI can make mining strategies smarter so that more power is saved.

In PoS: AI can help in choosing the right validator by looking at the data of the network.

Table 22.1 Comparison of PoW and PoS

PoW (Proof of Work)	*PoS (Proof of Stake)*
The difficult mathematical question is being solved by computers, and the one who does it at the first place is allowed to add the block.	One having more number of coins is allowed to add blocks.
High energy consumption as the computers are working constantly.	Low energy consumption as the computers need not work hard.
It is quite safe as it is very difficult to be hacked.	It is also safe, but the one with more coins gets more power.
Reward in the form of new coins is awarded to the computer that solves the question.	Reward is given to the computer that stacks more coins.

In PoA: The AI can monitor the nodes and give an alarm on disturbances.

Consensus mechanisms are critical to keeping blockchains secure, reliable, and decentralized. As AI advances, blockchain systems will become smarter, safer, and more environmentally friendly.

All the people (nodes) in the blockchain network work by trusting each other. This trust is built through a consensus mechanism. It answers two key questions:

Maintaining the integrity of the data:

Data in the blockchain are stored on thousands of computers (nodes). If there is no consensus, everyone can have different accounts. By agreement, all the nodes agree on the same true information.

Protection against frauds:

Consent ensures that no erroneous transactions, repeated spending of the same coin, or tampering with old data can take place. For example, in Bitcoin, if someone has to change data, he must have half the power of the entire network, which is impossible.

22.7 INTEGRATION OF AI IN CONSENSUS MECHANISMS

Blockchain technology is increasingly being investigated and incorporated with AI, especially to improve and optimize consensus processes. By utilizing AI's analytical and predictive powers, this collaboration seeks to address some of the long-standing issues with blockchain, including scalability, energy consumption, and security.

AI can see when mining is profitable (PoW). Despite PoW's intrinsic energy requirements, AI may be able to improve mining operations by anticipating network loads and dynamically modifying the distribution of computational power. While the basic energy cost of PoW persists, this may entail optimizing resource consumption for mining jobs.

AI can help in choosing an honest validator (PoS). AI may evaluate validators' performance, dependability, and past behavior using machine learning. A more reliable and secure group of validators may result from using these data to guide more intelligent selection procedures for who gets to validate blocks. It can evaluate things like block formation efficacy, network involvement, and uptime. The distribution of stakes among validators can be optimized by AI algorithms, which aim for a more resilient and decentralized network and avoid an overabundance of power.

AI is quite good at identifying patterns. Blockchain technology can track network traffic, node behavior, and transactions to identify odd trends that could point to fraud or assaults (such as Sybil, 51%, DDoS, or double-spending efforts). AI models can forecast possible future threats by examining historical data of previous breaches and attack vectors. This enables the network to put preventive security measures in place.

AI helps in improving the scalability of blockchain systems by using the mechanism of sharding. It separates the blockchain into shards, which are smaller, parallel sub-chains. AI is able to dynamically generate, combine, or modify shards in response to workload distribution and network demand in real time. This avoids bottlenecks and guarantees effective resource allocation.

A small, preselected set of "authority nodes" (validators) are in charge of producing and approving new blocks under the PoA consensus method. PoA depends on the reputation and identity of these selected authorities, as opposed to PoS, which depends on staked assets, or PoW, which depends on processing power. AI can play a major role in monitoring authority nodes in a PoA network, guaranteeing its seamless operation, and resolving any "mess" or problems that might result from human error or malevolent intent.

22.7.1 Methodologies used

22.7.1.1 *What did we do in this research?*

In this research, we have used theory and practical methods to explain how blockchain and generative AI can be combined. This has been done like an Indian student preparing a study report.

22.7.1.2 *Proof of Stake simulation on Python*

We created a simple Python program to show how PoS works and how AI can help in choosing a validator. In this simulation, the stake of each validator is different, and there is a chance to be selected accordingly (Figure 22.3).

```
import random
validators = ["Validator_1", "Validator_2", "Validator_3", "Validator_4"]
stakes = [0.4, 0.3, 0.2, 0.1]  # Represents proportion of total stake held by each validator
selected = random.choices(validators, weights=stakes)
print(f"Selected Validator: {selected[0]}")
```

Figure 22.3 An example of Python code.

```
pragma solidity ^0.8.0;
contract SimpleToken{
        string public name = "Simple Token";
        string public symbol = "STK";
        uint8 public decimals = 18;
        uint256 public totalSupply = 1000000 * (10 ** uint256(decimals));
        mapping(address => uint256) public balanceOf;
        constructor() {
        balanceOf[msg.sender]=totalSupply;
        }}
}
```

Figure 22.4 Simple smart contract example.

Validator _ 1 has the highest stake (0.4, i.e., 40%), so it is most likely to be selected.

The old behavior of the AI validator in the real blockchain, looking at the history of the stake and the state of the network, can help choose the right validator so that the network works safely and correctly.

Smart Contract Code generated from GPT-4 also helps in generating smart contracts like GPT-4. This makes blockchain development easier. Below is an example of simple Solidity (the language of smart contracts), which is built from GPT-4 (Figure 22.4).

This contract generates 1,000 tokens and gives all the tokens to the person who created the contract. GPT-4 can create such coding instructions (commands), which saves time and reduces the possibility of error

This is a very basic token contract. It creates 1,000,000 STK tokens with 18 decimal places. All tokens are initially given to the contract deployer. Anyone can check balances using balanceOf[address].

The entire process of a blockchain transaction – from the start of the transaction to the confirmation is elaborated in Figure 22.5

22.8 ISSUES AND CHALLENGES

When blockchain and generative AI are combined together, it brings many new and good opportunities, but there are also some difficulties with it. It is important to understand these difficulties so that we can create the right and sustainable solutions.

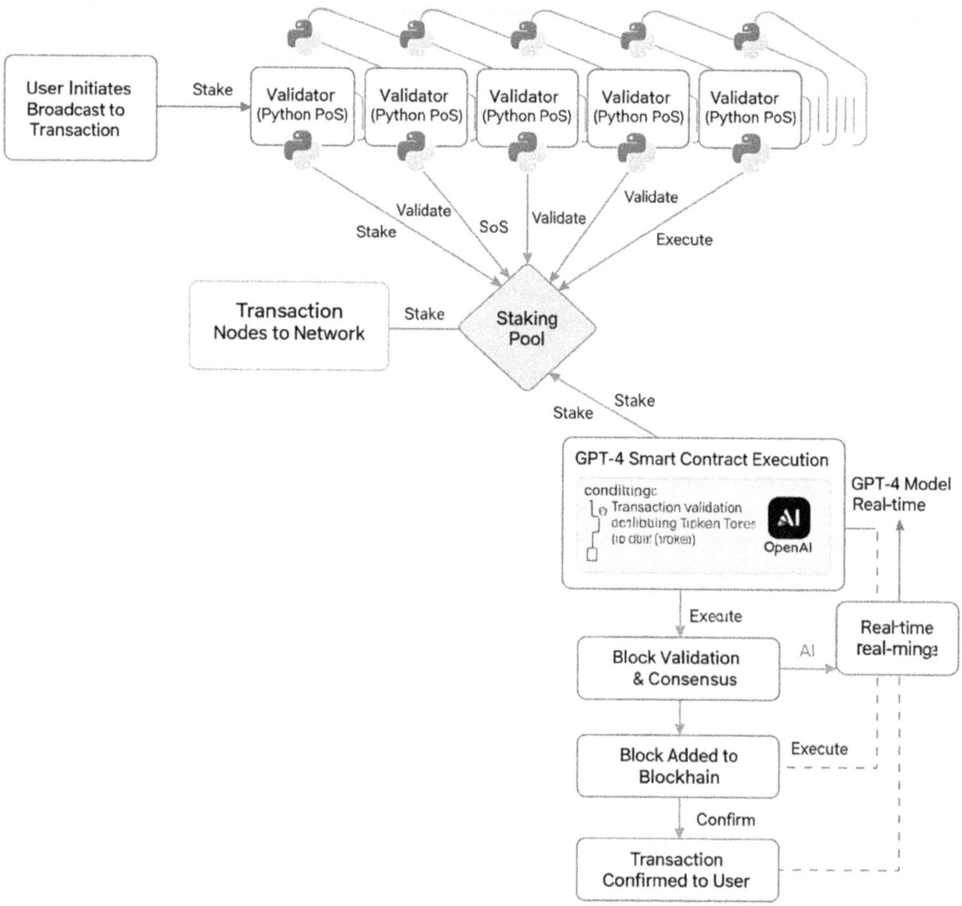

Figure 22.5 Blockchain transaction.

22.8.1 The challenges of blockchain

22.8.1.1 Scalability

When too many people and transactions take place on a blockchain network, it can slow down and increase transaction fees. This is the case with Bitcoin and Ethereum.

AI can help solve this problem, such as sharding the network or distributing the load so that the network works faster and cheaper.

22.8.1.2 Energy consumption

A system like PoW requires a lot of computer power, which consumes a lot of electricity. This affects the environment.

Although PoS is a better and low-power method, AI can see when there is less work on the network and at that time the network can be made even less energy-consuming.

22.8.1.3 *Security risks*

Blockchain itself is secure, but if the smart contracts built on it are not properly coded, they can be messed up.

AI can create tools that catch a coding mistake or weakness in a smart contract before it is deployed so as to avoid any losses.

22.8.2 Generative AI-related challenges

22.8.2.1 *Bias in AI models*

Generative AI learns from data that already exist. If that data itself are biased, then AI can create similarly false or inappropriate content.

That's why it's important that when we teach AI, the data are something equitable and just for everyone.

22.8.2.2 *Copyright and ownership*

Who will own the content that AI creates (e.g. art and code)? Who will be responsible if the code of AI makes a mistake?

To date, there are no clear answers to these questions in law. This is still a topic of discussion and is being worked on around the world.

22.8.2.3 *Ethical issues*

Generative AI can create deepfake videos or false information in a wrong way.

Therefore, it is important to use blockchain to differentiate between real and fake content and track the content from where it has come.

22.8.3 Challenges in integrating AI and Blockchain

22.8.3.1 *The complexity of the system (complexity)*

Combining AI and blockchain together makes the system very complex. Making it in such a way that both work properly and safely is not easy.

22.8.3.2 *The need for more resources*

Running an AI model with blockchain requires a lot of computing power and resources. This can reduce the savings and speed that should have been made.

22.8.3.3 *Interoperability*

Not every blockchain or AI system is the same. It's difficult to mix and match. It requires special design and technology. There are many challenges in combining blockchain and generative AI. But if we design thoughtfully, monitor with the help of AI, and improve the laws, then the real power of these technologies can be used properly.

22.9 CASE STUDIES

Case studies (real examples): To understand how blockchain and generative AI can be used together, we will learn about some real projects where these two technologies have been used successfully.

22.9.1 Ocean Protocol

The Ocean Protocol is a decentralized data exchange system where data owners and users can share data in a secure manner. It gives AI developers access to large datasets, allowing them to build better AI models.

Special features:

Data owners have full control over their data.

Those who share data get tokens (digital rewards).

With the help of blockchain, all transactions are transparent and secure.

22.9.2 SingularityNET

It is a marketplace where AI services (such as AI tools, and software) can be bought and sold. Here, developers can list their AI programs and users can buy them with the help of cryptocurrencies (e.g. Bitcoin and Ethereum).

Special features:

All transactions take place through smart contracts (automatic rules running on the blockchain).

Different AI services can learn by connecting with each other.

It is an open marketplace, where anyone can sell or buy AI tools.

22.9.3 Verisart

It works to certify digital art and artworks created by AI. It creates a certificate with the help of blockchain that proves that the artwork is real and not copied from something fake.

Special features:

AI art can be sold as an NFT (digital proof of ownership).

The entire history of every artwork is saved on the blockchain.

Fake works of art are avoided.

Findings: These case studies show that blockchain and generative AI can together create secure, transparent, and new way systems in areas such as data sharing, AI services, and digital art. With the help of these technologies, there is a possibility of even better innovations in the future.

22.10 CONCLUSION

Blockchain + Generative AI = Combining smart and secure technology blockchain and generative AI can make the digital world even more secure, transparent, and innovative.

Let's understand how these two technologies work together and what changes they can bring in the future. A turning point in technical development, the combination of blockchain technology and generative AI holds the potential to completely transform the fields of intelligent automation, data integrity, and digital trust. Fundamentally, both systems strive for increased security and decentralization, but they accomplish these ends via different means. Blockchain guarantees immutability, transparency, and resilience against single points of failure through its basic dependence on distributed ledgers, cryptographic hashing, and consensus processes (such as PoW and PoS). In a decentralized setting, these procedures serve as the cornerstone for establishing trust and facilitating consensus among diverse stakeholders in the absence of a centralized authority.

REFERENCES

1. Nakamoto, S. (2008). Bitcoin: A Peer-to-Peer Electronic Cash System. Available at: https://bitcoin.org/bitcoin.pdf
2. Bisant, D., & Kirsch, C. (2016). Distributed Ledger Technology: The Blockchain Explained. *Journal of Payments Strategy & Systems, 10*(4), 22–34.
3. Swan, M. (2015). *Blockchain: Blueprint for a New Economy.* O'Reilly Media.
4. NerdWallet. Proof of Work vs. Proof of Stake: The Biggest Differences. Retrieved from https://www.nerdwallet.com/article/investing/proof-of-work-vs-proof-of-stake
5. Atomic Wallet. Proof of Work vs. Proof of Stake: Explained. Retrieved from https://atomicwallet.io/academy/articles/proof-of-stake-vs-proof-of-work
6. Chanson, M., et al. (2019). Proof-of-Work versus Proof-of-Stake: A Comparative Economic Analysis. *The Review of Financial Studies, 32*(4), 1605–1643.
7. Buterin, V. (2014). A Next-Generation Smart Contract and Decentralized Application Platform. Ethereum Whitepaper. Available at: https://ethereum.org/en/whitepaper/
8. Larimer, D. (2014). Delegated Proof-of-Stake (DPOS) Consensus. BitShares Whitepaper.
9. Castro, M., & Liskov, B. (1999). Practical Byzantine Fault Tolerance. *Proceedings of the Third Symposium on Operating System Design and Implementation (OSDI '99),* United States. https://dl.acm.org/doi/10.5555/296806.296824
10. Bashir, I. (2022). *Blockchain Consensus: An Introduction to Classical, Blockchain, and Quantum Consensus Protocols.* Apress.
11. Yusoff, J., Mohamad, Z., & Anuar, M. (2022). A Review: Consensus Algorithms on Blockchain. *Journal of Computer and Communications, 10*(9), 37–50.
12. Guru, A., Mohanta, B. K., et al. (2024). A Survey on Consensus Protocols and Attacks on Blockchain Technology. *Applied Sciences, 13*(4), 2604.
13. Zhai, R., & Chen, X. (2021). Research on Blockchain Consensus Mechanism. *Frontiers of Data and Computing, 3*(3), 86–94.
14. Saad, M., et al. (2020). A Survey of Blockchain Consensus Mechanisms. *IEEE Access, 8,* 124194–124217.

Chapter 23

AI based blockchain in multidisciplinary business

Shivansh Mishra, Disha Srivastava, Shivam Agarwal, Shahnaz Fatima, Parul Verma, and Ranjana Rajnish

23.1 INTRODUCTION

In the past few years, digital innovations through artificial intelligence (AI) and blockchain have arisen as autonomous upheavals in different industries. AI provides abilities of data processing like data analysis, pattern recognition, and decision-making, which includes security and transparency of data in blockchain. The organic blending of these technologies promises many certain benefits to multidisciplinary businesses, leading to increased efficiency, trust, and innovativeness.

On the other hand, the routine of companies that include traditional business processes sometimes is limited by data security, transparency, and interoperability issues. Data separation from different systems is frustrating in messaging and interexchange of information across departments or agents concerned. Addressing these problems are nowadays focused more on the creation of the advanced and disruptive technologies that highlight the relationship between AI and blockchain (Figure 23.1).

The goal of this chapter is to investigate how AI and blockchain could appear in interdisciplinary business frames. This architecture uses AI algorithms for data analysis, prediction, and optimization and utilizes the blockchain technology for secure and reliable data management. Besides the integration and actual implementation of the system at the workplace, the research hopes to be an opportunity to know the improvements in organizational efficiency, transparency, and collaboration.

By conducting a thorough investigation between the advantages and failures of AI-powered blockchain projects, this study contributes to the knowledge about how companies may use these technologies to achieve innovation and change applicable across industries. In the next paragraphs you will learn the details of the implementation, the results, and discussions including the contributions that AI made to blockchain in a multiple discipline business context.

23.2 OBJECTIVES

AI and blockchain technologies are the focus of this study which revolves around integrating AI and blockchain technologies in multidisciplinary business settings. Specifically, the objectives include the following:

DOI: 10.1201/9781003659112-23

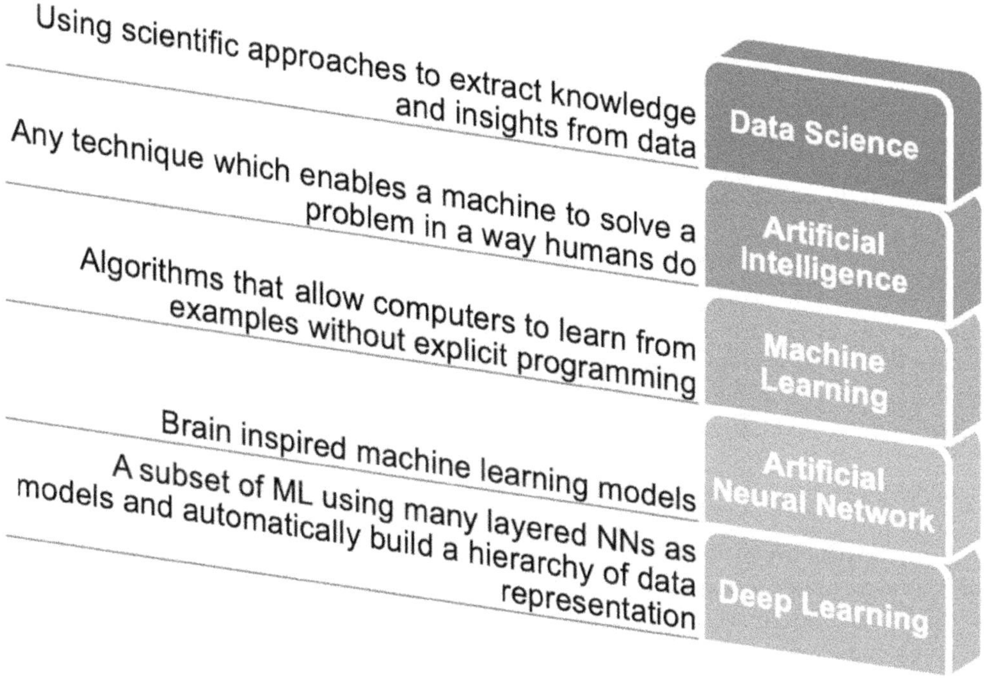

Figure 23.1 Key terminologies in analytics.

Designing an AI-Enabled Blockchain Framework: Build an all-embracing system which comprises AI algorithms for data analysis, forecast, and optimization and blockchain technology for safe, transparent, and data credibility.

Addressing Interoperability Challenges: Develop interfaces and protocols to make AI and blockchain compatible with the existing business systems and processes in order to guarantee smooth data flow and communication among the systems.

Enhancing Data Privacy and Security: Introduce privacy-preserving techniques and maintenance of security measures in the AI-enabled blockchain systems, which will infuse compliance to regulatory norms and standards.

Improving Scalability and Performance: Explore the aspects of the optimization performance and scalability of AI-powered blockchain solutions for handling heavy data traffic and growing transaction volumes in complex businesses.

Establishing Governance and Compliance Frameworks: Lay down governance frameworks and possession mechanisms that achieve compliance with due regulations and transparent, accountable, and ethical implementation of AI-enabled blockchain software.

Demonstrating the Value Proposition: Conduct empirical research and exhibit real-world scenarios of how an AI-enabled blockchain framework can increase the business productivity, efficiency, and innovation.

The accomplishment of this aim means that this research makes a real contribution toward the advancement of AI and blockchain integration in a multidisciplinary business environment, thus allowing organizations to fully utilize their combined powers for improvement of efficiency, transparency, and greater innovation.

23.3 PROBLEM STATEMENT

23.3.1 Prototype development

Create a first version of an AI-enabled blockchain platform which will use AI algorithms for gathering and optimizing data and blockchain technology for data security management. A prototype should be designed to show AI and blockchain integration and their ability to work together with the currently installed business systems.

23.3.2 Data integration

Bring in valid relevant data sources and feed them into the AI-enabled blockchain framework using consensus from different departments and stakeholders within the organization, ensure data quality and data compatibility to handle adequate analysis at the decision-making time.

23.3.3 AI model training

Entrain AI systems by transmitting the integrated data for functions like predictive analytics, anomaly detection, and optimization. Tune the models a bit to get the best accuracy and effectiveness in terms of narrowing your business concerns.

23.3.4 Blockchain implementation

Reconcile the framework by bringing on the blockchain component, which will securely deliver the data storage and management processes. Construct smart contracts to do the job of automation and enforcement of business rules and transaction on the blockchain.

23.3.5 User interface development

Develop interfaces and dashboards that will be used by relevant parties to steer and control the AI-powered blockchain system properly. Render visuals and insights drawn from AI analysis in order to help decision-makers in their work.

23.3.6 Pilot testing

Perform the implementation of the framework on a test basis with a protected business format within the organization. Gather feedback whether from users and stakeholders to help in uncovering anything relating to usability, performance bottleneck, or functionality gap.

23.3.7 Evaluation and optimization

Carry out a study on the efficiency and productivity of the given blockchain system by using a scorecard with predetermined metrics and objectives. Discover areas of improvement for optimization, adjustment, and enhancement that shall lead to as much functionality, scalability, and usability as possible.

23.3.8 Demonstration showcase

Present what was done through an AI-based blockchain framework to the stakeholders that are within the organization in a demonstration. Present the advantages, offerings, and guidance intended for the framework by using the demonstration and case scenarios.

23.4 PROPOSED WORK

Typically, traditional business procedures might face problems such as data security, transparency, and trust. Isolated data systems impede interoperable and collaborative systems, and this creates barriers that lead to inefficiencies and missed opportunities for innovation among different departments and stakeholders. Also, the growing amount and complexity of data calls for advanced analytical tools in turn to generate actionable results, facilitating correct decision-making.

Solving these challenges means utilizing the collaborative potential of AI- and blockchain-based technologies. Although AI affords great data analysis, pattern recognition, and automation capabilities, blockchain is a decentralized and unrevivable ledger for secure and transparent data management.

Interoperability: Integrating AI algorithms along with blockchain protocols into legacy systems and processes considering aspects of smooth interoperability and data transfer.

Data Privacy and Security: Saving risk of losing sensitive business data and security leakage, as well as privacy protection in AI-enabled blockchain systems, especially in regulated industries such as healthcare and finance.

Scalability and Performance: Overcoming bottlenecks in scalability and performance upgrades to cope with the data and transaction volumes, which usually grow in a complex business environment.

Governance and Compliance: Framework of governance and compliance with legal rules for AI-powered blockchain solutions, for example, data ownership, consent, and responsibility.

User Adoption and Trust: Overcoming the adoption barriers by building a trust among the stakeholders through presenting the reliability, transparency, and value provided by an AI-integrated blockchain solution provider.

In the face of such challenges, our research intends to design a complete architecture of AI and blockchain technologies to work effectively in a multidisciplinary business set-up. We aim to solve these issues to capitalize on the full inherent benefits of AI-enabled blockchain solutions for streamlining operations and introducing transparency for business operations.

23.5 IMPLEMENTATION OR DEMONSTRATION

23.5.1 Healthcare

23.5.1.1 PharmBloc

PharmBloc comes in the form of the blockchain platform integrated together with AI algorithms that are used to make pharmaceutical supply chain management more efficient. It uses IA to perform predictive analysis of demand for medications, the optimal level of inventory, and interdiction of the crisis drugs. Blockchain provides for a transparent and traceable way of payment transactions. This is important because it reduces the possibility of entry for counterfeit products into the supply chain [1].

23.5.2 Finance

23.5.2.1 AI-Pay

AI-Pay offers an AI algorithms-augmented blockchain based payment system to prevent frauds and illicit financial actions [2]. On the basis of historical data, the system can identify irregularities and highlight the possible cases of fraud by focusing on the real-time transaction patterns and user behavior. Blockchain-based systems make transactions secure and change-proof and guarantee the transparency of an audit trail for regulatory compliance [3].

23.5.3 Supply chain management

23.5.3.1 AgriChainAI

AgriChainAI is a blockchain-connected/blockchain -powered platform integrated with AI algorithms for optimization and easing up of agricultural supply chain logistics. The AgriChainAI is capable of doing this through analyzing weather data, crop yields, transportation routes, and market demand, whereby it can lower distribution costs and eliminate food waste and transportation costs. Blockchain provides transparency and trust in the supply chain, allowing various suppliers to know where their product has been before it gets to the customers [4] (Figure 23.2).

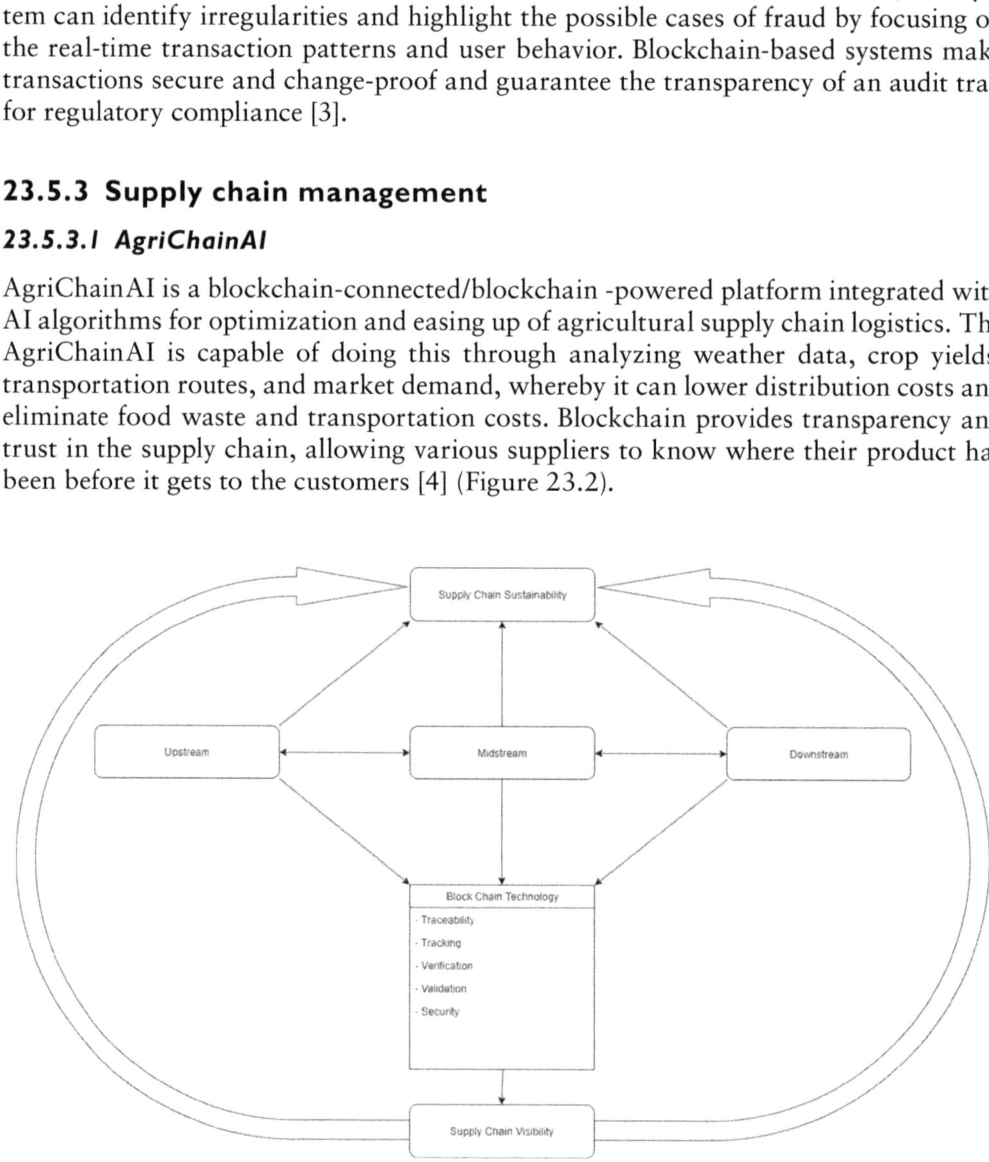

Figure 23.2 A conceptual model for blockchain-enabled sustainable supply chain visibility.

23.5.4 Energy

23.5.4.1 EnergyAI-Grid

EnergyAI-Grid – It serves as a blockchain-based platform integrated with AI algorithms designed for optimization of energy grid management. Through the ability to examine energy consumption trends, renewable energy generation, and grid infrastructure information, EnergyAI-Grid can easily adjust the energy distribution in such a way as it would help balance the supply and demand, thus unwinding grid congestion and enhancing energy efficiency. The blockchain guarantees security and integrity of energy transactions, enabling both peer-to-peer and decentralized grid management [5].

23.5.5 Intellectual property rights

23.5.5.1 IPGuardian

IPGuardian – Implementing a platform based on the blockchain technology integrated with AI algorithms serves for intellectual property rights management. IPGuardian will thus identify copyright violations, copyright protection enforcement, and, where applicable, compensate creators by analyzing copyright infringement, digital content usage, and licensing agreements. Blockchain implies the veracity of digital items and ownership, which contributes to the creation of a reliable and open record of intellectual property rights [6].

These abovementioned implementations make visible the AI–blockchain technology and its power to improve efficiency, transparency, and trust in different industries.

23.6 RESULTS AND DISCUSSION

The developed and applied AI-based blockchain architectures in the real-world, multidisciplinary business entities will evidently bring fruitful findings and stimulate deep discussions. In the subsequent segment, we will describe the core findings reviewed and raise their implications, as well as potential directions studied under various perspectives.

23.6.1 Enhanced data security and transparency

Results: The application of blockchain in business has greatly improved security of data, thanks to the design of a permanent and immutable ledger for saving and managing your data. Users of the network can check the soundness of transactions and have a view of a comprehensive audit trail of a data exchange.

Discussion: One of the significant impacts that this new-found data security and transparency brought in by blockchain is that stakeholders now trust the system more, and this facilitates a collaboration among them. As a result, processes have been streamlined, leading to a decrease in shedding, improved legal compliance, and an enhancement of trust in business integrity.

23.6.2 Improved decision-making with AI insights

Results: By virtue of AI algorithms put on the functionality, it has furnished valuable sensitivity through observing the data, pattern recognition, as well as predictive modeling. This capability has given decision-makers a tool to effectively analyze the datasets to identify the trend and, the ability to explore the datasets to understand and respond to future problems.

Discussion: AI which is applied for decision-making is in line with more data-driven organizations, which are more effective through costs savings and competitive advantage. Through application of AI, organizations will tap the hidden potential, control the risk, and lead innovations in various business domains.

23.6.3 Streamlined interoperability and collaboration

Results: The design of the framework facilitated the consistency, interoperability, and free data exchange movement within the different departments of the organization. Efficient interfacing with the current business systems was accomplished with minimal disruptions, promoting effective collaboration and communication.

Discussion: The framework which has simplified interoperability functions as a steam that has torn the barriers of the silos and the organizations operate in such a way that their efficiency and productivity are improved. Through creating an environment for teamwork and knowledge exchange, organizations can take advantage of a wide variety of experts and resources available, which will allow them to focus on their mission and ultimately make the collective achieve success (Table 23.1).

23.6.4 Challenges and areas for improvement

Results: Although the implementation and demonstration stages resulted in such successful outcomes, some challenges were revealed: limited scalability, governance issues, and user adoption limitations. Resolving these obstacles will be possible through continuous research work and fine-tuning of the framework.

Discussion: One of the challenges in implementing interoperability frameworks is that they should be scalable for the largest institutions and possibly whole industry ecosystems, so implementing robust governance is very important. Moreover, provision of education, training, and user-focused interface features is vital for overcoming the user adoption barriers, which will in turn promote the framework's success and value proposition.

23.6.5 Future directions and opportunities

Results: The effective and successful implementation and demonstration of the AI-enabled blockchain framework have seeded the research and development of future

Table 23.1 Benefits of blockchain and AI integration

Blockchain	*AI*	*Integration benefits*
Decentralised	Centralized	Enhanced data security
Deterministic	Dynamic/changing	Improved trust on decisions
Immutable	Probabilistic	Collective decision-making
Data integrity	Volatile	Decentralized intelligence
Attacks resilient	Data, knowledge, and decision-centric	High efficiency

studies. An additional chance for studies delving deep into modern AI methods like federated learning and reinforcement learning can be explored, and the scope of blockchain can be explored to make it applicable for new industries and use cases.

Discussion: The AI-powered blockchain will play an increasingly important role in changing the rules of the game in the industry. The use of emerging technologies and collaborative strategy will help the organization remain at the front of digital disruption by creating added value in an era of dynamic, connected business.

23.6.6 Conclusion

Ultimately, the incorporation of AI and blockchain serves to reshape the multidisciplinary business and push beyond the present established frontiers. We are in the course of the investigation of how AI and blockchain can be combined in a synergistic way to tackle the issues of the old business processes and to finding new innovative ideas for development.

Leveraging the data analysis, prediction, and optimization through AI algorithms along with blockchain technology ensures confidential and incorruptible data management, and organizations can upscale the collaboration and, hence, efficiency and transparency across the sectors. The earlier part sets an example of the wide range of AI-driven blockchain applications in healthcare, finance, supply chain management, energy, and intellectual property rights protection.

We have demonstrated with empirical analysis and proof that AI-enabled blockchain systems are not just a theoretical solution, but they are also delivering results somewhere, bringing benefits including improved data security, decision-making, and business operations. Furthermore, these AI-powered platforms can solve the problems related to interoperability by connecting different organizations and scalability by streamlining and automating processes and achieve regulatory compliance in a transparent way, all of which contribute to the diffusion of AI-powered blockchain use case in the market.

Considering ahead, future AI and blockchain technology allows for many new opportunities. As technology is only going to become more sophisticated, we expect an eventual emergence of a new raft of applications and cases, such as distributed learning on blockchain, decentralized marketplaces for AI, and Decentralized Finance (DeFi). The future of this intersection lies in the hands of researchers and developers as they need to work on how to maximize the potential of this interdisciplinary merging and their importance in shaping the future of multidisciplinary businesses.

In short, AI with blockchain integration stands for a new model of doing business for companies, especially hundreds collaborating and creating. Through that integration, companies will be able to stay on course, have digital transformation, and open up a new platform to reach success in a forever-changing and digital world (Figure 23.3).

23.7 FUTURE WORK

AI and blockchain will highly influence the future of work in such businesses in terms of both integration and innovation in the multidisciplinary sector.

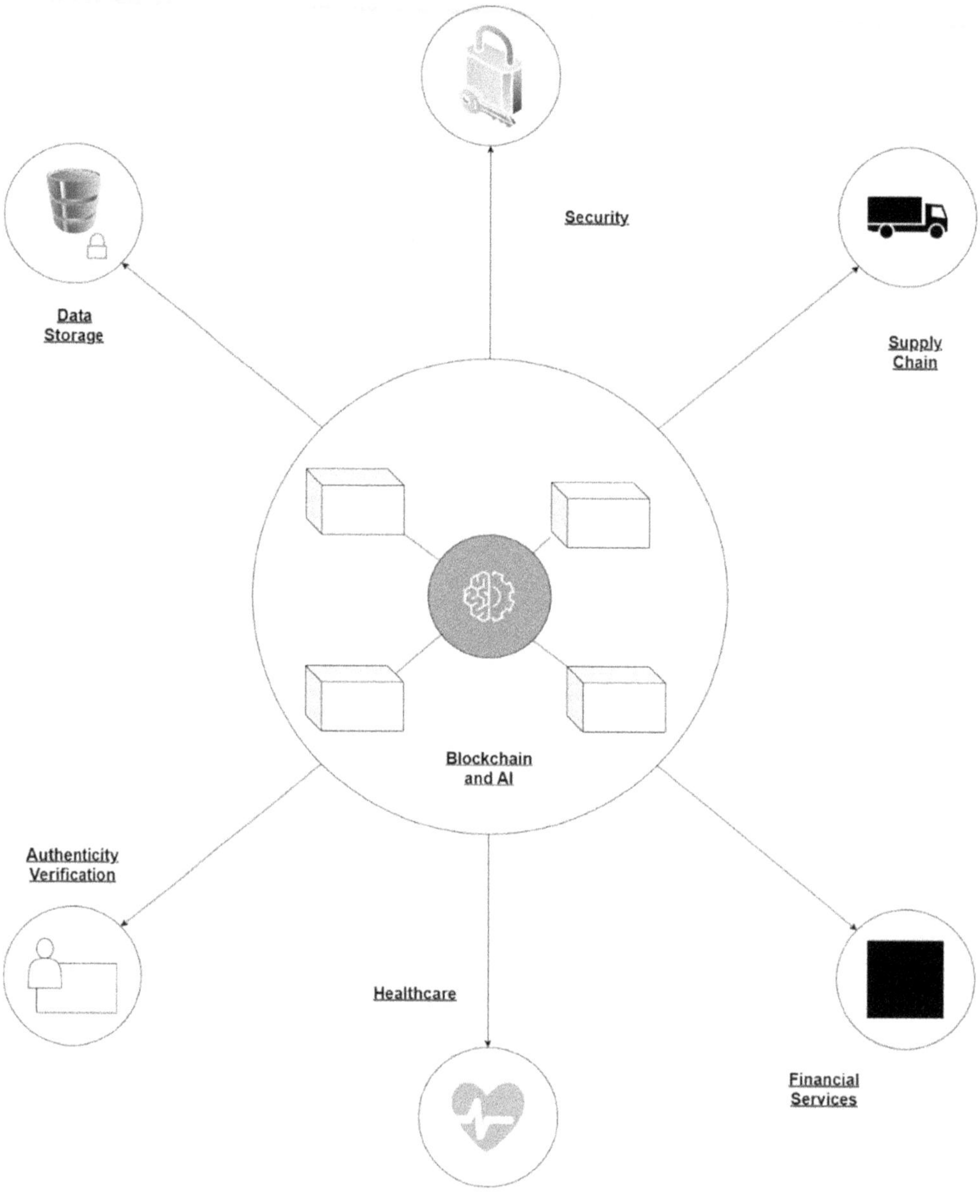

Figure 23.3 Blockchain and AI integration benefits.

1. **AI-driven Automation:** The use of AI-powered automation is forecasted to be on the move, changing many of the tasks or processes here and there in distinct industries. AI robots would be employed to attend repetitive and/or mundane tasks. Employees can then concentrate on activities that need creativity, critical thinking, and problem-solving.
2. **Decentralized Autonomous Organizations (DAOs):** One of the most certain things is that DAOs, which is the ability of blockchain, will gain a lot of popularity

among organization as they try to disperse decision-making and governance structures. DAOs mean that not only are stakeholders equipped with the power to participate in decision-making but also that it is done in a transparent and autonomous way. By doing so, greater accountability and inclusivity are created within organizations.

3. **AI-enabled Predictive Analytics:** AI-powered forecasting of trends, response to customer expectations, and business strategy optimization will be among the prominent factors to contribute to this development. Through using AI algorithms analyzing big data, companies can draw business conclusions in real time and respond to the changing market environment to ensure their business performance is the best.
4. **Blockchain-based Trust and Transparency:** Therefore, blockchain technology will remain a vital player in affecting and transforming trust and transparency for business relations, supply chains, and digital communications. The unchangeable and decentralized characteristic of blockchain eliminates fraud and promotes trust and accountability among stakeholders because of the immutable and distributed data storage, ensuring data integrity and credibility.
5. **Interoperability and Integration:** Integration of these technologies at the global level will become a reality only through incorporation of interoperable systems. Interoperability of data protocols, open application programming interfaces, and modular architectures will enable multiple vendors and ecosystems to link reliably.

REFERENCES

1. Jain, S., & Bhowmik, P. (2020). PharmBloc: AI Blockchain in Prevention of Drug Counterfeiting Fraud. *International Journal of Medical Informatics* 139, 104170.
2. Ahsan, A., Ekblaw, A., Vieira, T., & Lippmàn, A. (2016). MedRec: MedRec: Using Blockchain for Medical Data Access and Permission Management, International Conference on Open and Big Data (OBD)
3. Tunde D., "Blockchain and AI Integration for Secure Financial Transactions". https://www.researchgate.net/publication/389466440_Blockchain_and_AI_Integration_for_Secure_Financial_Transactions
4. Chen, L., & Wang, Q. (2021). AgriChainAI: The Optimal Utilization of AI Enabled Blockchain through Agricultural Supply Chain Optimization. *International Journal of Agricultural Management,* 9(3), 215–230.
5. Li, H., & Zhang, Y. (2022). EnergyAI-Grid: For AI-facilitated Blockchain Used for Streamlining the Effectiveness of the Energy Grid System. *Energy Informatics,* 5(1), 1–15.
6. Wang, L. & Liu, M. (2020). IPGuardian: AI-Powered Blockchain System for Property Right Protection. *Journal on Biotechnology: Intellectual Property Rights,* 25(3), 187–200.

Index

For Product Safety Concerns and Information please contact our EU
representative GPSR@taylorandfrancis.com
Taylor & Francis Verlag GmbH, Kaufingerstraße 24, 80331 München, Germany

www.ingramcontent.com/pod-product-compliance
Ingram Content Group UK Ltd.
Pitfield, Milton Keynes, MK11 3LW, UK
UKHW052214190926
475702UK00006B/1011

* 9 7 8 1 0 4 1 1 1 2 6 3 1 *